Trace Element Analytical Chemistry
in Medicine and Biology

Trace Element Analytical Chemistry in Medicine and Biology

Volume 5

Proceedings of the Fifth International Workshop
Neuherberg, Federal Republic of Germany,
April 1988

Editors
Peter Brätter · Peter Schramel

Walter de Gruyter · Berlin · New York 1988

Editors

Peter Brätter, Professor, Dr. rer. nat.
Hahn-Meitner-Institut Berlin GmbH
Research Group – Trace Elements in Health and Nutrition
Glienicker Straße 100
D-1000 Berlin 39
Germany

Peter Schramel, Univ. Doz. Dr. phil.
Gesellschaft für Strahlen- und Umweltforschung mbH
Institut für ökologische Chemie
Ingolstädter Landstraße 1
D-8402 Neuherberg
Federal Republic of Germany

Library of Congress Cataloging in Publication Data

Trace element analytical chemistry in medicine and biology.
Includes bibliographical references and indexes.
Contents: [v. 1.] Proceedings of the first international workshop, Neuherberg, Federal Republic of Germany, April 1980 -- v. 2. Proceedings of the second international workshop, Neuherberg, Federal Republic of Germany, April 1982 -- [etc.] -- v. 5. Proceedings of the fifth International Workshop Neuherberg, Federal Republic of Germany, April 1988.
1. Trace elements in the body--Congresses. 2. Trace elements--Analysis--Congresses. 3. Biological specimens--Analysis--Congresses. 4. Chemistry, Analytic--Congresses. I. Brätter, Peter, 1935- . II. Schramel, Peter.
QP534.T7 616.07'56 80-26803
ISBN 3-11-008357-4 (v. 1)

Deutsche Bibliothek Cataloging in Publication Data

Trace element analytical chemistry in medicine and biology :
proceedings of the ... internat. workshop / ... Internat. Workshop on Trace Element Analyt. Chemistry in Medicine and Biology. – Berlin ; New York : de Gruyter.
NE: International Workshop on Trace Element Analytical Chemistry in Medicine and Biology
Vol. 5 Neuherberg, Federal Republic of Germany, April 1988. – 1988
ISBN 3-11-011340-6

 Printing: Druckerei Gerike GmbH, Berlin. – Binding: Dieter Mikolai, Berlin. – Printed in Germany.

Preface

The fifth workshop on "Trace Element Analytical Chemistry in Medicine and Biology", organized by the GSF (Gesellschaft für Strahlen- und Umweltforschung mbH) was held, as in previous years, in Neuherberg. This year the workshop was combined with the annual meeting of the German Society for Minerals and Trace Elements e.V.

The fourth workshop had been devoted mainly to nutritional and clinical aspects. Therefore in the present meeting the subject matter was more oriented towards the state of the art of trace element analytical techniques.

The invited experts reported on newer aspects of analytical methods (G. Tölg, M. Stöppler, B. Welz), recent developments in preanalytical treatment of biological samples (G. Knapp) and the combination of analytical methods for speciation analysis (L. Duneman, P. Brätter). Aluminum and platinum (F. Alt) were elements of special interest. Attention was also paid to the trace element levels in body fluids and tissues (A.M. Reichlmayr-Lais) as well as to the role of trace elements in metabolic processes (P. J. Aggett) and in human nutrition (P.J. VanSoest).

The 74 papers accepted by the scientific committee illustrate the variety of analytical methods in use, together with their application in medical, biological and nutritional studies.

It was, once again, our aim to complete the book as soon as possible. Therefore the manuscripts were not extensively edited. The views and style of the contributions remain the responsibility of the named authors. The subject index reference is to the first page of a particular paper.

Our thanks are due to all participants who contributed by presenting a paper or by taking part in the valuable interdisciplinary discussions and to the scientists who acted as chairmen or referees for the poster sessions.

September 1988

P. Brätter
P. Schramel

Contents

NEWER ASPECTS OF ANALYTICAL METHODS

Plenary lectures given by:

G. Tölg

M. Stoeppler

B. Welz

PREANALYTICAL STEPS: SAMPLING AND SAMPLE TREATMENT

Plenary lecture given by:

G. Knapp

SPECIATION ANALYSIS

Plenary lectures given by:

L. Dunemann

P. Brätter

<u>TRACE ELEMENT LEVELS (AND REFERENCE DATA) IN BODY FLUIDS AND TISSUES</u>

Plenary lecture given by:

A.M. Reichlmayr-Lais

SPECIAL ELEMENTS: ALUMINIUM AND PLATINUM

Plenary lecture given by:

F. Alt

NEWER ASPECTS OF ANALYTICAL METHODS

Plenary Lectures given by:

G. Tölg

M. Stöppler

B. Welz

Chairmen:

W. Fresenius
G. Knapp
R. Schelenz

Poster Reports:

F. Alt, P. Brätter, G. Knapp, R. Schelenz
P. Schramel, G. Tölg, N.I. Ward, B. Welz

WHERE IS ANALYSIS OF TRACE ELEMENTS IN BIOTIC MATRICES GOING TO?

Günther Tölg

Institut für Spektrochemie und Angewandte Spektroskopie Dortmund
und Laboratorium für Reinststoffanalytik am Max-Planck-Institut für
Metallforschung Stuttgart, Bunsen-Kirchhoff-Straße 11-13,
D-4600 Dortmund 1

Abstract

Accomplishing the act of balancing our technological progress with the accompanying risks for safety and health in our life quality, calls for modern analytical science. It should serve as an indicator for the correct balance of forces in the substantial sphere. However, today such an indicating role is based on rough estimations, or on insufficient or unconfirmed information with respect to concentrations, binding forms and local distributions of toxic or of essential substances within a sample. Moreover, many complex mechanisms of synergetic and antagonistic physiological interactions have not yet been clarified - and consequently, we have to take into account severe misjudgements of risks.

In addition, one has reached the limits of financial means required for the increasing control and survey tasks of daily analytical routine. Accordingly, only a long-term planned strategy for the development of more powerful, more reliable and more economic analytical methods, which moreover guarantee a better local distribution (microdistribution analysis), are the prerequisites for an improvement of this situation in analytical sciences.

In view of the future tasks and the ultimate limits of trace- and micro-distribution analysis of the elements, the present state and an outlook on reaching the limit of analysis are critically discussed. Main emphasis is placed here on the possibility of improving conventional determination methods such as AAS, OES, XRFA, MS, NAA with regard to better power of detection and reliability. But also innovative analytical principles such as laser atomic spectroscopy (RIS, LEI, LIF) are introducted. They promise to develop into an essential basis for micro and trace element analysis of tomorrow.

Trace Element Analytical Chemistry in Medicine and Biology

As instrumental methods are always the last step in an analytical procedure, a brief reference will be made to the problem of sample preparation, mainly with regard to the sources of systematic errors. As for trace analysis at the ng/ml- or pg/ml-levels there are no reliable or certified standard reference materials available up to now, multistep procedures are still necessary. They must combine decomposition, preconcentration and determination methods in an optimal way to minimize systematic errors. The state-of-the-art of such multistep procedure in extreme elemental trace analysis will also be presented.

Motivation for an innovative trace analysis

In the search for a sensible balance of forces between our indispensible technological progress and the inevitable risks for health and safety resulting thereof, the question of the importance of Analytical Chemistry becomes one of actual urgency in every respect. Only a few of us are aware that in fact it is Analytical Chemistry which forms the basis for the acquirement of all informative data which are absolutely necessary in many fields of natural sciences, medicine, technology and environmental research. Informations thus gained, form the key to the cognitive connections between substantial concentrations of concentration changes and the recognition of physical, chemical and physiological properties and effects or changes taking place on the macro- and micro-scale [1]. Closely related of course is the question, whether or not analytical methods of today are sensitive enough for solving all substantial problems with which we will increasingly be faced in the future and which will be of an interdisciplinary nature in most cases.

The plain answer to this is: Without an innovative analytical science there cannot be envisaged any improvements in our living standards, in the substantial field of technology and in our natural environment. The emphasis of analytical progress is more specificly explained with regard to trace analysis by the following statements:

Power of detection

Analytical science must grow in the sense of higher power of detection [2].

An occasional dispute of this argument is, that the power of detection in analytical science is already far too high and that this contributes to the growing number of environmental disasters. Such argumentation is naive, but it does make evident that a powerful analytical science - like any other matter in this world - is subject to popular abuse. This is the reason why Analytical Chemistry must be placed into the hands of responsible and critically thinking experts. It goes without speaking that an uncritical acquisition and interpretation of data may have devastating results: the gravest of all in this respect is the increasing uncertainty and accompanying fear spreading in our society. Our aim must be to achieve the contrary, namely, the early recognition of risks and, whenever possible, the early minimization of them. In addition, one should strive to evaluate more objectively and by more detailed investigations those risks which are already known, irrespective of positive or negative results.

A further and more realistic discussion on the foreseeable limits of more powerful detection methods in the field of trace element analysis can be based on the omnipresent concentrations of the elements in all natural matrices. These concentrations correspond to the element abundance in the crust of the earth or in the biosphere.

The natural concentration levels practically determine the limit of the lowest detectable concentrations of elements in a biotic matrix. This limit lies for most elements at concentration levels above 1 ng/g, a level which is already analytically accessible today. The only analytical problem encountered here are the rare elements of which the omnipresent concentrations lie at the pg/g level and lower, as is the case, e.g. for platinum metals. In this connection, one must consider that in the above statement, it is not primarily a question of the improvement of the relative power of detection in concentration units (e.g., g/g or g/l). It is more a question of improvements of the absolute detection power (in mass units) and of techniques for the determination of the binding forms of the elements.

In biological systems, the analytical interest will increasingly focus on the micro distribution of the elements and their binding forms in tissue compartements, in individual cells and even in cell membranes. In field of micro trace analysis - which means trace analysis with very limited sample amounts/volumes - and even more so in the field of in-situ micro distribution analysis, one has therefore to aim at absolute limits of detection at the femtogramme level, as the following roughly sketched explanation should illustrate. Natural omnipresent concentrations in the blood serum may lie below the ng/g level, even for relatively abundantly occurring elements such as Al, V, Cr. At the present state-of-the-art, it is necessary for the determination of such elements at very low concentration levels at which they may possibly be physiologically effective, that greater sample amounts/volumes have to be available for analysis, sometimes amounts of more than 1 g. And in this manner, one only comes to integral conclusions. But if more detailed statements, for example, on element distributions within the individual protein fractions in human blood are required, a method with a 10-100 times greater power of detection would be necessary to perform such a task. And if an element concentration of 1 ng/g is to be quantitatively determined, one would require an absolute power of detection in the femtogramme region (1 fg = 10^{-15} g).

It is at this point, that one has principly to put the question what would have to be the minimum amount/volume of an element which must be available in order to be determined with a certain statistical error. From a statistical point of view (law of root N), there should be at least 10 000 atoms within the volume, if statistical errors are not to be greater than 1%. For an atom with an atomic weight of 60, this corresponds to a mass of 1 attogramme (10^{-18} g) [3].

In my later comments I shall expound the prognosis that there exists a good chance to realize this theoretically attainable power of detection with the aid of laser spectroscopy.

Reliability and economy

Our euphoric view on the improvement of the analytical power of detection is rapidly dimmed, if we return to the present state-of-the-art in trace analysis of the elements in the ng/g region.

Here, a long learning process resulting from interlaboratory comparative analyses taught us a bitter truth. Improvements in the relative, but particularly in the absolute power of detection are accompanied by an increase of systematic errors, both of chemical and physical nature. And consequently, the problem of reliability of analytical results is finally one of paramount importance in the extreme trace analysis [4].

This problematic situation with respect to its causes, as well as to the recognition and the avoidance of systematic errors in the extreme trace analysis is the subject of numerous publications [e.g., 4-7] and should therefore not be discussed in detail here. Today, one may proceed from the assumption that in common bulk analysis of biotic matrices - even at natural concentration levels - it is possible to obtain results with sufficient reliability for most elements at ng/g levels, when applying the appropriate expert knowledge and the necessary critical capability. The remaining insufficient or lacking results in routine analysis are in most cases less of a principal nature, but more a sign of insufficient qualification of data-producing laboratories. Here, it is a question of further education towards critical ability and continuation studies which should range in the foreground. But unfortunately, such an urgently demanded learning process is often hampered by very uncritical advertisements of the instrument manufacturers as well as by the not widely known complexity of analytical strategies which are necessary. It must also be emphasized, that the presently reliably determinable elements, classified as "unproblematic elements" may become problematic elements again, if they must be determined in micro regions with lower absolute limits of detection.

Often, the procedure for getting reliable results in the ng/g analysis is still very uneconomic, so that in pracitical applications, the arising costs become a point of central interest. However, one should keep in mind, that wrong analytical results gained by inexpensive procedures are at least just as uneconomic as correct results attained by more expensive methods. Most of the more cost-saving instrumental direct methods, which should self-evidently be employed in routine analysis, are highly matrix-dependent relative methods which must be calibrated with certified standard reference materials. This has to be done in order to compensate for elemental interferences caused by other

Table 1. Capability of determination methods in elemental trace analysis [8].

Method	Technique	LD Order of magn $ng \cdot mL^{-1}$ ≥ 1 ≤	Matrix effects	Multielement determination	Speciation analysis
AAS	flame (W-loop)	■(□)	●	no	
	furnace	□	●●●	no	
	HG (HGC)	□ (□)	●●	no	possible
ZAAS	furnace	□	●	no	
OES-DCP	pneum. nebuliz.	■	●●●	yes	
-ICP	pneum. nebuliz.	■	●	yes	
	ETV	■	●	yes	
	HG (HGC)	■ (□)	●●	yes	possible
-MIP	pneum. nebuliz.	■	●●●	limited	
	ETV	■ (□	●●	limited	
	HG (HGC)	■ (□	●●	limited	
- HC	volatilization	■	●●	limited	possible
FANES		□	●●●	limited	
AFS	laser ICP/ICP	■	●●	yes	
	HC / ICP	■	●	yes	
	laser/furnace	□	●●●	yes	
CFS	furnace	■	●●●	yes	
Optogalvanic		□		limited	
XRS	XRF	■	●●●	Z > 6	
	PIXE	■	●●●	Z > 14	limited
	TR-XRF	□	●	Z > 14	
MS	ID-TI	□	●●	limited	
	ID-ICP	■	●	yes	
	ID-FD	□	●●	limited	yes
NAA	INAA	■	(●)	yes	
	RNAA	□	(●)	limited	possible
Voltammetry	DPASV	□	●●●	limited	yes
	DPCSV	□	●●●	limited	yes
Chelat-GC	ECD	□	●●●	limited	yes
-HPLC	UVD	■	●●●	limited	yes
Spectrophotom.		■		no	yes
Fluorimetry		□	●●●●	no	yes

■ for a high number of elements; □ only for special elements; ● low, ●● medium; ●●● high; and ●●●● very high matrix effects [8] components in the sample during analytical signal generation. But such standards are not yet available in the majority of cases. And the situation will not improve much during the coming years, because the production of standards will be very slow due to high costs.

Accordingly, it may be concluded that the following three significant methodological quality criteria - power of detection, reliability and economy - will not loose in actuality in future innovative elemental analysis. The strategy for the optimization of these three interconnected criteria is very complicated, indeed, and it is not referred to in any great detail in this contribution.

2 Atomic spectroscopical determination methods of today and tomorrow

In addition to neutron activation analysis (NAA), which will not find elaborate mention here, because its possibilities and limitations are widely known, AAS, OES, XRFA and MS are the most significant determination principles available for biological trace analysis today [8,9], see Table 1. The manifold special methodological techniques based on these principles will only be referred to by way of example on the most important points with respect to the analysis of biotic matrices.

Atomic absorption spectrometry (AAS)

Today as in the past, this method is employed preferably for direct and indirect determinations of most of the trace elements in biotic matrices. Although until today mainly single elements can only be determined sequentially, this disadvantage is compensated for by high sensitivity. The latter reaches the lower pg-region as a result of the introduction of the electrothermal atomization in furnaces (ETAAS or furnace-AAS, preferably with graphite tube furnaces, GFAAS) and of the background compensation by the Zeeman effect. Exceptions are only the carbide forming elements and several non-metals, e.g., C, N, O, P, S and the halogens. The power of detection of AAS is only surpassed in a very few cases by the application of other principles (e.g., inverse voltammetry and chelate GC).

The power of detection of GFAAS is by 2-3 orders of magnitude higher compared to that of flame-AAS (FAAS). But one must not overlook the fact that in FAAS the interferences or systematic errors are far less. Therefore, FAAS should always be given the preference to GFAAS, if the detection power (1-100 ng/ml) is sufficiently high. Over the past few years, the absolute power of detection of FAAS has been improved by reducing the sample consumption, which is usually 1-2 ml, by the injection technique [10], the micro-wire loop technique [11], the "collection tube techniques" [12] and the application of high pressure nebulization [13] increasing the absolute power of detection by a factor of 10-50.

It is also possible to carry out a direct FAAS of solid samples, especially when rather volatile elements (e.g. Cd, Zn, Pb and Tl) are to be determined. When analyzing biotic matrices, the microsample (some mg) is decomposed in a separate decomposition device and subsequently the analyte is evaporated and fed directly into the flame gases in form of an aerosol (Fig.1). This technique in FAAS [14] allows it to avoid systematic errors of solution techniques, similarly as with the more complicated and problematic solid sampling GFAAS, which will be mentioned later.

In GFAAS [15, 16] a significant progress in the reduction of systematic errors has been realized by the introduction of the Zeeman background compensation and by the treatment of the graphite furnace surfaces, e.g. by coating with pyrographite or metal oxides, by using glassy carbon tubes or by operation with controlled and constant temperatures during atomization. The latter was e.g., achieved by the employment of platform techniques, tube-in-tube techniques and most promising, twin-furnace techniques, Fig. 2 [17]. Partial success also would be obtained by an empirical use of matrix modifiers [16]. The detection limit can be lowered by a factor of about four for all elements by using the signal summation [18]. But finally, it was a separate measurement of the analytical signal and the background signal by the electronical recording of time-resolved signals [19], which led to the application of optimal methodological parameters and thus, to the discovery of insidious systematic errors [20].

By the introduction of the Zeeman compensation and the time-resolved measurement of signals, the direct furnace AAS for solid samples has become very attractive. It became a powerful technique for the determination of a series of elements with unproblematic chemical behaviour during atomization (e.g., for Cd, Zn, Cu, Sb and with reservations, also Tl and Pb) in biotic matrices [21, 22]. Direct solid sampling in AAS offers several advantages. Wet chemical sample preparation involving the risk of contamination is avoided and the sample throughput can be extensively increased. The only time-consuming steps are the sample homogenization and the weighing of the sample for analysis. With the present commercially available instruments, however, the sample amounts used only may range from the upper micogramme to the lower milligramme, the power of detection is rather poor as compared to the one of common furnace techniques using solutions. The small amounts of sample used on the one hand necessitate it to homogenize the samples for particle diameters of <100µm, in order to facilitate representative analysis of larger samples [23]. On the other hand, they make it possible to perform microdistribution analyses (e.g., for the medical examination of histological tissue specimens), for which, however, TXRFA in most cases is more helpful (referred to later on in text).

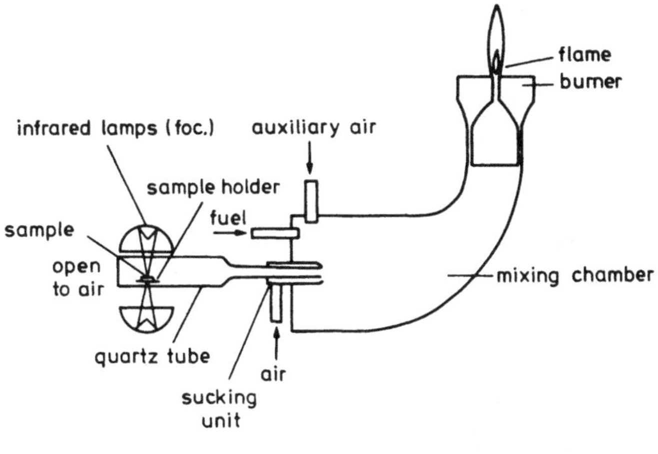

Fig. 1. Solid sample pre-burning technique in FAAS [4]

In addition to the well known cold vapour techniques as used in the determination of Hg, hydride generation AAS (HG-AAS) represents a special combination of chemical separation and pre-concentration with AAS determination [16]. It results in an improved power of detection and accuracy for all elements which can be quantitatively reduced to volatile hydrides such as As, Se, Sb and Sn. For example, with a special condensation tube technique (HGC-AAS) the Se from the generated H_2Se may be determined with excellent reliability down to the 1 pg region. With this technique, for instance, the Se content in segments of a single hair may be reliably determined with a relative standard deviation of < 5% [24].

Optical emission spectrometry (OES)

OES is the oldest multi-element atomic spectrometric method and originally involved the use of flame, electric arc or spark excitation. However, immense new activities took place after the introduction of various new plasma excitation sources like inductively coupled plasma (ICP), direct current plasma (DCP), capacitively coupled microwave plasma (CMP) and microwave induced plasma (MIP). Most experience for the analysis of biotic matrices has been gathered with the ICP. With pneumatic nebulization, the power of detection lies somewhere between that of flame and furnace AAS techniques [25].

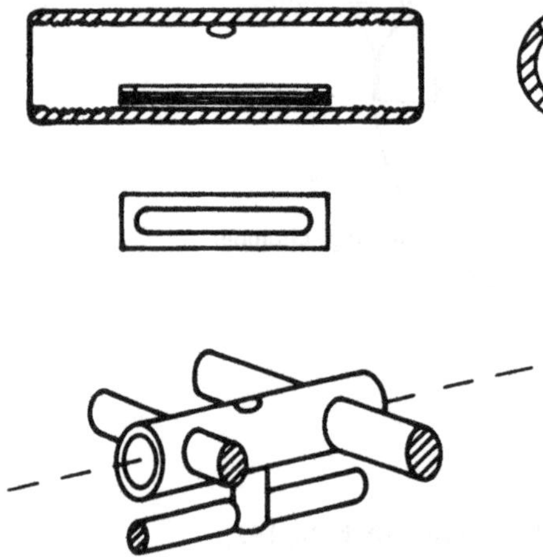

Fig.2. Isothermal atomization techniques in GFAAS. Upper part Graphite tube with L'vov platform, lower part Twin-furnace technique [8]

Especially for elements with thermally stable oxides (e.g. lanthanides), the limits of detection are low. This is in contrast to elements with high ionisation and most sensitive analytical lines with high excitation potential. Accordingly, Cd, Tl and Pb can be determined by AAS with significantly higher powers of detection. The powers of detection of ICP-OES and of DCP-OES methods are almost identical and are similar to those of spark-OES and conventional XRF techniques. According to empirical results obtained, interferences of alkali metals are greater in DCP-OES than in ICP-OES. An enhancement of the power of detection by one order of magnitude could be realized both by the use of ultrasonic nebulization or by employment of the high pressure aerosol injection and furnace techniques [26], as known from AAS. None of these special techniques are commercially available up to now. While CMP-OES offers no advantage over ICP-OES and DCP-OES, MIP-OES is a low-cost micro-method for solutions [27-29] with excellent absolute power of detection. In combination with GC and HPLC, it represents a sensitive, element-specific detection principle for speciation analysis [30-32].
After HPLC separation, we were able to detect different organomercury compounds via the Hg signal with a high power of detection compared to that of a conventional UV-detector [33].

When MIP-OES is combined with suitable decomposition and separation methods, the absolute limit of detection for Hg in organic matrices is less than 1 pg [34].
ETA-MIP-OES in combination with suitable decomposition and separation methods may also be a very useful principle for very sensitive simultaneous determination of many other elements like Fe, Ni, Co, Zn, Cd, Tl, Bi [36,37]. The main handicap for the improvement of the detection power in OES, but also in AAS and AFS, is that the atomization and the generation of the optical signal happen in the same energy source. While the atomization step requires high energy to crack chemical bonding, the generation of the optical signal needs lower energy to attain an optimal signal-to-background ratio. This problem can only be overcome by new concepts, where both steps of the procedure are separated and can be appropriately optimized. In AAS, the twin-furnace technique comes nearest to optimal operating conditions. A further way may be the principle of FANES - a combination of sample atomization in a furnace and analyte excitation in a glow discharge lamp [38,39]. However, also volatilization of solid samples with a focussed laser

beam (laser ablation) may bring advantages in micro-trace and in in-situ micro distribution analysis [40].

Laser spectroscopy

Monochromaticity, high spectral energy density and directivity are the specific properties of laser light. If tunable lasers are applied in analytical spectroscopy, low detection limits and excellent selectivity can be obtained. Figure 3 shows the basic principles of analytical laser spectroscopy. When directing the laser beam into the cloud of an atomized sample and tuning the frequency of the laser to resonance of an anlyte, there are three ways for quantitative analysis. One may measure the absorption (Laser Atomic Absorption Spectroscopy - LAAS) by the analyte, its fluorescence (Laser Induced Fluorescence - LIF) or the ionization products (ions or electrons) of the selectively excited analytes after appropriate ionization processes. We distinguish ionization by thermal collisions with other particles (Laser Enhanced Ionization - LEI), by electric fields (Field Ionization Laser Spectroscopy - FILS) and by photoionization (Resonance Ionization Spectroscopy - RIS). Reviews on LIF, LEI, FILS and RIS can be found in [41-43] and [44], respectively. In order to realize isotope selectivity RIS has been combined with mass spectrometer (Resonance Ionization Mass Spectroscopy - RIMS), see [45].

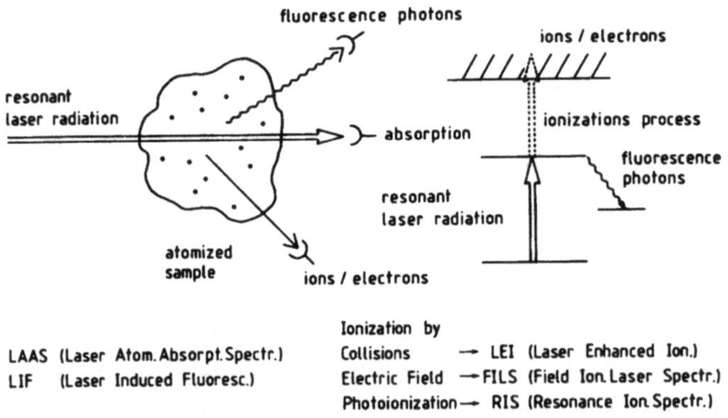

Fig. 3. Principles of excitation in laser atom spectroscopy (after K. Niemax)

Usually the tunable lasers applied in laser spectrochemistry are pulsed dye lasers with relatively broad line widths. A different approach was choosen in our institute where narrow band width continuous wave lasers are applied in LEI [46]. The longer interaction time of analytes with the continuous laser field compared with pulsed excitation allows to obtain lower detection limits. On the other hand, the small line width of cw lasers offers the possibility of applying high-resolution (Doppler-free) techniques to spectrochemistry.

Figure 4 shows a resonant Doppler-free two-photon spectrum of calcium recorded with two cw lasers [46].

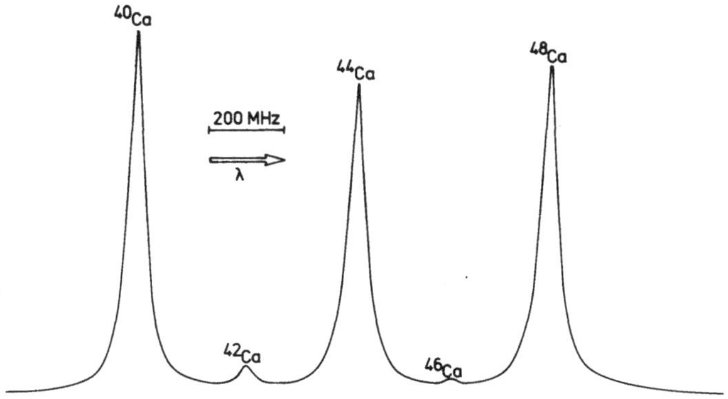

Fig. 4. Doppler-free 2-photon spectrum of the natural abundant isotopes of calcium (after K. Niemax)

The isotope components of the transition are resolved and, accordingly, isotope-selective laser spectrochemistry without a mass spectrometer can be performed. Note, that the most abundant isotop (^{40}Ca: 97%) as well as the rare ^{46}Ca (33 ppm) can be analyzed in one scan. The spectrum was obtained with a thermionic diode, an extremely sensitive detector [47]. With this detector it is possible to measure down to 10^3-10^4 atoms/cm^3 in the presence of 10^{13} atoms/cm^3. Calibration can be carried out, e.g. by isotope dilution as in mass spectrometry.

Currently, the most serious disadvantage in the practical utilization of laser spectroscopy is the very high price of the laser system which, however, may be substituted soon by far less expensive semiconductor diode lasers. With such laser diodes, it will also be possible to substitute the hollow cathode lamps in AAS, which will open the way to multielement AAS determinations in a very simple and cheap way [48].

Coming back to the very important fact that complete separation of the atomization and excitation steps in spectroscopic methods is the basis for a most significant innovation with respect to better sensitivity and reliability, it is to be expected that sample atomization with a focussed laser beam followed by laser excitation of the produced aerosol will be a very promising approach. This combination may open a new area of optical atomic spectroscopy, whereby most of the analytical goals can be reached, namely high absolute detection power (about 1 ag) for nearly all elements, microprobe work by laser focusing with a lateral resolution of about 1 µm, internal calibration by isotope dilution and a very cheap instrumentation. This could be realized provided, that laser diodes are available to 400 nm (today, about 700 nm). Then the entire optical wavelength spectrum to about 200 nm (today 350 nm) could be covered by doubling the laser frequency with the aid of nonlinear effects.

X-ray spectrometry (XRF)

The possibilities and limitations of conventional wavelength- and energy-dispersive XRF are very well known, in [49,50]. Their absolute power of detection is highest for elements with an atomic number around 30 (Zn). Here, the detection limits are about 0.1 µg and they become poorer for higher and lower atomic numbers. The main limiting factor for the power of detection is the Compton and Rayleigh scattering of the primary beam. Furthermore, interelement effects necessitate it to calibrate with standards that must be very similar in composition to the sample [51]. Better powers of detection and accuracy can only be achieved by decreasing the signal-to-noise ratio and the matrix effects. This can be done in three ways [52]:

1. By application of the total reflection technique (TXRF) [53-57]. Here, a very small angle of incidence of the exciting X-ray beam so that the latter is totally reflected at the target surface, Fig.5. When the target surface is optically flat (as is possible, e.g., with quartz, glassy carbon or polyacryl)

and even when it is homogeneously coated with a very thin film of sample material, the power of detection of TXRF is a factor of up to 1000 (pg-region) better than in conventional energy-dispersive techniques. The simultaneous measurement of the element peaks and the correlation of peak heights and concentrations is no longer a problem, because the interelement effects are negligible in contrast to conventional XRF with thick samples. Therefore, calibration can be done by using a single element as internal standard. This element must not be determined and its concentration needs not to be observable in the sample spectrum. The main problem lies in a sample preparation on the micro-scale. This starts with cleaning the sample carrier and distributing the elements to be determined as an homogeneous thin layer of less than µm thickness on a target area of approx. 8mm in diameter. Solution evaporation, cathodic sputtering or electrolytic deposition techniques can be used for this aim. In the case of biological materials and especially in medical studies, a great advantage is offered by the direct analysis of histological microtome-sections [58,59]. They permit the simultaneous determination of most of the elements of physiological interest, in e.g., lung tissue (Fig.6).

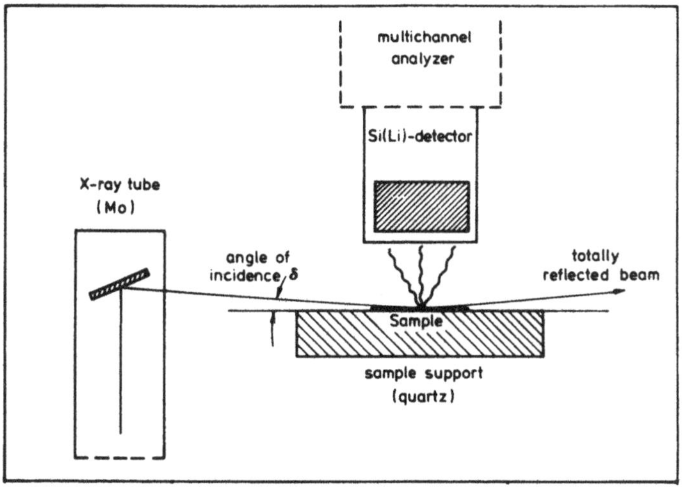

Fig. 5. Principle of TXRF [59]

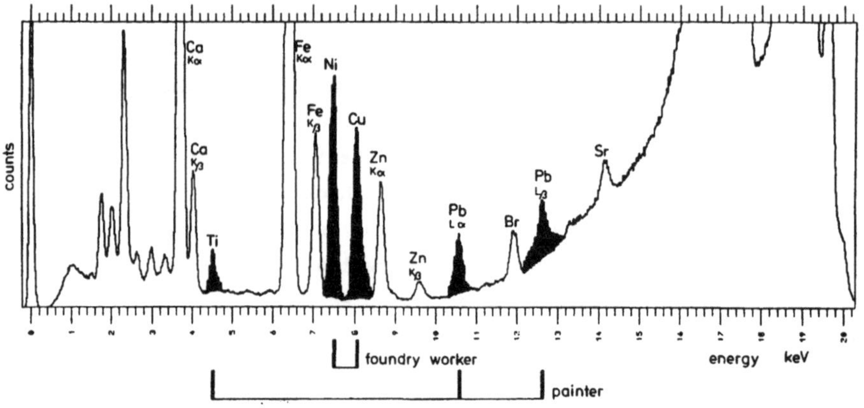

Fig. 6. TXRF-spectra of microtome sections of lung tissue of exposed and non-exposed persons [59].

Further XRF techniques using thin-film samples which enable detection limits in the pg-range recently have been described [60].

2. With respect to the power of detection, TXRF is superior to the proton-or more general the particle-induced X-ray emission (PIXE) method [61]; PIXE, however, yields a higher absolute power of detection than conventional XRF. The advantage of the much more expensive PIXE, which requires a particle-(proton-)accelerator, lies in the fact that the proton beam can be focussed to about 1 µm in diameter, so that multi-element microdistribution analysis can be carried out in the µg/g-range, e.g. across the diameter of a human hair. Many other very interesting applications with respect to biotic matrices are described in literature [62,63], sometimes, however, in a very uncritical way with respect to costs and accuracy.

3. Further very promising, but again much more expensive analytical tools stem from the development of synchrotron excitation sources for XRF (SXRF). Synchrotron radiation can be focussed to beam diameters between 1-10 µm which still possess sufficient intensity to penetrate through and excite thin samples [64-66]. The detection limits lie in the pg-range. Better power of detection may be expected here also when a totally reflecting excitation beam is used [67].

These three excitation sources in XRF open useful ways for very sensitive simultaneous determinations of microdistribution of most elements in their essential and toxic concentration levels in biotic matrices.

Mass spectrometry (MS)

Considering the principle, MS is the most universal, sensitive instrumental method for multi-element determination. In comparison to OES, the spectra show few lines which can be interpreted easily. The resolution of two isobaric isotopes of different elements is no technical problem so that the isotope dilution technique can be applied to obtain precise but in particular accurate quantitative results (with exception of single-isotope elements).

For practical applications, a wide-spread utilization of these advantages could recently be realized by the availability of high-performance low-cost quadrupole filters and by a considerable progress in the development of ion sources such as hf-plasma (ICP, MIP), glow discharges (GD) and laser sources. Vast experiences made in ICP-OES and GD-OES could be transferred to MS and enabled a rapid progress in terms of reliability and economic aspects.

For ICP-MS as well as for GDMS, calibration can be improved by application of isotope dilution, even in multielement procedures. Thus can be predicted for both techniques that they will lead to new lines of approach in routine analysis of solutions, extending the capabilities of the old thermionic technique with its impressing absolute power of detection down to the pg range [68,69].

In ICP-MS, the power of detection for most elements is at levels of 0.1 to 1 ng ml^{-1} [70,71], which is about one order of magnitude better than in ICP-OES. Experiences made so far show a much narrower variation range of detection limits for ICP-MS than for ICP-OES where they vary over a range of nearly five orders of magnitude. Furthermore, with ICP-MS analytical procedures can be realized which are closer to an "ideal multi-element analysis" than OES procedures usually are. Recent progress in applications of the ICP-MS techniques indicates that the optimistic perspectives given here will be confirmed in routine analysis.
GDMS is a very promising tool for direct bulk and thinlayer analysis of solids

and may offer a good chance for the analysis of solutions, too. As our first results for an application of GDMS to micro analysis [72] indicate, a substantial improvement in the absolute detection power can be expected, at least in comparison with the ICP as ion source.

A new generation of LAMMA and TOF-SIMS in combination with a secondary excitation of the sputtered neutral particle from the sample might lead to new and very sensitive micro-probe techniques with lateral resolution in the μm-range.
A very promising principle to improve the detection power can also be seen in the resonance ionization mass spectrometry (RIMS), as the first activities in this field [73] let expect.

Non-atomic spectroscopical determination methods

It must clearly be pointed out that numerous other classical or more chemical determination methods, like spectrophotometry, fluorimetry, chelate GC, chelate HPLC and voltammetry, have their firm place in the toolbox for trace analysis, not only with respect to high power of detection (Table 1), but also in view of simpler instrumentation and less problems in the calibration of such solution methods. The main reason why they will be substituted more and more by atomic spectroscopical methods in routine analysis is that they require more special analytical experience and knowledge in chemistry, sometimes they are more time-consuming or not easy to automize. Also, most of these methods allow only the determination of a single element or of some elements simultaneously. These solution methods are hardly suitable for micro-distribution analysis.
The limitations of most of these methods are well known. Their power of detection which enables extreme trace analysis, is mostly limited by blank problems and other sources of systematic errors (adsorption, desorption, volatility a.o.). Innovations are more a question of broader applications which is only of a practical sense, if better criteria of power of detection, selectivity, reliability and economy are to be obtained or new, more efficient multi-step procedures are to be created with their support.

3 The role of chemical decomposition and preconcentration methods in the future of extreme trace analysis

Considering all the progress in instrumental determination methods as mentioned above, one may gain the impression that chemical methods of sample preparation have no longer a future. This, however, would prove to be short-sighted with respect to any progress in elemental analysis. The reasons can clearly be defined [7].

- All direct instrumental methods have to be calibrated. But calibration causes great difficulties, except in very few cases such as in INAA, in XRFS, when analyzing very thin samples, or in SNMS which is neither depending a lot on matrix effects. In most methods, widely varying elemental cross interferences due to matrix or concomitant elements are encountered. So, as long as no confirmed mathematical functions between signal intensities and element concentrations exist, one has to refer to reliable standard reference materials which do not yet exist in most cases of practical analysis. Therefore, a solution technique with standard addition is the method of choice, in order to compensate for matrix effects during calibration. But this way is not always successful, either. Indeed, systematic errors may occur, if the calibration function over the entire concentration range is not linear, the background level is not known and the bonding form of the element to be determined is not the same as the one used in the standard solution.

- In order to achieve optimal power of detection and reliability, the elements to be determined should be isolated on a smallest possible target area or in a minimum volume. The matrix effects can be eliminated to a great extend; however, the concentration results in higher analytical signals, only if chemical operations are carried out on microscale (μl, mg).
- The reproducibility of the result of an instrumental method gives no information about its accuracy. In extreme trace analysis, the accuracy of a result can only be confirmed by employing two or more procedures basing on fully independent principles. This means that they have to differ from one another in each analytical step, such as sample preparation and determination. Consequently, only when different procedures come to the same results, one can assume that systematic errors are absent. In this context, multi-step procedures open an efficient strategy to recognize systematic errors. A further more cost-consuming way of gaining information concerning the accuracy of procedures is by interlaboratory comparisons.

From conclusions made with respect to these points and various other provisions for analysis in the ng/g- and pg/g-range, it follows that special attention has to be paid to chemical methods which allow the complete decomposition of a sample and the isolation and preconcentration of the elements to be determined prior to the actual determination step.

During the past years, vast experience has been gathered in the development of such combined solution procedures and their optimization regarding minimal systematic errors [5-7]. Care must mainly be taken in view of the minimization of the danger of contamination and to the avoidance of losses by adsorption and volatilization.

In this context, the problem of the decomposition of organic matrices could be solved in an optimal manner. Of the various successful principles available [74-76], the high pressure decomposition in a quartz bomb permitting the complete mineralization of organic samples [77,78], is the most expensive, but also the most efficient method. Today one can say that by this principle, decomposition of organic matters imposes no longer a problem as it existed over a period of many years in extreme elemental trace analysis [79].

Moreover, there are so many principles regarding chemical separation and preconcentration methods [80-85] available, that there is hardly any more need for new ones in practical trace analysis. Much more important seems to be the combination of the different steps in closed, on-line systems according to the flow injection principle (FIA) [86,87] and the development of efficient micro techniques. The latter must enable it to concentrate the solutions with the enriched elements to be determined drop by drop on as small a target area as possible for intensive signal generation. These sample preparation techniques will help to improve the power of detection and to lower systematic errors drastically. Their time-consumption can be extensively decreased by automation.

In my opinion the question "Where is analysis of trace elements going to?" must necessarily lead to the answer:
"Into a future with so many challenges for Analytical Chemistry as never before in its history". They only may be coped with if new methodological principles and analytical strategies can be introduced by a closer fusion of chemical and physical ways of thinking in our discipline and by the awareness that there exists an urgent demand for more criticism both for the production

and the interpretation of more detailed analytical information. In consequence, not only the analytical methods, but also the analysts should become more "sensitive".

Acknowledgements. The work has been supported by the "Ministerium für Wissenschaft und Forschung des Landes Nordrhein-Westfalen" and by the "Bundesministerium für Forschung und Technologie".

References

1. Tölg G., Garten R. (1985) Angew. Chem. (Int.Ed.Engl.) 24:485
2. Tölg G. (1988) Fresenius Z. Anal. Chem. 329:735
3. Tölg G. (1975) In: Svehla G. (ed.) Wilson's and Wilson's comprehensive analytical chemistry, vol. III. Elsevier Scientific, Amsterdam, pp 1-184
4. Tölg G. (1976) Naturwissenschaften 63:99
5. Tölg G. (1977) Fresenius Z. Anal. Chem. 283:257
6. Tschöpel P., Tölg G. (1982) J. Trace Microprobe Technol. 1:1
7. Tölg G., Tschöpel P. (1987) Anal. Sci. 3:199
8. Tölg G. (1987) Analyst 112:365
9. Broekaert J.A.C., Tölg G. (1987) Fresenius Z. Anal. Chem. 326:495
10. Berndt H., Jackwerth E. (1975) Spectrochim. Acta 30 B:169
11. Berndt H., Messerschmidt J. (1979) Spectrochim. Acta 34 B:241
12. Berndt H., Messerschmidt J. (1982) Anal. Chim. Acta 136:407
13. Berndt H. (1988) Fresenius Z. Anal. Chem. (in print)
14. Berndt H. (1984) Spectrochim. Acta 39 B:1121
15. Dittrich K. (1988) In: Fresenius W. et al, (eds.) Analytiker Taschenbuch, Bd. 8. Springer, Berlin, Heidelberg, New York, London, Tokyo, in press
16. Welz B. (1985) Atomic absorption spectrometry, 2nd engl. edn. VCH Verlagsgesellschaft, Weinheim
17. Frech W.E., Lundberg E., Cedergren A. (1985) Progr. Anal. At. Spectrosc. 8:257
18. Berndt H., Schaldach G., Klockenkämper R. (1987) Anal. Chim. Acta 200:573
19. Baasner J., Berndt H., Eiermann R. (1986) In: Welz B. (ed.) Fortschritte in der atomspektrometrischen Spurenanalytik, Bd. 2. VCH Verlagsgesellschaft, Weinheim, pp 387-395
20. Berndt H., Sopczak D. (1987) Fresenius Z. Anal. Chem. 329:18
21. Langmyhr F.J., Wibertoe G. (1985) Progr. Anal. At. Spectrosc. 8:193

22. 2nd International Colloquium: Solid Sampling with Atomic Spectroscopic Methods, Wetzlar (1987) Fresenius Z. Anal. Chem. 328:315
23. Stoeppler M., Kurfürst U., Grobecker K.H. (1985) Fresenius Z. Anal. Chem. 322:687
24. Piwonka J., Kaiser G., Tölg G. (1985) Fresenius Z. Anal. Chem. 321:225
25. Fassel V.A. (1971) In: XVI Coll. Spectrosc. Intern., Heidelberg, Plenary Lectures and Reports, Hilger A., London, pp 63-93
26. Broekaert J.A.C. (1987) Anal. Chim. Acta 196:1
27. Matousek J.P., Orr B.J., Selby M. (1984) Progr. Anal. At. Spectrosc. 7:275
28. Kollotzek D., Tschöpel P., Tölg G. (1984) Spectrochim. Acta 39 B:625
29. Haas D.L., Caruso J.A. (1984) Anal. Chem. 56:2014
30. Risby, T.H., Talmi Y. (1983) Crit. Rev. Anal. Chem. 14:231
31. Deruaz D., Mermet J.M. (1986) Analusis 14:107
32. Cammann K., Lendero L., Feuerbacher H., Ballschmiter K. (1983) Fresenius Z. Anal. Chem. 316:194
33. Kollotzek D., Oechsle D., Kaiser G., Tschöpel P., Tölg G. (1984) Fresenius Z. Anal. Chem. 318-485
34. Kaiser G., Götz D., Tölg G., Knapp G., Maichin B., Spitzy H. (1978) Fresenius Z. Anal. Chem. 291:278
35. Noijiri Y., Otsuki A., Fuwa K. (1986) Anal. Chem. 58:544
36. Volland G., Tschöpel P., Tölg G. (1981) Spectrochim. Acta 312:901
37. Aziz A., Broekaert J.A.C., Leis F. (1982) Spectrochim. Acta 37 B:381
38. Falk H., Hoffmann E., Lüdke Ch. (1981) Spectrochim. Acta 36 B:767
39. Falk H. (1988) Crit. Rev. Anal. Chem. 19:29
40. Moenke-Blankenburg L. (1986) Progr. Anal. At. Spectrosc. 9:335
41. Omenetto N., Winefordner J.D. (1979) Progr. Anal. At. Spectrosc. 2:1
42. Travis J.C., Turk G.C., Schenk J.R., van Dijk C.A. (1984) Progr. Anal. At. Spectrosc. 7:199
43. Letokhov V.S. (1987) Laser photoionization spectroscopy. Academic Press, Orlando, p 215
44. Hurst G.S., Payne M.G. (1988) Spectrochim. Acta B, in press
45. Fassett J.D.D., Travis J.C. (1988) Spectrochim. Acta B, in press
46. Niemax K. (1987) Naturwissenschaften 74:474
47. Niemax K. (1985) Appl. Phys. B 38:147
48. Hergenröder R., Niemax K. Spectrochim. Acta B, submitted
49. Van Grieken R., Markowicz A., Torök S.Z. (1986) Fresenius Z. Anal. Chem. 324:825

50. Klockenkämper R. (1982) In: Ullmans Encyklopädie der technischen Chemie, Bd. 5, Verlag Chemie, Weinheim. pp 501-518
51. William K.L. (1987) An introduction to X-Ray spectrometry. Allen & Unwin, Boston
52. Klockenkämper R. (1987) Spectrochim. Acta 42 B:423
53. Yoneda Y., Horiuchi T. (1971) Rev. Sci. Instrum. 42:1069
54. Wobrauschek P., Aiginger H. (1975) Anal. Chem. 47:852
55. Schwenke H., Knoth J. (1983) Nucl. Instrum. Methods 193:239
56. Michaelis W., Knoth J., Prange A., Schwenke H. (1985) Adv. X-Ray Anal. 28:75
57. Prange A. (1987) GIT Fachz. Lab. 6:513
58. Von Bohlen A., Klockenkämper R., Otto H., Tölg G., Wiecken B. (1987) Int. Arch. Occup. Environ. Health 59:403
59. Von Bohlen A., Klockenkämper R., Tölg G., Wiecken B. (1988) Fresenius Z. Anal. Chem., in press
60. Ruch C., Rastegar F., Heimburger R., Maier E.A., Leroy M.J.F. (1985) Anal. Chem. 57:1691
61. Garten R.P.H. (1984). In: Fresenius W. et al. (eds.) Analytiker Taschenbuch, Bd. 4, Springer, Berlin, Heidelberg, New York, pp 259-286
62. Li H.K., Malmqvist K.G., Carlson L.E., Akselsson K.R. (1984) Nucl. Instrum. Methods B 3:347
63. Li H.K., Malmqvist K.G. (1985) Nucl. Instrum. Methods B 12:257
64. Gilfrich J.V., Sketton E.T., Quadri S.B., Kirkland J.P., Nagel D.J. (1983) Anal. Chem. 55:187
65. Iida A., Goshi Y. (1984) Adv. X-Ray Anal. 28:61
66. Ketelsen P., Knöchel A., Petersen W. (1986) Fresenius Z. Anal. Chem. 323:807
67. Iida A., Yoshinaga A., Sakurai K., Goshi Y. (1986) Anal. Chem. 58:394
68. Heumann K.G. (1982) Int. J. Mass Spectrom. Ion. Phys. 45:87
69. Heumann K.G., Beer F., Weiss H. (1983) Mikrochim. Acta 1983:95
70. Gray A.L., Date A.R. (1983) Analyst 108:1033
71. Houk R.S. (1986) Anal. Chem. 58:97A
72. Jakubowski N., Stuewer D., Tölg, G., to be published
73. Donohue D.L., Young P., Smith D.H. (1982) Int. J. Mass Spectrom. Ion. Phys. 43:293
74. Tschöpel P. (1982) Pure Appl. Chem. 54:913
75. Tschöpel P. (1980) In: Ullmann Encyklopädie der technischen Chemie, Bd. 5, Verlag Chemie, Weinheim, p 27

76. Knapp G. (1984) Fresenius Z. Anal. Chem. 317:213

77. Schramel P., Hasse S., Knapp G. (1987) Fresenius Z. Anal. Chem. 326:142

78. Knapp G. (1985) Int. J. Environ. Anal. Chem. 22:71

79. Tölg G. (1983) Pure Appl. Chem. 55:1989

80. Mizuike A. (1983) Enrichment techniques for inorganic trace analysis. Springer, Berlin, Heidelberg, New York

81. Mizuike A. (1986) Fresenius Z. Anal. Chem. 324:672

82. Bächmann K. (1981) Crit. Rev. Anal. Chem.12:1

83. Zolotov Yu.A. Bodnya V.A., Zagruzina A.N. (1982/83) Crit. Rev. Anal. Chem. 14:93

84. Van Grieken R. (1982) Anal. Chim. Acta 143:3

85. Cassidy R.M. (1981) In: Lawrence J.F. (ed.) Trace analysis, vol 1. Academic Press, New York, pp 121-192

86. Möller J. (1988) In: Fresenius W. et al. (eds.) Analytiker Taschenbuch, Bd. 7. Springer, Berlin, Heidelberg, New York, London, Paris, Tokyo, pp 198-275

87. Tyson J.F. (1985) Analyst 110:419

DETERMINATION OF SELENIUM IN BODY FLUIDS USING HYDRIDE-GENERATION AND GRAPHITE FURNACE ATOMIC ABSORPTION SPECTROMETRY

B. Welz, M. Schubert-Jacobs, G. Schlemmer and M. Sperling
Department of Applied Research, Bodenseewerk Perkin-Elmer & Co GmbH,
D-7770 Überlingen, FRG

Abstract

Hydride-generation atomic absorption spectrometry (HG AAS) offers the advantages of excellent sensitivity for the determination of selenium and relatively simple instrumentation. Therefore it is well suited for routine analysis. The accuracy of the determination, however, depends critically on the acid digestion procedure, and severe systematic errors may be introduced when an improper procedure is used. A gradual destruction with nitric, sulphuric and perchloric acids to a final temperature of 310 °C in special flasks with long necks was found to be optimum for the subsequent determination of selenium using HG AAS.

Graphite furnace atomic absorption spectrometry (GF AAS) is frequently preferred over other techniques because it usually requires only minimal sample pretreatment or none at all. This is particularly true when up-to-date technology such as the stabilized temperature platform furnace (STPF) concept and Zeeman-effect background correction is applied. Selenium, however, due to interaction with carbon and the formation of gaseous molecules tends to cause more difficulties than other elements. Careful selection of the modifier and subtle details in the temperature program play an important role in eliminating matrix interferences.

Both techniques, HG AAS and GF AAS, can provide accurate results in the determination of selenium in body fluids at all concentrations of interest. It is important, however, to be aware of the traps and to avoid the pitfalls.

Trace Element Analytical Chemistry in Medicine and Biology

Introduction

Selenium has gained substantial interest as an essential trace element for humans as well as for some animals. The accurate determination of this element is, however, still a major challenge for the analyst. Acid-digestion fluorimetry was widely used for the determination of selenium (1-4) but atomic absorption spectrometric (AAS) techniques gained more and more importance because of their sensitivity and specificity. There was some doubt, however, about the accuracy of the results obtained, particularly when HG AAS was used (5, 6). This was shown to be predominantly caused by improper acid digestion procedures prior to hydride generation and determination by AAS (7-9). Many selenium compounds are volatile and can be lost during an inadequately controlled decomposition procedure (10). On the other hand, acid-resistant organoselenium compounds, such as selenomethionine, selenocysteine and the trimethylselenonium ion, are not fully decomposed, and thus low results are obtained (7, 8, 11). An acid digestion with nitric, sulphuric and perchloric acids at a temperature up to 310 °C, originally used for the decomposition of marine biological tissue samples (12) was applied successfully for the analysis of various human body fluids (8, 13).

GF AAS is the technique preferred by many analysts for trace element determination because it frequently permits direct analysis of body fluids without pretreatment steps such as acid digestion. The volatility of many selenium compounds, however, necessitates the use of a chemical modifier (14) in order to avoid losses in the pyrolysis stage. Sample atomization from a platform using STPF conditions (15) is also essential in order to reduce vapour phase interferences to a minimum. Erroneous results may be obtained, however, even when these measures are taken (9, 16). This is at least in part due to insufficient stabilization of some oxidation states of selenium by commonly used chemical modifiers, such as nickel (17). Other modifiers, such as a mixture of copper and magnesium nitrate (17, 18), or of palladium and magnesium nitrate (19) were found to be much more effective in stabilizing the various selenium compounds.

Spectral interferences due to iron and/or phosphate are another potential source of error in the determination of selenium when continuum-source background correction is used (16, 20, 21, 22). This interference is not observed in instruments with Zeeman-effect background correction (22-25), and this type of correction is therefore mandatory for an accurate direct determination of selenium in body fluids. Results of interlaboratory trials will be given in order to demonstrate the capability of the proposed methods and techniques.

Experimental

For HG AAS, a Perkin-Elmer Model 4000 atomic absorption spectrometer, equipped with an electrodeless discharge lamp for selenium, operated from an external power supply at 6 W, was used. A spectral slit-width of 2.0 nm was used to isolate the 196.0-nm resonance line. The signals were recorded with a Perkin-Elmer Model 56 strip chart recorder. A Perkin-Elmer Model MHS-20 hydride system was operated at a quartz cell temperature of 900 °C. The time settings at the controller were 40 s PURGE I, 8 s REACT (during this time, sodium tetrahydroborate solution is injected into the sample solution) and 30 s PURGE II. The acid digestions were carried out in an aluminium heating block. The temperature was controlled by a thermostat from a Perkin-Elmer F 20 gas chromatograph.

Sodium tetrahydroborate solution, 3 % (m/V), was prepared by dissolving 3 g of sodium tetrahydroborate powder (Riedel-de-Haen) in 100 mL deionized water and stabilizing with 1 g of sodium hydroxide. The solution was filtered and could be stored for only a few days. Hydrochloric acid (32 % m/V), perchloric acid (70 % m/V), sulphuric acid (96 % m/V) and nitric acid (65 % m/V) were of suprapure quality (Merck).

For sample digestion, graduated quartz flasks with a long neck and a nominal volume of 40 mL as shown in Figure 1 (Kürner Analysentechnik, Rosenheim, FRG) were used. It is important that the flasks have long necks which act as air coolers so that no selenium is lost during the decomposition (26).

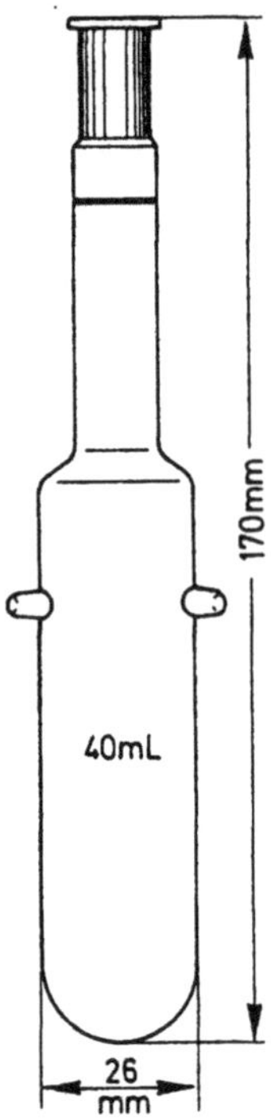

Fig. 1: Acid Digestion Flask, 40 ml with Gradiation Mark and Ground Glass Joint

0.5 mL of serum, whole blood or urine was placed into the digestion flask and 1.0 mL of nitric acid was added. The digestion flask was placed into the aluminium heating block, heated slowly (~ 15 min) to 140 °C and held at this temperature for about 20 min. After cooling to room temperature 0.5 mL of sulphuric acid and 0.2 mL of perchloric acid were added. The temperature was raised slowly to 140 °C, 200 °C, 250 °C and 310 °C, and held at each of these temperatures for 15 min before the next increase. The final temperature of 310 °C was held for at least 20 min so that most of the perchloric acid was removed, and the volume of the digestion solution was decreased to about 0.5 mL. After cooling to near room temperature, 20 mL of 5 mol/L hydrochloric acid was added to the digestion solution, heated to 95 °C and held at this temperature for 20 min in order to reduce all hexavalent selenium to the tetravalent form. The solution was allowed to cool to room temperature and diluted to volume (40 mL) with deionized water. 5 mL of the digestion solution (corresponding to 62.5 µL body fluid) and 15 mL of water were placed into the reaction flask of the hydride system for the determination of selenium. Larger aliquots or the total volume of the digestion solution may be transferred to the reaction flask and used for the determination if higher sensitivity is required.

For GF AAS, a Perkin-Elmer Model Zeeman/3030 atomic absorption spectrometer equipped with an HGA-600 graphite furnace and an AS-60 autosampler was used. Pyrolytically coated graphite tubes (Perkin-Elmer Part No. B010-9322) with an inserted L'vov platform (Perkin-Elmer Part No. B010-9324) were used exclusively. The graphite furnace temperature program used for the determination of selenium in body fluids is given in Table 1.

Copper nitrate solution, 1000 mg L^{-1} Cu, was prepared by dissolving 100 mg of copper metal powder in 1 mL of concentrated nitric acid and diluting to 100 mL with deionized water. Palladium nitrate solution, 3000 mg L^{-1} Pd, was prepared by dissolving 300 mg of palladium metal powder (Alfa Ventron Products No. 400030) in 1 mL of concentrated nitric acid with the addition of 10 µL of concentrated hydrochloric acid under gentle heating and diluting to 100 mL with deionized water. Magnesium nitrate solution, 2000 mg L^{-1} $Mg(NO_3)_2$, was prepared by dissolving 350 mg of magnesium nitrate hexahydrate (suprapure, Merck No. 5855) in deionized water and diluting to 100 mL. The copper nitrate-magnesium nitrate and palladium

Table 1: Graphite Furnace Temperature Program for the Determination of Selenium in Body Fluids Using Oxygen Ashing and Copper Nitrate-Magnesium Nitrate or Palladium Nitrate-Magnesium Nitrate Modifier.

Step	1	2	3	4	5	6	7	8
Temp (°C)	90	140	500	500	900	2100	2600	20
Ramp Time (s)	1	10	10	1	1	0	1	1
Hold Time (s)	40	10	30	30	10	3	5	5
Read						on		
Argon Flow (mL/min)	300			300	300	0	300	300
Oxygen Flow (mL/min)		300	300					

nitrate-magnesium nitrate mixed modifiers were prepared by mixing equal volumes of the above-mentioned solutions. Ten µL of the mixed solution were added to the body fluid in the graphite tube. All body fluids were diluted with a 0.15 % (m/V) Triton X-100 solution, and aliquots of 10 to 20 µL were used for determination.

Among the materials investigated were Seronorm Trace Elements Serum, Urine, and Whole Blood (NYCOMED AS, Oslo, Norway), and NBS Reference Material No. 8419, Bovine Serum (National Bureau of Standards Office of Standard Reference Materials, Washington, DC). Also investigated were several individual blood, serum and urine samples obtained from various sources.

Results and Discussion

1. Hydride-Generation AAS

In the hydride-generation technique, by the addition of sodium tetrahydroborate to the acidified sample solution, gaseous hydrogen selenide (H_2Se) is formed, stripped from the solution using an inert gas and atomized in a heated quartz cell. For this reaction it is essential that selenium is present in its tetravalent ionic form, i.e. as the selenite. Organic selenium compounds must be decomposed completely in order to apply this technique successfully.

In 1983 an IUPAC interlaboratory trial on the determination of selenium in lyophilized human blood serum reference materials was organized (6). Of the 27 participating laboratories, 4 applied HG AAS for the determination of selenium after acid digestion of the serum. The results obtained with this technique are depicted in Fig. 2, With the exception of one laboratory, the values exhibit large negative deviation relative to the accepted mean as well as substantial within-laboratory variances. These results indicated the existence of systematic errors, and all HG AAS data were therefore excluded in estimating the selenium level. An IUPAC project was started to investigate the sources of error in HG AAS determination of selenium in body fluids and to establish a reliable method based on this technique.

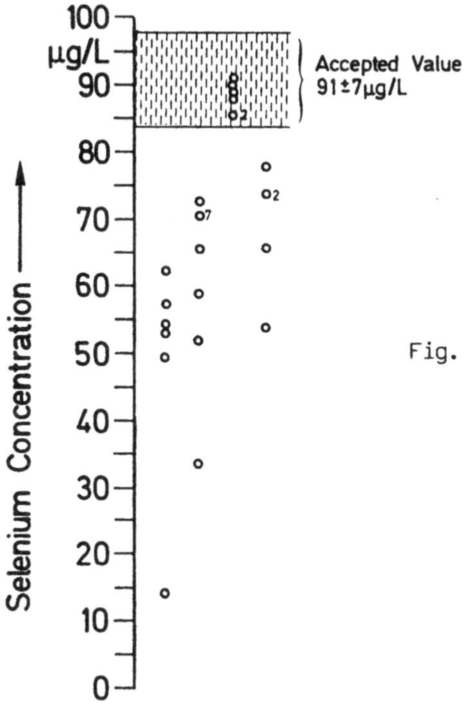

Fig. 2: Results Obtained for Selenium in Lyophilized Serum Using HG AAS (4 Laboratories, 6 or 12 Independent Determinations each) and Accepted Range. IUPAC Interlaboratory Trial, 1983.

In a first set of experiments several common acid digestion procedures were investigated with respect to their ability to mineralize organoselenium compounds. Selenomethionine is among the most acid-resistant compounds found in human body fluids (7, 11). Figure 3 shows the signals obtained from aqueous solutions of this compound after different acid digestions in comparison to those from a reference solution prepared from tetravalent ionic selenium (selenite). Without acid digestion, no signal for selenium is found for selenomethionine. Using a nitric acid digestion, 6 h on a boiling water bath, or 1 h at 160 °C under pressure, the recovery is 10 % and 50 %, respectively. Only after a nitric, sulphuric and perchloric acids digestion with a maximum temperature of 310 °C is full recovery obtained for selenium from selenomethionine. Essentially the same behaviour was found for selenomethionine added to whole blood. The recovery was 35 % and 77 %, respectively, for the two nitric acid digestion procedures, whereas it was always around 100 % after the nitric, sulphuric and perchloric acids digestion.

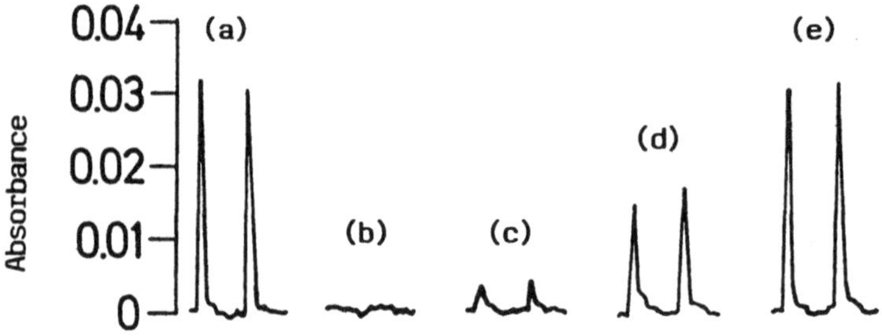

Fig. 3: Recovery of Selenium for Selenomethionine Depending on the Acid Digestion Procedure:
(a) 8 ng Se as the Selenite;
(b) - (e) 7.9 ng Se as Selenomethionine;
(a) and (b) no Acid Digestion,
(c) Nitric Acid, 6 h at 95 °C;
(c) Nitric Acid, 1 h at 160 °C under Pressure;
(e) Nitric, Sulfuric and Perchloric Acids, 310 °C.

After the proper acid digestion procedure was found and the working conditions were set, an interlaboratory trial was organized to demonstrate that the parameters could be transferred and accurate results be obtained for selenium in body fluids using HG AAS. Fourteen laboratories, nine of which had followed the recommended procedure outlined in the Experimental section, submitted their results. Three laboratories submitted additional results that were obtained using modified acid digestion procedures, and five laboratories reported only data that were obtained using alternate acid digestion or an alternate analytical technique.

All data reported for selenium in human serum, blood and urine and for bovine serum are depicted in Figures 4 to 7. The human serum and the urine had been analyzed previously in another IUPAC trial by 15 laboratories using 6 inherently distinct analytical methods. The mean values found in that interlaboratory trial were 90.7 $\pm$ 1.1 µg L^{-1} and 49.4 $\pm$ 0.7 µg l^{-1}, respectively (27). The 9 laboratories using the recommended procedure in this interlaboratory trial reported mean values between 84.2 µg L^{-1} and 100.7 µg l^{-1} with an average of 91.7 $\pm$ 1.7 µg L^{-1} for the serum sample. In the analysis of the urine sample, one laboratory was excluded because of inconsistency with the other laboratories. The other 8 laboratories using the recommended procedure reported mean values between 45.8 µg L^{-1} and 54.0 µg L^{-1} within an average of 49.9 ± 1.2 µg L^{-1}. The results for both materials are in excellent agreement with the selenium values established for these materials in previous interlaboratory trials.

Also distributed in the interlaboratory trial was an acidified aqueous reference solution with a nominal selenium content of 50.0 µg l^{-1} which was to be carried through the same acid digestion procedure as were the body fluids. One laboratory (the same as for the urine) had to be excluded because of inconsistency with the other laboratories. The other 8, which had used the recommended procedure, reported mean values between 44.7 µg L^{-1} and 54.6 µg L^{-1} with an average of 49.7 ± 1.0 µg L^{-1}. The similarity of the results obtained for this aqueous reference solution and those obtained for the urine was striking with respect to both accuracy and precision.This means that the urine matrix had no influence on the performance of the proposed procedure and that urine, if a proper acid digestion procedure is used, can be analyzed as accurately and precisely as an aqueous selenium reference solution.

Very little information was available for the whole blood sample, except for a value of 94 ± 1 µg L^{-1} obtained using acid digestion fluorimetry. The 9 laboratories using the recommended procedure reported mean values between 82.8 µg L^{-1} and 98.5 µg L^{-1} with an average of 93.0 ± 1.6 µg L^{-1}.

The NBS Reference Material bovine serum was of particular interest because its low selenium content of 16 ± 2 µg L^{-1} is close to that which can be expected in cases of severe selenium deficiency. The results of this study may therefore reflect the capability of the proposed procedure for the measurement of low selenium concentrations in human body fluids. The data of three laboratories had to be excluded because of inconsistency with other laboratories and/or large within-laboratory variance. This may be an indication that some laboratories apparently had difficulties in analyzing samples with a very low selenium content. The remaining 6 laboratories which used the recommended procedure reported mean values ranging from 14.0 µg L^{-1} to 19.0 µg L^{-1} with an average of 15.8 ± 0.8 µg L^{-1}. This was in excellent agreement with the recommended value for this material. The

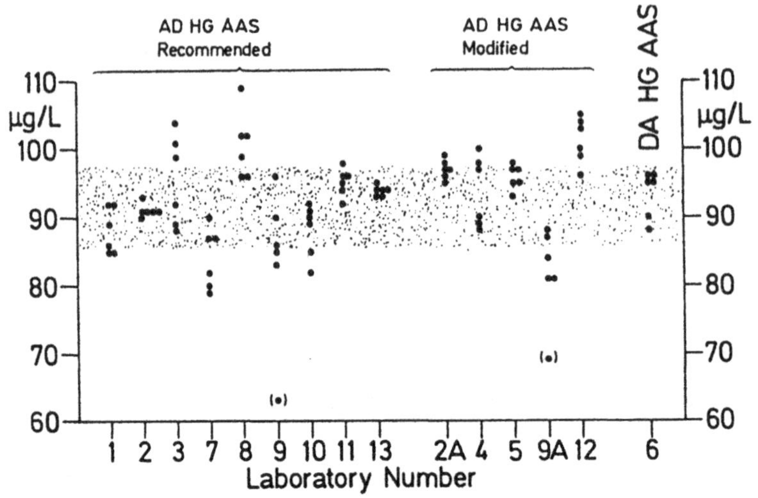

Fig. 4: IUPAC Interlaboratory Trial - Selenium in Seronorm Serum (Batch 105). DA HG AAS = Dry Ashing HG AAS.

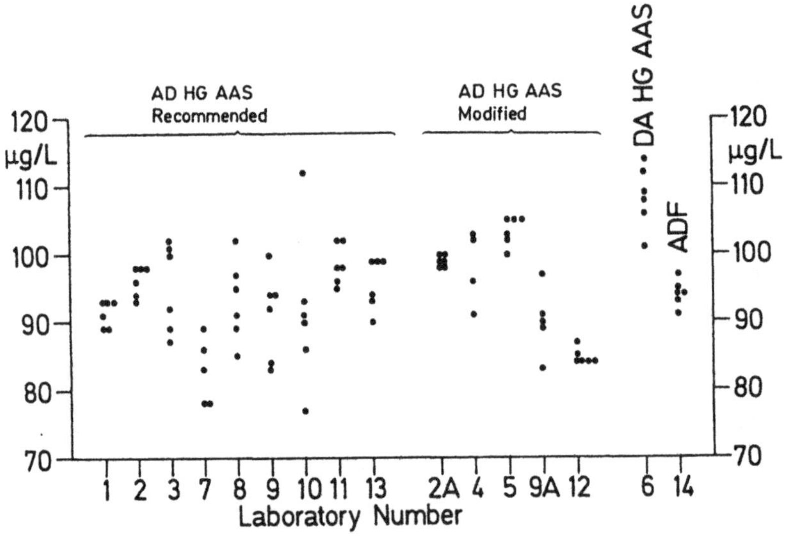

Fig. 5: IUPAC Interlaboratory Trial - Selenium in Seronorm Human Blood. ADF = Acid Digestion Fluorimetry.

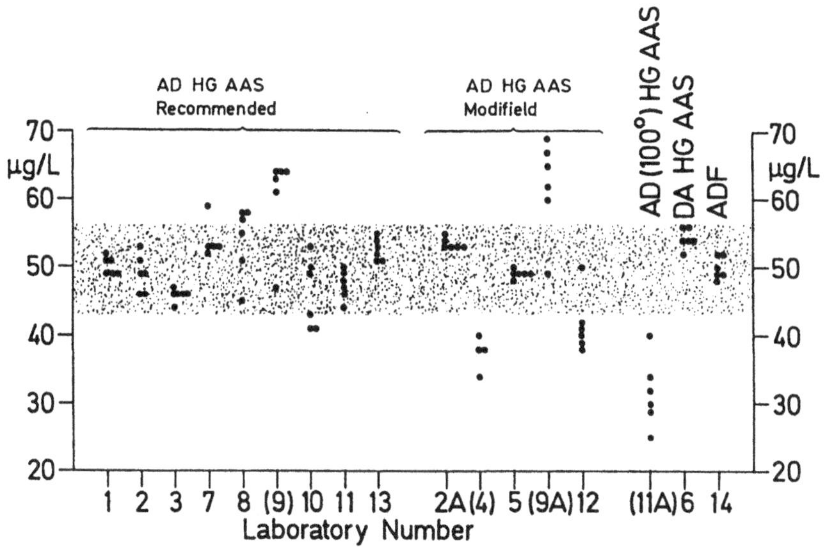

Fig. 6: IUPAC Interlaboratory Trial - Selenium in Seronorm Urine (Batch 108) AD (100°) = Acid Digestion at 100 °C only.

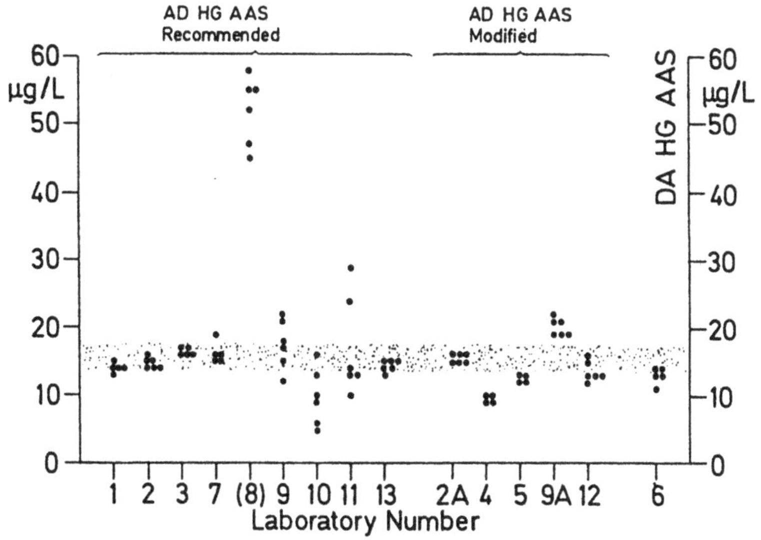

Fig. 7: IUPAC Interlaboratory Trial - Selenium in NBS Bovine Serum.

majority of the participating laboratories apparently were able to determine this low selenium value with good accuracy and precision using the recommended procedure. Three of the 4 laboratories which used other procedures, i.e. those which altered the sample treatment or acid digestion slightly, reported mean values below 14.0 µg L^{-1}, suggesting that the recommended procedure had to be followed more closely at low selenium concentrations.

This interlaboratory trial showed that systematic errors which were observed in earlier trials for HG AAS could be avoided if the sample decomposition procedure described in this study was applied. The recommended acid digestion procedure could be reproduced in other laboratories and, at least for low selenium levels, resulted in a better accuracy than other procedures which were well established in the participants' laboratories.

2. Graphite Furnace AAS

One of the distinct advantages of GF AAS is that liquid samples, such as body fluids, can be analyzed directly so that sample pretreatment can be reduced to a minimum. The sample is injected into the graphite tube, dried and pretreated in a pyrolysis stage in order to destroy and/or remove the majority of the sample matrix prior to the atomization stage. Particularly for the determination of volatile elements, such as selenium, careful control of the pyrolysis stage is a very important part of the analysis. Stabilization of the analyte with a "modifier" is essential in order to prevent volatilization of the analyte element at the temperatures required for an effective pyrolysis and removal of the matrix components. Mixtures of copper and magnesium nitrate or palladium and magnesium nitrate were found to be most effective in stabilizing all the oxidation states of selenium, as well as organic compounds, to pyrolysis temperatures of at least 1000 °C. Use of pyrolytic graphite coated tubes and sample deposition on a L'vov platform mounted in the tube were also found to be important parameters for accurate determination of selenium.

A substantial improvement in sensitivity and in long-term stability of the selenium signals was brought about by using "oxygen ashing", as is depicted in Fig. 8. When oxygen was introduced into the graphite tube during the pyrolysis stage at a temperature of 500 °C, the organic matrix was ashed and not only charred, as it was in an inert gas. While this had no influence on the selenium signal from an aqueous solution, it substantially enhanced the signal for selenium in blood and other body fluids.

When all these measures were taken, including the use of STPF conditions (15, 19) and of Zeeman-effect background correction (22-25), as was outlined in the Introduction, the peak area sensitivity for inorganic and for organically bound selenium was the same, within experimental error. Nevertheless we found consistently a somewhat lower sensitivity for selenium in body fluids compared to that for selenium in aqueous solutions containing

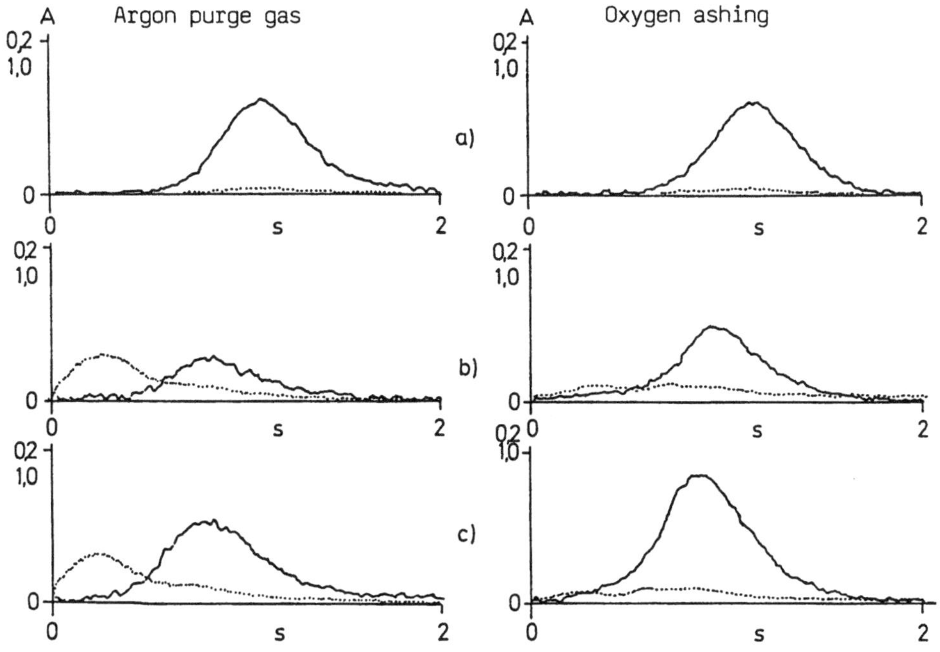

Fig. 8: Influence of Oxygen Ashing on the Signal for Selenium Using the Palladium Nitrate-Magnesium Nitrate Modifier.
a) Aqueous Standard Solution; b) Blood Sample; c) Blood Sample with Added Selenium

the same amount of modifier. This is illustrated in Figs. 9 and 10 for serum and urine, respectively. The slope of the additions curve was always lower than that for the matrix-free solutions, and it was different not only for serum and urine but also for a series of body fluids of the same type. This made it essential that the method of additions be applied routinely to the determination of selenium in body fluids. The results obtained with this technique, however, were accurate and in good agreement with those obtained with other techniques as can be seen in Table 2.

It should be mentioned in this context that the slope of the working curve for aqueous standards and for selenium added to the various body fluids was always identical when the samples were acid-digested. The type of acid digestion was hereby of little importance, and a simple digestion with hot nitric acid was absolutely sufficient. All determinations could in this case be made against a working curve established using aqueous standards.

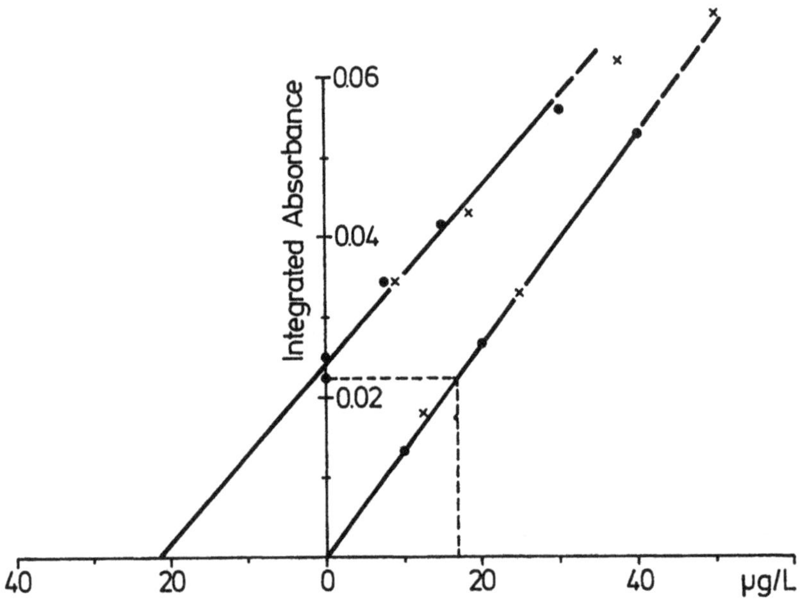

Fig. 9: Determination of Selenium in Serum Using Standard Calibration and the Method of Additions. Found: 88 µg L^{-1}; Reference Value 91 µg l^{-1}.
● Selenite; + Selenomethionine

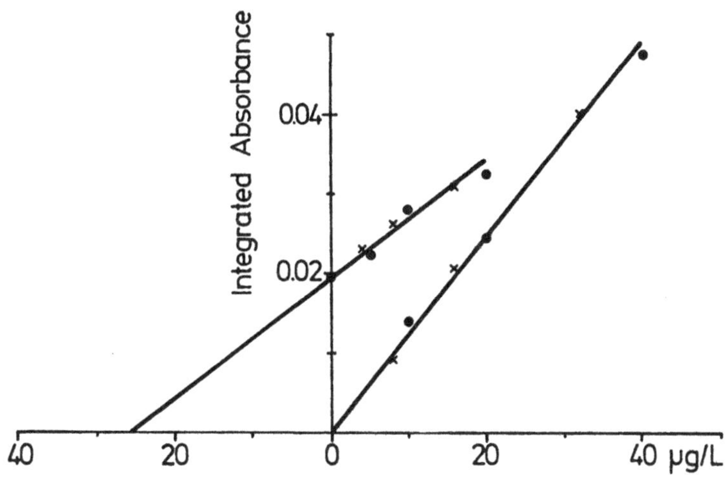

Fig. 10: Determination of Selenium in Urine Using Standard Calibration and the Method of Additions. Found 47 µg L^{-1}; Reference Value: 49.6 ± 2.8 µg l^{-1}.
● Selenite; + Trimethylselenonium Iodide.

Table 2: Determination of Selenium in Serum Samples Using GF AAS and the Method of Addition. Reference Values Obtained with Isotope Dilution Mass Spectroscopy (IDMS) Provided by C. Veillon, USDA, Beltsville, MD, USA.

Sample No.	Selenium Content (µg L^{-1})	
	GF AAS	IDMS
1	165	176
2	362	340
3	43	46.3
5	226	234
6	368	323
7	64	70
8	315	307
9	153	147
10	183	181

Conclusion

Both procedures investigated in this study, GF AAS and HG AAS, gave accurate results for selenium in body fluids when used properly. This means for HG AAS that an acid digestion with nitric, sulphuric and perchloric acids to a maximum temperature of 310 °C was mandatory to completely mineralize all organic selenium compounds. For GF AAS the requirements were the use of Zeeman-effect background correction, oxygen ashing in the tube and of a powerful modifier mixture such as palladium nitrate and magnesium nitrate. When body fluids were analyzed directly, the method of additions had to be applied for evaluation. Acid digested samples could be determined directly against aqueous standards.

References

1. Watkinson, J.H. 1966. Anal. Chem. 38, 92.

2. Haddad, P.R., L.E. Smythe. 1974. Talanta 21, 859.

3. Olsen, O.E., I.S. Palmer, E.E. Cary. 1975. J. Assoc. Off. Anal. Chem. 58, 117.

4. Chan, C.C.Y. 1976. Anal. Chim. Acta 82, 213.

5. Ihnat, M., B.K. Thompson. 1980. J. Assoc. Off. Anal. Chem. 63, 814.

6. Ihnat, M., M.S. Wolynetz, Y. Thomassen, M. Verlinden. 1986. Pure and Appl. Chem. 58, 1063.

7. Verlinden, M. 1982. Talanta 29, 875.

8. Welz, B., M. Melcher, J. Nève. 1984. Anal. Chim. Acta 165, 131.

9. Welz, B., M. Melcher, G. Schlemmer. 1984. In: Trace Element - Analytical Chemistry in Medicine and Biology, Vol. 3 (P. Brätter and P. Schramel, eds.) W. de Gruyter, Berlin, p. 207.

10. Robberecht, H.J., R.E. van Grieken, P.A. van den Bosch, H. Deelstra, D. van den Berghe. 1982. Talanta 29, 1025.

11. Nève, J., M.Hanocq, L. Molle, G. Lefebvre. 1982. Analyst 107, 934.

12. Welz, B., M. Melcher. 1985. Anal. Chem. 57, 427.

13. Welz, B., M.S. Wolynetz, M. Verlinden. 1987. Pure and Appl. Chem. 59, 927.

14. Ediger, R.D. 1975. At. Absorpt. Newslett. 14, 127.

15. Slavin, W., D.C. Manning, G.R. Carnrick. 1981. At. Spectrosc. 2, 137.

16. Welz, B., M. Melcher, G. Schlemmer. 1983. Fresenius Z. Anal. Chem. 316, 271.

17. Welz, B., G. Schlemmer, U. Voellkopf. 1984. Spectrochim. Acta 39B, 501.

18. Nève, J., L. Molle. 1986. Acta Pharm. et Toxicologica 59, Suppl. VII, 606.

19. Schlemmer, G., B. Welz. 1986. Spectrochim. Acta 41B, 1157.

20. Manning, D.C. 1978. At. Absorpt. Newslett. 17, 107.

21. Saeed, K., Y. Thomassen, F.J. Langmyhr. 1979. Anal. Chim. Acta 110, 285.

22. Welz, B., G. Schlemmer. 1986. J. Anal. At. Spectrom. 1, 119.

23. Fernandez. F.J., W. Bohler, M.M. Beaty, W.B. Barnett. 1981. At. Spectrosc. 2, 73.

24. Fernandez, F.J., R. Giddings. 1982. At. Spectrosc. 3, 61.

25. Bauslaugh, J., B. Radziuk, K. Saeed, Y. Thomassen. 1984. Anal. Chim. Acta 165, 149.

26. Kotz, L, G. Kaiser, P. Tschöpel, G. Tölg. 1972. Fresenius Z. Anal. Chem. 260, 207.

27. Ihnat, M., Y. Thomassen, M.S. Wolynetz, C. Veillon. 1986. Acta Pharm. et Toxicologica 59, Suppl. VII, 566.

population of the level l and g is the degeneracy. This means that the more excited atoms we have in a hot environment the more ions we will get. Also, the higher the atoms are excited the more easily they will be ionized. If we can selectively excite atoms from only one element to a high energy level we will get an amount of charges that is proportional to the number of atoms of that element in the interaction region. By applying a voltage over the interaction region the charge increase can be detected as a current increase [1-12].

The selective excitation can be achieved with laser light, which has all the features needed for that task, high intensity, sharp frequency and easy wavelength tuning. As mentioned, excitations to high levels is needed to get a large increase in the ionization. That can be made two ways, either by using one laser with short wavelength or by succesive excitation from two lasers with more moderate wavelengths (Two colour LEI).The first way is useful for many elements but not for those with the highest ionization limits. It also has the drawback that one can get strong molecular two-photon ionization in some wavelength regions. Most of those problems are solved by using two colour LEI. This method also gives further advantages which will be discussed later. A schematic picture of the different excitation methods can be seen in fig 1.

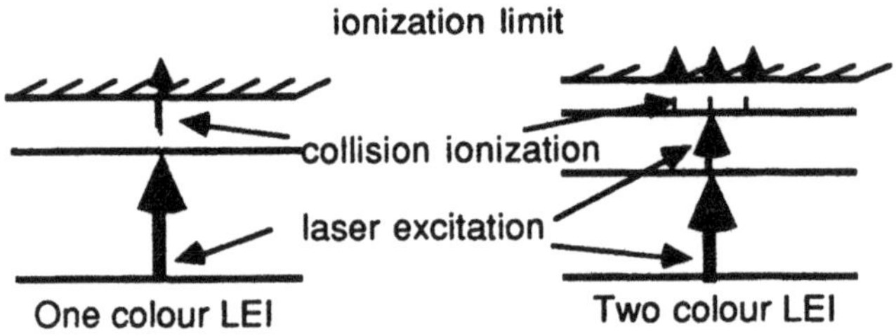

Fig 1. The physical mechanism of LEI

Experimental

A schematic view of the experimental equipment for two-colour LEI in flames is shown in figure 2. It consists mainly of four parts; a flame, a laser system, a charge collection system and a detection system. The flame, in which the solution containing

LASER ENHANCED IONIZATION SPECTROSCOPY IN FLAME AND FURNACE AS A TOOL FOR TRACE ELEMENT ANALYSIS

Mats Lejon, Sten Sjöström, Halina Rubinsztein-Dunlop, Ove Axner
Department of Physics
Chalmers University of Technology
S-412 96 Göteborg, Sweden

Abstract

Laser enhanced ionization spectroscopy (LEI) is an extremely sensitive method for detecting atoms and molecules in flames or furnaces. The LEI method is based on the enhanced rate of thermal (collisional) ionization which follows when the analyte element is selectively excited by light from a dye laser tuned to an appropriate wavelength. This enhanced rate of ionization is detected electrically by applying a voltage across the flame (furnace) and observing the changes in the electrical current.

Using LEI spectroscopy as an ultra-sensitive method for trace element analysis we have determined detection limits for more than 30 elements. The obtained detection limits, using a flame as an atomizer, are for many elements far below those which can be reached with conventional flame based methods. *For many of the investigated elements the detection limits are in the ppt-region.*

Introduction

Laser Enhanced Ionization is based on the fact that atoms in a hot (normally atmospheric pressure and thus collision-rich) environment has a certain probability of being ionized through collisions.This rate of ionization R is approximately proportional to:

$$R = n_1 \, g \, e^{-(E_i - E_1) / kT}$$

where E_i is the ionization energy for the element, and E_1 is the energy of the level from which the collision takes place, k is the Boltzman factor, T is the temperature, n_1 is the

Trace Element Analytical Chemistry in Medicine and Biology

the sample is aspirated, is supported on a single slot commercial atomic absorption burner and it is fed with acetylene and air. An excimer pump laser (Lambda Physik EMG 102) simultaneously pumps two dye-lasers (Lambda Physik EMG 2002). The dye-lasers can produce light from 220 nm to 1 µm (using second harmonic generation crystals for UV-light production). The energy of the laser light is about 1-10 mJ per pulse in the visible, 0.01-1 mJ in UV, and the duration of the laser pulse is 20 ns. The maximal repetition rate for the pump laser is 100 Hz. The charge collection system consists of a pair of stainless steel plate electrodes (held at -1500 volt) positioned at both sides of the flame and a burner head (connected to ground via a resistor) on which the signal is measured. The electrical detection system consists of a fast amplifier, a boxcar integrator and a recorder. We are also working with a graphite furnace (Perkin-Elmer HGA-72) as an atomizer. The experimental set-up for the graphite furnace is similar to the above described.

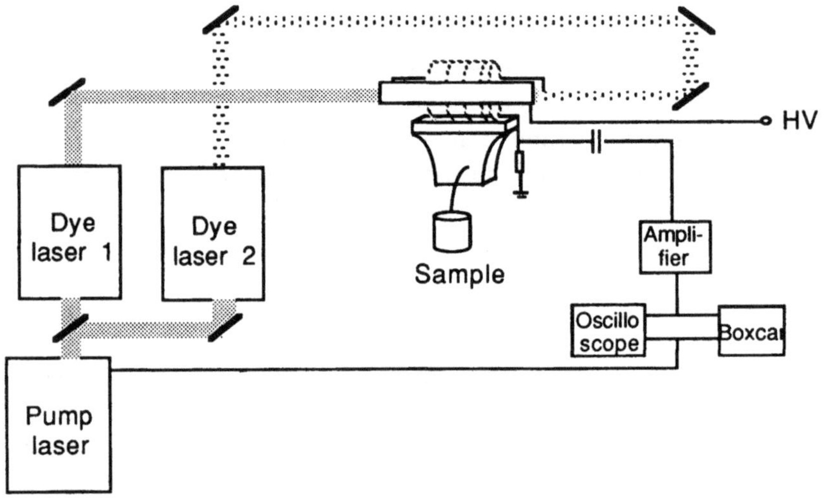

Fig 2. Experimental setup for LEI in flame.

Reduction of spectral interferences with two colour LEI

All methods for trace element analysis in realistic samples suffer from spectral interferences. This occurs if there is a matrix element present in the sample which has an absorption line close to that of the analyte element. Problems associated with spectral interferences of the above described type can be almost totally eliminated using a background correction technique based on two-colour LEI [12,13]. This background correction is made by doing three measurements, the two-colour and two one-colour. The difference between the two-colour and the sum of the one-colour measurements can then be used as a true representation for the concentration of the analyzed element.

Detection limits for Laser Enhanced Ionization in flame

Na — element
0.001 — det.limit one step LEI (ng/ml ppb)
0.0006 — det.limit two step LEI (ng/ml ppb)

H																	He
Li 0.005	Be											B	C	N	O	F	Ne
Na 0.001 0.0006	Mg 0.003 0.0001											Al 0.2	Si	P	S	Cl	Ar
K 0.1	Ca 0.006	Sc	Ti	V	Cr 0.2	Mn 0.04 0.0002	Fe 0.08 0.01	Co 0.06 0.04	Ni 0.02	Cu 3	Zn	Ga 0.03	Ge	As 3000	Se	Br	Kr
Rb	Sr 0.003	Y	Zr	Nb	Mo	Tc	Ru	Rh	Pd	Ag	Cd 0.2 0.01	In 0.001	Sn 0.4	Sb 50	Te 5	I	Xe
Cs	Ba 0.2	La	Hf	Ta	W 300	Re	Os	Ir	Pt 1	Au 1.2	Hg	Tl 0.006	Pb 0.2 0.003	Bi 0.2	Po	At	Rn
Fr	Ra	Ac															

Ce	Pr	Nd	Pm	Sm	Eu	Gd	Tb	Dy	Ho	Er	Tm	Yb 1.7 0.1	Lu
Th	Pa	U	Np	Pu	Am	Cm	Bk	Cf	Es	Fm	Md	No	Lr

Table 1. Detection limits for LEI in flames

Multielement capability and low detection limits.

Although LEI spectroscopy is a method detecting one element for each excitation wavelength used, it is a multi-element method in the sense that many elements can be investigated in rapid sequence with only a minimum of experimental rearrangement [9,10]. This is due to the recent development of new dyes tuneable over wide wavelength ranges combined with the large number of optical transitions available to LEI spectroscopy for each element, which makes it possible to investigate many elements using just one dye (or one combination of dyes in two-colour LEI). The time needed for changing wavelength in this case is just a few seconds and the total time for measuring all elements will therefore merely be determined by the sampling time.

A summary of the detection limits for one- and two-colour LEI in flames is given in Table 1. Some of the two-colour results are not significantly better than corresponding one-colour results. The main reason for this is that whenever the sensitivity of the one-colour measurement is so high that the detection limits are determined by contamination in the laboratory the improved sensitivity that follows from two-colour excitation will not lower the detection limits.Despite this it may still be very advantageous to use two-colour excitation in many cases due to a better selectivity and better background correction

LEI in graphite furnace yields pg analysis.

LEI in graphite furnace shows an even greater potential than does LEI in flames [11,14,15]. One important advantage with the graphite furnace compared to the flame is that micro-analyses can be performed. The detection limits today are in the pg range, see Table 2, but can probably be significantly lowered if the measurements are performed with clean-room facilities. However, LEI in furnace is not as well developed as LEI in flames and more research is needed before it can be considered as a reliable and easy-to-use method for trace element analysis.

Table 2. Detection limits with LEI in graphite furnace

Element	λ_1 (nm)	λ_2 (nm)	Det. lim.(pg) absolute	Det. lim.(ppb) in 50 μL water
Co	298.72	480.82	5	0.1
Cr	298.60	483.59	5	0.1
Mg	285.21	none	1	0.02
Mn	280.11	472.97	1	0.02
Na	285.30	none	5	0.1
Ni	294.39	497.67	7	0.14
Pb	283.31	471.99	0.5	0.01

REFERENCES

1. R. B. Green, R. A. Keller, P. K. Schenck, J. C. Travis, and G.G. Luther, 1976, J. Am. Chem. Soc. 98, 8517.

2. P.K.Schenck, W. C. Mallard, J. C. Travis, and K. C. Smyth, 1978, J. Chem. Phys. 69, 5147.

3. G. C. Turk, J. C. Travis, J. R. DeVoe, and T. C. O'Haver, 1979, Anal. Chem. 51, 1890.

4.* J. C. Travis, G. C. Turk, and R. B. Green, 1982, Anal. Chem. 54, 1006A.

5. G. J. Havrilla, S. J. Weeks, and J. C. Travis, 1982, Anal. Chem. 54, 2566.

6. W. G. Mallard and K. C. Smyth, 1982 Comb. and Flame 44, 61

7. T. Berglind, S. Nilsson and H. Rubinsztein-Dunlop, 1987, Physica Scripta 36 246

8. N. Omenetto, T. Berthoud, P. Cavalli, and G. Rossi, 1985, Anal. Chem. 57, 1256.

9. O. Axner and I. Magnusson, 1985, Physica Scripta 31, 587.

10. O. Axner, I. Magnusson, J. Petersson, and S. Sjöström, 1987, Anal. Spec. 41, 19.

11. I. Magnusson. O. Axner, I. Lindgren, and H. Rubinsztein-Dunlop, 1986, Appl. Spec. 40, 968.

12. I. Magnusson, O. Axner, and H. Rubinsztein-Dunlop, 1986, Physica Scripta 33, 429.

13. O. Axner, M. Lejon, I. Magnusson, H. Rubinsztein-Dunlop, S. Sjöström, 1987, Appl. Optics 26 3521.

14. A. S. Gonchakov, N. B. Zorov, Yu. Ya. Kuzyakov, and O. I. Matveev, 1979, Anal. Lett. 12, 1037.

15. I. Magnusson, S. Sjöström, M. Lejon, and H. Rubinsztein-Dunlop, 1987, Spectrochimica Acta vol. 42B,713-718

* review article

QUANTITATIVE ENERGY DISPERSIVE X-RAY FLUORESCENCE ANALYSIS (EXRFA) OF BIOLOGICAL SAMPLES USING INFORMATIONS FROM THE BACKGROUND CONTINUUM

V. Cercasov, H. Schreiber
Institute of Physics, University of Hohenheim,
Postbox 70 05 62, D-7000 Stuttgart 70 (FRG)

Abstract

A simple semi-empirical correction method based on the determination of two interdependent parameters: effective mass and effective thickness of the samples is presented. The parameters are calculated from the background intensities in two different energy regions of the continuum, usually at 5 and 15 keV. These energies are adequate for the information depth of the characteristic radiations of the interesting elements and account therefore in a representative manner for the mean extinction effects in biological samples of irregular shape. This method was applied on samples of fir and spruce needles in form of pellets and of individual needles as well, and on the feathers of birds.

Introduction

Biological samples are characterized by containing only relatively small amounts of interesting elements in a matrix of low atomic number (H,C,N,O). Often it is impossible to gain an ideal geometrical shape, e.g. plane pellets of infinite thickness. On the other hand for EXRFA the high counting rate caused by thick samples reduces the efficiency of the determination of elements with low and medium X-ray energies.
Thereof results the need to develop a calibration procedure for samples with arbitrary shape. In recent years several approaches have been proposed to account for matrix extinction effects using informations from spectral background or from the

Trace Element Analytical Chemistry in Medicine and Biology

scattered radiation of the tube. For example in (1), (2) and (3) the informations are taken from the coherently and incoherently scattered excitation radiation peaks; additionally a parameter called "effective thickness" of the sample can be derived. O'Reilly and King (4) combine the informations from scattered radiation with the fundamental-parameter method. Characteristic of these methods is the different information depth of the scattered radiation and of the energies used for the analysis. Deviations from the homogeneity lead to inconsistent corrections for the really analysed part of the sample depending on the characteristic energies in relation to the sample thickness.

The use of background radiation in the energy regions strongly affected by extinction could reduce this discrepancy. However only in samples consisting of a light matrix with small contents of higher Z elements the disturbance of the background radiation spectrum by absorption effects is sufficiently low to permit a suitable extinction correction. Biological materials like leaves, needles or feathers meet these conditions.

Experimental

The measurements were done with a tube excited (Rh) EXRFA device (Finnigan Inc.) using an observation angle of 135° with respect to the exciting beam direction. The tube voltage was usually 40 kV.

The investigated samples were

a) layers of dust particles on filters
b) thin pellets of plant material
c) irregular shaped biological objects: needles, pieces of bark, feathers

with thickness varying between 0.1 and 20 mg/cm^2. Hence in the interesting energy range (2 to 15 keV), very different sample configurations result, from "infinite thick" to "infinite thin". As a consequence the deviations from the homogeneity in the sample material have extremely different effects on interdependent parameters like mean extinction coefficient or mean concentration value. For our next considerations, we assume that the biological matrix, consisting essentially of light elements, is homogeneous and is responsable for the extinction processes. A homogeneous distribution of heavy elements is not required, if additional informations about this distribution exist. In this case an appropriate extinction correction can be applied.

Calculation Method

The intensity of the characteristic radiation as well as of the background for a small sample depends on the position of the sample with respect to the exciting beam and detector aperture. We assume for both kinds of radiations the same geometrical dependence. Therefore for large samples of different shapes and positions the ratio between scattered and characteristic radiations remains the same (desregarding extinction effects). Our measurements confirmed this assumption for lateral deviations up to a position, where the relative efficiency was reduced to only 1% (10 mm from center), and for height variations up to 8 mm. That means, for a given sample the undisturbed scattering radiation represents a useful parameter for the relative efficiency.

For a fixed position of the sample the intensity of the scattered radiation depends on the sample mass and its elemental composition. In typical biological materials of limited thickness we find high energy regions, where this intensity depends predominantly on the mass of the sample. For an appropriate weighing procedure we choose the energy region between 13.7 keV and 13.9 keV, which can be disturbed at most by high contents of Rb and Sr. At this energy, for sample thickness of 30 mg/cm^2 we have an intensity loss of only 10% by extinction. The correction is achieved with the mean extinction coefficient for biological matrices, assuming an exponential law.

Equivalent to conditions, that are frequently met in practice, we defined the following standard status: uniform lateral mass and element distribution over the whole aperture, and plane shape at reference height "zero". To ascertain the relative insensitivity to variations in the mean atomic number of biological samples, we measured under the described standard conditions materials with different composition. The results are presented in Fig. 1.

Informations on the "effective thickness" of the sample can be derived from the shape of the scattered radiation spectrum. At high energies the sensitivity to thickness variations is small, at low energies the scattered radiation has its origin only in the outer layer of the sample. Consequently we used a medium energy region, corresponding to the information depth for elements from Ca to Fe, in order to calculate a parameter D called "effective thickness". This region should fulfill the important requirement of being little disturbed by characteristic radiation. For most biological materials with small

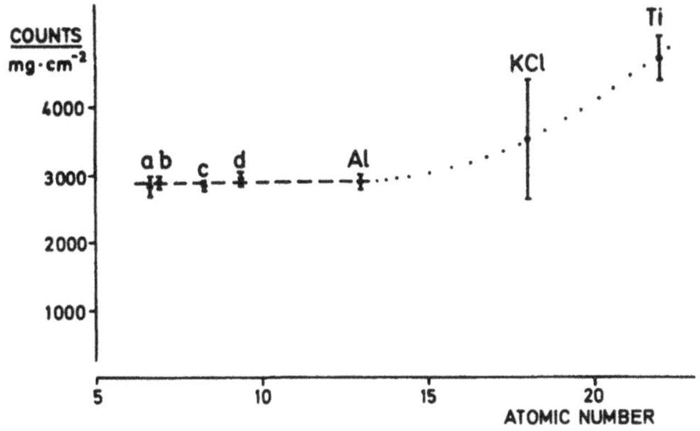

Fig. 1. Specific yield of scattered radiation at 13.8 keV for materials with different average atomic number (a: cellulose, b: cellulose nitrate, c: teflon, d: cellulose filter impregnated with KCl)

contents of V and Cr, the background intensity near 5.3 keV can be selected, however after applying a linear correction procedure, accounting for these elements. It is possible to use a simple exponential law, containing a mean value of the extinction coefficient for biological materials at this energy. For higher K and Ca contents of 6-8% by weight, the influence of these elements cannot be neglected. The extinction coefficient of the matrix has to be modified. It can be sufficiently approximated by an empirical linear function, depending on the counting rate between 3.1 keV and 4.1 keV.
Neglecting the above mentioned correction procedure, it can be stated: the counting rate C_2 at the higher energy (near 13.8 keV) furnishes the "effective mass" of the sample; the counting rate at the lower energy C_1 in relation to C_2 represents a model parameter for the "effective thickness". Because of the interrelationship, caused by the correction procedures, it is useful to apply an iterative computation method, alternating between approximations for thickness and for mass. Now having knowledge on the exact value of thickness, it is possible to correct for the extinction of the characteristic radiation. The energy dependent correction function, necessary for the conversion of the whole measured spectrum into a spectrum of an "infinite thin" sample, has been determined empirically by transmission measurements. This treatment makes it possible to apply successfully the X-ray software "Super ML" of Tracor Northern, Inc. (5) (6), using a reference library with "infinite thin" standards.
It is another profit of the method to yield concentration values not depending on variations of the tube current.

Results

To test the accuracy of the method, we used cardboard pieces of different shape e.g. large and small circles, strips. They were placed at various positions in the beam and arranged in single

and multiple layers of 3 to 60 mg/cm^2. Even for K and Ca, where the extinction correction factors attain more than 10, the consistency of the results was satisfactory.
The correction of the influence of different K and Ca contents could be tested by using pellets of appropriate Standard Reference Materials (NBS 1572 and 1575, IAEA/V-10).
These results encouraged us to apply this method on samples of single needles of fir and spruce since 1984.
The results in Table 1 show an agreement for the elements Fe, Zn and Sr, but a systematic manner of deviation for Ca, K and S, that is for elements with lower characteristic energies. Here the homogeneous model becomes insufficient: the extinction in the cuticle of the single needle samples has to be accounted for. Using the measured extinction coefficient for Ca, K and S, the ratios (in the last column of Table 1) lead to an identical "effective thickness" value of the not mineralised cuticle of 20 μm for these 3 years old needles. Thus measuring a needle sample in these two preparation forms offers an additional information for applying a more realistic two-layer model for a needle.

Table 1. Single Needle Results and Pellet Values (mg/g)

Element	Individual needles (mean of 10 needles)		Pellets (70 needles)	Ratio: Pellets / Individuals
	Tip part	Bottom part		
Sr	0.158	0.131	0.147	1.02
Zn	0.106	0.103	0.109	1.04
Fe	0.167	0.146	0.150	0.96
Ca	16.9	18.1	20.5	1.17
K	3.93	3.87	5.90	1.51
S	0.90	0.92	1.88	2.07

References

1. Van Espen, P., L. Van't dack, F. Adams and R. Van Grieken. 1979. Anal.Chem. 51, 961-967.

2. Van Dyck, P.M. and R.E. Van Grieken. 1980. Anal.Chem. 52, 1859-1864.

3. Török, Sz., P. Van Dyck and R. Van Grieken. 1984. X-Ray Spectrom. 13, 27-32.

4. O'Reilly, T. and B.-S.W. King. 1987. In: Advances in X-Ray Analysis, Vol. 30 (C.S. Barrett et al., eds.). Plenum Press, New York and London, pp. 165-174.

5. Schamber, F.H.. 1977. In: X-Ray Fluorescence Analysis of Environmental Samples (T.G. Dzubay, ed.). Ann Arbor Science, Ann Arbor. pp. 241-257.

6. Wodke, N.F. Jr. and F. Schamber. 1978. Super ML: Operation and Program Description. Tracor Northern, Inc.

AUTOMATIC RADIOCHEMICAL NEUTRON ACTIVATION ANALYSIS PROCEDURE FOR THE SIMULTANEOUS DETERMINATION OF THE TRACE ELEMENTS As, Au, Cd, Cs, Cu, Hg, Mo, Rb, Se AND Zn IN HUMAN SERUM

D. Van Renterghem and R. Cornelis

Laboratory of Analytical Chemistry, Institute for Nuclear Sciences, Rijksuniversiteit Gent, Proeftuinstraat 86, B-9000 Gent, Belgium

Abstract

In view of the simultaneous determination of the trace and ultra-trace elements As, Au, Cd, Cs, Cu, Hg, Mo, Rb, Se and Zn in human serum with radiochemical neutron activation analysis, a destruction-separation procedure has been developed. It consists of the combustion of an irradiated serum sample in the apparatus Trace-O-Mat (1-2), the dissolution of the ash and the separation of the elements by means of ion exchange.

Introduction

Neutron activation analysis is an accurate and sensitive method for the simultaneous determination of As, Au, Cd, Cs, Cu, Hg, Mo, Rb, Se and Zn in human serum, provided radiochemical separations are included. In this work, the organic ion exchanger Dowex 1X8 is employed to eliminate the major radioactivities of ^{82}Br ($T_{1/2}$ = 35.30 h), ^{42}K ($T_{1/2}$ = 12.36 h) and ^{24}Na ($T_{1/2}$ = 14.96 h) and to group the elements of interest in such a way as to avoid γ-spectral interferences from predominant radionuclides.
The separation is preceded by the destruction of the lyophilized, irradiated sample in the apparatus Trace-O-Mat. The combustion is maintained by the O_2-stream, while the loss of volatile elements is prevented by the liquid N_2-cooling. The residual ash is dissolved in an acid, suitable for the subsequent separation.

Trace Element Analytical Chemistry in Medicine and Biology

Experimental

1. Reagents

The strong basic organic anion exchanger Dowex 1X8, 100-200 mesh, Cl^--form, is purchased from Serva. The technical grade product exhibits the same properties as the more purified resin. This is of primary importance, because the quality of the resin drastically influences the behaviour of the elements As, Cu and Se.
All liquids used are of analytical grade. The products are very pure or pro-analysis.

2. Apparatus

The sample is pressed to a pellet with a Parr® pellet press.
The apparatus Trace-O-Mat and the peristaltic pump for transport of the fluids are operated by a control unit.
A glass column with an internal diameter of 8 mm is packed with the resin to a height of 75 mm.

Procedure

The destruction-separation procedure has been developed with the aid of the radiotracers ^{76}As ($T_{1/2}$ = 26.32 h), ^{198}Au ($T_{1/2}$ = 64.68 h), ^{82}Br ($T_{1/2}$ = 35.30 h), ^{115}Cd ($T_{1/2}$ = 53.46 h), ^{137}Cs ($T_{1/2}$ = 30.1 a), ^{64}Cu ($T_{1/2}$ = 12.70 h), ^{197}Hg ($T_{1/2}$ = 64.10 h), ^{42}K ($T_{1/2}$ = 12.36 h), ^{99}Mo ($T_{1/2}$ = 66.02 h), ^{24}Na ($T_{1/2}$ = 14.96 h), ^{86}Rb ($T_{1/2}$ = 18.66 d), ^{75}Se ($T_{1/2}$ = 119.77 d) and ^{65}Zn ($T_{1/2}$ = 244.0 d).

1. Preparation of the pellet

About 10 µl of the tracer solutions was spotted on 500 mg freeze-dried human serum. After drying, the doped serum was wrapped in a piece of a precision wipe and pressed to a pellet with a diameter and a height of approximately 10 mm.

When it will come to irradiated serum, it will have to be pressed after an irradiation for 7 h in a thermal neutron flux of 4.3 x 10^{12}n cm^{-2} s^{-1}.

2. Destruction of the sample

The pellet is placed on the sample holder in the apparatus Trace-O-Mat (figure 1). The H_2O-, N_2 (liquid)- and O_2 (1 l min^{-1})- supplies are opened. The halogen - infrared - reflecting lamps are turned into the highest position and switched on at maximum power for 15 s.

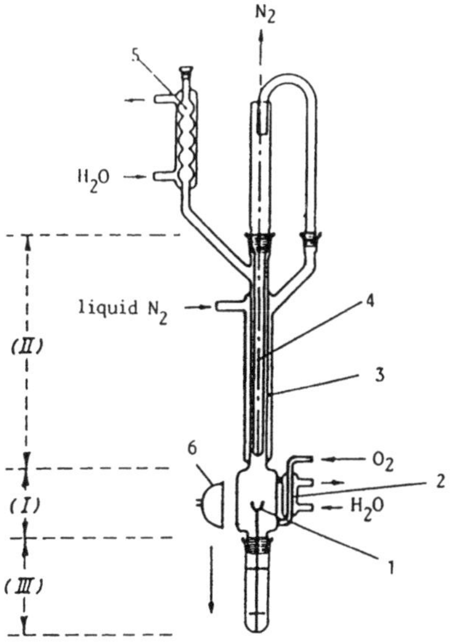

I. Combustion chamber
II. Cooling system
III. Quartz test tube

1. sample holder
2. oxygen inlet
3. cooling mantle
4. cold finger
5. condenser
6. halogen - infrared - reflecting lamps

Figure 1 : Trace-O-Mat

Two to three minutes later, when the sample is completely ashed, the N_2 (liquid)-supply is closed, the lamps are lowered and operated at half power. The O_2-stream is reduced, but a gentle flow is pertained to prevent the gases from entering the O_2-inlet during the reflux process.
The dissolution of the ash is accomplished by the following procedure :
- reflux during 1 h with the carrier solution. The carrier solution, 2 ml 9M HBr + 2.7% H_2O_2, contains 100 µg of the eluted and 10 µg of the retained elements

- reflux during 15 min with 2 ml 9M HBr
- rinse with 2 ml 9M HBr
- turn the sample holder upside down
- reflux during 5 min with 4 ml 10M HCl
- rinse two times with 4 ml 10M HCl.

3. Separation

The separation scheme for the elements of interest is presented in figure 2.

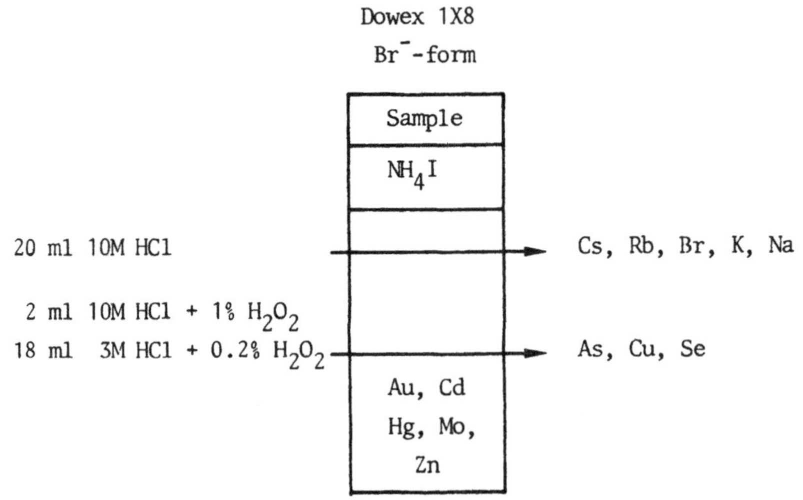

Figure 2 : Separation Scheme

The anion exchanger Dowex 1X8, originally in the Cl^--form, is transformed to the Br^--form by washing the resin with 30 ml 9M HBr. This transformation, as well as the dissolution of the ash in HBr are prerequisites to prevent the adsorption of disturbing amounts of Br.
Control of the oxidation state of the elements As, Br, Cu and Se is achieved by addition of 100 µl of a saturated NH_4I-solution in H_2O on the top of the column. Subsequently, a peristaltic pump transfers the sample in small aliquots from the apparatus Trace-O-Mat onto the column. The adsorption and elution speed are adjusted to 0.5 ml/min.

The first eluate, 20 ml 10M HCl, contains the radioisotopes ^{134}Cs and ^{86}Rb, that will be measured after decay of the major activities of ^{82}Br, ^{42}K and ^{24}Na.

The second fraction, 2 ml 10M HCl + 1% H_2O_2 and 18 ml 3M HCl + 0.2% H_2O_2, comprises the elements As, Cu and Se. The H_2O_2 transforms As(III) to As(V), dissolves the red Se-precipitate and keeps the Se in a soluble form (3).

The elements Au, Cd, Hg, Mo and Zn remain on the resin.

All fractions and standards are counted on a Ge(Li)-detector.

Results and Discussion

The results of the tracer experiments throughout the radiochemical procedure are given in table 1.

TABLE 1 : Experimentally determined recovery in the eluate and on the resin

Element	Recovery in eluate, %		Retention on resin, %
	10M HCl	3M HCl	
Cs	> 99	0	0
Rb	> 99	0	0
Br	55 - 70	0	0.5 - 1.5
K	> 99	< 0.1	< 0.1
Na	> 99	< 0.1	< 0.1
As	0	>98	0 - 1
Cu	< 1	>95	1 - 2
Se	0	75 - 78	20 - 22
Au	0	0	> 95
Cd	0	0	100
Hg	0	0	100
Mo	0	0	> 99
Zn	< 0.2	< 0.5	> 98

The elements Au, Cd, Hg, Mo and Zn are quantitatively retained on the resin, while the others, except Se, are eluted for more than 95%.
The behaviour of Se is more complicated. About 75% is eluted with 3M HCl + 0.2% H_2O_2, while the rest remains on the resin. This does not constitute a drawback, because the sum equals nearly 100% and the calculation of the concentrations is based on the k_0-standardization method (4-5).
The effectiveness of the separation scheme has been confirmed by the analysis of an irradiated serum sample.

Conclusion

The experiments with tracers and with irradiated serum prove that the proposed destruction-separation procedure will allow a precise and accurate determination of the stated trace elements in human serum by neutron activation analysis.

The authors acknowledge with gratitude the financial support given by "The National Fund for Scientific Research", "The Inter-University Institute for Nuclear Sciences" and "The Ministry of Scientific Research, Prest Program".

References

1. Knapp, G., S.E. Raptis, G. Kaiser, G. Tölg, P. Schramel and B. Schreiber. 1981. Fresenius Z. Anal. Chem., 308, 97.

2. Heng-Bin Han, G. Kaiser and G. Tölg. 1981. Anal. Chim. Acta, 128, 9.

3. Tjioe, P.S., J.J.M. De Goeij and J.P.W. Houtman. 1977. J. Radioanal. Chem., 37, 511.

4. Simonits, A., F. De Corte and J. Hoste. 1975. J. Radioanal. Chem., 24, 31.

5. De Corte, F. 1987. "Hoger Aggregaat" Thesis, State University of Gent.

PREANALYTICAL STEPS: SAMPLING AND SAMPLE TREATMENT

Plenary Lecture given by:

G. Knapp

Chairman:

B. Welz

Poster Reports:

B. Welz

NEW DEVELOPMENTS IN THE DECOMPOSITION OF BIOLOGICAL MATERIAL

G. Knapp
Institute for Analytical Chemistry, Micro- and Radiochemistry
Graz University of Technology
Technikerstraße 4, A-8010 Graz, Austria

Abstract
The requirement for high performance sample decomposition methods becomes increasingly demanding in the field of trace element analysis in biological materials.
Five powerful methods will be discussed in this lecture; three techniques for wet chemical decomposition in closed systems, one plasma ashing technique and one combustion method. Finally two examples for the decomposition of mg-amounts of biological materials will be given.

Despite a remarkable progress in the field of measuring techniques, decomposition methods for biological materials are still important. The main task of decomposition methods still is the destruction of the organic sample matrix and the transformation of solid materials into an aqueous solution suitable for measurement.

In the course of the last few years the following improvements have been made:

- Reduction of systematic errors
- Multipurpose application to different sample materials
- Mechanization of decomposition process
- Simple handling
- Reduction of decomposition time

Reduction of systematic errors

The most serious systematic errors in decomposition methods are

- contamination caused by decomposition reagents,
- contamination caused by components of the vessel material or by adsorbed impurities,
- losses of elements caused by adsorption on the surface of the vessel or by reaction with vessel material and
- losses of elements caused by volatilization.

Invited Lecture Presented at the 5th International Workshop
Trace Element Analytical Chemistry in Medicine and Biology, 15th to 18th April 1988
Neuherberg, Federal Republik of Germany

The contamination by decomposition reagents can be reduced enormously by using only small quantities of highly pure reagents.
Liquid decomposition reagents (e.g. nitric acid) in a highly pure quality can be achieved by the so-called subboiling distillation[(1)]. In *Figure 1* a subboiling distillation apparatus is represented schematically. The results on *Table 1* show that even very

Fig. 1. Drawing of a sub-boiling still.
(1) distillation chamber,
(2) heating filament,
(3) cooling finger,
(4) inlet funnel,
(5) bottle with purified acid,
(6) acid to be distilled.

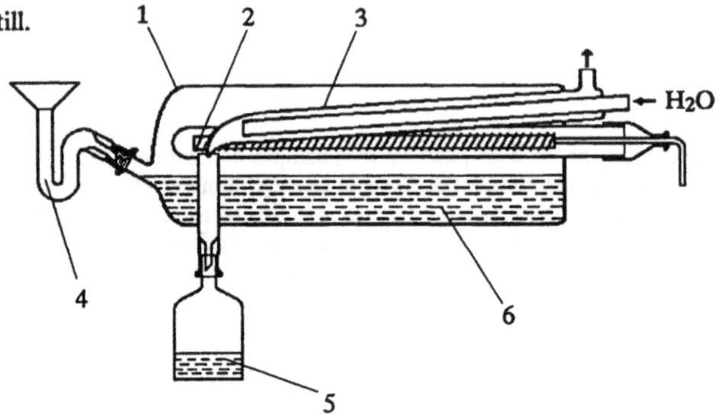

expensive, highly pure special reagents are more contaminated than "subboiling"-reagents. In addition the contamination is increased by repeated opening of the reagent bottle. The amount of decomposition reagents needed will be reduced enormously by using closed systems. These pressure-decomposition methods have additional advantages which will be mentioned later on.
Using oxygen as decomposition reagent is another possibility to reduce blanks.

Tab.1. Residual impurities in water and different acids [ng/ml]

			Cd	Cu	Fe	Al	Pb	Mg	Zn
H_2O		subboiling	0.01	0.04	0.32	<0.05	0.02	<0.02	<0.04
HCl	10M	subboiling	0.01	0.07	0.6	0.07	<0.05	0.2	0.2
HCl	10M	suprapure	0.03	0.2	11	0.8	0.13	0.5	0.3
HCl	12M	p.A.	0.1	1.0	100	10	0.5	14	8.0
HNO_3	15M	subboiling	0.001	0.25	0.2	<0.005	<0.002	0.15	0.04
HNO_3	15M	suprapure	0.06	3.0	14	18	0.7	1.5	5.0
HNO_3	15M	p.A.	0.1	2.0	25	10	.05	22	3.0
HF	54%	subboiling	0.01	0.5	1.2	2.0	0.5	1.5	1.0
HF	40%	suprapure	0.01	0.1	3.0	1.0	3.0	2.0	1.3
HF	54%	p.A.	0.06	2.0	100	5.0	4.0	3.0	5.0

Oxygen can be obtained practically free of blanks and only small amounts of highly pure acid are necessary to dissolve the combustion residue.

Contamination from the vessel material can be reduced by using appropriate materials and applying suitable cleaning methods. *Table 2* shows the element contents of materials used in trace element analysis(2).

Tab.2. Element contents of different materials [ng/g]

Element	Glassy carbon Sigradur G®	PTFE Teflon®	Quartz Heralux®	Quartz Suprasil®	Borosilicate-glass
B	100	–	100	10	main
Na	350	25000	1000	10	main
Mg	100	–	100	100	$6x10^5$
Al	6000	–	30000	100	main
SI	85000	–	main	main	main
Ca	80000	–	800-3000	100	10^6
Ti	12000	–	800	100	3000
Cr	80	30	5	3	3000
Mn	100	–	10	10	6000
Fe	2000	10	800	200	$2x10^5$
Co	2	2	1	1	100
Ni	500	–	–	–	2000
Cu	200	20	70	10	1000
Zn	300	10	50	100	3000
As	50	–	80	0.1	500-22000
Cd	10	–	10	–	1000
Sb	10	0.4	2	1	8000
Hg	1	10*	1	1	–

*) strongly dependent on storage conditions

If it is not necessary to work with hydrofluoric acid, quartz glass has proved to be the best material for trace element analysis. Steaming with nitric acid or hydrochloric acid is the best possible way for cleaning vessels made of quartz glass or PTFE(3). In *Figure 2* such a steaming apparatus is presented schematically.

Losses of elements due to adsorption on the vessel surface usually are smaller with quartz glass than with other materials. It has been observed that elements, above all mercury, diffuse into the surface of PTFE and on the other hand be leached, which again causes losses and contamination.

Losses caused by volatilization of elements or compounds can be avoided by decomposing in closed systems or by collecting the volatile components. Both processes are applied to the most recent decomposition methods.

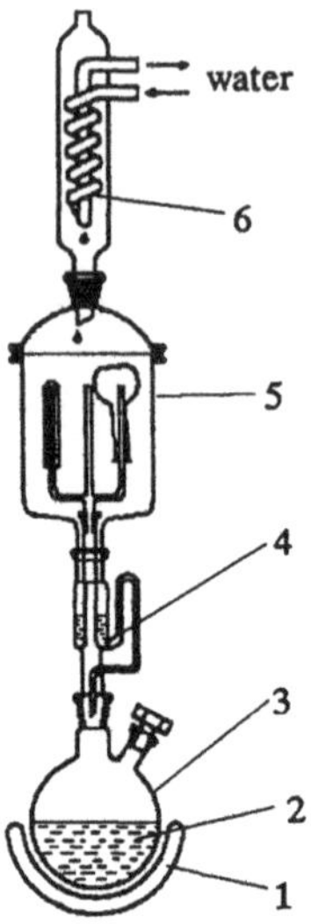

Fig.2. Drawing of a steaming apparatus.
(1) heater,
(2) acid or water,
(3) round flask,
(4) siphon,
(5) steam chamber,
(6) reflux condenser.

Decomposition methods of the most recent generation

State of the art for the decomposition of biological materials are the ashing methods, shown below. This does not preclude, of course, that some of these methods still have teething troubles. The most recent generation comprises the following techniques and devices:

1. Wet chemical decomposition in a closed system, with nitric acid or mixed acids consisting of nitric acid, hydrochloric acid, perchloric acid, hydrogen peroxide and hydrofluoric acid. The following devices are available.

1.1. Decomposition in closed Teflon® vessels at medium pressure (up to 8 bar) with microwave heating(4);

- Microwave Digestion System MDS-81D®, CEM Corporation, USA,*)

1.2. Decomposition in closed Teflon® vessels at high pressure (up to 85 bar) with microwave heating;

- Microwave Acid Digestion Bomb®, Parr Instrument Company, USA,*)

1.3. Decomposition in closed quartz- or glassy carbon vessels at high pressure (up to 120 bar) with conventional heating(5, 6, 7);

- High-Pressure-Asher HPA®, Anton Paar Company, Austria, *) **)

2. Low temperature ashing with radio frequency induced oxygen plasma;

2.1. • Cool-Plasma-Asher CPA-4®, Anton Paar Company, Austria, *) **)

3. Combustion in a dynamic system with complete recovery of all volatile elements;

3.1. • Trace-O-Mat VAE-II®, Anton Paar Company, Austria, *) **)

1. In the present comparison of modern pressure decomposition systems I have not taken into consideration the classical PTFE-decomposition bomb, as this system is not state of the art any longer. If one is willing to accept the well-known disadvantages of PTFE as vessel material, a combination with micro-wave heating will be advantageous.

In *Table 3* the most important decomposition data of the three decomposition systems are compared with each other. The main differences lie in the kind of heating and the vessel material. The decomposition methods of CEM-Corporation and Parr Instrument Company apply microwave heating, whereas the High-Pressure-Asher® has a conventional heating system.

Microwave heating enables us to decrease the decomposition time by the factor 3 to 10. Looking at the decomposition times for organic and inorganic samples, it can

Tab.3. Comparison of the analytical data of the three pressure decomposition systems.
A) Microwave Digestion System MDS-81D®,
B) Microwave Acid Digestion Bomb®,
C) High-Pressure-Asher HPA®;

	A)	B)	C)
inorganic sample (dry weight)	1 - 2 g	1 - 2 g	0.2 - 0.5 g
organic sample (dry weight)	0.1 - 0.5 g	0.2 g	1.0 - 1.5 g
maximum temperature	200° C	250° C	320° C
maximum pressure	8 bar	85 bar	120 bar
ashing time	2 - 60 min	2 - 60 min	45 - 150 min
bomb volume	120 ml	45 ml	85 ml
vessel material	Teflon®	Teflon®	quartz

*) Distributor:
Hans Kürner Analysentechnik
P.O. Box 171
D-8200 Rosenheim, FRG

**) Distributor:
Questron Corporation
P.O. Box 2387
Princeton, NJ 08543-2387, USA

be seen that the microwave heating has adventages for inorganic samples. For decomposing silicate minerals only one to five minutes are needed, 1 to 2 hours are necessary when applying a High-Pressure-Asher®. For organic samples, ranging from food to plastics, 15 - 60 minutes are necessary when applying micro-wave decomposition and 60 - 120 minutes when using a High-Pressure-Asher®. Teflon is used as vessel material for both microwave decomposition systems, whereas High-Pressure-Asher vessels made of pure quartz glass are applied for all acids except hydrofluoric acid. In the case of decompositions with hydrofluoric acid in the High-Pressure-Asher® glassy carbon is applied.

When comparing the fields in which these three decomposition methods are applied the following picture emerges. For the analysis of inorganic sample materials the microwave systems with their very short decompostion times are advantageous. As in this field the element concentration is comparatively high, systematic errors by adsorption or diffusion into the vessel material - as it can be observed when using Teflon® - do not become evident to a significant extent. When analyzing biological samples, which very often have an extremely low element content, significant systematic errors may occur when using Teflon as material for decomposition vessels. In this field the High-Pressure-Asher® with its decomposition vessels made of highly pure and inert quartz glass proves to be advantageous. A further benefit of the High-Pressure-Asher® is the complete destruction of all organic compounds with nitric acid at 320°C[8, 9]. At 200°C, the maximum temperature of microwave-ashing with Teflon®-vessels, not all of the organic compounds are destroyed[10]. This leads to systematic errors with the determination of arsenic and selenium by Hydride-AAS or Hydride-ICP-OES[11] and with voltammetric determination of heavy metals[8, 9]. The slightly longer decomposition time should be accepted in favour of the lower error rate.

2. The ashing of organic samples by means of a high frequency induced oxygen plasma is not new, but only KAISER and TÖLG described, how a plasma ashing device has to be designed[12]. The scheme in *Figure 3* shows our version of this technique. A quartz test tube with a cooling finger serves as a digestion vessel. Volatile elements including arsenic and selenium which are partially lost in traditional plasma ashers are retained on the cooling finger. Mercury however is lost. When decomposition is complete the sample vessels are taken out of the plasma ashing device and inserted in a reflux heater. The ash and elements adsorbed onto the cooling finger are dissolved by refluxing with acid in a separate device[13]. A largely automized plasma decomposition apparatus with four independent ashing units, the Cool-Plasma-Asher CPA-4®, has been developed in cooperation with the Anton Paar Company.
This method is suitable for extreme trace element analysis of biological materials. Up

Fig.3. Cool-Plasma-Asher®. Flow diagram of the process. (I) Sampling, (II) ashing, (III) refluxing, (IV) measurement. (1) sample, (2) cooling water, (3) vacuum, (4) oxygen, (5) HF-coil, (6) magnetic stirrer, (7) heating block, (8) measurement of the solution.

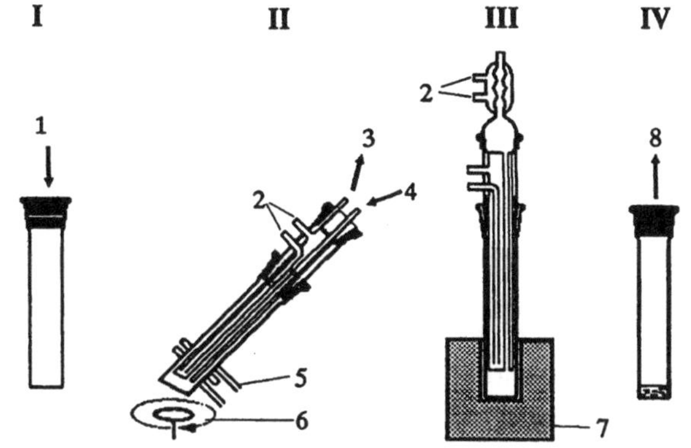

to 1 gram of sample material is ashed and dissolved in 2 ml highly pure acid. Therfore a comparatively high element concentration in the decomposition solution can be achieved with extremely low blanks. However, the long ashing periods of several hours have to be mentioned.

3. As the last one in the series of high performance decomposition methods I would like to mention the second generation of a proven combustion technique for extreme trace element analysis, namely the combustion in a dynamic system[14, 15, 16].

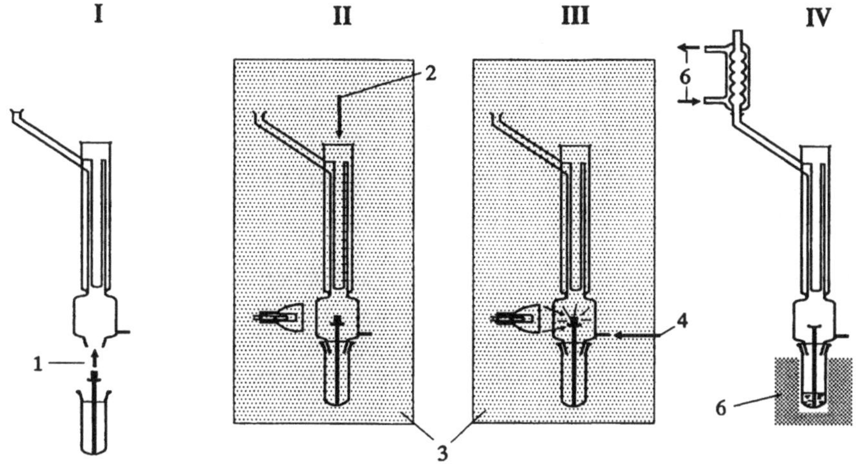

Fig.4. Trace-O-Mat II®. Flow diagram of the process. (I) Sampling, (II) cooling, (III) burning, (IV) refluxing. (1) Sample transfer into the burning chamber, (2) liquid nitrogen, (3) burning apparatus, (4) oxygen, (5) refluxing heater, (6) cooling water.

In *Figure 4* a flow chart of this method is shown. The sample is burned in a stream of oxygen. The burning system is entirely made of quartz glass. An essential feature is the cooling system above the combustion chamber which is filled with liquid nitrogen. This condenses all volatile element traces together with the combustion products CO_2 and H_2O. After combustion the residual ash as well as the condensed volatile elements are dissolved in 2 ml acid by refluxing. We call this burning device Trace-O-Mat®. The advantage of the second generation of this device, the Trace-O-Mat II®, is the higher sample throughput of 10 samples per hour. As can be seen in *Figure 4*, several independent combustion units made of quartz glass are applied, passing through the process one by one. First of all the combustion unit is loaded with the sample and transfered to the burning device. Then the combustion takes place after the quartz unit has been filled with liquid nitrogen. Finally the elements are collected by refluxing in a special device.
As with the Cool-Plasma-Asher®, comparatively high element concentrations in the decomposition solution can be obtained with the Trace-O-Mat®.

Decomposition of micro-quantities of samples

The decomposition of mg-sample-quantities is of great importance in the biological field. Recently we have been dealing with element determination in biopsy samples. In this special case tissue samples with a wet weight between 5 and 20 mg had to be decomposed without systematic errors. Finally I would like to give a brief report on our experiences in this field.
The analytical multistage procedure used in our laboratory consists of wet decomposition in closed vessels and element measurement by means of Zeeman-AAS. As the majority of trace elements are bound to proteins, phosphorus is used as internal standard and the dry weight will not be taken into consideration.
For the pressure decomposition of biopsy samples small quartz vessels are used. 5 - 20 mg of sample material are decomposed with 100 µl of concentrated nitric acid. The High-Pressure-Asher® and the microwave apparatus microdigest®300 from PROLABO Company, France, were compared.

The microwave decomposition is performed in a 10 ml quartz tube closed with a PTFE screw-cap. In the microdigest®300 the tightly sealed sample vessel is heated for 20 minutes at 40 % of the microwave's power capacity. After cooling down, the decomposition solution which is adhered to the screw cap, is collected in the vessel by centrifugation. Then the vessel is opened and the solution is made up to 500 µl with water.

When decomposing with the High-Pressure-Asher® 2 ml quartz vessels are applied. These vessels are built in such a way that they fit into the sample tray of the Zeeman-

AAS equipment[17]. This process fulfills an important trace analytical principle – using one and the same vessel from sampling to measurement. The quantities of the samples and reagents are the same as with the microwave decomposition. In the High-Pressure-Asher® the samples are heated for 30 minutes at 300°C. The following treatment of the samples is the same as with the micro-wave decomposition.

It is a short matter to compare the two methods. During high-pressure-ashing much smaller quartz vessels can be applied, because the inside pressure can rise up to 120 bar. As these vessels fit exactly into the sampler of the graphit-furnace-AAS, no transfer of the decomposition solution with all its possible errors is necessary. In contrast to this, microwave ashing can be performed in less time.

References

1. Kuehner, E.C., R. Alvarez, P.J. Paulson, T.J. Murphy. 1972. Anal.Chem. 44, 2050.
2. Tölg, G., P. Tschöpel. 1987. Analyt.Sci. 3, 199.
3. Tschöpel, P., L. Kotz, W. Schulz, M. Veber, G. Tölg. 1980. Fresenius Z. Anal. Chem. 302, 1.
4. Kingston, H.M., L.B. Jassie. 1986. Anal. Chem. 58, 2534.
5. Knapp, G. 1984. ICP Information Nwsltr. 10, 91.
6. Knapp, G. 1984. Trends in Anal. Chem. 3, 182.
7. Knapp, G., A. Grillo. 1986. American Lab. 3, 76.
8. Schramel, P., S. Hasse, G. Knapp. 1987. Fresenius Z. Anal. Chem. 326, 142.
9. Würfels, M., E. Jackwerth, M. Stoeppler. 1987. Fresenius Z. Anal. Chem. 329, 459.
10. Würfels, M., E. Jackwerth, M. Stoeppler. 1988. Fresenius Z. Anal. Chem. 330, 160.
11. Welz, B., M. Schubert-Jacobs, G. Schlemmer, M. Sperling. 1988. Lecture at the 5th International Workshop Trace Element Analytical Chemistry in Medicine and Biology, 15th to 18th April 1988, Neuherberg, Federal Republic of Germany.
12. Kaiser, G., P. Tschöpel, G. Tölg. 1971. Fresenius Z. Anal. Chem. 302, 1.
13. Raptis, S.E., G. Knapp, A.P. Schalk. 1983. Fresenius Z. Anal. Chem. 316, 482.
14. Knapp, G., S.E. Raptis, G. Kaiser, G. Tölg. 1981. Fresenius Z. Anal. Chem. 308, 97.
15. Heng-Bin Han, G. Kaiser, G. Tölg. 1981. Anal. Chim. Acta 128, 9.
16. Raptis, S.E., G. Kaiser, G. Tölg. 1982. Anal. Chim. Acta 138, 93.
17. Knapp, G. 1985. Intern. J. Environ. Anal. Chem. 22, 71.

ALKALI-DIGESTION METHOD USING TMAH FOR DETERMINATION OF TRACE ELEMENTS CONTAINED IN BIOLOGICAL MATERIALS

Shumpei Shimizu
Moses Lake Industries, Inc.
780 Randolph Road, Moses Lake, Washington 98837, U.S.A.

Tsurahide Cho, Yukio Murakami
Tama Chemicals Co., Ltd., 5-36-2 Kamata, Ota-Ku, Tokyo, Japan

Abstract

TMAH, mainly used as developer in LSI industry, has possibly been produced on large scale with ppb level contaminant (US Pat. 4,634,509). This study to utilize it as solubilizer of biological materials and to analyze trace elements in it was done by one experimenter with GF-AAS, ICP-AES and ICP-MS. The analytical results of some standard materials (biological) showed satisfactory agreement with the authorized values. The proposed TMAH-solubilization method was proven to be safe, easy and time saving procedure with high accuracy and good precision.

Introduction

The presence of trace elements in biological material is spotlighted as essential element in nutrition or toxic one in environmental or toxicological problems. Accordingly, concentration range of them especially attracts attention in realizing highly accurate results from these standpoints. Development and automatization of analyzing instruments have remarkably made rapid progress to meet this demand, but still serious problems remain in sample pretreatment, reagent contaminant and decomposition method of biological materials.[1)] Under such circumstances, the standard reference materials are recommended by the authorities of various countries to check for accuracy and precision in analyzing methods and results.[2] It is evident that preparation of highly homogenized samples[3] as well as solubilization of trace elements into homogeneous system, which can feed to automatized analytical instruments, is urgently required to determine trace elements in biological materials. Instead of conventional ashing methods (dry or wet),[4] many improved methods are proposed to eliminate any loss of elements and to render the sample into homogeneous system. Still, some problems remain for such elements as As, B and others which are essentially likely to be lost in their treatment. To cover this defect, radioactivation analysis[5] is regarded as a supplement, but it is tedious and in restricted use.

This study proposes solubilization method of biological materials by tetramethylammonium hydroxide(TMAH), a strong organic base equal to NaOH or KOH, which our team has succeeded in the mass production with ppb levels of contaminants (US Pat.4,634,-509). Thus, our team is now looking for a variety of analytical uses in solving of some problems. In this study, fundamental conditions of solubilization of biological materials with TMAH are examined. One experimenter determined many elements by using TMAH-solubilized samples with GF-AAS, ICP-AES and ICP-MS. The evaluation of the proposed method was made by analyzing three standard reference materials and by checking if the results obtained were consistent or not with those authorized values.[2]

Experimental

A. Tetramethylammonium hydroxide TMAH, $(CH_3)_4NOH$, M.W.91.15 Water solution decomposes at 130-140°C. and forms carbonate by absorbing CO_2 in the air. Its purity and viscosity are shown here.

Table 1 Typical analytical results of impurities detected by GF-AAS ng/ml

Al	As	Ca	Co	Cr	Cu	Fe	Mg	Ni
<1	<5	<1	<1	1	<1	<1	<1	<1

B. Analytical instruments used

1. GF-AAS: Perkin-Elmer Mod.5000 with HGA-500 GF.
2. ICP-AES: Seiko Elec. Mod.1200

C. The proposed TMAH-solubilization methods of biological materials

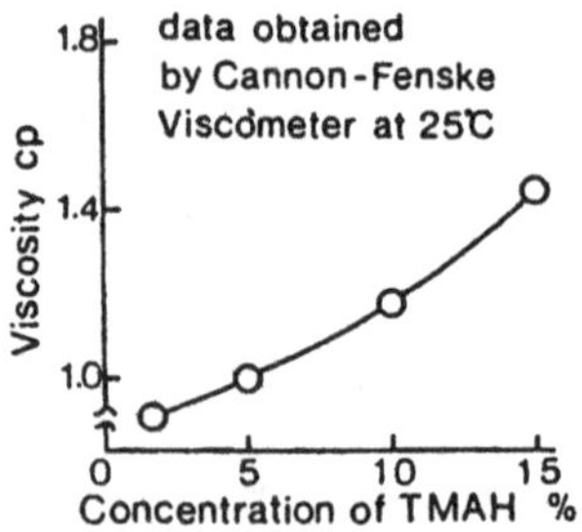

Fig.1 Viscosity of TMAH solution

Accurately weigh 200-500mg and add 5ml of 10-25% TMAH solution and if necessary add about 10ml of water. The solution is then left at room temperature for all night. When much of insoluble residues are observed, gently heat to 65-85°C for 1-2 hours to hasten dissolution. Use this solution as solubilized sample solution. Colored state can be tolerated.

Results and Discussion

Solubilized status is not always identical depending on chemical nature of individual sample. For instance, mussel can become transparent solution by heating to low temperature with 10-15% TMAH, but even not-transparent solution, being kept at room temperature with TMAH 10% or less, resulted in a good agreement with the certified values. Milk powder can easily dissolve with 1% TMAH solution but some precipitates are formed with higher concentration of TMAH. Hair was not completely solubilized even in TMAH 25% with heating at 85°C, but satisfactory analytical results could be obtained.
When analyzing with GF-AAS or CIP-AES, though TMAH completely decomposed, the absorbance or the light emission intensity tends to decrease for some elements with the increase of TMAH, respectively. But, the detection limits (define to S/N=2 with GF-AAS and 3σ of BG with ICP-AES) obtained are not so low compared with that of 0.68% HNO_3 sample solution. The concentration effect on detection limits of some elements are shown in Table 2. The detection limits of several elements, when TMAH is less than 5%, did not make much difference. Then, it is

Table 2 Concentration effect of TMAH solution on the detection limit of some elements

GF-AAS

Element	Wave-length nm	Concentration effect of TMAH 0.5	1.0	2.5	5.0	7.5	10%	HNO_3 0.68% (pg)
		on the detection limit (pg)						
As	193.7	210	210	150	180	180	180	44
Sb	217.6	36	36	32	44	44	48	36
Sn	224.6	28	28	28	32	28	28	32
Cd	228.8	1.1	1.0	1.1	0.9	0.7	0.6	0.7
Co	240.7	13	13	15	18	19	16	13
Pb	283.3	14	10	10	9	9	9	12
Mo	313.3	200	22	22	210	210	230	76
Sr	460.7	14	130	100	110	110	76	56

ICP-AES

Element	Wave-length nm	Concentration effect of TMAH 0.5	1.0	2.5	5.0	7.5	10%	HNO_3 0.68% (µg/ml)
		on the detection limit (µg/ml)						
Sn	189.9	71	69	93	110	110	120	84
As	193.7	104	79	140	140	110	180	120
Mo	202.0	16	16	14	16	20	20	17
Sb	206.8	70	75	80	90	100	140	54
Ni	221.6	10	14	10	10	12	13	11
B	249.6	15	12	14	18	18	20	15
Mn	257.6	2	2	2	2	3	4	2
Cr	267.7	7	7	7	7	10	8	7
V	292.4	15	10	11	10	10	14	11
Ti	334.9	3	2	4	4	8	6	3
Ba	455.3	1	2	2	2	2	4	2

recommended that solubilized sample solution should be used with TMAH 5% or less. Then, in this study, standard addition method was mainly used. In dissolving, it is necessary to find optimum concentration of TMAH for sample weight taking into account the content of the element to be determined. In case of mussel, the ratio of the estimated content (ug/ml) of the element to detection limit (ug/ml) are shown in Table 3 (ICP-AES). This ratio is helpful to examine if the said sample weight is correct or not. In Sr and Mn where this ratio is 2 figures or more, the sample weight can be reduced or increased solution volume. In Ti, B and Cu, this condition can be used as it is. In Cd, Cr and As, it is suggested to double the sample weight or to decrease the solution volume to 25ml. While in Pb, larger sample weight or smaller solution volume shall be used.

Table 3 Ratio of estimated content (µg/ml) of the element to detection limit (µg/ml), analyzing with ICP-AES

Sr	Mn	Ti	B	Cu	Cd	Cr	As	Pb
80	35	8.5	3.6	2	0.55	0.45	0.4	0.04

Sample Weight 250mg, Sample volume 50ml TMAH 5%

With GF-AAS, as pyrocoated tube cannot be used for long time when TMAH concentration is 1% or more, normal tube was used instead of it despite the resulting lower sensitivity. However, its lowered sensitivity is still higher than ICP-AES and, accordingly, the said condition can be applied to most of the elements. It may be said that it is rather difficult to obtain satisfactory analytical results of all the elements by

Table 4 Analytical results of Mussel, Hair and Bovine Liver

Ele-ment	GF-AAS n=5	ICP-AES n=5	Certified value
Mussel(NIES No.6)			μg/g
As	9.0 ± 0.5	11 ± 0.73	9.2 ± 0.5
B	ND	12 ± 1.3	not given
Ba	ND	0.52 ± 0.05	not given
Cd	0.87 ± 0.01	0.78 ± 0.03	0.82 ± 0.03
Co	0.38 ± 0.01	0.36 ± 0.04*1	(0.37)
Cr	0.55 ± 0.02	0.75 ± 0.13	0.63 ± 0.07
Cu	5.0 ± 0.10	5.0 ± 0.10	4.9 ± 0.3
Mn	17 ± 0.81	16 ± 1	16.3 ± 1.2
Mo	0.49 ± 0.04	0.90 ± 0.06	not given
Ni	0.85 ± 0.04	0.95 ± 0.04*2	0.93 ± 0.06
Pb	0.79 ± 0.06	<10	0.91 ± 0.04
Sn	0.67 ± 0.14	<5	not given
Sr	13 ± 0.81	16.4 ± 1.3	(17)
Ti	ND	6.8 ± 0.31	not given
V	<1	0.59 ± 0.07	not given
Hair(NIES No.5)			
As	0.86 ± 0.06	<5	not given
B	ND	4.2 ± 0.44	not given
Ba	ND	2.5 ± 0.10	(3.2)
Cd	0.24 ± 0.016	<1	0.20 ± 0.03
Co	0.16 ± 0.015	<1	(0.10)
Cr	0.99 ± 0.02	1.8 ± 0.2	1.4 ± 0.2
Cu	16 ± 0.36	16.7 ± 0.7	16.3 ± 1.2
Mn	5.1 ± 0.14	5.0 ± 0.36	5.2 ± 0.3
Mo	<1	<5	not given
Ni	1.9 ± 0.17	<2	1.8 ± 0.1
Pb	6.3 ± 0.20	<10	(6.0)
Sn	2.8 ± 0.20	4.2 ± 0.60	not given
Sr	2.1 ± 0.12	2.5 ± 0.27	2.3 ± 0.2
Ti	ND	17.8 ± 1.64	(22)
V	<1	<1	not given
Bovine Liver(NBS 1577a)			
As	<1	<5	0.047 ± 0.006
B	ND	0.69 ± 0.04	not given
Ba	ND	0.11 ± 0.01	not given
Cd	0.36 ± 0.02	0.49 ± 0.06	0.44 ± 0.06
Co	0.20 ± 0.01	<1	0.21 ± 0.05
Cr	0.12 ± 0.02	<0.5	not given
Cu	179 ± 5	128 ± 6	158 ± 7
Mn	10.1 ± 0.37	9.5 ± 0.7	9.9 ± 0.8
Mo	3.5 ± 0.12	3.3 ± 0.2	3.5 ± 0.5
Ni	<1	<2	not given
Pb	0.19 ± 0.01	<10	0.135 ± 0.015
Sn	<0.5	<5	not given
Sr	0.11 ± 0.01	0.15 ± 0.02	0.138 ± 0.003
Ti	ND	0.26 ± 0.03	not given
V	<1	0.10 ± 0.01	0.099 ± 0.008

Remarks: 1. ND stands for Non-Detectable.
2. Figures in parentheses are Reference values.
3. *1 Sample weight 1g. *2 2g.

only one experiment depending on chemical nature of samples.

The analytical results of mussel and hair (NIES) and bovine liver (NBS) by the proposed method with GF-AAS and ICP-AES are shown in Table 4 together with the authorized values.

The obtained results fairly agree with the authorized values with the exception of As of mussel by ICP-AES. This means that the proposed method assures high accuracy and good precision for the biological materials.

As some elements in mussel, such as Co, Sr, B, Ti and Ba, were not given the certified values, correlation between the sample weight and the content of the element was examined analyzing with GF-AAS and ICP-AES. As Table 5 shows, high correlation coefficient of 0.9 or more is

Table 5 Correlation coefficient between sample weight and content of the elements

Method	Ele-ment	Sample g	TMAH %	Correlation coefficient
GF-AAS	Co	0.05 – 0.15	0.5 – 5.0	0.933
	Sr	0.05 – 0.15	0.5 – 5.0	0.976
ICP-AES	B	0.05 – 5.0	0.5 – 20	0.991
	Ti	0.05 – 5.0	0.5 – 20	0.989
	Ba	0.2 – 0.7	10 – 20	0.918

obtained in fairly wider range of them. Thus, it may be said that the results obtained by this method are highly reliable ones.

It is made clear by the results in Tables 4 and 5 that the standard addition method proved to eliminate any influence of TMAH and that determination of trace elements in ppb level first becomes possible by highly purified TMAH together with analytical instruments having enough high sensitivity.

In conclusion, the proposed TMAH-solubilization method for biological materials was proven to be the most suitable method, which can give accurate and precise results, even for volatile elements such as As, B and others, with rapidness, safety, easiness and without danger of explosion and introducing any contamination. The effectiveness of this method on applying with ICP-MS was also supported without giving any significant effect on mass spectra. The spectra of purity of TMAH and of trace elements in TMAH-solubilized mussel sample were displayed on the poster session. This study will be detailed in another report because of shortage of space. The solubilization of biological materials with TMAH gives reaction products which are usually soluble in water and are convenient to feed to the applying instruments. The detailed report on it will also appear in another journal.

Reference

1. Versieck, J. 1986, J. Res. of Natl. Bur. Stand., 91 No.2, 87-92.
2. Claude Veillon 1986, Ana. Chem., 58 No.8, 851-866A.
3. Parr, R.M. 1986, J. Res. of Natl. Bur. Stand., 91 No.2, 51-57.
4. Lonnerdal, B., et al. 1980, in Trace Element Analytical Chemistry in Medicine and Biology, P.619.
5. Advisory Group, IAEA, 1984, Ana. Chim. Acta 165 1-29.

SEMI-MICRO ACCESSORY TO AN AUTOMATED WET DIGESTION SYSTEM FOR ASHING SMALL SAMPLE AMOUNTS

A. Frank

Department of Chemistry, National Veterinary Institute
Box 7073, S-750 07 Uppsala, Sweden

Abstract. In order to digest limited sample amounts (< 300 mg d.w.) in small ashing tubes, resulting in sample volumes appropriate for trace element analysis, an accessory to an automated wet ashing apparatus was developed. Aluminium pieces were constructed with outer dimensions of the normal ashing glass tubes to fit into the holes of the aluminium block and were provided with drilled inner holes to facilitate the use of smaller ashing tubes. By using time and temperature controlled standardized - wet digestion, the ashing is simple. Further treatment and analysis can be performed in the same tube. The reliability of the digestion procedure was checked by analysis of 17 elements including selenium in standard reference materials.

Introduction

By automated wet digestion the preparation of biological materials, e.g. in trace element analysis, is safe and reproducible. Ashing time and temperature is controlled continuously. When performed overnight, it saves time and personnel resources. A commercially available apparatus (Tecator Digestion System, model 40; Tecator AB, Höganäs, Sweden) is equipped with borosilicate glass tubes of 80 ml volume. In case of limited sample amounts the use of smaller glass tubes is preferable. For this purpose an accessory to the above instrument was developed.

Description of the equipment and accessory

The apparatus consists of two units: an electrically heated

Trace Element Analytical Chemistry in Medicine and Biology

aluminium block with a capacity of 40 borosilicate glass tubes (80 ml) and a microprocessor for programming and control of temperature and time.

Forty pieces of aluminium were turned on a lathe so that they had the round-bottom profile of the glass tubes. The length of the pieces was 3 mm longer than that of the holes in the aluminium block and their diameter was adapted to fit easily into the holes. Holes were drilled in the aluminium pieces so that the smaller borosilicate ashing tubes with a total volume of 28 ml could be inserted into them. The dimensions of the glass tubes were: outer diameter 17 mm, inner diameter 12.7 mm, length 210 mm. Two small holes, opposite each other, were drilled in the uppermost part of the aluminium pieces to facilitate easy removal from the block with the aid of a tool (Fig. 1.).

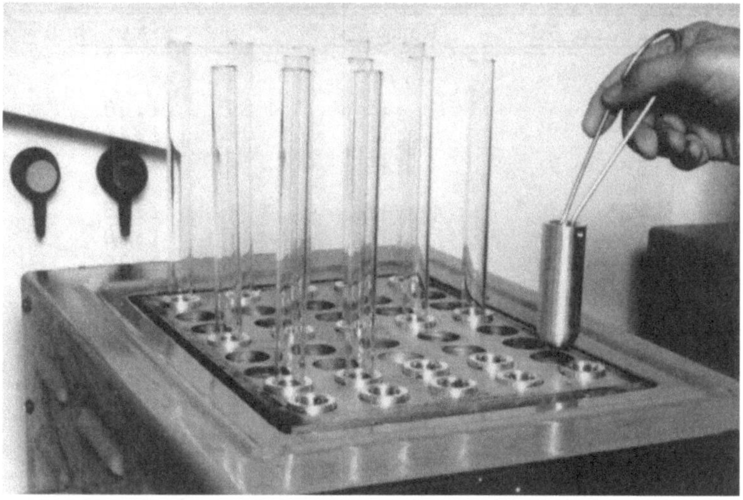

Fig. 1. The aluminium heating block with the semi-micro accessory and ashing tubes.

Experimental

Biological materials of animal or plant origin up to 300 mg d.w. (1.5 g w.w.) are digested with a mixture of 65% nitric acid and 70% perchloric acid (7:3 by vol.) (1). The material should not contain more than 150 mg fat (2). After addition of the acid mixture (5 ml), the samples are kept at room temperature for 3-5 hours. The digestion is performed overnight according to the standard ashing program shown in Table 1. This program is usually applied with decomposition of biological materials in 80 ml glass tubes. The amount of the material should not exceed 5 g w.w. (1 g d.w.) with a maximum lipid content of 0.5 g and is digested with the oxidising acid mixture (15 ml).

Table 1. Standard ashing program for wet digestion of organ tissue, max 1.5 g wet weight, fat content max 150 mg, 5 ml acid mixture (small ashing tubes)

Temperature °C	Ramp, h	Time, h
50	:15	1:00
70	:15	1:45
100	:15	1:45
120	:15	1:15
138	1:15	1:00
150	1:30	1:30
160	:45	:45*
180	:15	:45
225	:30	2 - 4

*It is advisable to inspect the solutions before increasing temperature to 225°C. Dark coloured samples indicating incomplete digestion, should be removed from the block and treated separately. After the addition of a few ml nitric acid the ashing procedure is repeated.

The reliability of the ashing procedure in the semimicro ashing tubes was checked by determination of Ca, Cd, Co, Fe, Cu, K, Mg, Mn, Na, P, Sr, V, and Zn in standard reference materials such as bovine liver (NBS SRM 1577a) and, in addition to the mentioned elements Cr, Pb, and Ni in oyster tissue (NBS SRM 1566). The dry residue was dissolved in 3.0 ml ionic buffer and analysed by simultaneous multi-element analysis using d.c. plasma-atomic emission spectrometry (3). The results are shown in Table 2.

Table 2. - Concentration of 16 elements in NBS standard reference materials (N = 6)

Element	Bovine liver (1577a) Certif. (µg/g)	Found (µg/g) $\bar{x} \pm SD^a$	Oyster tissue (1566) Certif. (µg/g)	Found (µg/g) $\bar{x} \pm SD^a$
Ca	120 ± 7	102 ± 6	1500 ± 200	1400 ± 120
Cd	0.44 ± 0.06	0.41 ± 0.03	3.5 ± 0.4	3.3 ± 0.07
Co	0.21 ± 0.05	0.22 ± 0.01	0.4[b]	0.29 ± 0.01
Cr			0.69 ± 0.27	0.55 ± 0.05
Cu	158 ± 7	148 ± 4	63.0 ± 3.5	59.5 ± 1.0
Fe	194 ± 20	175 ± 7	195 ± 34	185 ± 7
K	9960 ± 70	10040 ± 290	9690 ± 50	9620 ± 514
Mg	600 ± 15	576 ± 16	1280 ± 90	1240 ± 26
Mn	9.9 ± 0.8	10.7 ± 0.3	17.5 ± 1.2	18.5 ± 0.4
Na	2430 ± 130	2410 ± 21	5100 ± 300	4870 ± 85
Ni			1.03 ± 0.19	0.92 ± 0.06
P	11100 ± 400	12190 ± 36	8100[b]	7640 ± 119
Pb			0.48 ± 0.04	0.44 ± 0.06
Sr	0.138 ± 0.003	0.111 ± 0.004	10.36 ± 0.56	10.80 ± 0.54
V	0.099 ± 0.008	0.095 ± 9.009	2.8[b]	2.8 ± 0.03
Zn	123 ± 8	102 ± 6	852 ± 14	816 ± 15

[a] 95 percent confidence interval
[b] Non-certified value

Selenium was determined in bovine liver (NBS SRM 1577a) animal muscle (IAEA H-4) and milk powder (IAEA A-11), serum (Nyegaard Seronorm 105), and urine (Nyegaard Seronorm 108) by FIA-hydride atomic absorption spectrometry with an electrically heated quartz cuvette. The perchloric acid residue (~1 ml) was diluted with hydrochloric acid to a suitable measuring volume (4). The results are shown in Table 3.

Table 3. - Selenium concentration in standard reference materials

Standard reference material	sample weight, g	Certif.[a]	Found[a]	
bovine liver (NBS 1577a)	0.1	0.71 ± 0.07	0.70 ± 0.04	µg/g
milk powder (IAEA A-11)	0.3	0.0339 ± 0.0072	0.0307 ± 0.005	"
animal muscle (IAEA H-4)	0.2	0.280 ± 12%	0.300 ± 0.008	"
serum (Seronorm 105)	0.5	0.090 ± 0.006[b]	0.090 ± 0.004	µg/ml
urine (Seronorm 108)	1.0	0.049	0.048 ± 0.004	"

[a] 95 percent confidence limit of the overall means

[b] $\bar{x}$ ± SD

The concentrations of the different elements analysed were generally in good agreement with the certified values.

Conclusions

A semi-micro accessory to an automated wet digestion system has been developed and proven to be efficient. The small ashing tubes allow digestion of limited sample amounts. The ashing residue can be dissolved in small volumes. Then, the analytical work is accomplished in the same tube or is easily transferred elsewhere for further chemical treatment.

References

1. Frank, A. 1976. Fresenius Z. Anal. Chem. 279, 101.

2. Frank, A. 1983. Kemisk Tidskrift 85, (8), 63.

3. Frank, A. and L.R. Petersson. 1983. Spectrochim. Acta 38B, 207.

4. Galgan, V. and A. Frank. 1988. In: Trace Element Analytical Chemistry in Medicine and Biology (P. Schramel, ed.). Walter de Gruyter, Berlin · New York.

AUTOMATED SYSTEM FOR DETERMINATION OF SELENIUM IN BIOLOGICAL MATERIALS.

Automated wet digestion, hydride generation by flow injection analysis and atomic absorption spectrometry

V. Galgan and A. Frank

Department of Chemistry, National Veterinary Institute
Box 7073, S-750 07 Uppsala, Sweden

Abstract. An automated system for the determination of selenium in biological materials consisting of automated wet digestion, continuous hydride generation by flow injection analysis and atomic absorption spectrometry is described. Samples up to 5 g wet weight are digested by $HNO_3/HClO_4$. Atomisation is performed in an electrically heated quartz cuvette. The detection limit of the method is 0.070 µg/l (0.5 ml injection volume). The characteristic concentration is 0.150 µg/l (A=1%) and the capacity 120 measurements per hour. The accuracy of the method was checked by analysing different standard reference materials.

Introduction

Selenium is determined routinely in biological materials at the National Veterinary Institute. The selenium poor environment in the Scandinavian countries and the demand for analysis of a great number of samples focused the interest on the development of an analytical method with high capacity, including automated wet digestion, automated selenium determination and computer aided calculation.

Materials and Method

Reagents.

Nitric acid, 65%; perchloric acid, 70%; hydrochloric acid, 37%; all reagents of analytical grade (Merck, FRG);

Trace Element Analytical Chemistry in Medicine and Biology

sodium tetrahydroborate solution, 4.4 M in 14 M sodium hydroxide solution (Novakemi, Netherlands); selenium standard solution, (BDH, UK); nitrogen gas (AGA, Sweden).

Apparatus:
Automated wet digestion (Tecator Digestion System, model 40, Tecator, Höganäs, Sweden), borosilicate ashing tubes (total volume of 28 of 80 ml).
Flow-injection system consisting of :
FIA-star -5020 Analyzer; -5007 Auto sampler, -5102-001 Injection valve "V-100", -5020-032 Chemifold SR III; PTFE reaction coils and connection tubes (Tecator AB, Höganäs, Sweden);
Atomic absorption spectrometer (Varian, model AA-1475, Australia, Perkin-Elmer 403, FRG)
Gas-liquid separator (Varian, Australia), electrically heated quartz cuvette (Varian and according to (7)).
Selenium hollow cathode lamp (Hamamatsu, Photonics K.K. Japan) or selenium Photron Super Lamp and power supply (Photron PTV, Australia), selenium electrodeless discharge lamp (Perkin-Elmer, FRG)
Reduction and needle valves (Brooks Instruments,Netherlands)
Computer: Commodore PC 20 (FRG) and printer.

Procedure

Samples (up to 5 g wet weight or 1 g dry weight) are digested with a mixture of 65% nitric acid and 70% perchloric acid (7:3 by vol.) in an automated wet ashing system. Selenium standards are treated in the same way. Forty samples can be digested overnight by a microprocessor-controlled programme of temperature and ashing time (1). The last digestion step at 225 °C should not last longer than 30 min to prevent evaporation to dryness and loss of selenium.

Selenium is reduced from Se^{VI} to Se^{IV} by heating 30 min at 110 °C in 2.4 M hydrochloric acid (5/10ml) and diluted in the same tube with 2.4 M hydrochloric acid to a suitable volume (5, 10, 25 or 50 ml). Seleniumhydrogen is generated by the addition of sodium tetrahydroborate in a flow injection system with a hydrochloric acid solution as the carrier. Samples are picked up from an autosampler and injected into the carrier stream. The flow injection manifold is shown in Fig. 1.

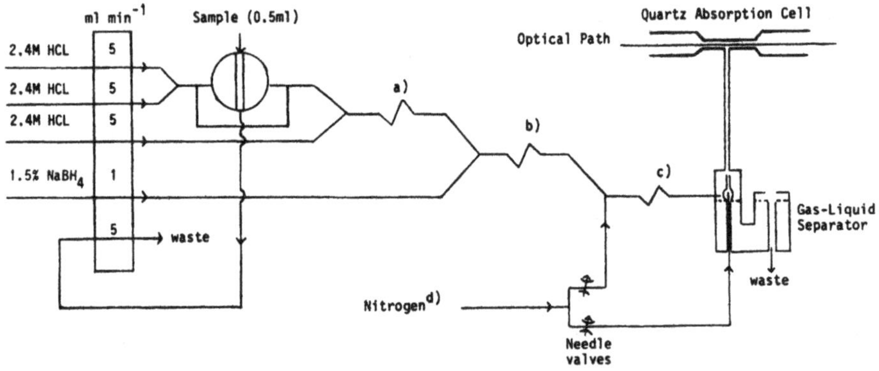

a) 12cm length; 0.7 mm i.d.
b) 30cm " ; 0.7 mm i.d.
c) 40cm " ; 0.97mm i.d.
d) Nitrogen flow rate for both streams: 50ml min^{-1}

Fig. 1. - Flow injection manifold for the selenium determination by hydride generation and atomic absorption spectrometry (HG-AAS).

Atomisation is performed in an electrically heated quartz cuvette at maximum selenium absorption signal. Selenium is determined at 196 nm using a hollow cathode lamp or an electrodeless discharge lamp by atomic absorption spectrometry. To evaluate the signal from the instrument the spectrometer is connected to the Fia-instrument, which in turn is attached to a PC. At calibration maximum six standard concentration values may be entered. The selenium concentration in the biological materials is calculated and stored in the PC. Generally three injections per sample solution are performed.

Results and Discussion

The absorbance is linear to a selenium concentration of 12 µg/l. The precision (10 injections) was < 1 % RSD at the 10 µg/l level and better than 3 % RSD at 1 µg/l. The detection limit of the method (3 * standard deviation of the noise) was 0.070 µg/l when injecting 0.5 ml sample volume. The characteristic concentration was 0.150 µg/l (A=1%). The capacity is 120 - 240 measurements per hour depending on the flow rate of the sample solution through the valve and of its total volume available.

Depending on the sample amount, the limit of determination in biological materials is 1-2 µg and 10 µg Se/kg sample (sample weight being 1.5 - 5 g and 0.1 g w.w. respectively). Concentrations as low as 0.008 µg Se/kg plasma or serum of cow and sheep are not unusual in Scandinavian countries.

The accuracy of the method was checked by analysing Standard Reference Materials (Table 1.).

Table 1. - Selenium concentration in standard reference materials

Standard reference material	Certif.[a]		Found [a]			n
bovine liver (NBS 1577)	1.1	± 0.1	1.13	± 0.04	µg/g	5
bovine liver (NBS 1577a)	0.71	± 0.07	0.70	± 0.04	"	5
rice flour (NBS 1568)	0.4	± 0.1	0.38	± 0.06	"	5
milkpowder (IAEA A-11)	0.0339	± 0.0072	0.0307	± 0.005	"	5
animal muscle (IAEA H-4)	0.280	± 12%	0.300	± 0.008	"	5
serum (Seronorm 105)	0.090	± 0.006[b]	0.090	± 0.004	µg/ml	8
urine, (Seronorm 108)	0.049		0.048	± 0.004	"	5

[a] 95 percent confidence limit of the overall means
[b] $\bar{x} \pm SD$

Several serious interferences are mentioned in the literature in connection with selenium determination by HG-AAS. Interactions are described with As, Co, Cu, Ni, Sb, Sn etc. (2, 3, 4, 5).

However, these elements usually do not have any substantial importance in the biological material analysed at our Institute. The concentrations are often too low to cause any noticeable interaction with selenium during hydride generation. In addition, higher concentrations of interacting elements are tolerated because of the very short resistance time of selenium in the system (6,7).

However, there are a few exceptions. The liver of the sheep at chronic copper poisoning and that of the mute swan may have copper concentrations as great as 1000 mg/kg or more. In such cases a copper determination should precede selenium determination.

A system of automated wet digestion, hydride generation by flow injection and atomic absorption spectrometry has been proven to be suitable for determining selenium in a wide range of biological materials such as liver, kidney, muscle, blood, blood fractions, milk, yolk, yeast, grass, cereals etc.

References

1. Frank, A. 1988. In: Trace Element Analytical Chemistry in Medicine and Biology, Vol. 5 (P. Schramel ed.). Walter de Gruyter, Berlin . New York.

2. Welz, B., M. Melcher. 1984. Analyst 109, 569.

3. Welz, B., M. Melcher. 1986. J.Anal.At.Spectrom. 1. 23.

4. Meyer, A., Ch. Hofer, G. Tölg, S. Raptis, G. Knapp. 1979. Fresenius Z.Anal.Chem. 296, 337.

5. Piwonka, J., G. Kaiser, G. Tölg. 1985. Fresenius Z.Anal. Chem. 321, 225.

6. Fang, Z., S. Xu, X. Wang, S. Zhang. 1986. Anal.Chim.Acta 179, 325.

7. Åström, O. 1982. Anal.Chem. 54, 190.

ANALYSIS OF MANGANESE IN BLOOD SAMPLES FROM INFANTS AND CHILDREN

W.S. Dempster, H.de V. Heese, F.H. Pocock, H. Breuer

Department of Paediatrics and Child Health, Institute of Child Health, University of Cape Town, South Africa

Abstract

In childhood nutrition and paediatric disorders little is known of the importance of manganese (1). Published information discloses serious inconsistencies showing widely divergent levels. The small volume of blood available from infants for quantitation at nanogram level increases the difficulty of analysis. Recent reviews (2) suggesting that serum manganese levels are normally distributed around a mean of 0.6 ng/ml prompted us to investigate sources of error in measurement as previously we had reported a value of 1.2 ng/ml (3).

This paper describes the optimal conditions for analysis by electrothermal atomic absorption spectrometry. The effect of extraneous additions to blood samples from steel needles was quantitated by using a laser to produce small holes in a baboon's vein for blood collection. A series of investigations are reported in adult male volunteers to determine the flushing effect of drawing five 2 ml samples of blood through the same steel needle.

The results demonstrate a variable error from extraneous additions. The concentration in serum obtained using a laser was approximately half. Traces of haemolysis may greatly contribute to the confusion of measured values in serum and plasma.

Trace Element Analytical Chemistry in Medicine and Biology

Materials and Methods

Manganese was quantitated in serum and whole blood by electrothermal atomic absorption spectrometry using a Pye Unicam SP9 atomic absorption spectrometer equipped with automatic sampler graphite furnace, SP9 computer and a deuterium arc background corrector. Hollow cathode lamps and pyrolytically coated graphite tubes were employed. The instrument settings were wavelength 279.5 nm, slit 0.5 nm, lamp current 10 ma, background correction high and sample volume injection 15 μl.

OPTIMISED FURNACE PROGRAM

	TEMP (°)	TIME (secs)	RAMP
Phase 1 (drying)	110	60	9 (2°C/sec)
Phase 2 (ashing)	1000	30	9 (2°C/sec)
Phase 3 (atomisation)	2500	3	0
Phase 4 (tube clean)	2700	4	0

During the atomisation stage the functions of temperature control, autozero, gas stop and peak timer were selected. To reduce the risk of contamination from the laboratory environment the atomic absorption spectrometer was enclosed in a class A laminar airflow cabinet conforming to US Federal Code 209B. A similar cabinet was installed for sample and reagent preparation procedures. Both laminar airflow cabinets were situated in a room used exclusively for trace metal analysis.

Stringent precautions were taken to ensure that all materials used were free from manganese and the concentration was determined in serum and whole blood using the protocol of standard additions. Serum was diluted 1:4 and whole blood 1:5 in 10% ethylene glycol to which standard additions of 1, 2 and 3 ng/ml of manganese were made in the case of serum and 10, 20 and 30 ng/ml to the dilution of whole blood.

Results

Recovery Data:	91.7 to 94.2%.
Relative Standard Deviations:	within batch assay 0.7 - 2.3%.
	between batch assay 5.9%.
Sensitivity:	7.8×10^{-12}g.

Experiments to determine sources of contamination.

Polystyrene containers and sample diluent (10% ethylene glycol): The mean concentration of manganese in 10% ethylene glycol added to polystyrene containers was found to be 0.15 ng/ml. No manganese was detected in the diluent injected directly into the graphite tube.

Disposable Steel Needles: Using a laser beam to produce small holes in an anaesthetised baboon's vein, blood was collected without using a syringe. The baboon's forearm was shaved and the vein was surgically exposed. Following meticulous cleaning with alcohol and double distilled de-ionised water a small hole was made with a laser and four 2 ml samples of blood were collected into polystyrene containers. At the same time two 2 ml samples were taken using a syringe and disposable steel needle from a vein in the opposite forearm. Samples were taken from two baboons and the following table shows the manganese concentration in these samples.

	Baboon 1		Baboon 2	
Sample	Laser	Syringe & Needle	Laser	Syringe & Needle
1	0.9	1.75	0.8	1.70
2	0.6	1.20	0.75	1.10
3	1.85*		0.82	
4	0.7		0.68	

*Sample No 3 showed a trace of haemolysis.

Flushing Effect: The flushing effect of drawing blood through stainless steel needles was investigated by taking six 2 ml samples of human blood from adult male volunteers. These specimens were collected through the same needle into five different syringes. The results are shown in the following table.

	Volunteers					
Sample	No.1	No.2	No.3	No.4	No.5	No.6
1	2.60	2.80	1.90	1.70	1.70	1.40
2	0.87	0.80	1.65	0.90	0.90	0.96
3	1.02	0.85	2.30	0.65	0.70	1.14
4	0.91	0.85	1.80	0.70	0.70	0.68
5	1.05	0.50	1.75	0.90	0.90	0.77

The serum of volunteer 3 showed traces of haemolysis.

Discussion

The graphite furnace conditions for the analysis of manganese in serum and whole blood required a very slow drying and ashing phase with a linear ramp setting equivalent to a temperature increase of 2°C per second. This was found to greatly enhance the signal and produce good precision. This slow program, however, restricts the number of samples that can be processed in a normal working day.

Ethylene glycol as a diluent further improved the precision and signal where it functions as a surfactant during the drying stage forming a film which prevents the droplet from bursting at high temperature causing loss of element. Observation of the sample droplet in the graphite tube showed a slow and even reduction in size.

Serum and whole blood are relatively viscous fluids. Extremely poor precision was obtained if the furnace autosampler was improperly aligned or if the probe for injection of the sample was not optimally set to ensure deposition of the entire droplet on the bottom of the graphite tube.

The results demonstrate the excellent precision that can be achieved when precautions are taken to ensure these optimum settings.

In any discussion about the determination of manganese at ng/ml levels the problem of contamination cannot be overstated and for any analytical method it is necessary to obtain an accurate measurement of a blank sample. This was done by determining the concentration of manganese in the sample diluent in the same manner as for quantitation in serum and whole blood. The mean value was found to be 0.15 ng/ml. No manganese was detected in 10% ethylene glycol injected directly into the graphite tube indicating that 0.15 ng/ml of manganese is introduced into the sample during preparation. The appropriate concentration may therefore be subtracted for correction of the measured value.

To assess the significance of errors resulting from collection procedures the concentration from extraneous additions must be compared with the intrinsic level of elements in serum and whole blood. For manganese, where the mean serum level is approximately 0.6 ng/ml, error resulting from extraneous additions can be greater than the intrinsic concentration and is therefore important. For this reason we measured manganese in serum to obtain an indication of the degree of contamination resulting from collection procedures. Using a laser to obtain blood without using a syringe and needle we were able to demonstrate that the concentration of serum manganese was approximately half the level of serum taken from the same source using a stainless steel needle. The flushing effect of drawing 2 ml samples of blood through the same needle showed higher levels in the first sample. To eliminate this problem some workers have advocated discarding the first sample in volumes ranging from 10 ml to 40 ml. For obvious reasons this is not possible in small volume samples available from infants and children.

We suspect that haemolysis is as great a problem as sample contamination and may be responsible for the variation in reported levels in serum. We noted elevated values in samples containing minor traces of haemolysis, e.g. the level of manganese in serum from baboon no.1 was 0.6 ng/ml. A trace of haemolysis just detectable to the naked eye increased the concentration to 1.85 ng/ml. The effect of extraneous additions from steel needles make the quantitation of manganese in serum from infants and children extremely difficult. Extraneous additions to whole blood where the intrinsic level is in the order of twenty-fold higher than serum would have less effect. A small volume of sample is required for analysis and a portion of this sample could be used to flush contamination from the needle.

Acknowledgments

We wish to thank Mrs L.G. Makepeace for typing the manuscript. The study was supported by grants from the McCaul Bell Bequest and the South African Medical Research Council.

References

1. Shaw, J.C.L. Trace elements in the fetus and young infant. 1980. Am. J. Dis. Child. 134, 74-81.
2. Versieck, J. Trace element in human body fluids and tissues. 1985. CRC Critical Reviews in Chemical Laboratory Sciences. 22(2), 97-184.
3. Heese, H.de V., Lawrence, M.A., Dempster, W.S., Pocock, F. Reference concentrations of serum selenium and manganese in healthy nulliparas. 1988. S Afr Med J. 73, 163-165.

<u>SPECIATION ANALYSIS</u>

Plenary Lectures given by:

L. Dunemann

P. Brätter

Chairmen:

P.J. VanSoest
L. Dunemann

Poster Reports:

L. Dunemann, I. Lombeck

CHARACTERIZATION AND IDENTIFICATION OF ELEMENT SPECIES BY COMBINED ANALYTICAL METHODS

L. Dunemann, G. Schwedt
Institute for Inorganical and Analytical Chemistry of the TU Clausthal, Paul-Ernst-Straße 4, 3392 Clausthal-Zellerfeld

Abstract

In general, published data on trace element concentrations in biological and environmental material reflect only the total contents with no regard to the binding form of the element. In this article we present some examples of speciation analysis permitting more detailed statements about potential effects of the investigated element. In particular, the combination of separation and detection methods is described as a powerful tool in speciation analysis.

Introduction

If today we want to talk about the dangers of heavy metals or other elements we can no longer consider the total concentration alone. Although it is important to know the total content exactly, it is even more important to consider the binding form, i.e. the particular species of the element in a particular matrix. Only then can we make statements about mobility, resorption, bioavailability, accumulation, and toxicity of this element.

For this reason it is necessary for investigations on trace elements in environmental and biological samples to employ speciation analysis. Speciation helps us to distinguish different element species concentrations. These concentrations of

Trace Element Analytical Chemistry in Medicine and Biology

separated element species can give us much more information about the effects of the metals in a particular matrix. We now can come closer to answering the question whether or not a certain concentration of an element is available to an organism. For example, is the amount of a heavy metal in a food sample toxic, or is a plant able to utilise a nutrient element out of its soil matrix?

If we start with the simpliest matrix in the geobiochemical environment - a natural water - we already have to deal with a great variety of species. Some of the most common ones are shown in Table 1.

Table 1: Heavy Metal Species in Natural Waters

1. FREE IONS	$Me^{n+} \cdot aq$ in different oxidation states
2. INORG. COMPLEXES	$Pb(CO_3)_2^{2-}$, $CdCl^+$, $Al(OH)_4^-$, $Zn(OH)_3^-$
3. ORG. COMPLEXES	R-COO-Me, R-S-Me, fulvates
4. Adsorbed to 3.	humic acids, proteins, lipids
5. COLLOIDS	Mn-oxides, Fe-oxides/hydroxides, sulfides
6. Adsorbed to 5.	to clay minerals (as Me^{n+}, $MeCO_3$, MeS), to Fe/Mn-oxides ("secondary structures") = exchange-adsorption
7. OCCLUSIONS	Me occluded in lattice structures (from mineral genesis)

In matrices other than natural waters we have to take an even higher number of species into consideration, and they may differ substantially from the above scheme.

If we want to use speciation analysis, there is a wide spectrum of separation and detection methods available. But when we have to deal with a complex matrix, we have may have to use more than one of these methods. It is imperative to choose a number of such methods in order to get a clear view of the element species involved. Of course, it would be very time-consuming to apply only a few of them one by one

Table 2: Methods for Speciation Analysis

a) SEPARATION METHODS	
Extractive separation:	solid/liquid, liquid/liquid
Membrane separation:	ultra filtration, dialysis
Adsorptive separation:	adsorption onto ion exchange or chemically modified silica media
Chromatographic separation:	LC: gel filtration, ion exchange, adsorption, reversed phase, hydrophobic interaction, affinity including HPLC TLC: adsorption, reversed phase, including HPTLC GC: capillary and packed columns
Electrophoresis:	native, pore gradient, IEF
b) DETECTION METHODS	
Total, after digestion:	AAS, ICP-AES, photometry, voltammetry/polarography
Selective, according to oxidation or binding states:	photometry (ligand exchange), voltammetry, potentiometry with ion selective electrodes
Characterization of ligands:	molecular spectroscopy (IR, UV, fluorescence) and post-column derivatization of functional groups
c) COMBINED METHODS (with online or offline detection)	
Single step separation:	LC/UV for distribution patterns LC/AAS for metal detection LC/CFA,FIA for detection of metal and dominant functional groups
Multi step separation:	ultra filtration/ion exchange/AAS gel filtration/ion exchange/AAS gelfiltration/electrophoresis/AAS
Speciation schemes: (= multi step speciation)	combination of all pertinent separation and detection techniques

separately. It is much faster and more economical to use combined methods, e.g. by coupling a separation with a detection method. Table 2 gives a condensed overview about some important methods successfully used in our laboratories for speciation analysis.
The traditional separation of analytical chemistry into inorganic and organic analysis must be overcome for speciation analysis, because only the application of the whole range of analytical methods - and especially combinations of them - can lead us to satisfactory results.
Although some suggestions on speciation procedures have been made by analytical chemists as well as environmentalists, biologists and toxicologists we should always keep in mind to check a given method with reference to the matrix being investigated. This is particularly true for speciation schemes which are rarely applicable to more than one matrix or more than one problem in question (see Discussion).

Selected Examples

Plant uptake of heavy metals

If we want to follow the pathway of an element (e.g. a heavy metal) from the soil into the plant, and further into animals or humans we first of all should have a look at the soil solution, i.e. the interface between the plant and the soil. Only those heavy metal species which can be solvated or dispersed by the soil solution can be absorbed by the plant roots.
The cation exchange capacity (CEC) of plant roots is in direct proportion to the numbers of fixed anions in the cell walls. Ion distribution (cation exchange and anion repulsion) within the Donnan free space is the typical Donnan distribution which occurs also in soils at the surfaces of negatively charged clay particles (1). Although this is an explanation for the uptake of some common ions, it may not be true for all ionic species observed in such a system. It also seems reasonable to

suggest a more or less strong adsorption of cations to the fixed anions in the cell wall micropores; solvated anions from the soil solution could than pass through the micropores of the cell walls with little resistence.
Besides the charge, especially the net charge, the uptake depends upon the size of the species (see Table 3).

Table 3: Penetration of Metal Species Through Cell Wall Pores*

Pores in cell walls		<5.0 nm
Hydrated ions	normally	<1.0 nm
	K^+	0.66 nm
	Ca^{2+}	0.82 nm
Sucrose		1.0 nm
Metal chelates excluded		>10,000 Dalton (depending on the shape)

*Based on (1)

Larger species than approximately 10,000 Dalton, or those with a high net charge, cannot enter the inner cell as a whole molecule, they are restricted or prevented by the cell wall pores. In such a case it depends upon the stability of the bonds between the central ion and its ligands whether or not a certain complex can be broken down (e.g. by root exudates). If so the metal can be taken up as a free ion or it can be transformed (re-complexed) into a smaller complex with organic or inorganic ligands from the soil rhizosphere.
To summarize, the resorption of metal species (e.g. complexes) depends on the charge, the size, and the stability of the species. All these properties can be evaluated by a combination of speciation methods.

Speciation of soil equilibrium solutions. First of all it is neccessary to evaluate both the size and stability of metal species. A fractionation due to molecular size can be done very rapidly by ultra filtration and gel filtration. Ultra filtration is faster than the chromatographic method, but the assessment of the stabilities of binding forms is done more elegantly by the coupling of gel filtration to chemical reaction detectors. Chemical reaction detectors serve not only for the detection of metals and functional groups (see Figure 1) but also for the estimation of species stabilities. This double function allows us to gain information on distribution patterns within a single run on a chromatographic column.

Figure 1: Chemical Reaction Detector (CFA[1] System) for the Evaluation of Carboxylic Groups (2-4)*

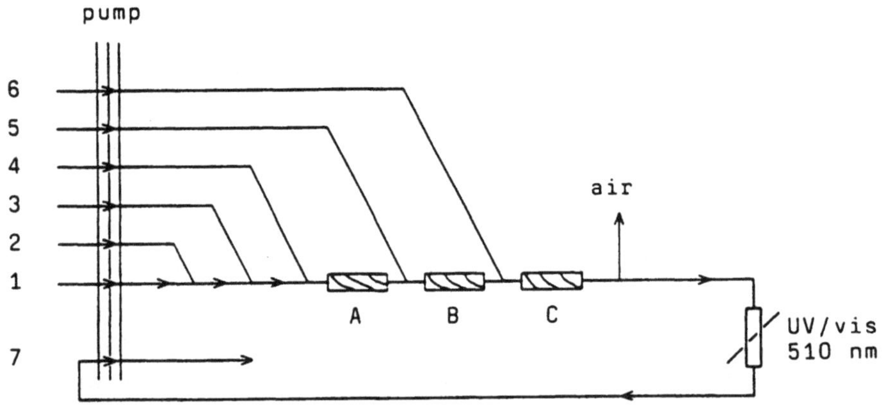

Detector:	UV/vis, 510 nm		
Pump:	Peristaltic, 8 channels, adjusted to a flowrate of 1 ml/min using a tubing with 1.02 mm i.d.		
Tubings:	1	From GFC column	(i.d. 1.02 mm)
	2	Air	(i.d. 1.02 mm)
	3	Hydroxylamin	(i.d. 0.76 mm)
	4	CMCTS[2]	(i.d. 0.76 mm)
	5	Ascorbic acid	(i.d. 0.76 mm)
	6	Fe(III)chloride	(i.d. 0.76 mm)
	7	Waste	(i.d. 1.14 mm)
Coils:	A	14 turns, thermostated to 37°C	
	B	28 turns, thermostated to 37°C	
	C	14 turns.	

[1] Continuous flow analysis
[2] N-Cyclohexyl-N'-(2-(N-methylmorpholino)ethyl-carbodiimide-p-toluolsulfonate

*From (5)

Correlations of distribution patterns of metals and functional groups can give distinct hints concerning the nature of the dominant ligands, which may play a potential role in metal complexation. We can ascertain potential binding partners without time consuming steps (6). From this work it is even pos-

Figure 2: Correlations of Fe with Dominant Functional Groups in a Soil Solution (for more details see (5,6))

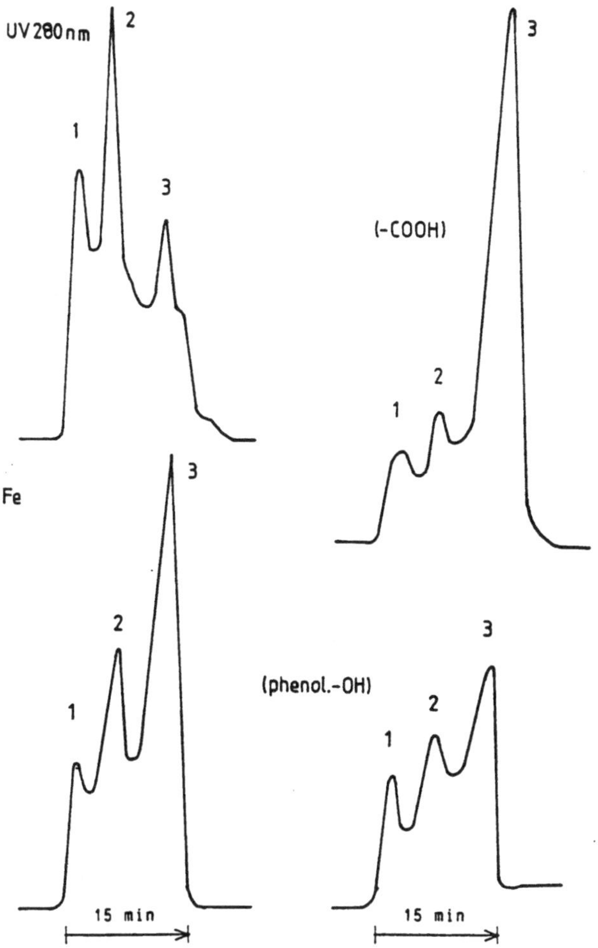

sible to make the statement that the simultaneous presence of carboxylic and phenolic groups have a major effect on iron complexation in soil solutions (see Figure 2). But for zinc this is not the dominant factor (6).
It is an important fact that dextrane based gel filtration media has a very weak ion exchange capacity, so that we can screen the amount of free metal ions simply by applying a sample to a very small column filled with a dextrane gel (Sephadex). Some metal ions, e.g. zinc, are retarded by the gel media and can be rinsed from the column with citric acid (6).

Bioavailability of metal species from soil solutions. Metal uptake by the roots is an important factor for metal bioavailability, but not the only one. Metal species can only enter the food chain if they are transferred to the eatable parts of the plant, mostly the shoot. In case of cereals the metal tranfer must go as far as the corn. During this passage the binding forms of the metals taken up by the roots may change. So the most important considerations are

a) the binding forms of those metals in the soil solution,
b) the total metal concentrations in the corn.

In addition to a vegetational pot trial reported elsewhere (7), we wanted to find out more about the bioavailability of heavy metal species in the soil solutions. So we collected the soil solutions from the pot trial and carried out a nutrient solution culture experiment.
We wanted to know if young plants were able to take up the metal species from the soil solutions. Therefore we compared the uptake rates of nickel species from soil solutions with those from adapted nutrient solutions containing inorganic nickel salts, either alone or in the presence of strongly chelating agents (e.g. DTPA). As shown in Table 4, nickel in the soil solution must be chelated a significant amount. Furthermore we found a strong negative correlation between the absorbance of the soil solution (at 280 nm) and the uptake rate of nickel by the oat plants in the nutrient solution trial (8). This leads

to the conclusion that a significant amount of nickel in the soil solution is organically complexed.

Table 4: Uptake Rates of Nickel Species by Oat Plants from Different Nutrient Solutions

Binding form	uptake rate
Free nickel	high
Ni from soil solutions	
Luvisol[1]	medium-high
Cambisol[2]	medium-low
Chelated nickel	low

↑ increasing availability of nickel

[1] Parabraunerde
[2] Pelosol

These results are in a good agreement with the results of the chemical speciation procedure, by the need of ultra filtration and ion exchange, also applied to the soil solutions (Table 5).

Table 5: Chemical Assessment of Nickel Species (Speciation) in the Soil Solutions

Size		Free species	
<500 D	40%	in Luvisol	20%
500 - 5000 D	40%	in Cambisol	10%
>5000 D	20%		

The presented experimental approaches are summarized in the speciation scheme in Figure 3.

Figure 3: Scheme for the Evaluation of the Availability of Heavy Metals in Soils

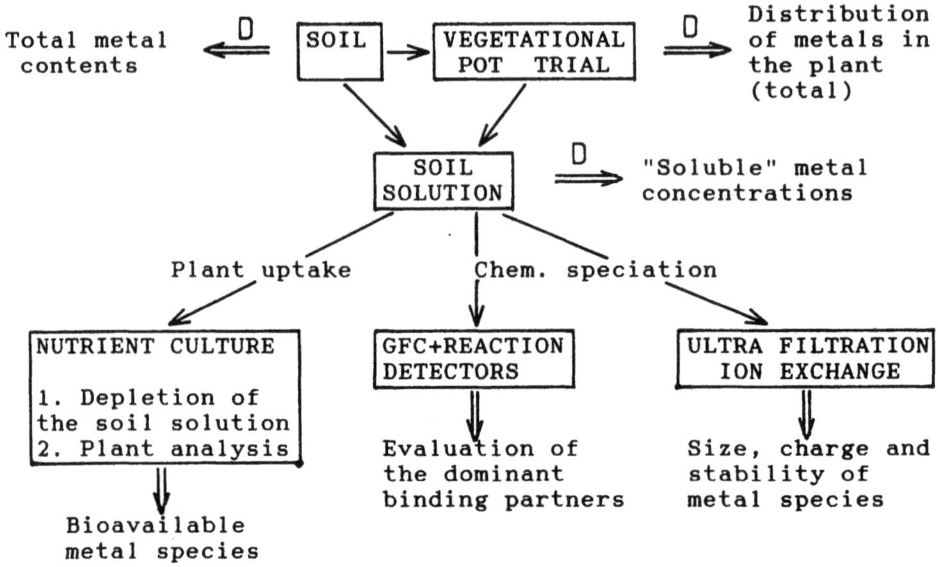

D = After complete sample digestion

Availability of metals in foodstuffs

In recent research work (9), we looked for a combination of separation and detection techniques in order to evaluate different metal species in soya bean flour extracts (Table 6).

Table 6: Total and Extractable Heavy Metal Concentrations in Soya Bean Flour (in mg/kg dry matter)

	Total	Water extractable	
Nickel	7.50 mg/kg	5.65 mg/kg	(75.3%)
Copper	15.2 mg/kg	12.8 mg/kg	(84.3%)
Zinc	41.2 mg/kg	24.5 mg/kg	(59.5%)

We chose gel filtration chromatography, because it is a fast and reliable method for the separation of proteins and other pertinent compounds in aqueous food samples (10). Electrophoretic techniques, such as polyacrylamide gel electrophoresis in pore gradients for the determination of molecular weights and isoelectric focussing for the determination of isoelectric points, were used as reference techniques.

Metal determinations were executed either by online or offline coupling with flame AAS (Figures 4-6). Online detection was only used for the fractionation via GFC (Figure 4), because the sample volumes applied to the electrophoretic gels were to small to allow detection by FAAS. Nevertheless metals could be determined after electrophoretic elution and concentration of the separated protein fractions (Figure 6).

As we supposed, the metal distributions of zinc, nickel, and copper differed significantly from the protein distribution (Figure 4). This fact supports the hypothesis, that metals are bound to special protein structures.

The addition of zinc (Figure 4) proves that spiking of an unknown matrix should be avoided, because it may alter the metal distribution within the matrix significantly. We suppose that the complexing ability of the potential binding partners in Fraction III (Figure 4) is already exhausted, whereas this is not true for Fraction I. This fraction can bind further metals because of an excess of free binding capacitiy.

The experiments showed that electrophoresis is superior to gel filtration chromatography with respect to the protein separation, because it gives a better performance with a greater number of separated fractions. Referring to the trace metal distribution patterns both methods reveal comparable results (Figures 5 and 6).

As in the soil solution experiment it is also important to determine the dominant binding partners of the investigated metals. We discussed this in an earlier publication (10).

To sum this up we present a speciation scheme for the evaluation of available metals in a food matrix (Figure 7).

Figure 4: Directly Coupled GFC-FAAS Showing Associated Zinc in Soya Bean Flour (unspiked and spiked with 2.5 μg/ml)

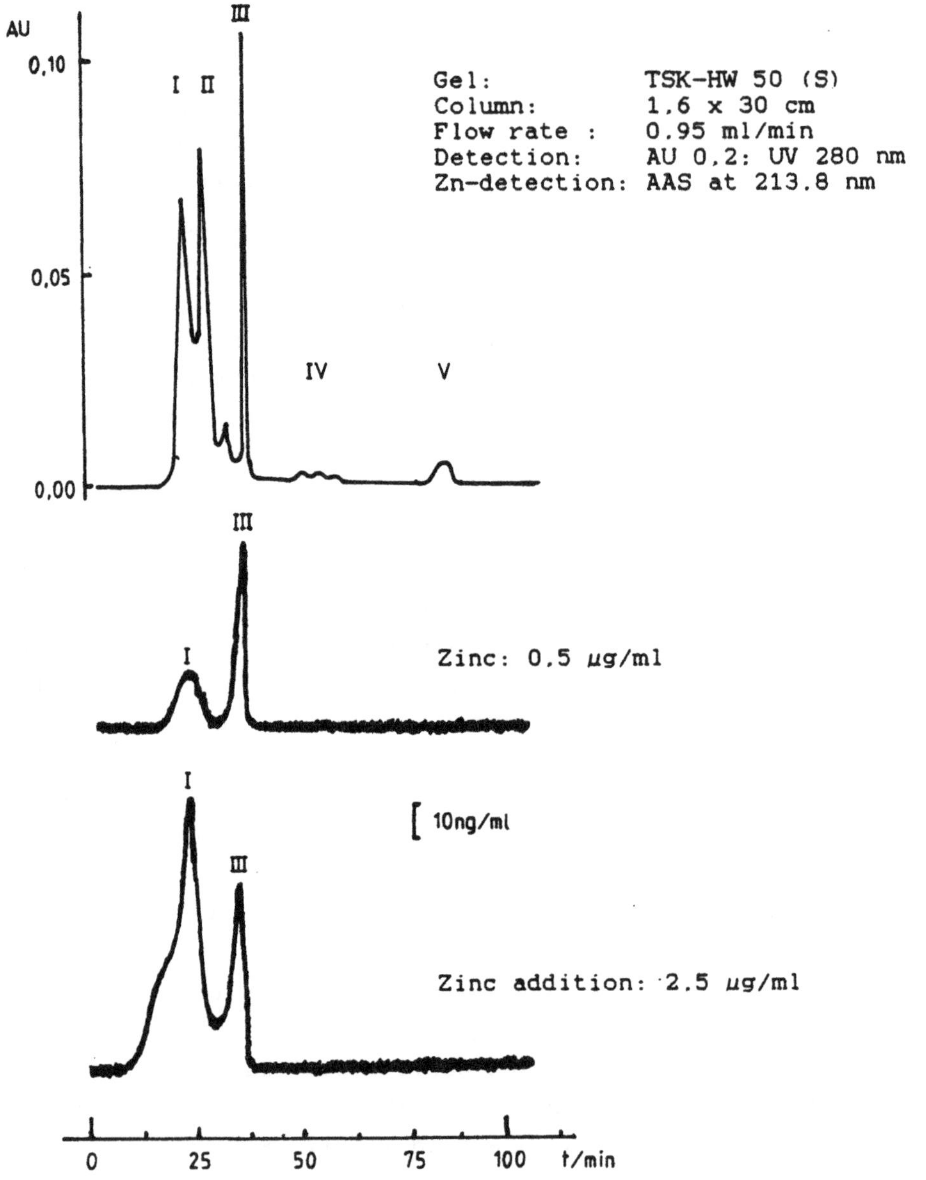

Figure 5: GFC Distribution Patterns for Nickel and Zinc in Soya Bean Flour

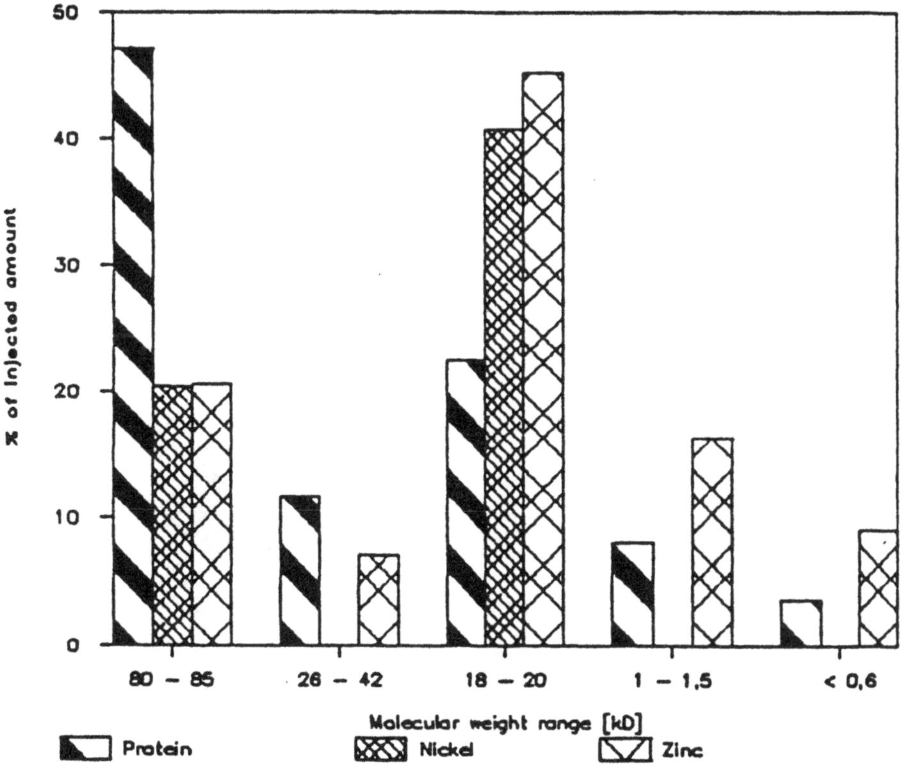

Relative Distribution:

	Protein [%]	Nickel [%]	Zinc [%]	MW [kD]
Fraction I	47.1	20.4	20.6	80 - 85
Fraction II	11.7	<8.8	7.1	26 - 42
Fraction III	22.4	40.7	45.2	18 - 20
Fraction IV	8.1	<8.8	16.3	1 - 1.5
Fraction V	3.6	<8.8	9.0	< 0.6
	----	----	----	
Recovery [%]	92.8	87.6	98.2	

Figure 6: Electrophoretic Distribution Patterns for Copper, Nickel, and Zinc in Soya Bean Flour

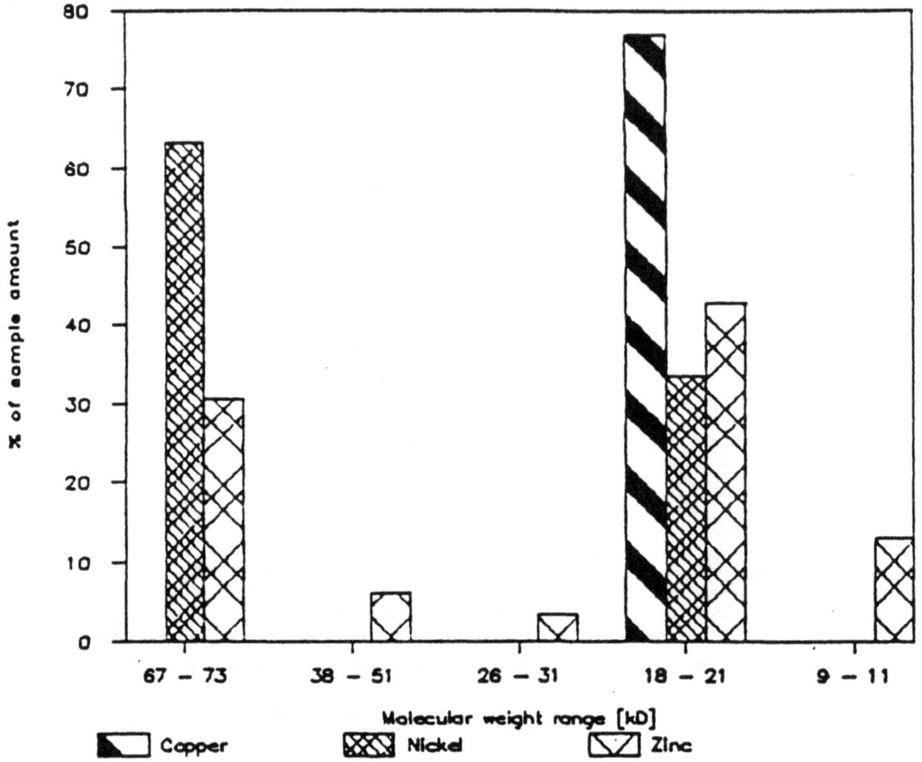

Relative distribution:

	Copper [%]	Nickel [%]	Zinc [%]	MW [kD]
Fraction I	<10	63.1	30.6	67 - 73
Fraction II	<10	<10	6.1	38 - 51
Fraction III	<10	<10	3.5	26 - 31
Fraction IV	76.9	35.5	42.9	18 - 21
Fraction V	<10	<10	13.9	9 - 11
	------	------	------	
Recovery	76.9	96.7	96.3	

Figure 7: Scheme for the Evaluation of Metal Availability in Foods

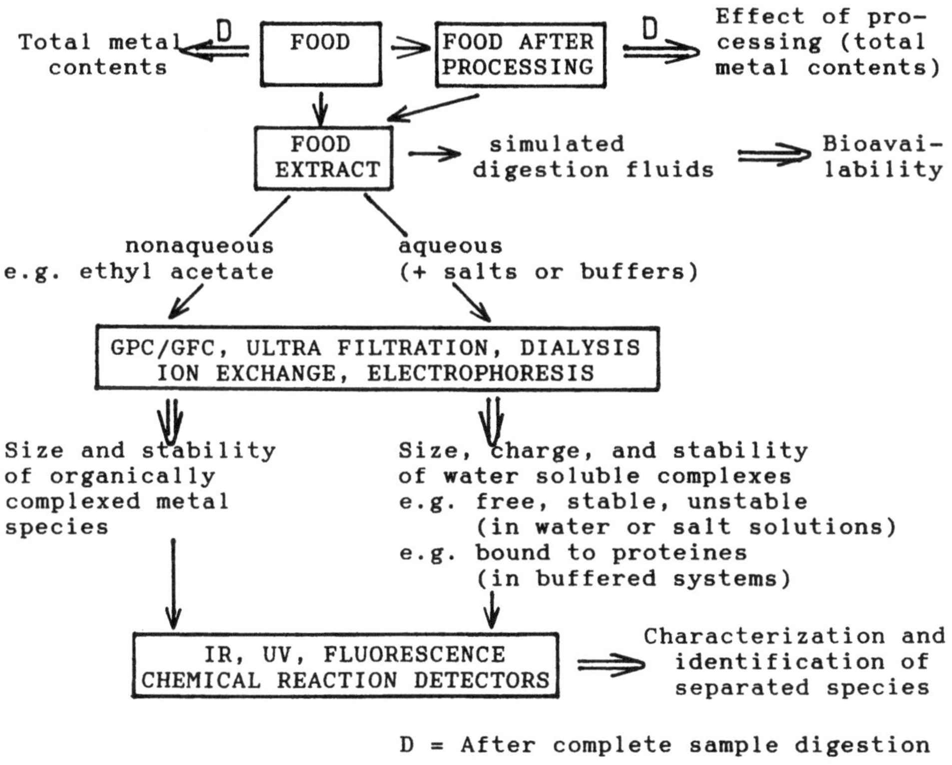

Discussion

For most applications of speciation analysis it is important to start with extraction procedures (11-13), followed by chromatographic or membrane saparation techniques (14). For the evaluation of species stabilities the use of ion exchange, ultra filtration, and dialysis is recommended (11, 14-16). Afterwards all pertinent spectroscopic methods can be used for the final characterization of the separated and partially characterized binding forms (17). For other investigations (oxidation states (17-18)) or applications to special samples see (17).

Besides finding the distribution of heavy metals in a plant part it is important to find the distribution within the plant cells (20). The next step must be the evaluation of the binding forms of these heavy metals. This can be done either by extraction procedures (20) or by other separation and detection methods mentioned before.

Addition of heavy metals as inorganic salts or combined with complexing agents (e.g. EDTA, DTPA) to a nutrient solution or a plant or food extract is often used, because of the low sensitivity of an applied detection method (e.g. 21-25). Nevertheless, it should be avoided, if we want to evaluate native binding forms or effects of element species in their original matrices. Changes in the distribution patterns may occur, which can alter the speciation results. Adding toxic concentrations of heavy metals is expecially harmful: they can change the metabolism in the plant cells (even morphological changes of plant components are observed (22)), and what may result is a dramatic change in metal species distribution. In real systems metals may mostly be taken up as metal complexes, but not as free ions (26).

Sometimes we cannot detect the concentration of a minor trace element due to dilution effects of separation steps or due to the low sensitivity of a detection technique. If there is no alternative to the speciation procedure, an enrichment step, e.g. adsorption to reversed phase media (27), should be preferred to the addition of the free metal salt.

On the other hand, addition of heavy metals as inorganic salts is a useful tool for investigations of the most stable binding form(s) (e.g. metallothionein). According to the experiments of Solomons et al. (28) a successive addition of metal salts is proposed to evaluate the total complexing ability of the matrix components. Most matrices have a much higher binding capacity for heavy metals than is needed for the amount already present. Although there is a dynamic equilibrium within

the native matrix between free and bound metal, this can be changed by the addition of free metal ions. After the successive addition of metal salt the system re-equilibrates until the complexing ability of the matrix is exhausted (19,28).
In special cases it may also be quite useful to add metal salts to the elution buffer of a chromatographic system (e.g. gel filtration) as a precaution to prevent losses of heavy metals during the separation process. But this attempt should be limited to the final characterization of a binding form with known properties (e.g. 16,24).
The use of calculation programms like GEOCHEM or CHELATE, often applied to natural waters and soil solutions (e.g. 29) is limited to problems where it is considered the system is in equilibrium and so thermodynamic data can be used, or when experimental methods are applied simultaneously (30-31). Robb et al. (32) used a computerized simulation technique to investigate metal species in digested food. They considered the food to be completely broken down into its amino acids and metal ions which afterwards take part in complex equilibrium reactions. It is still useful to have this technique for calculating the potential range of metal species that may be present in such matrices, because our knowledge about them is still limited. Like GEOCHEM and CHELATE this approach still depends only on thermodynamic data. But if we have to deal with kinetically controlled reactions the use of such computer programms is not recommended (33).

Conclusions

On its way from the soil via the soil solution into the plant and further into animals and humans, the original binding form of a metal undergoes several severe changes. It may be broken down either by certain properties of the soil solution (pH, redox potential, microorganisms, root exudates) or by the digestion fluids of animals or humans. On the other hand, the complexing abilities of most matrices are so strong that free metal ions are soon re-complexed, forming another binding form. If we want to follow this complex pathway and if we want to evaluate the mobility, resorption, bioavailability, accumulation, and toxicity of metals in their matrices we must use efficient speciation analysis methods.

Although the localization of metal accumulations in animal or plant components and cells is of fundamental interest, it is even more important to get information about the stability of the binding forms within such accumulations. This includes the knowledge of other potential binding ligands, because some binding forms in such a matrix may be quite unstable and this must result in a dynamic equilibrium, i.e. recomplexation reactions between the central ion and an outer complex ligand. It is of further importance to take into account the fact that there may be kinetically stable compounds (e.g. excretion products) too. This is why a mere consideration of thermodynamic properties is not enough for speciation analysis of complex matrices.

We should always be aware of the limitations of speciation analysis: low detection limits after separation procedures (dilution effects!); danger of contaminations or loss of small traces of metals in the low concentration range, and severe changes of the native binding forms.

Future Developments

It is imperative to use sensitive detection techniques for the speciation of minor trace elements in order to get better results after a separation step. This is particularly true for complete speciation schemes. Using sensitive detection methods we can avoid enrichment techniques producing the risk of artifacts or changes in the original species distribution.

If the maxima of UV detection and the detection of heavy metals sometimes do not correlate, it could be useful to carry out the UV detection at multiple wavelengths, best done by a diode array detector. This is especially useful if different sample components are present simultanously. For a screening procedure this is the most time-saving technique because not only can we save on lengthy separation steps, but also there will be minimum disturbance of the metal binding form equilibria.

The use of electrophoresis for metal speciation can be applied successfully for the separation of stable binding forms in food extracts. However, more work is neccessary before this method can be established for species analysis in a wide range of sample types.

References

1. Marschner, H. 1986. Mineral Nutrition of Higher Plants. Academic Press, London, p. 7
2. Kasai, Y., T. Tanimura, Z. Tamura. 1975. Anal.Chem. 47, 34
3. Takeuchi, T., R. Horikawa, T. Tanimura. 1980. Anal.Lett. 13, 603
4. Fishbein, W.N., J. Daly, C.L. Streeter. 1969. Anal.Biochem. 28, 13
5. Dunemann, L.: Fresenius Z.Anal.Chem. (in preparation)
6. Dunemann, L., G. Schwedt. 1984. Fresenius Z.Anal.Chem. 317, 394
7. Marschner, H., L. Dunemann, R. Schulz, P.H. Brown: Z.Planzenernähr.Bodenk. (in preparation)
8. Marschner, H., L. Dunemann, R. Schulz, P.H. Brown, N. v. Wirén. 1988. Poster presentation at the "1. Statuskolloquium" of the "Projekt Abfall-Wasser-Boden (PWAB)", Nuclear Research Center Karlsruhe
9. Dunemann, L., H. Reinecke: Fresenius Z.Anal.Chem. (in preparation)
10. Dunemann, L., G. Schwedt. 1986. Fresenius Z.Anal.Chem. 325, 121
11. Schwedt, G., M. Metschies, A. Schweizer, D. Zöltzer. 1987. Fresenius Z.Anal.Chem. 327, 142
12. Förstner, U., C. Sellhorn. 1986. In: Vom Wasser 67, Verlag Chemie, Weinheim, 159

13. Brümmer, G.W. 1986. In: The Importance of Chemical "Speciation" in Environmental Processes (M. Bernhard, F.E. Brinckman, P.J. Sadler, eds), Springer-Verlag, Berlin, p. 169

14. Schweizer, A., G. Schwedt. 1988. Fresenius Z.Anal.Chem. (in press)

15. Schwedt, G., A. Schweizer, G. Hendrich. 1988. Z.Lebensm. Unters.Forsch. (in press)

16. Meisch, H.U., I. Beckmann, J.A. Schmitt. 1983. Biochim. Biophys.Acta 745, 259

17. Bernhard, M., F.E. Brinckman, P.J. Sadler, eds. 1986. The Importance of Chemical "Speciation" in Environmental Processes. Springer-Verlag, Berlin

18. Thiele, A., G. Schwedt. 1985. Z.Lebensm.Unters.Forsch. 180, 126

19. Stoeppler, M., H.W. Nürnberg. 1984. In: Metalle in der Umwelt (E. Merian ed.). Verlag Chemie, Weinheim, p. 81

20. Schwedt, G., G. Jahns. 1987. Fresenius Z.Anal.Chem. 328, 85

21. Grill, E., E.-L. Winnacker, M.H. Zenk. 1986. Science 230, 674

22. Weigel, H.J., H.J. Jäger. 1980. Plant Physiol. 65, 480

23. Rauser, W.E. 1984. Plant Physiol. 74, 1025

24. Fujita, M., T. Kawanishi. 1986. Plant Cell Physiol. 27, 1317

25. Wagner, G.J., R. Yeargan. 1986. Plant Physiol. 82, 274

26. Fujita, M. 1985. Plant Cell Physiol. 26, 295

27. Weber, G., G. Schwedt. 1983. Deutsche Lebensm.Rundsch. 79, 13

28. Solomons, N.W., F. Viteri, T.R. Shuler, F.H. Nielsen. 1982. J.Nutr. 112, 39

29. Mullins, G.L., L.E. Sommers, T.L. Housley. 1986. Plant Soil 96, 377

30. Adams, T.M., J.R. Sanders. 1985. In: Chemical Methods for Assessing Bioavailable Metals in Sludges and Soils (R. Leschber, R.D. Davis, P. L'Hermite, eds.). Elsevier, London, New York, p. 68

31. Behel, D., D.W. Nelson, L.E. Sommers. 1983. J.Environ. Qual. 12, 181

32. Robb, P., D.R. Williams, D.J. McWeeny. 1986. Inorg.Chim. Acta 125, 207

33. Stumm, W., L. Keller. 1984. In: Metalle in der Umwelt (E. Merian, ed.). Verlag Chemie, Weinheim, p. 21

COMBINATION OF HPLC AND ICP-AES FOR THE SPECIATION OF SELENIUM AND OTHER TRACE ELEMENTS IN BODY FLUIDS.

P. BRÄTTER, B. GERCKEN, A. TOMIAK and U. RÖSICK

Hahn-Meitner-Institut Berlin, FRG
Res.Group: Trace Elements in Health and Nutrition

SUMMARY

On-line combination of gel filtration chromatography and ICP-AES has been applied for the fractionation and identification of metal-containing species in biological fluids including human blood serum, red blood cell lysate and mother's milk.
High-performance liquid chromatography columns were coupled to a sequentially operating ICP-AES system for multielement detection. During a chromatographic run the intensity data of each measuring sequence were transferred to a separate computer system and subjected to various evaluation procedures. Superimposed distribution profiles of the selected trace elements were obtained for each sample. Integration time as well as wavelength scan rate limits the number of elements to about six possible to be studied during one chromatographic run. The distribution of elements of clinical interest including copper, iron, zinc, magnesium or calcium could be simultaneously studied inserting a sample volume of between 250 and 1000 µl.
To increase sensitivity of Se-determination a procedure was developed to determine selenium protein profiles in body fluids via the on-line coupling of gel-permeation chromatography and ICP performing the acid digestion procedure and the formation of Se(IV) with the subsequent hydride generation in the connecting flow. Recovery studies have been carried out to examine and optimize the wet ashing conditions with the aid of various selenium compounds. To demonstrate the usefulness of this technique for the speciation of selenium its distribution in samples of human origin has been measured. In serum Se was found to be distributed among 3 different fractions. If human erythrocyte lysates were subjected to gel filtration the Se eluted with the glutathione peroxidase and hemoglobin. In breast milk four Se-peaks could be detected in the effluent at molecular weights of >600, 90, 25 and 10 kDa, respectively.

Trace Element Analytical Chemistry in Medicine and Biology

INTRODUCTION

Trace metals bound to specific proteins are subject to changes in connection with physiological or pathological processes in an organism. Measurements of the specifically protein bound elements in body fluids like whole blood or serum may be a more sensitive index in the assessment of trace element imbalances than the determination of their total amount present. With the increasing interest on the role of trace elements in living systems in recent years analytical work has been focussed on speciation studies. Various methods bave been described for coupling chromatographic techniques with element specific detection including liquid and gas chromatography for the fractionation and atomic absorption spectrometry (AAS) [1-5], neutron activation analysis (NAA) [6], PIXE [7] and plasma emission spectrometry [8-12] for metal identification. According to their known biological functions in clinical medicine and nutrition studies the distribution of the most interesting protein bound elements in serum or whole blood are copper, iron, selenium and zinc. The methods of choice for their separation have to meet a few clinical requirements:

The speciation has to be carried out normally on small sample volumes and the results of analysis are needed for diagnostic purposes on the same day. Despite of its sensitivity a drawback for the application of AAS techniques for metal speciation is the only single element detection property. NAA offers multielement detection capabilities but on the other hand its application is time consuming and depends on the availability of a high neutron flux density irradiation source.

Some recent works have demonstrated the application of plasma emission spectrometry for the identification of protein bound metals in liquid chromatographic effluents. With the developments in gel permeation high performance liquid chromatography (HPLC) simultaneous element detection including Cu, Fe and Zn can be achieved by direct coupling of a HPLC column to an inductively coupled plasma (ICP) polychromator system. The sequential mode of element detection by ICP offers some advantages in connection with the free choice of wavelengths. First it is easy to select the emission line of an element which is free of spectral interferences. Second, it is possible to measure the background variation as close as possible to the selected element channel. Clearly, sequential analysis limits the number of elements which can be detected in a fraction without substantial loss of information. That is due to the sequential integration of each channel over a fixed period within a measuring cycle. The analytical advantages of ICP include linear dynamic range of several orders, good reproducibility and detection capability for the elements of current clinical interest including Cu, Fe, Zn, Ca, Mg. The direct determination of Se by ICP-AES suffers from poor sensitivity. Its detection limit

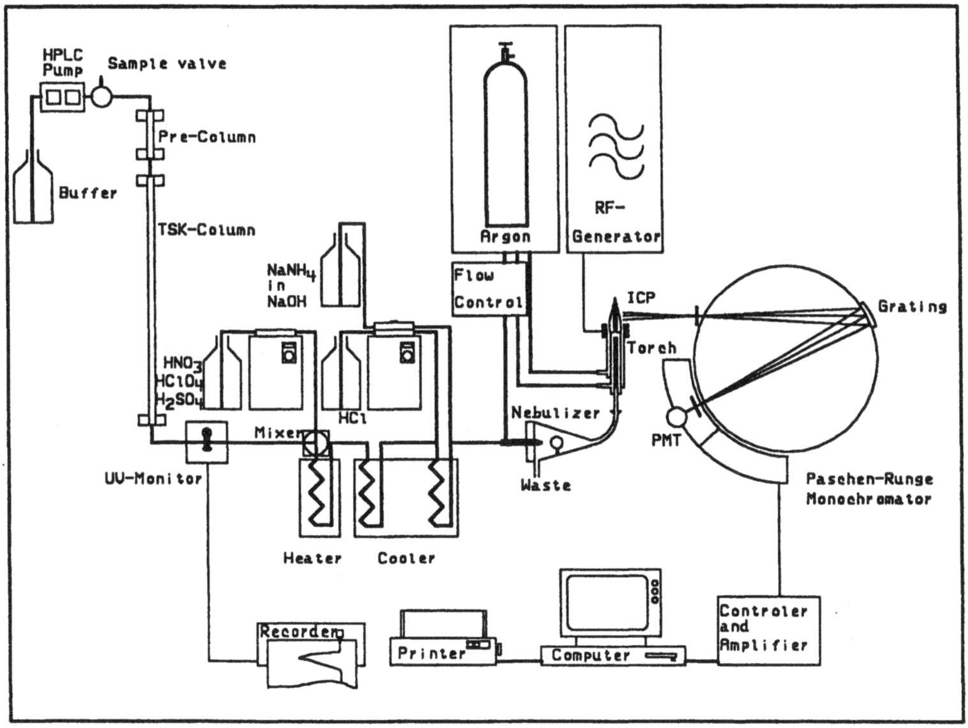

Fig. 1: Scheme of the trace element speciation in biological fluids by on-line coupling of gel filtration chromatography and ICP-AES.

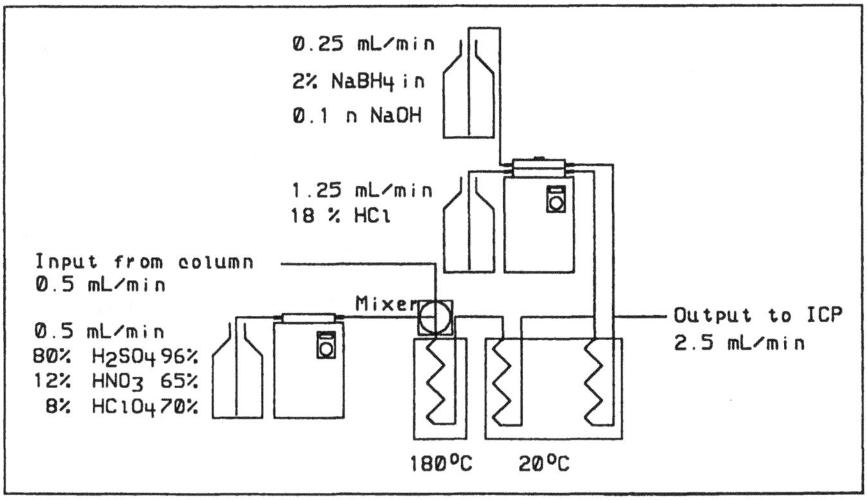

Fig. 2: Scheme of the on-line wet digestion and hydride generation system for the speciation analysis of selenium.

is in the range of 50 microgram per liter. Recently continuous generation of volatile hydrides becomes an interesting technique to improve the detection limit of selenium determination.

Detecting selenium via selenium hydride in body fluids simultaneously with the elements of clinical interest in one chromatographic run is an attractive aspect of speciation studies. With respect to the analytical requirements one has to recognize that selenium analysis of biological fluids via the hydride technique includes the acid digestion of the sample followed by the complete conversion of selenium (VI) to the oxidation state (IV). So far speciation of selenium was carried out in the batch mode analyzing each fraction collected after gel filtration. Restrictions in the application of these techniques has been attributed to the long time necessary to produce element protein profiles. Until now the on-line procedure has not been realized due to the difficulties of bringing the selenium into a form which is reactive with sodium tetrahydroborate. We developed a procedure to determine selenium protein profiles in body fluids via the on-line coupling of gel permeation chromatography and ICP-AES performing the acid digestion procedure and the formation of Se (IV) with the subsequent generation of selenium hydride in the connecting flow.

EXPERIMENTAL

Instrumentation :

The instrumental setup of the coupled analytical systems is shown in fig. 1. The chromatography was performed with a system consisting of a HPLC pump (LKB 2150), Reodyne injection valve , UV- monitor LKB 2138, and recorder LKB 2210. A short precolumn was placed after the sample injection valve to protect the analytical column. Protein was fractionated was on TSK G 3000 and TSK G 2000 columns (length: 600 mm. i. diameter: 7.5 mm).

Element detection was carried out by the sequential ICP atomic emission spectrometer system ARL 3520 (Applied Research Laboratories, Ecublens, Switzerland). Plasma and sample carrier gas flows were controlled via mass flow controller (Instruments S.A.).

Gel columns were calibrated with purified human milk and serum proteins including β-lactoglobulin B, α-lactoalbumin, serum albumin, IgA, casein, lactoferrin and transferrin. The proteins were assayed spectrophotometrically at 280 nm.

Fig. 3: Flow scheme of the instrumental procedure.

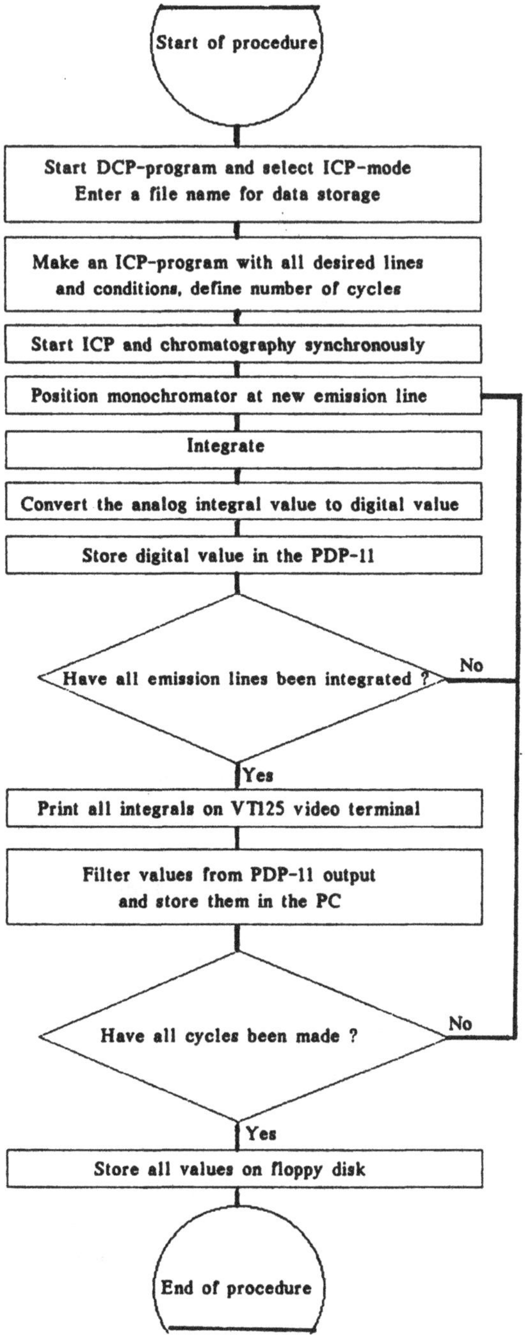

Operating parameters:

1) ICP-AES

		Wavelength [nm]
HF-generator power [W]	1400	Se 196.09
Frequency [MHz]	27.125	Se background 196.09±0.035
Plasma torch	Fassel type	Cu 324.76
Nebulizer	Meinhard type B3	Fe 259.86
Observation height [mm]	15	Mn 257.61
Photomultiplier gain [V]	1000	Zn 213.86
Nebulizer gas flow rate [l/min]	0.80	
Plasma gas flow rate [l/min]	1.00	

2) Gel filtration

Buffer	0.02 M Phospate + 0.1 M NaCl	pH = 7.4
Flow rate [ml/min]	0.5	
Sample volume [µl]	1000	Breast milk
	250	Serum, Erythrocyte-lysate
UV-wavelength [nm]	280	

Procedures :

In our experimental arrangement (fig. 2) the wet digestion of the effluent was intitiated by continuously adding a ternary mixture of nitric-perchloric-sulfuric acid into a stirring aparatus (Teflon, effective volume: 0.35 ml). The solution passed a Teflon coil (about 10m length, i.d. 0 .5 mm), which is placed in in a heater at a temperature of 180 ^{o}C. After cooling down the digested solution, 5 molar hydrochloric acid was continuously introduced into the liquid flow to reduce the Se (VI) to the oxidation state (IV). Subsequently selenium hydride was generated by introducing a solution of sodium tetrahydroborate. The liquid flow was delivered to the Meinhard type nebulizer of the ICP-system (flow rate: 2.5 ml/min) and the spray chamber was used as gas - liquid separator. The gaseous hydride was swept from the spray chamber into the plasma by an argon flow together with droplets containing the other elements under investigation. Compared to the direct technique the portion of selenium introduced into the plasma is increased considerably. Under the conditions we employed for the on-line analysis the detection limit was improved about 50 fold. A second advantage is that the other elements of interest could be detected together with selenium in the separated fractions.

Operation of the system and data collection:

The ICP-AES system was operated using the build-in procedures and the manufacturer's program. Each wavelength chosen was examined sequentially and the signal at each line integrated over 5 s (30 s for Se). This cycle of measurements was repeated automatically 70-85 times providing sufficient points to cover the whole chromatogram. The analytical data from the measurements were intercepted and transferred to a separate Texas Instruments professional computer system with an attached monitor. Superimposed plots of intensity versus integration points for each element were plotted on a Hewlett-Packard plotter (7475A). Then the data were subjected to various evaluation procedures. A flow scheme of the procedure is given in fig.3.

Problems arising:

The analytical problems we met in the development of the procedure are listed according to the main steps in table 1. In the following attempts to their solutions are considered.

On-line wet digestion and hydride generation

The wet digestion step is the most critical in the on-line procedure. The efficiency of wet digestion depends on several parameters including the temperature and the type of selenocompounds present in a sample. We studied some effects using pure organic selenocompounds including cystine, cystamine and methionine. The anorganic selenite was used as a reference. The results given in fig.4 show that selenomethionine was the most critical compound. A drop in the efficiency of the wet digestion procedure was observed moving from 160 to 200°C. We assume that this effect might be caused by bubble formation. At higher temperatures the increasing reaction rate seems to compensate this negative effect. Varying efficiency as a function of temperature may not seriously disturb the presence of selenopeaks in the chromatogram but it complicates a quantitative evaluation.

Contamination

A trace element chromatogram can be influenced by losses and/or contaminations during the fractionation on the column. Before use the buffer can be cleaned by passing through a Chelex 100 colunn. In case of the element zinc serious problems arose from the interaction between the metal ligand complexes and the gel matrix. We found that TSK gels bound zinc from human milk samples. On the other hand serum albumine picks up Zn from the column material during a chromatographic run. In order to maintain natural relationships between the different zinc complexes one has to carry out milk equilibration treatments. Fig. 5 illustrates that in the case of zinc speciation analysis of breast milk satisfying conditions were reached after four till five consecutive chromatographic runs.

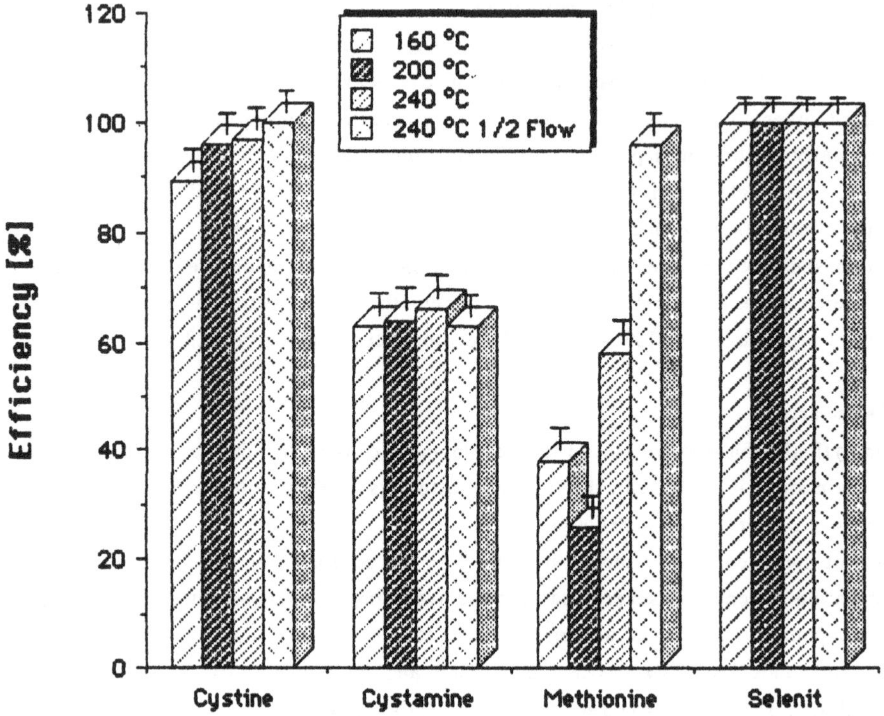

Fig 4: Dependence of the hydride formation efficiency on temperature and flow rate in the wet digestion of different selenocompounds.

STEPS	PROBLEMS	SOLUTIONS
FRACTIONATION	• CONTAMINATION	BUFFER PASSED A CHELEX 100 COLUMN
	• LOSSES • DISTURBANCE OF THE METAL-PROTEIN ASSOCIATION	EQUILIBRATION BY FIVE CONSECUTIVE CHROMATOGRAPHIC RUNS
	• IDENTIFICATION OF CONSTITUENTS	PURIFIED PROTEINS,
	• RESOLUTION	OPTIMIZATION OF INTEGRATION TIME, FLOW RATE, INJECTION VOLUME, IONIC STRENGTH OF THE BUFFER
WET DIGESTION	• DIFFERENT RECOVERY OF SELENO-COMPOUNDS	OPTIMIZATION OF TEMPERATURE, FLOW RATE, HEATING TIME
ELEMENT DETECTION	• SPECTRAL INTERFERENCES • BACKGROUND VARIATION • DETECTION LIMITS	CONSECUTIVE MEASUREMENT ADJACENT TO THE ELEMENT CHANNEL

Table 1: Problems arising in the main steps of the procedure.

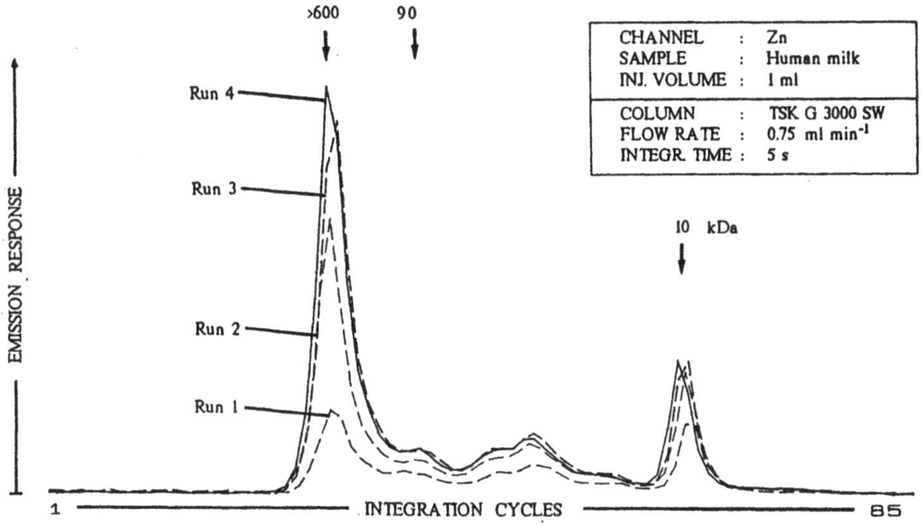

Fig. 5: Gel filtration chromatography of human milk: Effect of consecutive runs on the distribution of zinc.

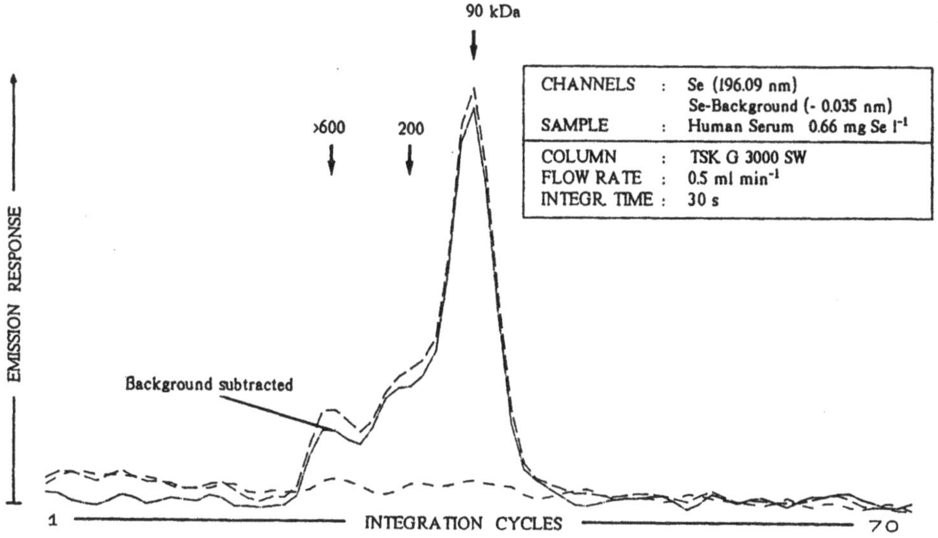

Fig. 6: Distribution of selenium in human serum after background subtraction.

Localisation of specific proteins

Gel filtration chromatography separates substances according to their molecular size. On TSK columns e.g. transferrin elutes after the albumin despite of its greater molecular weight. Compa rative runs using commercially available purified substances of human origin were found to be a suitable approach to indentify the position of the interesting components in human body fluids [15]. Compared to the UV-profiles of the effluents from the columns the protein-binding pattern of trace elements measured by ICP-AES showed a higher resolution. That can be used as an analytical tool in the characterization of the gelchromatographic properties and for easier localisation of specific proteins in the UV-profiles.

Background variation

Peaks can be simulated in a chromatogram by the variation of the background under the wavelength measured. This effect might occur when the mass load of the plasma varies due to a changing protein composition of the effluent or when the salt fraction appears at the end of the fractionation. The sequential ICP system enables the free choice of a background channel as close as possible to the element's wavelenght. By means of this setting the background can be controlled and, if necessary, subtracted as shown in fig.6 . Here Se speciation in serum was carried out and the background at 0.035 nm before the selenium line at 196.09 nm showed no variation. In serum at least 3 Se peaks could be localized at different molecular weight regions.

Quality control:

Assessment of the accuracy of the selenium determination using the ICP-AES with the hydride formation system was obtained by the analysis of certified reference materials. The results are shown in table 2. They indicate that our method provides results which are reliable for the speciation analysis carried out in this work on body fluids.

Table 2: Se concentration (μg/g) measured in standard reference materials.

SAMPLE	REFERENCE VALUE	VALUE OBTAINED [n = 5]
NBS Oyster tissue (SRM-1366)	2.1 ± 0.5	1.86 ± 0.20
NBS Bovine liver (SRM-1577)	1.10 ± 0.1	0.98 ± 0.15
IAEA Horse kidney (H-8)	4.67 ± 0.30	4.95 ± 0.80
Seronorm 103	0.090 ± 0.006	0.105 ± 0.04

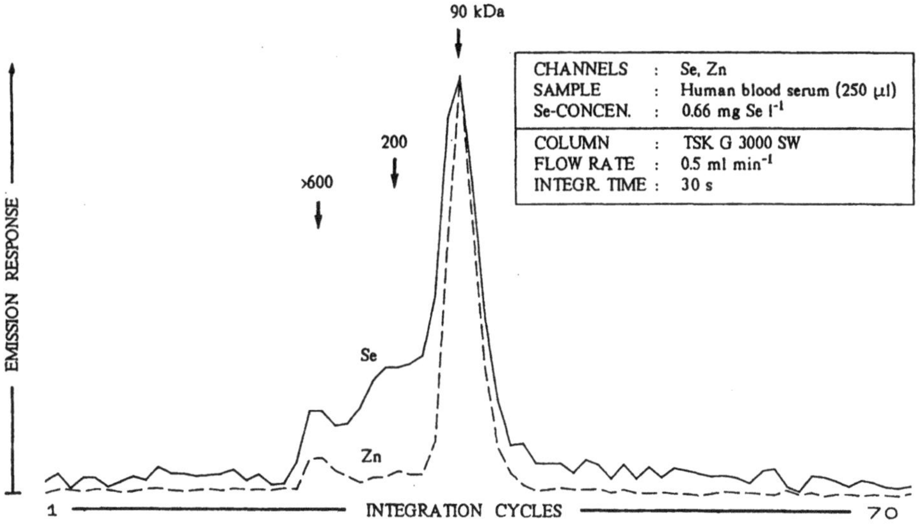

Fig. 7: Gel filtration chromatography of human blood serum: Distribution of selenium and zinc.

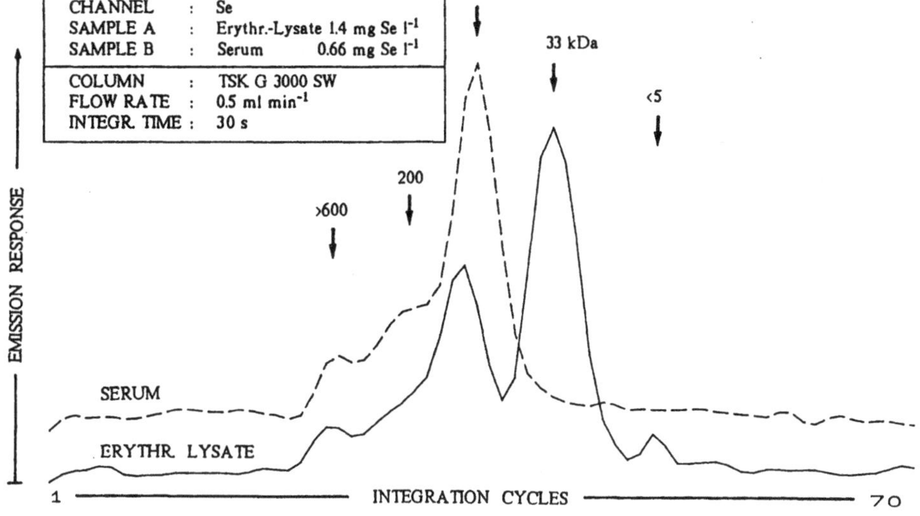

Fig. 8: Distribution of selenium in human serum and lysed erythrocytes.

APPLICATIONS

In order to demonstrate the feasibilties of the experimental approach the distribution of Se together with the other trace elements of interest was studied in human body fluids including serum, erythrocyte-lysate and breast milk.

Serum

Selenium was measured together with zinc in human serum in one chromatographic run shown in fig.7. The location of the Se-peaks seemed to be similar to those of the zinc. Zinc is bound in serum mainly to albumin. The highest selenium peak at 90 + 15 kDa is in the same region but shows a broader shape when compared to zinc. That might be due to the presence of the selenium containing enzyme glutathion peroxidase (molecular weight: 88 kDa) which elutes in the same region. The selenium distribution among the different protein fractions of serum was investigated in samples of 7 subjects. The percentage of selenium bound to the 90 kDa fraction was 77± 4%. 10± 3% was found in the 200 kDa range and 13± 5% bound to the high molecular weight exclusion fraction (>600 kDa). The selenium distribution in serum and erythrocyte-lysate of a subject is shown in fig. 8. The arrows indicate the positions of the selenium containing fractions.

Erythrocytes

Gel filtration chromatography of erythrocyte-lysate on a TSK column is shown in fig.9. Selenium eluted in two peaks which correspound to a molecular weight of 90 kDa and 33 kDa. The high molecular weight compond could be classed with the selenoenzyme GSH-Px. Together with selenium iron and zinc were monitored. Iron indicates the position of hemoglobin in erythrocytes and zinc the position of the enzyme carboanhydrase. With respect to the narrowness of these two markers a clear identification of the selenium binding complex at 33 kDa is not possible. According to Beilstein et al. [13] selenium is bound in erythrocytes to the globin fraction of the hemoglobin molecule as selenocysteine. It is known that the formation of the selenoenzyme glutathion peroxidase depends on the level of the selenium intake of subjects. Furthermore, with increasing selenium intake, this enzyme reaches a saturation state in erythrozytes. Therefore it was of interest to investigate the binding of that part of selenium in the erythrocytes which is not actually needed for enzyme formation. By aid of speciation studies using the developed on-line procedure information can easily be obtained concerning the effect of "excessive" selenium. In fig.10 the selenium distribution in erythrocyte-lysate is compared for two healthy persons having different nutritional selenium statuses. The selenium contents of both samples differs by about one order of magnitude. The high value represents subjects living in seleniferous areas of Venezuela [14] and the lower one approximately the mean of the German population. Figure 10 indicates a significant

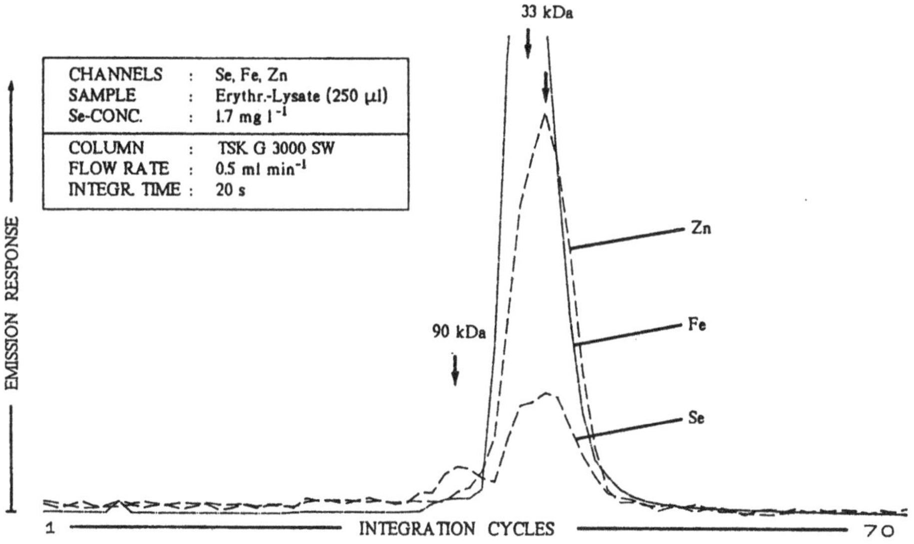

Fig. 9: Gel filtration chromatography of lysed human erythrocytes: Distribution of selenium, iron and zinc.

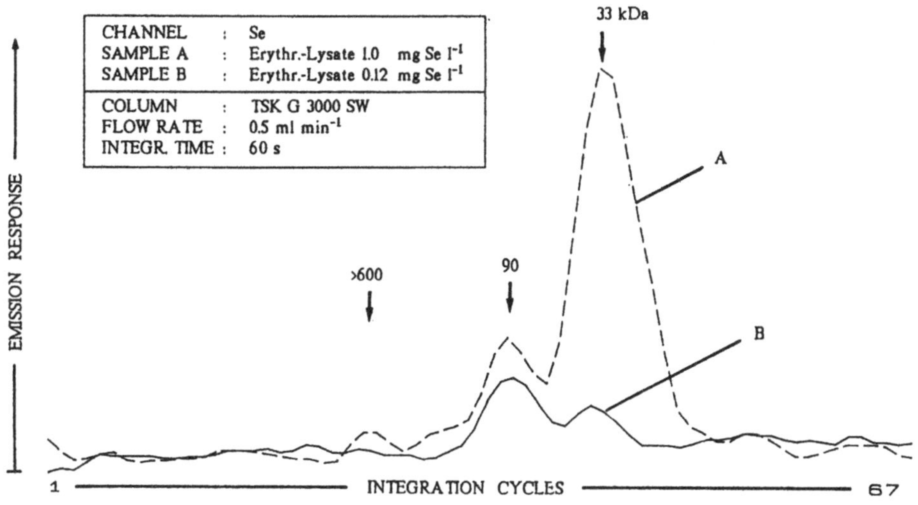

Fig. 10: Effect of different dietary selenium intake levels on the distribution of selenium in lysed human erythrocytes.

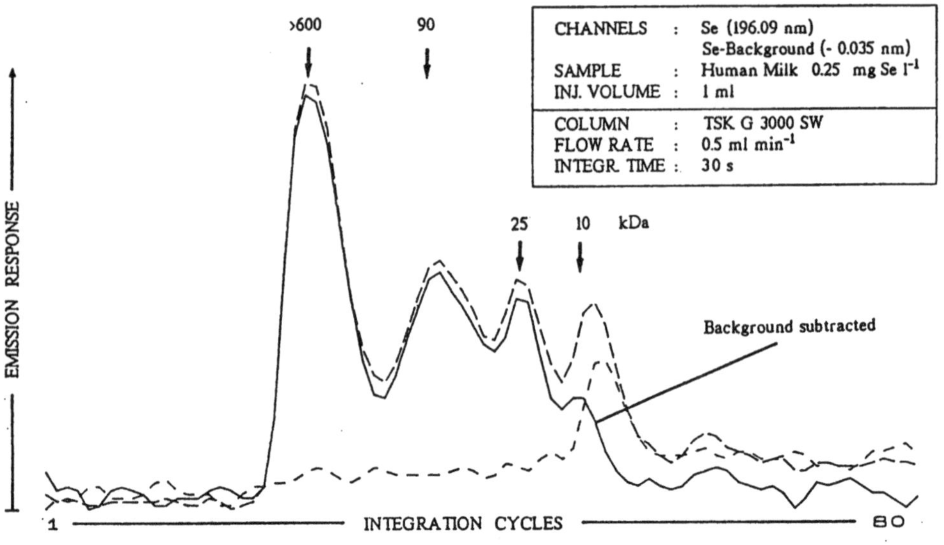

Fig. 11: Distribution of selenium in skimmed human milk after background subtraction.

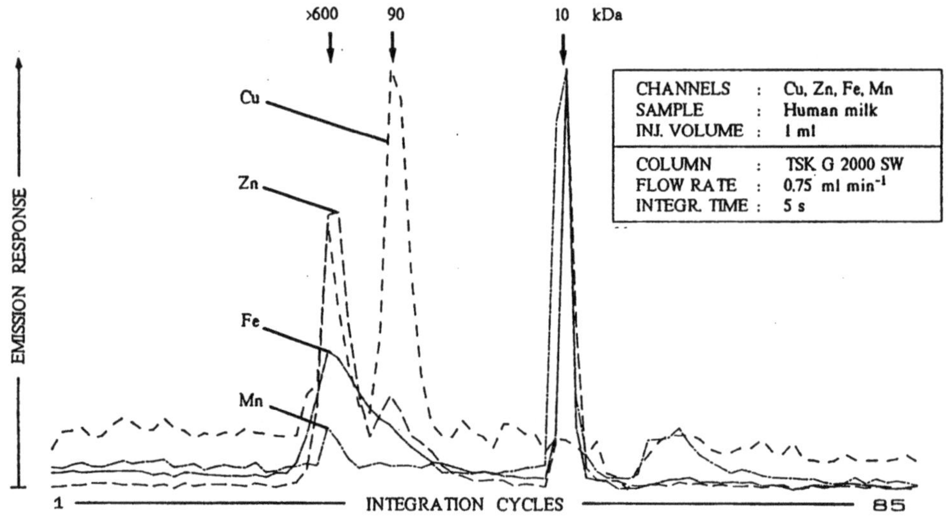

Fig. 12: Gel filtration chromatography of skimmed human milk: Distribution of copper, iron, zinc and manganese.

difference in the selenium fraction at 33 kDa. Therefore it can be assumed that the "excessive" selenium mentioned above is bound to the hemoglobin.

Breast milk

Applying the on-line combination of HPLC and ICP-AES measurement of the selenium distribution in skimmed human breast milk was carried out including background control and its subtraction. In fig.11 the importance of this treatment can be seen with respect to the selenium peak in the low molecular weight region. Subtraction yielded a selenium peak of lower intensity and its shift to a position around 10 kDa. In total 4 selenium peaks have been detected in mature breast milk. Another chromatogram of defatted human milk (fig.12) shows the distribution of copper, iron, manganese and zinc . The specific pattern of each element has been discussed elsewhere [15]. With respect to our findings for the low molecular weight region at about 10 kDa might be of special interest for selenium. At this position the elements iron, manganese and zinc elute exactly in the same fraction. For zinc the binding to citrate seems to be established [16]. From our gelchromatographic studies [15] we found some evidence that iron also binds to citrate in human milk. Further investigations are needed to confirm that manganese and selenium have similar binding behaviour in the low molecular weight region.

REFERENCES

[1] D .C. Chilvers, J.B. Dawson, M-H. Bahreyni-Toosi, A. Hodgkinson (1984)
Identification and determination of Cu- and Zn-protein complexes in blood plasma after chromatographic separation on DEA-Shepharose Cl-6B.
Analyst 109, 871-876

[2] P.E. Gardiner, E. Rösick, U. Rösick, P. Brätter, G. Kynast (1982)
The application of gel filtration, immunonephelometry and electrothermal atomic absorption spectrometry to study the distribution of Cu-, Fe- and Zn-bound constituents in human amniotic fluid.
Clinica Chimica Acta 120, 103-117

[3] P.E. Gardiner, P. Brätter, H. Gessner, M. Stöppler, H.W. Nürnberg (1984)
The distribution of zinc in human erythrocytes.
J.Clin.Chem.Clin.Biochem. 22. 159-163

[4] P.E. Gardiner, J.M. Ottaway G.S. Fell, R.R. Burns (1981)
The application of gel filtration and electrothermal atomic absorption spectrometry to the speciation of protein-bound Zn and Cu in human blood serum.
Analytica Chimica Acta 124, 281-294

[5] L. Ebdon, S. Hill, P. Jones (1987)
Application of directly coupled flame atomic absorption spectrometry - Fast protein liquid chromatography to the determination of protein-bound metals.
Analyst 112, 437-440

[6] D.J.R. Evans, K. Fritze (1969)
The identification of metal-protein complexes by gel chromatography and neutron activation analysis.
Anal.Chim.Acta 44, 1-7

[7] J. Pallon, P. Pakarinen, K. Malmquist, K.R. Akselsson (19879
Feasibility study of gel filtration as a separation method for blood serum proteins in combination with PIXE analysis.
Biol.Trace Element Res. 12, 401-409

[8] P.E. Gardiner, P. Brätter, B. Gercken, A. Tomiak (1987)
On-line sequential detection by inductively coupled plasma emission spectrometry of trace elements after liquid chromatography of biological fluids.
J.Anal.Atomic Spectr. 2, 375-378

[9] J.W. Carnahan, K.J. Mulligan, J.A. Caruso (1981)
Element selective detection for chromatography by plasma emission spectrometry.
Anal.Chim.Acta 130, 227-241

[10] J.P. McCarthy, J.A. Caruso, F.L. Fricke (1983)
Speciation of arsenic and selenium via anion-exchange HPLC with sequential plasma emission detection.
J. Chromatographic Science 21, 389-393

[11] P.E. Gardiner, P. Brätter, V.E. Negretti, G. Schulze (1983)
The use of a direct current argon plasma as a detector in gel filtration chromatography of biological fluids.
Spectrochim.Acta, Part B 38, 427-

[12] M. Thompson, B. Pahlavanpour, S,J. Walton, G.F. Kirkbright (1978)
Simultaneous determination of trace concentrations of As, Sb, Bi, Se and Te in aqueous solution by introduction of the gaseous hydrides into an inductively coupled plasma source for emission spectrometry.
Analyst 103, 568-579

[13] M.A. Beilstein, P.D. Whanger (1983)
Distribution of selenium and glutathione peroxidase in blood fractions from humans rhesus squirrel monkeys, rats and sheep.
American Institute of Nutrition 2138-2146

[14] P.Brätter, V.E. Negretti, U. Rösick, W.G. Jaffé, H. Mendez C., G. Tovar E. (1984)
Effects of selenium intake in man at high dietary levels of seleniferous areas of Venezuela.
in Trace Element Analytical Chemistry in Medicine and Biology (Eds. P. Brätter and P. Schramel) W. de Gruyter, Berlin- New York, Vol.3, 29-46

[15] P. Brätter, B. Gercken, A. Tomiak, U. Rösick (1988)
Speciation of trace elements in human milk and milk formulas.
in Trace Element Analytical Chemistry in Medicine and Biology (Eds. P. Brätter and P. Schramel) W. de Gruyter, Berlin-NewYork, Vol. 5. (in press)

[16] B. Lönnerdal, A.G. Stanislowski, L.S. Hurley (1980)
Isolation and identification of a low molecular weight zinc binding ligand from human milk.
J. Inorganic Biochemistry 12, 71-80

DISTRIBUTION OF SELENIUM IN COW'S AND HUMAN MILK.

P. Van Dael, H. Deelstra

Laboratory of Food Sciences, Dept. Pharm. Sciences,
University of Antwerp (UIA), B-2610 Wilrijk

G. Vlaemynck, R. Van Renterghem
Government Dairy Research Station, B-9230 Melle

Abstract

In the present study the distribution patterns of selenium in cow's and in human milk were compared. The whey fraction, obtained by ultracentrifugation, represented 40 % and 72 % of the total selenium content of cow's and human milk, respectively. The lipid fraction contained approximately 10 % of total cow's and human milk selenium. Cow's milk selenium concentration was approximately 50 % higher during winter than during summer, but did not influence the distribution pattern. After solubilization of milk fat globule membranes 61 % and 80 % of selenium was found in the outer fat globule membrane for cow's and human milk, respectively. Determination of the selenium content in individual proteins of cow's milk revealed that the highest selenium concentrations were present in β-lactoglobulin and in κ-casein.

Introduction

Since 1957 it has been proved that selenium is an essential element for a variety of mammals (1). Strong evidence for the essentiality of selenium in human beings comes from the pre-

Trace Element Analytical Chemistry in Medicine and Biology

sence in the enzyme glutathione peroxidase, which protects cells against oxidative damage by peroxides (2). Low selenium intakes have also been associated with Keshan disease, a cardiomyopathy, which has been virtually eliminated by supplementation with sodium selenite (3).

Babies, particularly neonates, are especially susceptible to selenium inadequancy, because of their dependence on milk or infant formulas, with low selenium content, as exclusive sources of food. However the total amount on itself does not ensure the overall utilisation or bioavailability of selenium (4). The chemical forms of selenium and its distribution in food are important determinants of selenium bioavailability.

This study was designed to determine the overall selenium content and its distribution in human and cow's milk.

For cow's milk the selenium distribution over the different protein fractions was determined.

Materials and Methods

Milk samples. Fresh mature human milk samples were obtained from apparently healthy women (n = 15). Mature cow's milk samples (n = 60) were collected at a local farm. All samples were collected in acid-washed plastic containers, pooled and stored frozen at -20°C. Major protein fractions of cow's milk were purchased from Sigma Chemical Company. A mixture of immunoglobulins and proteose-peptones were prepared from cow's milk by the method of Aschaffenburg and Drewry (5).

Fractionation procedure. Skimmed milk was prepared by centrifugation at 1500 x g for 30 min at 4°C. Whole milk was fractionated into fat, whey and casein fractions by ultracentrifugation (Beckman Model L 3-50) at 165,000 x g for 1 h at 4° C (6). In order to separate the lipid components, milk fat globule membranes were solubilized with 0.2 % sodium dodecyl sulphate (SDS) at 38 C for 1 h. Centrifugation at 4° C at 10,000 x g for 30 min separated the outer and inner fat globule membranes from the triglycerides (7). After separation all fractions were lyophilized and stored at -20°C.

Analysis of selenium. Samples were digested with a mixture of nitric and perchloric acids (8). The digests were analysed for selenium by hydride generation atomic absorption spectrometry (Perkin-Elmer 372 with an MHS-20 unit). Conditions for selenium determination have been described elsewhere (9). The method of standard addition was used throughout the whole procedure and each sample was analysed in threefold.

Results

The selenium content in whole and skimmed cow's and human milk are summarized in Table 1. The selenium concentration of cow's milk was proved to be higher during winter (approximately 50 %) than in summer.

Table 2 shows the results of the selenium distribution in milk fractions after ultracentrifugation. The whey fractions contained 40 % and 72 % of the total milk selenium collected from cows (average of summer and winter values) and women, respectively.

Table 1. Selenium concentrations of fresh and skimmed cow's and human milk (ng/ml).

Type of milk		Whole milk	Skimmed milk
Cow	Summer	11.3 ± 1.0	10.1 ± 0.4
	Winter	17.3 ± 4.5	16.0 ± 2.7
Human		14.7 ± 1.3	13.7 ± 1.8

Table 2. Selenium distribution in cow's and human milk (ng/ml).

Type of milk		Lipid fraction	Whey fraction	Casein fraction
Cow	Summer	1.2 ± 0.2	4.2 ± 0.2	5.9 ± 0.1
	Winter	1.9 ± 0.2	7.0 ± 0.3	8.2 ± 0.6
Human		1.3 ± 0.3	11.0 ± 1.1	2.7 ± 0.3

The proportion of the total milk selenium associated to the casein fraction was 50 % (average of summer and winter values) for cow's milk and 18 % for human milk. Regardless of source, approximately 10 % of total milk selenium was found in the lipid fraction. The distribution of selenium over the two fat globule membranes and the triglyceride fraction is shown in Table 3. The major portion of selenium within the cow's and human milk fat fractions was found to be associated with the outer membrane (61 % resp. 80 %), smaller fractions were found in the inner membrane (25 % resp. 13 %) and triglyceride fraction (14 % resp. 7 %).

Table 3. Distribution of selenium over cow's and human milk fat fraction (%).

Type of milk	Outer fat globule membrane	Inner fat globule membrane	Triglyceride fraction
Cow(*)	61 ± 6	25 ± 3	14 ± 3
Human	80 ± 8	13 ± 2	7 ± 1

(*) only values for winter.

The results of the selenium analysis of the different cow's milk proteins are summarized in Table 4. As may be seen from these values, whey proteins, especially -lactoglobulin, were found to be the most selenium-rich cow's milk proteins.

Table 4. Selenium content of the different proteins of cow's milk.

	% composition of milk (10)	ppb Se (ng/g)	ng Se/100 ml milk(*)
Whole milk	100	11 - 17(**)	1100 - 1700
Skimmed milk	95 - 96.5	10 - 16(**)	950 - 1550
Proteins	3.5 - 4.0		
#Caseins	2.4 - 2.8	130 - 240	312 - 672
α_s-casein	1.5 - 1.9	93 - 99	140 - 188
β-casein	0.9 - 1.1	144 - 160	130 - 176
κ-casein	0.3 - 0.4	426 - 435	128 - 174
#Whey proteins	0.5 - 0.7	490 - 660	245 - 462
α-lactalbumin	0.1 - 0.15	180 - 250	18 - 37
β-lactoglobulin	0.2 - 0.4	797 - 831	159 - 332
serum albumin	0.01 - 0.04	140 - 160	1 - 6
others (***)	0.12 - 0.28	175 - 190	21 - 40
proteose-peptones	0.06 - 0.18	200 - 245	12 - 44

(*) after recalculation; (**) ng/ml
(***) mainly immunoglobulins and proteose-peptones (5).

Discussion

The daily selenium intake of Belgian infants fed human milk is lower than the recommended dietary allowances proposed by the Food and Nutrition Board (11). Since infant formulas available in Belgium have even lower selenium concentrations than human milk (12), babies fed these formulas may be at increased risk for selenium deficiency. Therefore knowledge of the distribution patterns of selenium in cow's and human milk can be very useful in developing a formula diet rich in selenium.
The total selenium content of cow's and human milk reported in this study are much lower than those from high selenium areas (13, 14), but in agreement with previously published data for Belgium (12, 15). The higher selenium concentration of cow's milk during winter is probably due to differences in feed consumed (15).
In the present study considerable differences in the distribution of selenium in cow's and human milk were found. The content of this trace element in the lipid fraction was for both cow's and human milk approximately 10 %. This value is significantly higher than that published by Debski et al. (16), who reported only 3 % being associated with the lipid fraction. The findings that the major part of selenium in either cow's and human milk fat was localized in the outer fat membrane fraction, is possibly due to the more protein-rich nature of the outer membrane fraction and the presence of sulphur-containing amino acids in these proteins (18). The fraction found in the triglycerides may at least in part be due to difficulties

with separation of the inner membrane from the triglycerides, resulting in contamination of the latter fraction. A major association of other trace elements with the outer membrane fraction was also reported by other investigators (17).

Since human milk contains only one-tenth the quantity of casein of cow's milk, the 18 % found in the former was rather high.

Finally the whey fraction of human milk contained the major portion of total milk selenium (72 %), while the cow's whey fraction contributed for about 40 %. The selenium content in the casein and whey fraction probably depends upon nutritional factors. For total cow's milk selenium content within the casein fraction significantly different values were reported : Yoshida et al. (14) found 40 %, Debski et al. (16) 29 %, while Mathias et al. (19) reported approximately 60 % .

As can be seen from the results (Table 4) the whey proteins, especially the β -lactoglobulin, seemed to be the most selenium-rich cow's milk protein fraction. A possible explanation of this phenomenon could be the presence of a large quantity of sulphur-containing amino acids in these proteins (20).

Within the casein fraction, κ -casein revealed the highest selenium concentration.

As a preliminary conclusion addition of the more selenium-rich whey proteins of cow's milk, could be considered to enrich infant formulas. However bioavailability studies of these proteins are necessary for more definite conclusions.

Acknowledgement

Kathy Somers skillfully carried out part of the work.

References

1. Shamberger, R.J.. 1983. In: Biochemistry of Selenium (E. Frieden, ed.). Plenum, New York, pp. 31-58.

2. Ganther, H.E., D.G. Hafeman, R.A. Lawrance, R.E. Serfass, W.G. Hoekstra. 1976. In: Trace Elements in Human Health and Disease (A.S. Prasad, ed.). Academic, New York, pp. 165-234.

3. Chen, X., G. Yang, J. Chen, X. Chen, Z. Wen, K. Ge. 1980. Biol.Trace Elem.Res. 2, 91.

4. Alexander, A.R., P.D. Whanger, L.T. Miller. 1983. J.Nutr. 113, 196.

5. Aschaffenburg, R., J. Drewry. 1957. Biochemistry 65, 273.

6. Lönnerdal, B., C.L. Keen, L.S. Hurley. 1985. Am.J.Clin.Nutr. 41, 550.

7. Fransson, G.-B., B. Lönnerdal. 1983. Pediatr.Res. 17, 912.

8. Robberecht, H.J., R.E. Van Grieken, P.A. Van Den Bosch, H. Deelstra, D. Vanden Berghe. 1982. Talanta 27, 1025.

9. Verlinden, M.. 1981. University of Antwerp, PhD Dissertation.

10. Swaisgood, H.I.. In: Development in Dairy Chemistry (P.F. Fox, ed.). Elsevier Applied Science Publisher, pp. 1.

11. Food and Nutrition Board, National Research Council. 1980. Nutr.Rev. 38, 290.

12. Roekens, E., H. Robberecht, M. Van Caillie-Bertrand, H. Deelstra, R. Clara. 1985. Eur.J.Pediatr. 144, 45.

13. Hadjimarkos, D.M., C.W. Bonhorst. 1961. J.Pediat. 59, 256.

14. Yoshida, M., K. Yasumoto, K. Iwami, H. Tashiro. 1981. Agric.Biol.Chem. 45, 1681.

15. Deschuytere, M.A.. 1987. University of Antwerp, PhD Dissertation.

16. Debski, B., M.F. Picciano, J.A. Milner. 1987. J.Nutr. 117, 1091.

17. Fransson, G.-B., B. Lönnerdal. 1984. Am.J.Clin.Nutr. 39, 185.

18. Cawston, T.E., M. Anderson, G.C. Cheeseman. 1976. J.Dairy Sci. 43, 401.

19. Mathias, M.M., D.E. Hogue, J.K. Loosli. 1967. J.Nutr. 93, 14.

20. Maubois, J.L., G. Brule. 1982. Lait 617, 485.

SPECIATION OF TRACE ELEMENTS IN HUMAN MILK AND MILK FORMULAS

P. Brätter, B. Gercken, U. Rösick, A. Tomiak

Hahn-Meitner-Institut Berlin GmbH, Glienicker Str. 100, D-1000 Berlin 39

Abstract

An ICP atomic emission spectrometer coupled to a gel filtration chromatography system was used to characterize the species containing Cu, Zn, Fe, Mn, Mg and Ca in human milk and milk formulas based on cow's milk. The distribution profiles clearly indicate that the Cu, Zn and Fe protein binding in human breast milk and milk formulas is different. Investigations of protein bound trace elements in human breast milk before and after birth showed differences in Zn and Fe bound to citrate. An increase of citrate concentration occurred after birth at the first day of milk secretion with the factor of 25 ± 3.

Introduction

Human milk and/or milk formulas based on cow's milk are the main nutrient fluids of newborn infants. Assuming that the composition of breast milk may satisfy the growing demands of healthy infants during the early months of life it can be a reference to evaluate the nutritional value of alternative formulas. In an earlier study we compared the serum level of exclusively breast-fed and formula-fed infants with respect to essential trace elements [1]. The data obtained suggest that absorption and utilization of several trace elements may depend not only on the total amount in the nutrient fluid but also on the chemical form in which they occur. Therefore, the aim of the present work was:

1) to determine trace element containing species in human breast milk

2) to compare trace element profiles of human breast milk and cow's milk based infant formulas

Trace Element Analytical Chemistry in Medicine and Biology

3) to investigate the variation of protein bound trace elements in breast milk before birth and during the lactation period.

Instrumentation

For the chromatographic separation an LKB system (LKB Instruments, Freiburg, FRG) composed of a pump (LKB 2150), column, UV-detector (LKB 2138) and recorder (LKB 2210) was used. The element measurements were carried out with the sequential ICP-AES system ARL 3520 (ARL, Düsseldorf, FRG). The nebulizer and plasma gas flow rates were electronically controlled by a mass flow controller system (Instruments S. A., Unterhaching, FRG). The data were evaluated on a Texas Instruments professional computer using a program of our own development. A hard copy of the graphical display was plotted on a Hewlett-Packard plotter 7475 A.

Analyses

Speciation of trace elements in human milk was carried out by on-line coupling of GEL FILTRATION CHROMATOGRAPHY (GFC) with sequential INDUCTIVELY COUPLED PLASMA EMISSION SPECTROMETRY (ICP-AES) for element detection.
The block diagram of the analytical arrangement is shown in fig.1, the operating parameters are given in table 1.
Human milk was defatted by centrifugation at 10 000 g for 10 minutes and chromatographed on TSK-Gel-columns. The protein composition of the effluent was monitored with a UV-detector at 280 nm. To establish the identities of the UV-absorption peaks purified human milk proteins were used as shown in fig.2. The molecular weight association of the proteins was performed after calibration of the column with the following protein standards: thyroglobulin (669kD), ferritin (450kD), ceruloplasmin (150kD), transferrin (80kD), myoglobin (17,8kD) and vitamin B_{12}(1,355kD).
The element detection was carried out sequentially using an ARL 3520 ICP-AES system. During a measuring cycle the signals of up to six element channels were integrated for 5 seconds. Usually four elements have been detected in one chromatographic run repeating the measuring cycle automatically 85 times. The measured values

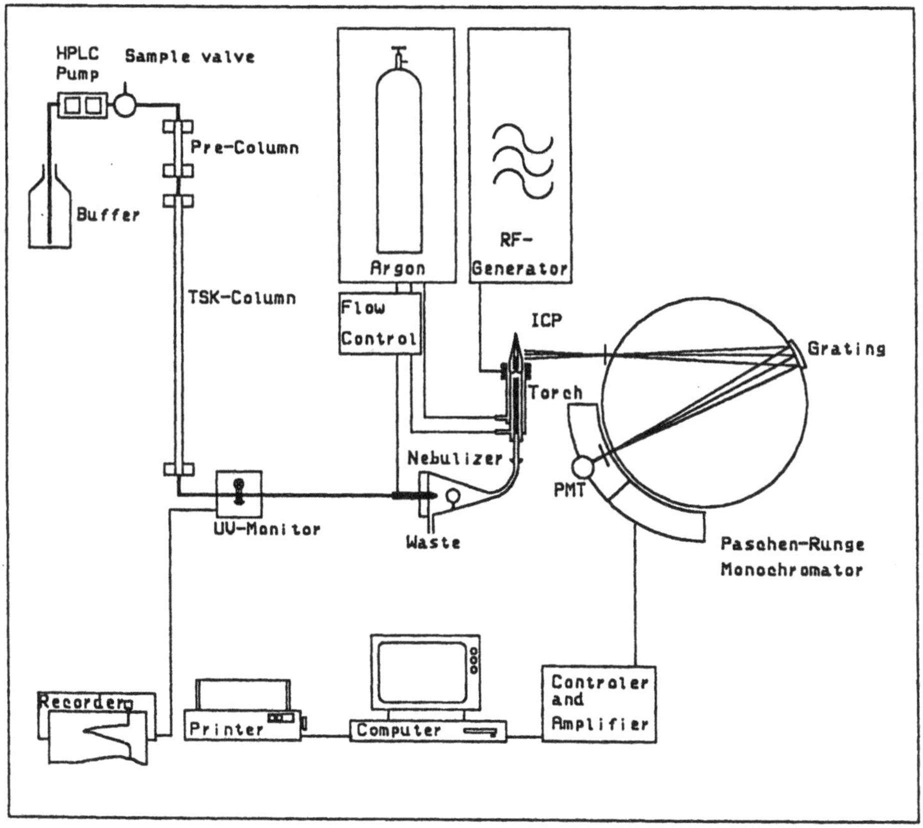

Fig. 1: Block diagram of the analytical arrangement

were transferred to a separate computer system and subjected to different evaluation procedures. For each element channel the minimum and maximum of the signal were taken and the difference was set to 100%. Typical plots of intensity vs. consecutive integration steps are shown in fig.3.
The total trace element contents of the milk samples were measured with the ICP-AES system after dilution of the samples with deionized water. The trace element content of each component can be calculated by distributing the total element content among individual components with respect to their corresponding peak areas.

Table 1: operating parameters

A. Gel filtration chromatography

Columns	TSK G 3000 SW GlasPac 600 x 8 mm
	TSK G 3000 SW 600 x 7.5 mm
	TSK G 2000 SW 600 x 7.5 mm
Buffer	0.1 M HEPES + 0.1 M NaCl, pH = 7.4
Flow rate [ml/min]	0.75
Sample volume [µl]	1000
UV detector wavelength [nm]	280

B. ICP-AES

Wavelength [nm]			
		Plasma torch	Fassel type
		Nebulizer	Meinhard type C2
Cu	324.76	Observation height [mm]	15
Zn	213.86	Photomultiplier gain [V]	1000
Fe	259.95	Power [W]	1200
Mg	279.55	Frequency [MHz]	27.125
Mn	257.61	Nebulizer gas flow rate [l/min]	0.80
Ca	393.37	Plasma gas flow rate [l/min]	1.00
		Integration time [s]	5

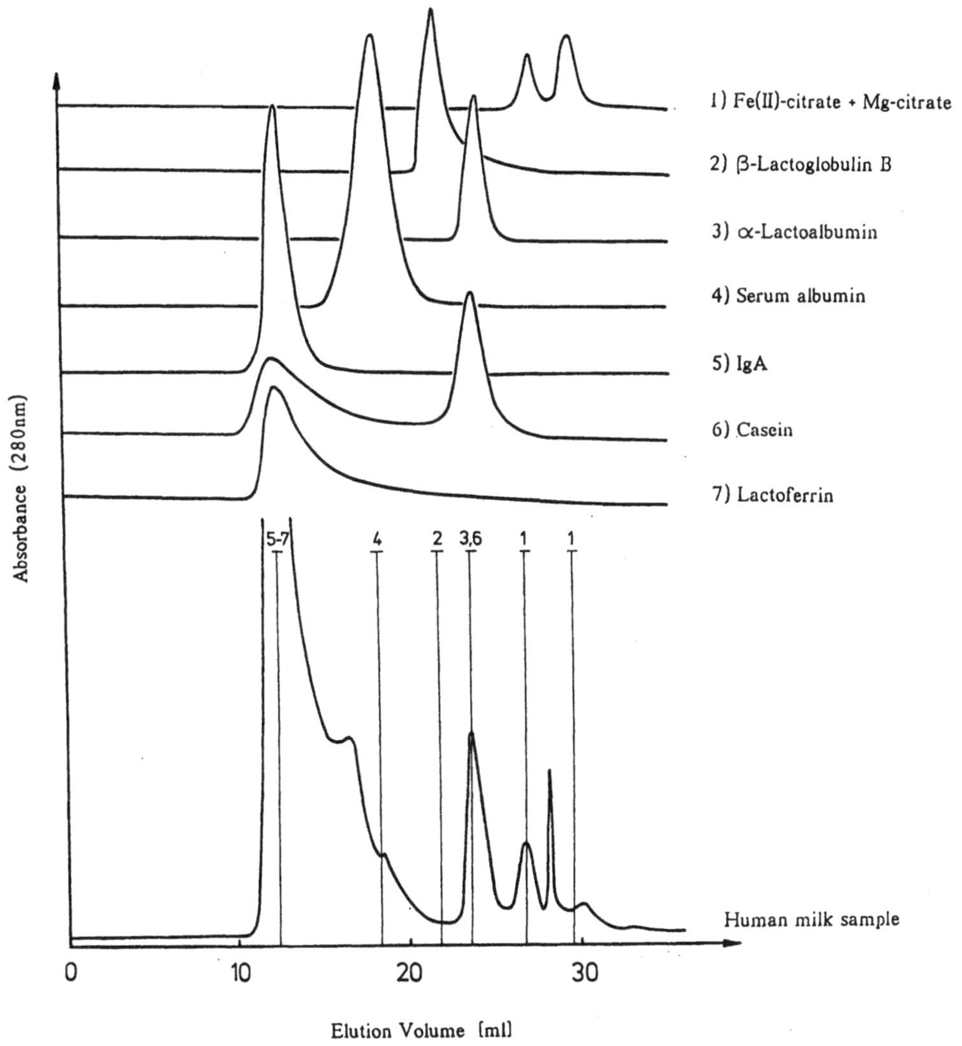

Fig. 2: Gel-filtration chromatogram of human milk, localization of compounds in the sample using purified substances

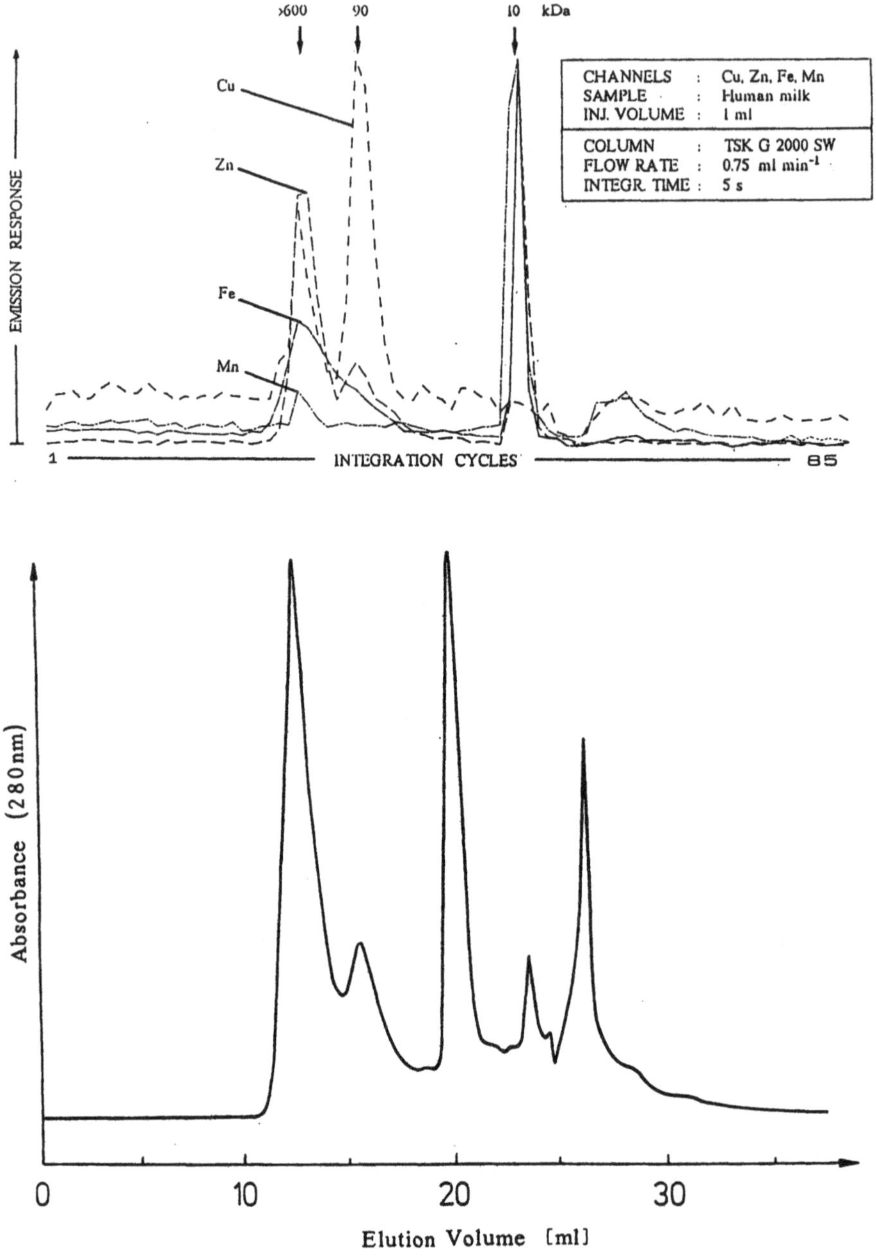

Fig. 3: Gel-filtration chromatogram of human milk
upper part: trace element distribution
lower part: UV-absorbance at 280nm

Results and Discussion

1) The results of speciation analyses of breast milk are given in table 2.

Table 2: Occurrence of trace elements in fractions of human milk after gel filtration chromatography

MW [kDa]	>600	90-80	30-20	10-9	<1
localized proteins	Casein Lactoferrin	Serum albumin	Casein α-Lactoalbumin	Citrate	Salts
Cu	+	+	+	-	-
Zn	+	+	+	+	-
Fe	+	-	-	+	-
Mn	+	-	-	+	-
Mg	-	-	-	-	+
Ca	-	-	-	-	+

We found Zn, Fe and Mn to be coeluted in a fraction of the low molecular weight region between 9 and 10 kDa. The Zn-peak of this fraction was identified by different authors as Zn-citrate [2, 3, 4]. Using an enzymatic test (Boehringer, Mannheim, FRG) we identified citrate in this fraction [5]. Fe-citrate in a pure form as well as added to milk samples yielded the same elution characteristics as the low molecular weight species in the sample.

2) Measuring the distribution of Cu, Zn and Fe among the various components in human milk and cow's milk infant formulas different patterns were obtained for Cu, Zn and Fe (fig. 4). Comparing both nutrient fluids, in human milk an additional Cu-peak was found under the serumalbumin peak (fig. 4b) and 3 additional Zn-bonding compounds (indicated by arrows) have been detected (fig.4c). In case of Fe human milk exhibited the major part in the fraction of high molecular weight, while cow's milk showed the main part in the fraction of low molecular weight (fig 4d).

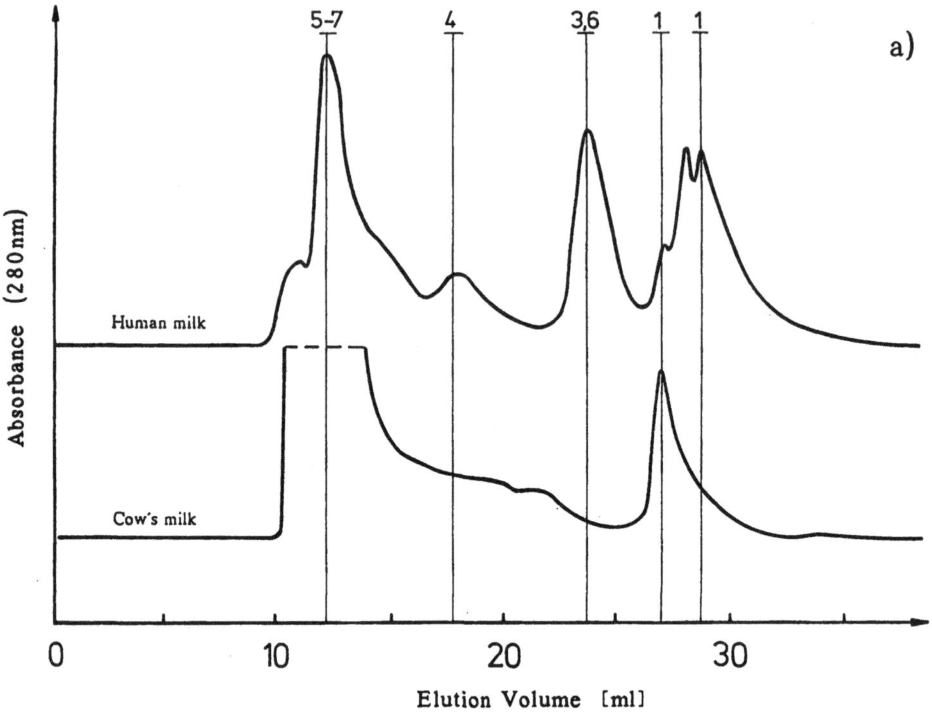
5-7
4
3,6
1
1
a)
Absorbance (280nm)
Human milk
Cow's milk
0
10
20
30
Elution Volume [ml]

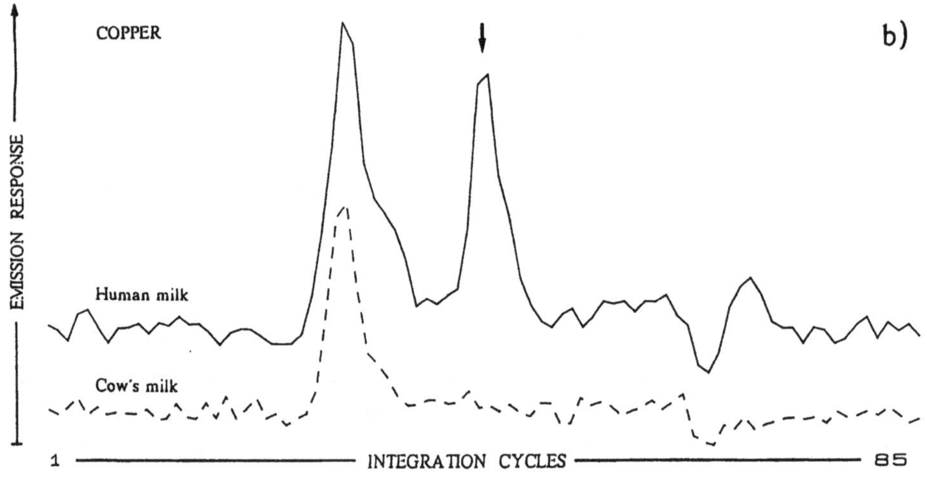
COPPER
b)
EMISSION RESPONSE
Human milk
Cow's milk
1
INTEGRATION CYCLES
85

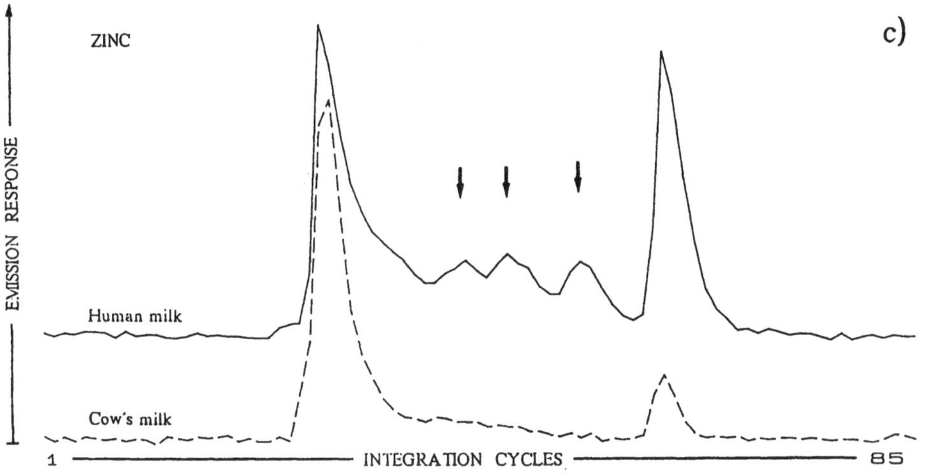

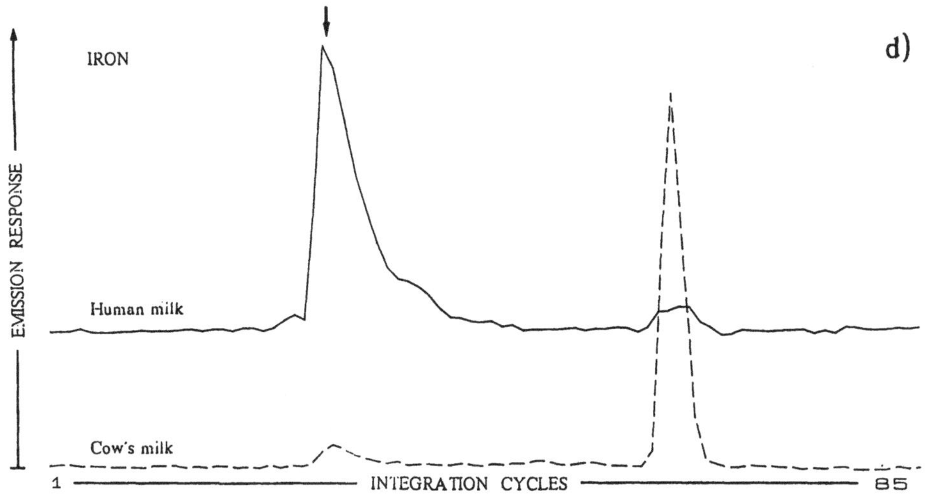

Fig. 4: Trace element and protein patterns of human milk and infant formula based on cow's milk

a) UV-absorbance at 280nm, peaks are indicated by numbers according to the localization procedure of fig. 2

b) Cu-channel

c) Zn-channel

d) Fe-channel

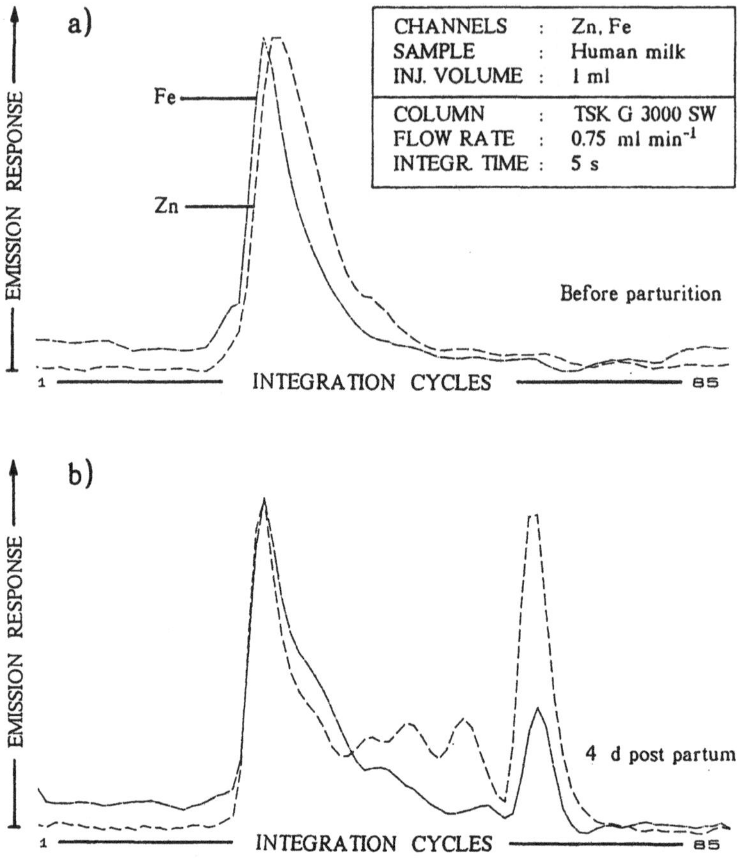

Fig. 5: Gel-filtration chromatogram of Zn and Fe derived from human milk a) before and b) after parturition Localization of citrate by c) UV-absorption (indicated by arrow) and d) fractional enzymatic analyses

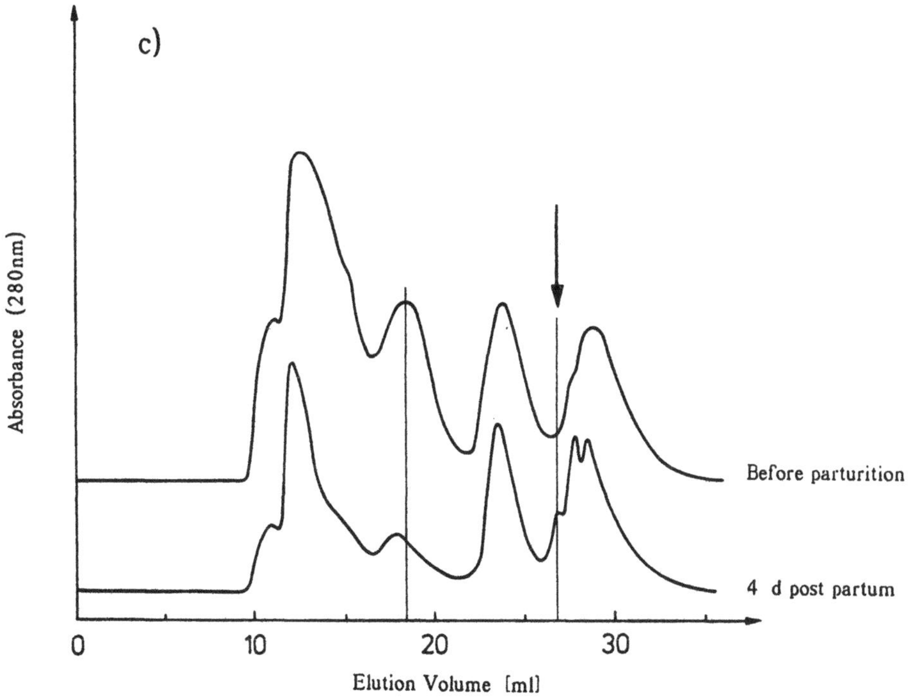
c)
Absorbance (280nm)
Before parturition
4 d post partum
0
10
20
30
Elution Volume [ml]

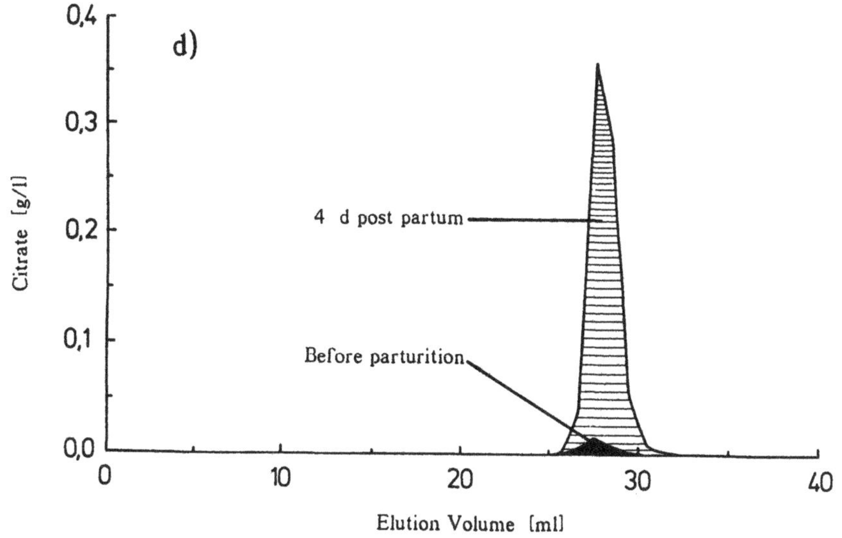
d)
Citrate [g/l]
0,4
0,3
0,2
0,1
0,0
4 d post partum
Before parturition
0
10
20
30
40
Elution Volume [ml]

3) Comparing samples of prenatal milk secretion and postpartum milk a considerable increase of the citrate concentration was found after delivery at the first day of milk secretion (factor 25 ± 3). Zn and Fe appeared simultaneously in the chromatogram (fig.5) at the same position as the elution volume of citrate was observed, indicating a specific binding. No similar peak characteristic was found for Cu.

Because Cu and Zn are essential for the development of infants, we assume that the occurrence of specific ligands in breast milk including citrate leads to an increased bioavailability of these elements in comparison to cow's milk. In the chromatograms of colostrum relatively high amounts of Zn and Fe appeared in the fraction of low molecular weight eluting together with citrate. We suggest that this ligand binding Zn and Fe may be important for the absorption by neonates.

References

1. Brätter, P.; Negretti de Brätter, V. E.; Rösick, U.; v. Stockhausen, H. B.
Trace element concentration in serum of infants in relation to dietary sources
In: Trace Element Analytical Chemistry in Medicine and Biology, Vol. 4 (1987)
(Brätter, P. & Schramel, P., eds.) pp 133-143, de Gruyter, Berlin, New York

2. Lönnerdal, B.; Stanislowski, A. G.; Hurley, L. S.
Isolation of a low molecular weight ligand from human milk
J. Inorg. Biochem. 12 (1980) 71-78

3. Blakeborough, P.; Salter, D. N.; Gurr, M. I.
Zinc binding in cow's and human milk
Biochem. J. 209 (1983) 505-512

4. Martin, M. T.; Jacobs, F. A.; Brushmiller, J. G.
Identification of copper binding and zinc binding ligands in human and bovine milk
J. Nutr. 114 (1984) 869-879

5. Harzer, G.; Kauer, B. S.
Binding of zinc to casein
Am. J. Clin. Nutr. 35 (1982) 981-987

QUANTITATIVE DETERMINATION OF PROTEINS USING POLYACRYLAMIDE GEL ELECTROPHORESIS AND NEUTRON ACTIVATION

Susan F. Stone

Department of Chemistry and Biochemistry, University of Maryland, College Park, Maryland 20742 and Center for Analytical Chemistry, National Bureau of Standards, Gaithersburg, Maryland 20899

Rolf Zeisler

Center for Analytical Chemistry, National Bureau of Standards, Gaithersburg, Maryland 20899

Glen E. Gordon

Department of Chemistry and Biochemistry, University of Maryland, College Park, Maryland 20742

Abstract

A combination of powerful and sensitive methods for separation and element determination is evaluated to accomplish trace element speciation. Polyacrylamide gel electrophoresis (PAGE) and neutron activation analysis (NAA) have been combined to determine the phosphorus associated with separated phosphoproteins by autoradiography and densitometry. The proteins themselves can then be quantified. Results are presented for several phosphoproteins and phosphoprotein-containing natural matrix samples. Phosphorus can be determined down to 0.2 μg. In addition, the expansion of the method to selenium-containing macromolecules is being studied. A quantitation limit of 0.25 nmole Se has been achieved.

Trace Element Analytical Chemistry in Medicine and Biology

Introduction

The emphasis on total trace element content in biological samples has begun to shift toward trace element speciation. This has resulted in the need for development of techniques for the accurate determination of the trace element species in complex biological samples. Approaches taken to accomplish these determinations have combined a good separation method with a sensitive method for element determination. The method required for elemental speciation must have a separation technique with sufficient resolving power, which will not alter or destroy the species, and a technique for elemental determination with sufficient sensitivity, which will not be influenced by the species being analyzed. Various separation techniques have been employed for these speciation studies. These include ultracentrifugation, thin-layer chromatography, gel filtration, high performance liquid chromatography, ion-exchange chromatography and, recently, polyacrylamide gel electrophoresis (PAGE). Some of these methods have been combined with detection via neutron activation analysis (NAA) (1,2).

Neutron activation analysis is sufficiently sensitive to meet the above requirements, allowing trace element association with macromolecules to become the subject of many studies. Many elements can be determined and, in general, the technique is independent of matrix and elemental species. Previous studies have demonstrated the feasibility of combining NAA with PAGE separations. PAGE has proven to be a powerful separation tool for proteins in complex biological samples (3). In the work combining NAA and PAGE, gels with separated proteins are activated, producing radionuclides of the elements associated with the protein bands on the gel. The detection of these radionuclides is done either by gamma-counting or by autoradiography. It was then possible to locate the separated proteins using one of these methods (4).

In this contribution, the extension of the detection of qualitative associations by PAGE and NAA towards quantitative determination is described. To apply this method for quantitative analysis, we have used as model compounds, selected phosphoproteins and phosphoprotein-containing natural matrix samples. In addition, the extension of the method to selenium-containing macromolecules is being studied.

Experimental Section

PAGE separation

One-dimensional PAGE separations (on a molecular weight basis) were performed, both with and without sodium dodecyl sulfate (SDS). The detergent SDS improves resolution in the protein separation; however, it might interfere with the subsequent NAA or destroy the original trace element species. Polyacrylamide slab gels, 20% non-gradient, (180 x 150 x 0.75 mm), with Tris/glycine running buffer and sample buffer, were prepared according to reagent amounts recommended by Tollaksen (5). Acrylamide, N,N'-methylene-bisacrylamide, and N,N,N',N',-tetramethylethylenediamine (TEMED) were from SERVA Fine Biochemicals. Glycine, Tris buffers, and ammonium persulfate were purchased from Sigma Chemical Company. Sodium dodecyl sulfate (SDS) was obtained from Pierce. All chemicals were evaluated for their trace element content (4). Electrophoresis, using a Hoefer 600 series Vertical Gel Slab unit, was run at a constant amperage of 20 mA per gel. Proteins were separated both under denaturing (with SDS) and non-denaturing (without SDS) conditions. All preparation and sample handling was performed under a Class 10 clean hood to minimize contamination. Wells for the application of protein solutions had been formed by the insertion of Teflon combs into the prepared slab gels prior to polymerization. Gels were fixed in 50% ethanol, 0.1% formaldehyde and either stored in 20% ethanol or stained with Coomassie Brilliant

Blue. If detergent removal was needed, the procedure of Olden and Yamada (6) was followed.

Irradiation

Gels to be irradiated were conditioned with acetic acid/glycerol and then dried on Whatman 41 filter paper at 80 °C for two hours under vacuum. Each gel was individually packaged in cleaned linear polyethylene bags. Irradiations took place at a specially designed facility at the University of Virginia reactor in Charlottesville, VA (3). The facility includes the use of an aluminum cassette, which is placed adjacent to the reactor core. The cassette is 20 cm square and holds a stack of samples up to 3 cm in thickness. The facility was operated at a fluence rate of $1.7 \cdot 10^{17} m^{-2} s^{-1}$. The length of irradiation ranged from 4 - 6 hours, followed by a one-week decay before unpackaging and shipment to the National Bureau of Standards for analysis.

Autoradiography and densitometry

Following irradiation, autoradiographs were taken of the gels, and densitometric measurements were made of the resulting bands. Autoradiographs of the gels were done by exposure to Kodak X-OMAT film, using Kodak cassettes. Exposure times varied from one day to 3 weeks, depending on the protein and the amount applied. For the selenium exposures, a Kodak cassette with an intensifying screen was used. The development process involved the use of Kodak GBX developer and fixer chemicals for 5 and 3 minutes, respectively. A laser densitometer (LKB 2202 UltroScan) was employed to take absorbance measurements of the autoradiographs at 633 nm. The measurements were taken through the middle of the protein bands.

Proteins and standards

The phosphoproteins, obtained from Sigma Chemical Company, contained various ratios of phosphorus: phosvitin - 10%, α-casein - 1%, and β-casein - 0.67%. Protein solutions of approximately 1 mg/mL were prepared from the lyophilized proteins dissolved in sample buffer. Sample applications ranged from 2 L to an approximate maximum sample loading of 50 μL. A range of phosphorus/protein amounts was used, from 5 μg P as phosvitin, down to 0.01 μg P as β-casein. Standard Reference Material Non-fat Milk Powder, SRM 1649 (7), was used as the natural matrix containing phosphoproteins. A solution of 2 μg/mL was prepared from this standard. The phosphorus standard used was a solution of $NH_4H_2PO_4$, prepared and pipetted at various concentrations onto dried gels. A solution of glutathione peroxidase from human erythrocytes (from Calbiochem) was used without modification. A Se tracer was prepared by irradiating a pure Se solution (Se shot, from Alfa Products, dissolved in nitric acid) pipetted onto a Whatman filter paper. A filter paper with 365 μg selenium was irradiated for 6 hours at $1.03 \cdot 10^{18} nm^{-2}s^{-1}$. The filter paper was dissolved in a small amount of nitric acid and the solution diluted to 50 mL to yield a tracer solution of 7.3 μg/mL. Solutions were prepared from the primary solution, ranging from 5 μg/mL to 0.07 μg/mL, with 2 μL of each solution pipetted onto filter papers. Iron foils were included to measure fluence distributions over the irradiation area.

Results and Discussion

Phosphoproteins

Quantitative determination of phosphoproteins was accomplished by irradiating and evaluating PAGE-separated proteins and standards. Calibration curves were obtained from the images of the standard solutions. These curves were with the separated proteins and, hence, the masses of each protein. Fluence variations over the gel length, measured by the iron foils, were significant and corrections have been made for the fluence differences between the location of the proteins and standards.

An example of absorbance measurements of autoradiographed protein bands from phosvitin versus the mass of applied phosphorus (0.1 - 0.5 μg phosphorus) is shown in Figure 1. Error bars represent the uncertainty in repetitive measurements of the same track of one protein band. This assessed only the reproducibility of the computer-supported densitometric measurements, and had a mean standard deviation (1σ) of 1.5%. Linear responses are demonstrated over the applied range of concentrations, both with and without the

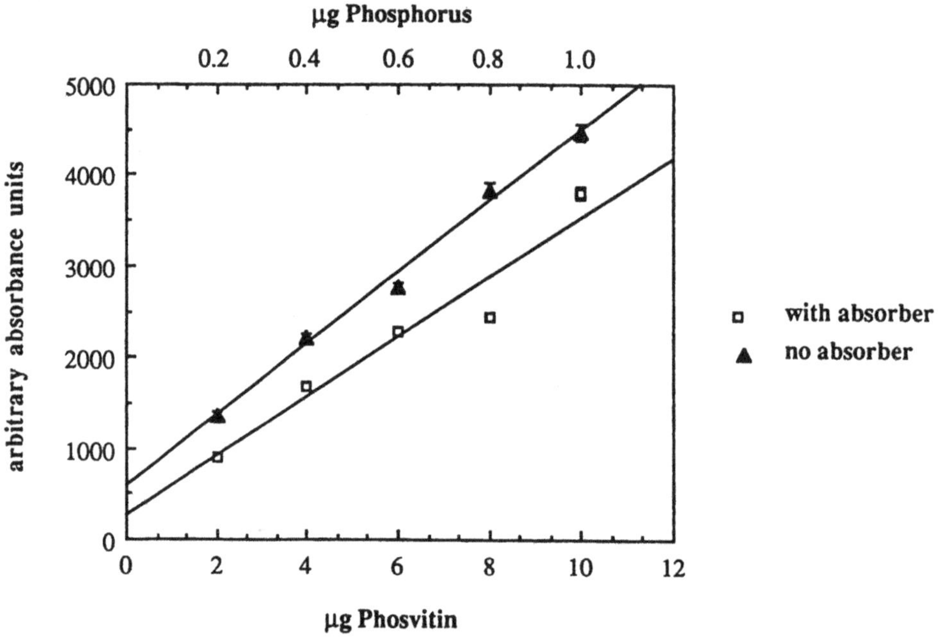

Figure 1. Results of PAGE/NAA method on Phosvitin (2 - 10 μg protein). Two different sets of results are presented; autoradiography with and without the use of an aluminum absorber. Error bars represent the uncertainty (1σ) in repetitive measurements of the same track of one protein band.

use of an aluminum absorber. The use of an absorber is desirable, because of narrowed bands in the autoradiograph and also the reduction of blank contribution. The latter is indicated by the decreased intercept of the calculated line for the measurements using the absorber. Linearity was also observed in the 1-5 μg phosphorus range.

Similar experiments were performed with α-casein and β-casein over the concentration ranges 0.02 to 0.10 μg P and 0.01 to 0.10 μg P, respectively. The latter is at the present limit of detection. Quantitation of the α-casein had uncertainties of 20-50% on individual determinations. Several conditions contributed to the difficulties in the casein detection and quantitation. Even with the use of SDS, the separations did not have the required resolution. A different preparation of samples and additional work with electrophoresis, e.g., the use of gradient gels, are needed. Also, the exposures need to be done under optimized conditions. The optimization methods described below for the selenium work are not suitable for these proteins; in addition, exposure times longer than two weeks have limited usefulness since the nuclide of interest (^{32}P) has a half life of 14.2 days. Other options, such as the use of high contrast film, will be explored.

These separation problems caused large uncertainties in the quantitative determination of caseins in SRM 1549 milk powder. A reasonable estimate of the most easily resolved α-casein component in milk powder is 11% (by weight), with a probable relative uncertainty of 20-50%. Improvements in sample preparation and electrophoresis, noted above, will be needed to achieve more accurate determinations. No data on the casein quantities in SRM 1549 are available at present. Comparisons with other techniques will be included as they become available to further assess the performance of the method.

The SDS detergent aids in providing good separation of the proteins. The lower energy sulfur β^- resulting from the SDS-PAGE gel autoradiographs can be screened out using the appropriate thickness of aluminum, but the Na present in the SDS causes problems because of the activity of ^{24}Na in the irradiated gel. The removal method for SDS from the gels proved to be reliable, with no observable scattering of ^{36}S β-spots from the SDS in the gels appearing on the autoradiographs. Therefore, SDS gel separation will be the method of choice in future determinations, unless the protein-element association of interest would be adversely affected.

Selenium

Initial attempts to detect glutathione peroxidase directly using the PAGE/NAA method were inconclusive. Applications from the prepared solutions of the selenoprotein at the maximum PAGE loading (about 50 μL) were not detectable. Radiotracer experiments were performed to assess the minimum detectable amounts for selenoprotein determination. A comparison of results from the different exposure conditions is shown in Figure 2. High sensitivity can be achieved by using an intensifying screen at -80 °C during exposure. The use of direct exposures or exposures with the intensifying screen at room temperature do not provide enough contrast and sensitivity. Under current conditions for gel irradiations, 0.25 nmole of selenium can be quantified. Spots representing levels below 0.25 nmole were qualitatively detected, but could not be quantified due to lack of contrast. Improvements in contrast are needed to lower the limit of quantitation. Nevertheless, the 0.25 nmole determination is within the working range of separated proteins in gel electrophoresis, and compares favorably to the determination, under extreme conditions, of one picomole of a purified selenoprotein containing one selenium atom in its molecule (8).

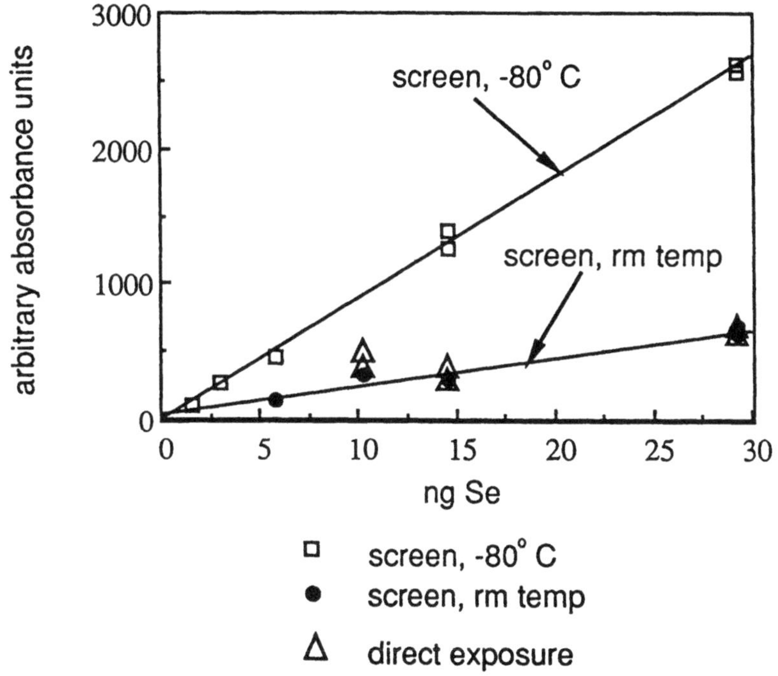

Figure 2. Comparison of results from Se radiotracer experiments using autoradiography under 3 conditions; direct exposure, exposure with an intensifying screen at room temperature, and exposure with an intensifying screen at -80 °C.

Our method, therefore, is applicable to protein samples within normal concentration ranges.

Acknowledgment

The assistance of Dr. D. Reeder and Dr. J. Edwards for advice and use of equipment is gratefully acknowledged, as are the personnel at the University of Virginia Nuclear Reactor Facility, especially B. Hosticka.

Certain commercial equipment, instruments, or materials are identified in this paper in order to adequately specify the experimental procedures. Such identification does not imply recommendation or endorsement by the National Bureau of Standards, nor does it imply that the materials or equipment identified are necessarily the best available for the purpose. This work will be included in the dissertation of S.F.S. to be submitted as a requirement for the Ph.D. degree from the University of Maryland.

References

1. Jayawickreme, C. K., Chatt, A. J. Radioanal. Nucl. Chem. in press.

2. Kwok, K. S., Snapka, R. M., Bernard, J. A., Harling, O. K., Varshavsky, A., 1985. Trans. Am. Nucl. Soc., 49, 157.

3. O'Farell, P. H., 1975. J. Biol. Chem. 250. 4007.

4. Stone, S. F., Hancock, D., Zeisler, R., 1987. J. Radioanal. Nucl. Chem., 112, No. 1, 95-108.

5. Tollaksen, S. L., Anderson, N. L., Anderson, N. G., 1981. Operation of the Iso-Dalt System, ANL-BIM-81-1, 6th. Ed., Argonne National Laboratory, 26.

6. Olden,K., Yamada, K. M., 1977. Anal. Biochem. 78. 483-490.

7. Standard Reference Material 1549, Non-Fat Milk Powder Certificate, 1985. National Bureau of Standards, Gaithersburg, MD.

8. Ching, W. M., deWeerd, B., Stadtman, T. C., 1985. Proc. Natl. Acad. Sci., 82. 347.

SEPARATION OF CHROMIUM(III) FROM CHROMIUM(VI) USING GRAPHITE FURNACE CHEMISTRY

Y. Schmitt, J.D. Kruse-Jarres
Institute of Clinical Chemistry and Laboratory Medicine
Katharinenhospital Stuttgart, FRG

Abstract

Chromium is ubiquitous in nature and can occur in different oxidation states from which Chromium(III) and Chromium(VI) are the most important. The main source for the intake is the food. Chromium is also found in waste water from industries which use Chromium for metallurgical purposes in the production of alloys and stainless steel, for important components of refractory materials and for colors. Chromium(III) was found to be an essential component of the glucose tolerance factor while Chromium(VI) can have acute and chronic toxic effects.

So it is necessary to find out ways of differentiating chromium(III) from chromium(VI). We used graphite furnace chemistry as the basis for metal speciation by electro-thermal atomic absorption spectrometry. After the addition of trifluoroacetylacetone, for example, chromium(III) can be removed from a graphite tube before atomization. Different authors reported different compositions of solutions and temperature programs for that purpose. In contrast to these authors reporting quantitative removing from water or urine samples we found under various modified analytical conditions a maximum of 50 - 60 % removing only. Already lots of problems arose when using water samples. So trifluoracetylacetone seems to react with the graphite leading to a decrease of chromium(III) being removed. In human plasma the results of removing chromium(III) was worse.

Trace Element Analytical Chemistry in Medicine and Biology

Introduction

The properties of the metal chromium which was discovered as an element in 1797 were quickly recognized to be of significant economic importance (1). Chromium is ubiquitous in nature and can occur in different oxidation states from which chromium(III) and chromium(VI) are the most important. The main source for man is the intake with food. Chromium is also found in waste water and airborne contaminations from industries which use chromium for metallurgical purposes in the production of alloys and stainless steel, for important components of refractory materials like glass or ceramics, and for colors. A source for the exposure of humans may be the domestic water supply, too (2).

The resorption in the gastrointestinal tract depends on different factors like oxidation state. Resorption does not exceed 1 to 5 %. The absorption by inhalation is not important for non exposed people (2).

Chromium was found to be an essential component of the glucose tolerance factor, chromium(III) being the essential trace element (3). Even high doses of orally given chromium(III) do not have harmful effects. In contrast hexavalent chromium can have acute and chronic toxic effects. The exposure can result in irritative lesions of the skin or the respiratory and gastrointestinal tract. Patients surviving acute intoxication were observed to have liver necrosis, tubular necrosis of kidneys and cerebral edema (2). Causing DNA damages the mutagenicity was shown later (4).

For this reason it is essential to separate chromium(III) from chromium(VI) in biological samples. Several reports have been published concerning the determination of the different oxidation states by chemiluminescence (5) or liquid chromatography (6), respectively. Other authors (7,8) reported methods using electrothermal atomic spectroscopy. This paper reports the results of separation experiments in human plasma following the investigations of (7,8) as a guideline which both had good results for the separation in aqueous solutions and urine although using different reagents and graphite furnace operating parameters.

Material and Methods

Zeeman 3030 atomic absorption spectrometer with graphite tube furnace HGA 600 and autosampler AS 60 from Perkin-Elmer Corporation and pyrolytically coated tubes without platform were used.

The two different graphite operating parameters according to (7,8) are shown in Table 1 and 2. The different compositions of the solutions for preatomization removal of chromium(III) are shown in Table 3. The heparinized plasma was taken from patients and stocked with solutions of chromium(III) and chromium(VI) up to a concentration of 25 µg chromium/l. Cr(III) was CrCl3 and C(VI) was K2Cr207 both from Merck Corp., FRG.

Table 1. Graphite Furnace operating Parameters according to (7):

	temperature in degrees c.	ramp time in s	hold time in s
drying	130	20	40
volatilization of chromium(III)	400	1	300
charring	1200	10	40
atomization	2700	1	10

Table 2. Graphite Furnace operating Parameters according to (8):

	temperature in degrees c.	ramp time in s	hold time in s
drying	80	5	50
	150	15	70
volatilization of chromium(III)	220	25	40
	500	10	10
charring	1000	1	10
atomization	2400	0	6

Table 3. Composition of Solutions for preatomization Removal of Cr(III)

A: according to (7)

10 ml TFA
20 ml of 25% tetramethylammonium hydroxide solution in methyl alcohol
20 ml of a 0.5 M aqueous sodium acetate solution

B: according to (8)

1 g urea
700 µl TFA
100 µl ammonia solution
filled to 10 ml with a 0.2 M aqueous sodium acetate solution

Experimental section and results

First of all aqueous solutions of 25 µg of chromium(III) and chromium(VI) per liter were used for the comparison with the results of the two above mentioned authors (7,8). 20 µl of the chromium solutions and 10, 20 or 30 µl of the solutions A and B were pipetted into the graphite tube.

The reduction of the signals received is shown in figure I. The solution A lead to a smaller loss of chromium(VI), but was not able to remove chromium(III) as well as the solution B when using 10 µl volume of the reagents. The higher the volume of the reagents (10, 20 or 30 µl) the bigger the reduction of chromium(III) was but also bigger the loss of chromium(VI). The loss of chromium(VI) increased more than the reduce of chromium(III), so that the difference between both became smaller. That difference was smaller for the solution A than for B.

Because of these preliminary results the method according to (8) was chosen for the experiments in plasma. The results are shown in figure II. The reduction of the chromium(VI) signal was 15 % using 10 µl volume. Increasing the amount of reagents to 20 or 30 µl led to an higher reduction of the chromium(VI) signal to 20 or 25 %. The chromium(III) signal was reduced by 35 % using 10 µl. Increasing that volume to 20 or 30 µl increased the reduction to 45 %.

The results of preatomization separation from chromium(III) and chromium(VI) show obviously that the difference between remaining chromium(VI) and chromium(III) was smaller than it was in aqueous solutions.

After that parameters like temperature and time during the graphite furnace program, pH, composition of the solution were altered. With regard to these it was not possible to improve the separation significantly.

Figure I. Relative signals in aqueous solutions (mean value of 5 determ.)

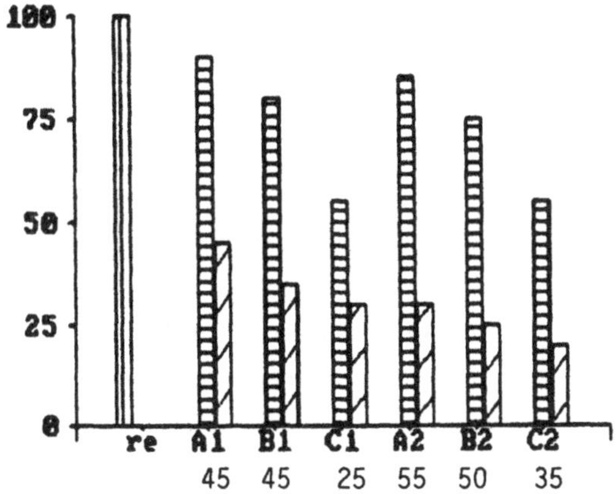

re: reference signal without preatomization separation
A1, B1, C1: solution according to (7) 10, 20 and 30 µl
A2, B2, C2: solution according to (8) 10, 20 and 30 µl

difference between both signals in percent

Figure II. Relative signals in human plasma (mean value of 5 determ.)

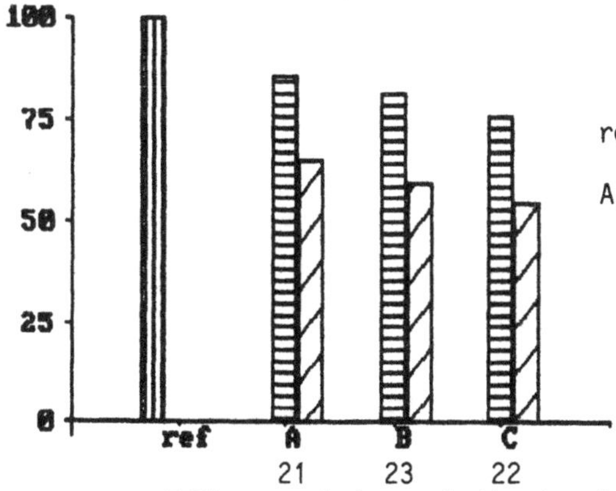

ref: reference signal without preatomization separation
A, B, C: solution according to (8) 10, 20 and 30 µl

difference between both signals in percent

As shown in figure III there was an important effect concerning the stability of the Cr(VI) signal. The longer the graphite tube was used for separation the smaller the stability of Cr(VI) became. The Cr(III) removal was not changed significantly, this signal decreased 6 % after 80 determinations. In contrast the Cr(VI) signal decreased about 30 %.

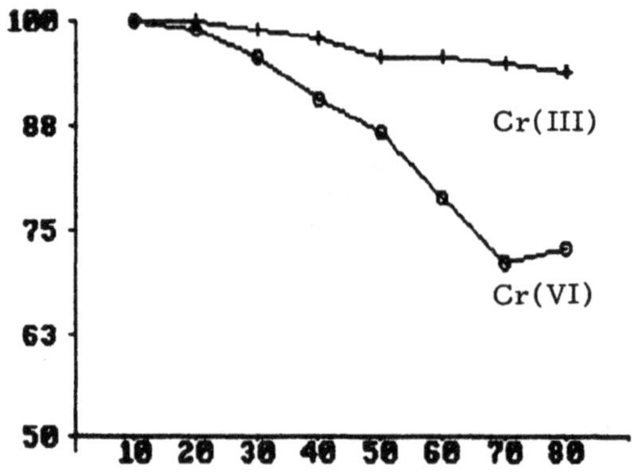

Figure III: Behavior of Cr(VI) and Cr(III) stability with increasing number of atomization in a single graphite tube.

Discussion

Already the separation in aqueous solutions was not as good as reported by (7,8), in human plasma it was distinctly worse. Using more volume of the reagent led only to a little better separation of the different chromium species. This was seen by (7) in urine, too. The values of repeated determinations in human plasma oscillate more than in aqueous solutions. There were oscillations between +/- 25 % in plasma in contrast to +/- 10 % in aqueous solutions.

(7) did not see a difference between uncoated and pyrolytically coated tubes with the L´vov platform concerning the removal of chromium(III), but differences between tubes with or without platform. This may be one reason for the yields in this work although the differences reported from (7) were only a few percent. He thought that one reason for the incomplete removal of Cr(III) was the higher porosity of the graphite and the formation of chromium carbides in tubes without platform. Chromium(VI) was retained in all types of tubes in the same manner.

The reaction between chromium or the solutions used or both together with the graphite in the tube has an influence on the stability of chromium(VI) during the drying, charring and atomization procedure. The influence was bigger in plasma than in aqueous solutions.

It was not clear whether the valence of chromium changes during the graphite furnace operating procedure as one more reason for these results. Reduction from chromium(VI) to chromium(III) was found in similar experiments in ammonium buffer solution (9). The reduced stability of chromium(VI) with the increasing number of atomization with the same graphite tube might have its reason in changing the valency.

As shown above good separation of chromium(III) from chromium(VI) could not be reached neither in aqueous solution nor in human plasma. Both reagents (7,8) did not work in that extent reported by these authors. Change of components and other factors could not improve the separation. Influences from the matrices in plasma are stronger than urine matrices. Further investigations especially on valency changing during graphite furnace operating will be done.

Literature

(1) Rinehart, W.E., S.C. Gad. 1986. Am Ind Hyg Assoc J 47 (11). pp. 696

(2) Hertel, R.F. 1986. IARC-Scientific publications. pp.63-77

(3) Mertz, W., K. Schwarz. 1959. Am J Physiol 196. pp. 614-618

(4) Leonard, A., R.R. Lauwerys. 1980. Mutat Res 76. pp. 227-239

(5) Bhargava, O.P., H.E. Bumsted, F.I. Grunder, B.L. Hunt, G.E. Manning, R.A. Riemann, J.K. Samuels, V. Tatone, S. Jeffwaldschmidt, P. Hernandez. 1983. Am. Ind. Hyg. Assoc. J. 44, p. 433

(6) The Japan Association of Industrial Health. 1985. The recommendation of TLV. Jpn. J. Ind. Health 27. p. 195

(7) Arpadjan, S., V. Krivan. 1986. Anal Chem. pp. 2611-2614

(8) Dungs, K., H. Fleischhauer, B. Neidhart. 1985. Fresenius Z Anal Chem. pp. 280-289

(9) Sawatari, K. 1986. Industrial Health 24. pp. 111-116

SEPARATION AND ISOLATION OF COPPER-METALLOTHIONEINS BY FAST PROTEIN LIQUID CHROMATOGRAPHY (FPLC)

H.Nederbragt, Moniek M.L. de Wit

Department of Pathology, Faculty of Veterinary Science, University of Utrecht, Utrecht, the Netherlands

Abstract

Copper metallothioneins have been isolated and separated using FPLC methods in which a Superose 12 column was used for gelfiltration and a Mono Q column for ion-exchange chromatography. Properties of isometallothioneins may be changed by conditions of preparation and storage as revealed by ion-exchange chromatography of our samples.

Introduction

Metallothionein (MT) is a low molecular weight protein with a high number of cysteine molecules that is able to bind metals of which Zn, Cd and Cu are the most important (see ref.1 for an extensive review of details). Usually, MT's are isolated from fractions obtained from gelfiltration on Sephadex G75, containing proteins in the 10 kDa range. The proteins from these fractions can be separated further by an ion-exchange procedure for which mainly DEAE A25 gels are used from which different isoMT's can be eluted with a salt gradient. The proteins thus obtained have been shown to be pure and have been used for biochemical studies and production of antibodies. Our interest in CuMT was raised by its proposed function in protection against the deleterious effects of Cu in cells and the probability of strain differences in rats with respect to the toxic effects of Cu (2, 3). However, the isolation and separation of MT's down to the level of isoMT is a time-consuming procedure taking at least 2 days per sample, and is inappropriate for analyzing a large number of samples.

Trace Element Analytical Chemistry in Medicine and Biology

An alternative for MT analysis was developed by Suzuki (4) who used an HPLC system in which the eluate of the column was nebulized directly and continuously into the flame of an atomic absorption spectrophotometer (AAS). Now, several comparable HPLC systems, differing in column material, have been described (5,6).
We describe here some preliminary results with a method in which a Fast Protein Liquid Chromatography (FPLC) system, containing Superose 12 for gelfiltration and Mono Q for ion-exchange chromatography, is compared with classical Sephadex and DEAE chromatography. Although FPLC may be somewhat slower than HPLC, it combines the advantage of speed with the possibility of semi-preparative isolation of MT's. Due to our interest in Cu toxicity we have directed ourselves mainly to CuMT.

Methods

Male random-bred Wistar rats (200 g) were injected i.p. with 3 mg Cu/kg bw and killed 18 h later. Livers were homogenized in a 10 mM Tris HAc buffer, pH 7.4, containing 2mM 2-mercaptoethanol and 0.01% NaN_3. The homogenate was centrifuged 10 min. at 6,000 g and the resulting supernatant at 100,000 g, both at 4^oC. Samples of the latter supernatant were used for chromatography on Sephadex G75 or Superose 12 after storage at 4^oC. Presumed MT containing fractions (low E280, Cu/Zn containing proteins at Ve/Vo = 2) and other proteins from Sephadex elution as indicated, were concentrated under N_2 using Amicon YM-5 filters; these or non-concentrated samples were used for gelfiltration on a Superose 12 column or for ion-exchange chromatography on a DEAE A25 or a Mono Q column. Properties of all columns and conditions of

Table 1. Properties of the Elution Systems and Conditions Used in Comparing Different Methods of Gelfiltration and Ion-Exchange Chromatography

	Sephadex G75	Superose 12	DEAE A 25	Mono Q
column	2.5 x 90 cm	1.0 x 30 cm	1.6 x 25 cm	0.5 x 5.0 cm
elution rate	30 ml/h	30 ml/h	20 ml/h	60 ml/h
sample	5 ml	200 ul	100-200 ml	10-20 ml
fractions	5 ml	0.5 ml	5 ml	0.5 ml
elution time	18 h	1.25 h	25-40 h	1.5 h
Tris concentr.	10 mM	100 mM	10 mM (init)	10 mM (init)

elution are given in table 1. All elutions were carried out at room temperature and eluates were recorded continuously for absorption at 280 nm (not shown). All buffers contained mercaptoethanol routinely. Cu and Zn concentrations of fractions were determined by AAS, using an air-acetylene flame.

Results and discussion

Gelfiltration of the liver supernatant on Sephadex G75 resulted in the Cu and Zn profiles shown in fig.1. Four Zn peaks (one with a prominent shoulder) and three Cu peaks (including a small near the void volume) were detected. The large Cu and Zn containing peak D was supposed to be MT. When a sample of the same supernatant was eluted on Superose 12 (fig.2) two Cu-containing peaks could be found, the void volume peak probably being to small to be detected. The Zn pattern on Superose was different from that on Sephadex and showed three peaks and a shoulder. This difference can be explained by : a. the difference in the molecular weight range in which

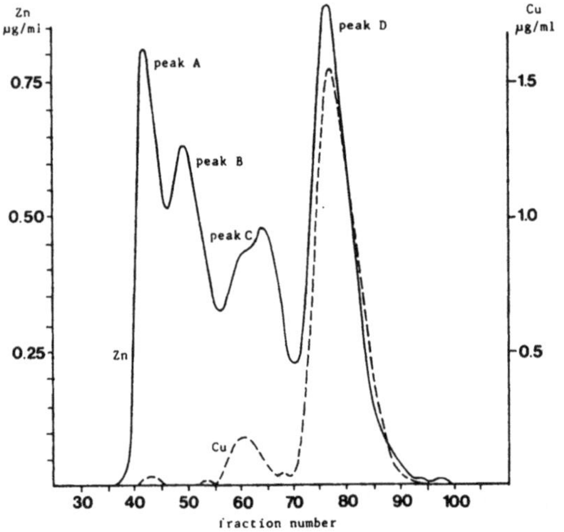

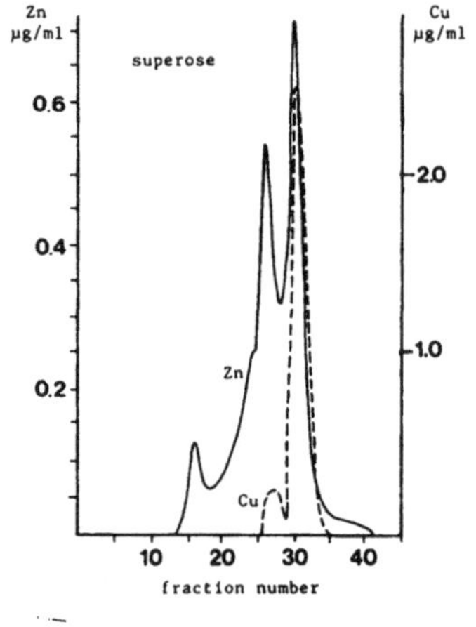

Fig.1 (left). Chromatography of 100,000 g supernatant of a liver of a Cu-jected rat on Sephadex G75, showing the profiles of Zn and Cu concentrations in the fractions.
Fig.2 (right). Chromatography of the same supernatant as in fig.1 on Superose 12.

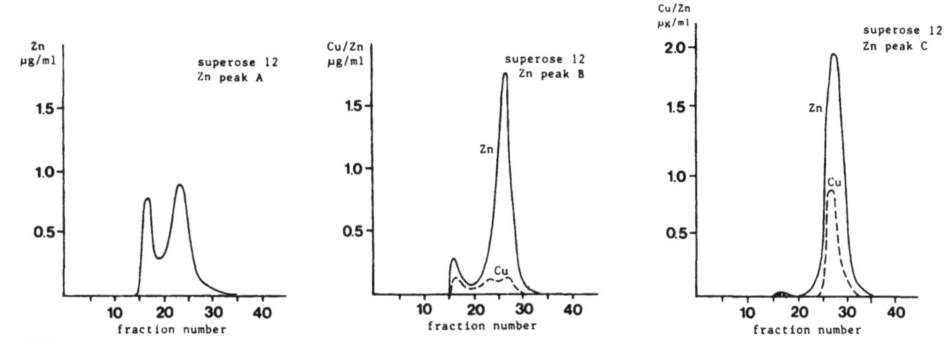

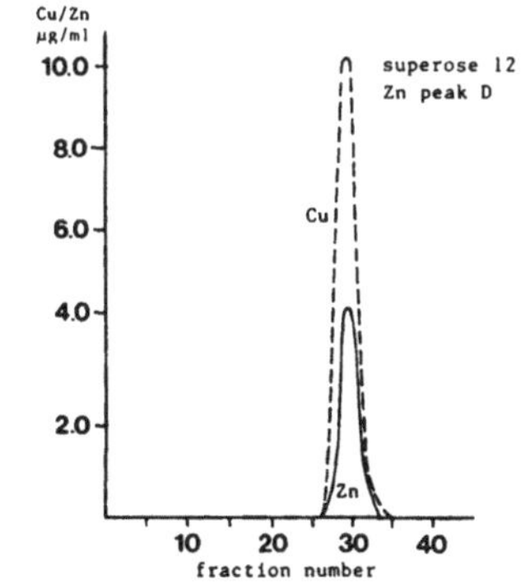

Fig.3. Fractions obtained by Sephadex G75 chromatography of fig.1 containing peaks A, B, C and D eluted separately on Superose 12. The Zn and Cu profiles are shown; no Cu could be detected in A.

separation of proteins is achieved (Sephadex G75: 5-60 kDa; Superose 12: 5-250 kDa); this could be illustrated by eluting concentrated Sephadex peaks A and B on Superose (fig.3A and B): these peaks both contained Zn proteins eluting in fractions 16 (void volume) and 25 of the Superose column. b. the less efficient separation of proteins in the MW range from 20-60 kDa on Superose; this could be seen when peak C was eluted from the Superose column (fig.3C): this shoulder containing peak on Sephadex eluted as a single peak from the Superose. The MT containing peak D also eluted as a single peak from the Superose (fig.3D). As was the case in the Cu and Zn profile after elution of the complete supernatant (fig.2) the top of the MT peak eluted in fraction 31.

The Sephadex fractions containing MT (peak D in fig.1) were brought directly onto a DEAE A25 column or, after concentration, onto a Mono Q column. Elution was performed with a 10 mM Tris buffer, followed by a Tris gradient from 10 to 400 mM. In the figures the full lenght of the run is not drawn completely.

The isoMT pattern after elution on DEAE A25 is shown in fig.4. A sharp Zn

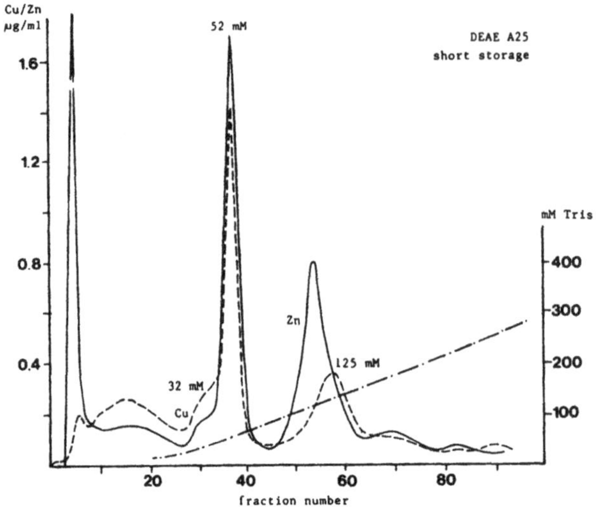

Fig.4. Chromatography of peak D from fig.1 on Sephadex DEAE A25 showing the profiles of Zn and Cu concentrations of the fractions; the Tris concentration at which the top of the Cu-containing peaks are eluted is given in the graph.

peak and a broad Cu peak were eluted before starting the gradient. In addition three Cu and Zn containing proteins were found: a shoulder at 32 mM and two peaks at 52 and 125 mM. The isoMT pattern after elution on Mono Q for CuMT is shown in fig.5. The large Cu peak obtained before starting the gradient was the result of overloading the column and could be reduced by loading a smaller sample that did not essentially affect the other peaks. These peaks eluted at Tris concentrations of 35 mM (shoulder), 53 and 121 mM, all three in very good agreement with those obtained from DEAE A25. In addition, a new Cu peak at 74 mM was found.

Considerable problems arose when we tried to reproduce these findings. It appeared that the storage of the samples caused changes in the properties

Fig.5.Chromatography of peak D from fig.1 on Mono Q, showing the Cu concentration of the fractions; the Tris concentration at which the top of the peaks eluted is given in the graph.

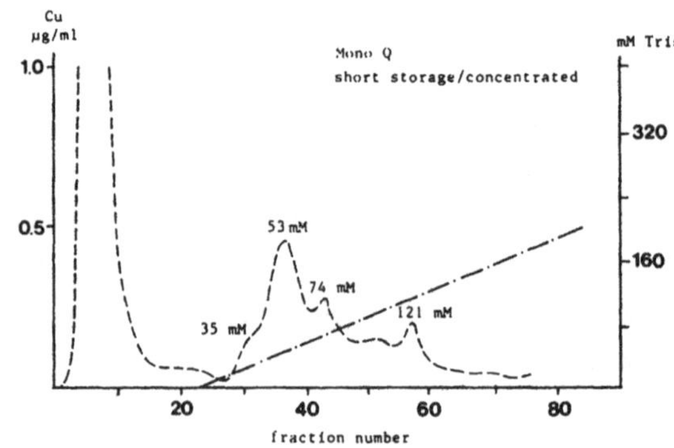

of the presumed isoMT's which were expressed in their behaviour on ion-exchange columns. This was illustrated by the effect of longer storage at 4°C of a Sephadex peak D sample, resulting in disappearance of former and appearance of new peaks with a higher charge after elution over DEAE A25. When new rats were injected with Cu and a supernatant and Sephadex peak D were prepared immediately followed by chromatography over Mono Q, Cu peaks with a lower charge were obtained.
Although our results suggest that Mono Q is a fast and reliable alternative for DEAE A25 ion-exchange chromatography the final confirmation of its use in isolation of CuMT's has to await the solution of the problem of artifacts occurring during preparation and storage, such as has been described before (7). The use of Superose 12 columns in the isolation is advantageous particularly when a large number of samples has to be analyzed for MT e.g. in cultured cells.

References

1. Kägi,J.H.R. and Y.Kojima (eds). 1987. Metallothionein II. Birkhäuser Verlag, Basel.

2. Nederbragt,H. and A.J.Lagerwerf. 1986. Hepatology 6, 601-607.

3. Nederbragt,H., M.Gijbels and A.J.Lagerwerf. 1987. In: Trace Element-Analytical Chemistry in Medicine and Biology, Vol.4 (P.Brätter and P.Schramel,eds). Walter de Gruyter, Berlin-New York. pp 589-596.

4. Suzuki,K.T. 1980. Anal.Biochem. 102, 31-34.

5. Klauser,S., J.H.R.Kägi and K.J.Wilson. 1983. Biochem.J. 209, 71-80.

6. Lehman,L.D. and C.D.Klaassen. 1986. Anal.Biochem. 153, 305-314.

7. Suzuki,K.T. and T.Maitani. 1981. Biochem.J. 199, 289-295.

THE EFFECTS OF COOKING AND SIMULATED GASTRO-INTESTINAL DIGESTION ON THE SPECIATION OF CADMIUM IN PIG KIDNEY

R.C. Massey, H. Crews

Ministry of Agriculture, Fisheries and Food, Food Science Division, Queen Street, Norwich NR2 4SX, UK

L. Ebdon, J.R. Dean

Department of Environmental Sciences, Plymouth Polytechnic, Drake Circus, Plymouth PL4 8AA, UK

Abstract

The speciation of cadmium in retail pig kidney has been examined by size exclusion chromatography (SEC) using an inductively coupled plasma-mass spectrometer (ICP-MS) as detector. Approximately 35% of the total cadmium in uncooked kidney was soluble at pH 8.0 and SEC/ICP-MS revealed three discrete peaks whose retention times corresponded to molecular weights of 1.2×10^6, 7.0×10^4 and 6 to 9×10^3 daltons. In the cooked pig kidney some 30% of the total cadmium was soluble and this was all associated with the 6 to 9×10^3 daltons peak. Following simulated gastric digestion at pH 2.5 the majority of the cadmium was solubilised and of less than 1000 daltons. On simulated intestinal digestion at pH 6.8 a single peak was observed which corresponded to 20% of the total cadmium originally in the kidney and which co-eluted with the cadmium binding compound extracted from the cooked kidney.

Introduction

The occurrence of cadmium in foodstuffs is of concern due to the well established toxicological properties of the metal (1). Fortunately the large majority of ingested cadmium is excreted in the faeces and in the case of orally administered ionic cadmium only approximately 6% is absorbed from the intestinal tract(2). When cadmium is consumed as part of

Trace Element Analytical Chemistry in Medicine and Biology

a meal the degree of uptake is generally rather lower and in the range 2-3% (3, 4). Investigations into the chemical speciation of cadmium in food have revealed that the metal is not present in the ionic form but complexed to various binding proteins (5, 6). It is therefore possible that the chemical form of cadmium in the foodstuff influences its behaviour on gastro-intestinal digestion and subsequent uptake from the intestine. One of the factors that has hindered an understanding of the possible role of speciation in this area has been the lack of suitably sensitive and selective analytical instrumentation. As a result it has in the past been necessary to artificially enhance the cadmium content of foods prior to speciation studies with the attendant risk of altering the chemical form of the metal. With the recent introduction of ICP-MS(7) and its interfacing to high performance liquid chromatography (8) it is now feasible to examine the speciation of cadmium at the levels at which it is present in normal retail foods. We now wish to report the changes in the chemical speciation of cadmium that occur during cooking and simulated gastro-intestinal digestion of pig kidney.

Experimental

Pig Kidney. The pig kidney was purchased at a retail outlet in Norwich in 1986 and homogenised and divided into two equal portions. One half was placed in a Pyrex dish, covered with aluminium foil and baked in a domestic electrically heated oven set at 190°C for 1.5 hours. The cooked kidney together with associated juices was rehomogenised and stored at -15°C with the uncooked sample pending analysis.

Total Cadmium Determination. Sub-samples of the uncooked and cooked kidneys were digested with nitric and sulphuric acids followed by hydrogen peroxide. After chelation and extraction the cadmium was determined by graphite furnace atomic absorption spectroscopy. The method is described in detail elsewhere (9). The wet weight total cadmium content of the uncooked and cooked pig kidney were 0.22 mg/kg and 0.23 mg/kg respectively. Each sample was analysed in triplicate and the repeat determinations were within ±5% of the corresponding mean.

Extraction of Cooked and Uncooked Kidney. The method employed was similar to that devised for the extraction of metallothionein (10) and is described in detail elsewhere (9). Briefly, the sample was homogenised in Tris-HCl buffer (0.02M, pH 8.0) containing sucrose (0.25M). Following centrifugation and treatment with Lipoclean (freon, Hoechst UK Ltd, Hounslow, UK) the extract was filtered (0.22 um) and analysed by SEC/ICP-MS.

Simulated Gastro-Intestinal Digestion of Cooked Kidney. A two stage enzymolysis procedure which is described in detail elsewhere (11) was used. Briefly, the cooked sample (25 g, wet weight) was treated with pepsin (1%, 50 ml) in 154 mM saline, acidified to pH 1.8 with hydrochloric acid and maintained at 37°C for 4 hours in a "Stomacher" bag (Seward Laboratory, London) in a shaking water bath. An aliquot was then removed, readjusted to pH 2.5 and centrifuged and filtered prior to SEC/ICP-MS analysis. Following simulated gastric treatment the bulk of the digest was adjusted to pH 7.0 with saturated sodium bicarbonate and intestinal juice (50 ml) added comprising equal volumes of (i) pancreatin (30 mg/ml) and amylase (10 mg/ml) in saline and (ii) bile salts (1.5 g/litre) in saline. The digest was maintained at 37°C for 4 hours and then centrifuged and filtered prior to SEC/ICP-MS analysis.

SEC/ICP-MS. The molecular weight speciation of cadmium was studied using a Superose 12 SEC column (Pharmacia, Milton Keynes, UK) interfaced directly to a Plasmaquad ICP-MS (VG Isotopes, Winsford, UK) monitoring ^{114}Cd as described elsewhere (8). Aliquots (100 ul) of the sample extracts were injected onto the column and eluted with a mobile phase (0.75 ml/min) comprising Tris-HCl buffer (0.12 M, pH 7.5) except for the gastric digest which was eluted wtih Tris-HCl buffer (0.12 M, pH 2.5). The column was regularly calibrated and a plot of log molecular weight against Kav yielded a straight line for the following standards: vitamin B_{12}, myoglobin, β-lactoglobulin, ovalbumin, bovine serum albumin, gammaglobulin and thyroglobulin.

Results

During preliminary investigations of the chemical form of cadmium in the uncooked kidney extract it became apparent that there were a number of problems specifically associated with speciation studies on trace amounts of the metal in a complex biological matrix. Of major concern was evidence of changes in speciation during chromatographic analysis and also anomalous recoveries of cadmium injected onto the SEC column. At pH 2.5 ionic cadmium was completely eluted from the column at the retention time anticipated for lower molecular weight species. Ionic cadmium did not elute from a clean column at pH 7.5 and this was presumably due to binding of the cation to residual carboxylate groups on the stationary phase. However, if metallothionein had previously been injected onto the column subsequent injections of ionic cadmium at pH 7.5 resulted in a peak of retention time corresponding to metallothionein and magnitude equivalent to the amount of ionic cadmium injected. This phenomenon suggests that a fraction of the previously injected metallothionein remains bound at the top of the SEC column and is desorbed and eluted on complexation with subsequent injections of ionic cadmium. The reverse situation was also observed to occur if even low levels of ionic cadmium, from the mobile phase for instance, were allowed to adsorb onto the SEC column at pH 7.5. Under these conditions injection of an extract of uncooked kidney resulted in the elution of a number of high molecular weight peaks whose combined magnitude corresponded in some cases to over 4 times the amount of cadmium injected onto the column. This behaviour was attributed to proteins with a high affinity for cadmium complexing and desorbing the column-bound cation. To resolve these difficulties the presence of extraneous cadmium in the chromatograph was reduced to a minimum by using an all glass/teflon system (FPLC, Pharmacia) and pretreating the mobile phase with an in-line Chelex-100 (Bio-Rad Laboratories Ltd, Watford, UK) scrubber column. To prevent additional artefacts caused by interaction with components of previous samples adsorbed on the column a new guard column packed with Superose 12 (prep grade) was used for every injection of the kidney extract and the analytical column was regularly cleaned as described elsewhere (9). Employing these techniques consistent chromatography was obtained on repeat injection of the extracts and enzymic digests of the kidney samples and, additionally, acceptable recoveries (75 to 100%) of the injected cadmium were achieved.

The results of the SEC/ICP-MS speciation studies are shown in Table 1. Of the total cadmium (0.22 to 0.23 mg/kg, wet weight) in the uncooked and cooked pig kidney some 30 to 35% was solubilised on extraction at pH 8.0. Three peaks were present for the uncooked sample and the retention times corresponded to molecular weights of 1.2×10^6, 7.0×10^4 and 6.3×10^3 daltons. The identities of these species have yet to be established but it seems likely that the first peak was associated with blood protein aggregates such as α_2 macroglobulin (c.a. 9×10^5 daltons) and the second with plasma albumen (c.a. 6.9×10^4 daltons). The third and largest peak had a molecular weight similar to that anticipated for metallothionein (c.a. 1×10^4) as estimated by SEC(10). On cooking the pig kidney under normal domestic conditions the two higher molecular weight species were destroyed whilst the third peak was unaffected. The survival of the latter further indicates that it is a metallothionein-like compound in view of the well established thermal stability of this class of protein (12, 13). The retention time of this putative metallothionein corresponded to a molecular weight of 8.9×10^3 daltons which is somewhat higher than 6.3×10^3 daltons estimated for the low molecular weight peak in the uncooked kidney extract. Co-injection of a mixture of the cooked and uncooked extracts revealed the presence of the two high molecular weight uncooked species at 1.2×10^6 and 7.0×10^4 daltons and a single peak at 6.9×10^3 daltons. There was no evidence from the profile and width of the latter peak to suggest the presence of two poorly resolved components. It would thus seem that the same low molecular weight compound is present in the cooked and uncooked extract.

TABLE 1

CHANGES IN SPECIATION OF CADMIUM AS EFFECTED BY COOKING AND SIMULATED GASTRO-INTESTINAL DIGESTION OF PIG KIDNEY

SAMPLE	MOLECULAR WEIGHT* (DALTONS)	CONCENTRATION OF CADMIUM (ng/ml) IN EXTRACT/ DIGEST	EXTRACTED CADMIUM SPECIES AS % OF TOTAL CADMIUM IN UNCOOKED KIDNEY
Uncooked	1.2×10^6	2.1	7
	7.0×10^4	2.5	8
	6.3×10^3(a)	6.5	20
Cooked	8.9×10^3(a,b)	10.2	30
Gastric digest of cooked kidney (pH 2.5)	$< 1.0 \times 10^3$	25	60
Intestinal digest of the above gastric sample (pH 6.8)	1.2×10^4(b)	7.5	20

*; as estimated by size exclusion chromatography

a, b; peaks co-elute on co-injection (see text)

Simulated gastric digestion of the cooked pig kidney with pepsin under acidic conditions resulted in solubilisation of some 60% of the cadmium. A similar phenomenon has been observed on gastric digestion of crab meat (14) and was found to be due predominantly to acid mediated dissociation of cadmium from the binding proteins. SEC/ICP-MS analysis of the gastric digest revealed the presence of a single peak whose retention time corresponded to cadmium species of less than 1×10^3 daltons. Simulated intestinal digestion of this gastric digest resulted in 80% of the cadmium binding to the residual insoluble fraction. The 20% remaining in solution was associated with a single peak whose retention time corresponded to a molecular weight of 1.2×10^4 daltons.

Co-injection of a 50:50 mixture of this intestinal digest with the cooked kidney extract revealed a single peak whose retention time corresponded to 1.2×10^4 daltons. As mentioned previously injection of the cooked kidney extract by itself resulted in a single SEC peak with retention time equivalent to 8.9×10^3 daltons. The elution of this peak at a somewhat shorter retention time corresponding to 1.2×10^4 daltons on co-injection suggests that components of the intestinal digest cause the metallothionein-like protein in the cooked kidney extract to elute fractionally earlier from the SEC column. The fact that these two peaks co-elute indicates that a similar cadmium binding protein is present in both the pH 8.0 extract and the gastro-intestinal digest of the cooked kidney. If this is the case it would appear that the metallothionein-like component in the kidney is resistant to both simulated gastric and intestinal enzymolysis and that the cadmium rebinds to this protein when the pH is raised to 6.8 after gastric digestion. The presence of stable, sample-derived, cadmium species in the intestinal digest suggests that the form of the metal in the foodstuff may influence the extent of its uptake from the digestive tract.

Characterisation studies are currently in progress to firmly establish whether, as now seems likely, the cadmium binding protein in the simulated gastro-intestinal digest is derived from the metallothionein-like component extracted from the cooked kidney.

References:

1. Samarawickrama G.P.. 1979. In: The Chemistry, Biochemistry and Biology of Cadmium (M Webb, ed). Elsevier, Amsterdam, pp 341-421.

2. Rahola T., A. Ritva-Kaarina, J.K. Miettinen. 1973. Ann. Clin. Res. 5, 214.

3. Newton D., P. Johnson, A.E. Lally, R.J. Pentreath, D.J. Swift. 1984. Human Toxicol. 3, 23.

4. Heard M.J., A.C. Chamberlain. 1982. Human Toxicol. 1, 411.

5. Wagner G.J.. 1984. Plant Physiol. 76, 797.

6. Wiedow M.A., T.J. Kneip, S.J. Garte. 1982. Environ. Res. 28, 164.

7. Houk R.S., V.A. Fassel, G.D. Flesch, H.J. Svec, A.L. Gray, C.E. Taylor. 1980. Anal. Chem. 52, 2283.

8. Dean, J.R., S. Munro, L. Ebdon, H.M. Crews, R.C. Massey. 1987. J. Anal. Atom. Spectrom. 2, 607.

9. Crews H.M., R.C. Massey, L. Ebdon, J.R. Dean. Fd. Chem. Toxicol. to be submitted.

10. Webb M.. 1979. In: The Chemistry, Biochemistry and Biology of Cadmium (M. Webb, ed) Elsevier, Amsterdam pp 195-266.

11. Crews H.M., J.A. Burrell, D.J. McWeeny. 1985. Z. Lebensm. Unters Forsch. 180, 405.

12. Cherian M.G.. 1979. Environ. Health Perspect. 28, 127.

13. Nordberg M., Y. Kohima. 1979. In: Metallothionein (J.H.R. Kagi, M. Nordberg, eds). Birkhauser Verlag, Basel pp 41-99.

14. Massey R.C., J.A. Burrell, D.J. McWeeny, H. Crews. 1986. Toxicol. Environ. Chem. 13, 85.

CHANGE OF SPECIATION OF FUNGUS BOUND CADMIUM DURING DIGESTION IN THE RAT

G.A. Jackl, D. Berg and W.E. Kollmer

Institut für Strahlenbiologie, Gesellschaft für Strahlen- und Umweltforschung München, D-8042 Neuherberg, F.R.G.

G. Reidel
Nuklearmedizinische Klinik und Poliklinik rechts der Isar der TU München, Ismaninger-Str.22 ,D-8000 München 80, F.R.G.

Abstract

The major shifts of Cd between the different fractions of food on its way through the gastrointestinal tract was examined by radioactive ^{109}Cd. Generally a major shift was observed from the soluble fraction to the solids and from the low molecular to the high molecular compounds of the soluble fraction. At pH 3 the element in the soluble fraction of the stomach was practically completely in the form of the free ion. At neutral pH it appeared in the region of high molecular weights in the stomach contents and persisted onto the faeces if the tracer was administered as a chloride. If it was administered bound to the protein of the mushroom *Agaricus arvensis* the major peak was observed in the medium range in the stomach contents and in the small intestine. In the faeces it appeared in the same region as the major peak in the chromatographs of the samples obtained from the animals that had received $CdCl_2$.

Introduction

Cadmium is a widespread contaminant in the environment. It is known as a hazard to man mainly because of its accumulation in

Trace Element Analytical Chemistry in Medicine and Biology

the kidney during long periods of chronic intake.
In environmental exposure the ingestion along with the diet is a main path of intake. (1). Normally only a small portion of its content in the food is resorbed in the intestine (2,3). However, this portion may change depending on the quality of the food or the nutritional status of the subject (4-8). Therefore it is of interest if the nutrient to which Cd is bound exerts an influence on its resorption in the intestine. One of the nutrients which may contain a high amount of Cd is the mushroom *Agaricus arvensis* (Schiff. ex.Fr.) (9-11). This study was undertaken to investigate what happens to Cd bound to the protein of this mushroom on its passage through the intestinal tract.

Material and methods

Two experiments were carried out: One for determining the radioactivity in the stomach, the major segments of the intestine and in the faeces in the fractions a) free, b) soluble c) solid and another where the soluble Cd binding compounds were analyzed by chromatography. Experimental animals were adult male rats of a Wistar derived strain of approximately 300 g body weight in the first and 220 g in the second experiment. They were kept at an artificial day-night rhythm (12/12 hrs.). Two weeks before the intubation of the tracer they were fed a solid semipurified diet (Altromin C-1000-R). Tap water was supplied *ad lib.*. Dried fruit bodies of the mushroom *Agaricus arvensis* were extracted and the extracts were labelled with radioactive ^{109}Cd as described before (12). At the same time a solution of $CdCl_2$ was also labelled with ^{109}Cd. Each administered dose of the fungus extract and the cadmium chloride solution contained 200 µg stable cadmium per ml. In the first experiment the specific activity was 9.19 MBq/mg, in the second 55.5 MBq/mg. In each experiment 2 rats received under a light ether anesthesia 0.5 ml of the solution of the labelled

$CdCl_2$ or the labelled fungus protein by way of a stomach tube followed by 0.5 ml of water. In both experiments one animal of each groups was sacrificed under ether anesthesia 1 h after this administration. The stomach and the small intestine were recovered and their radioactivity determined. In the first experiment the caecum and the large intestine were also recovered. On the following day the other two animals were sacrificed, the same segments of the intestine recovered and the faecal pellets collected. Then the whole contents of the stomach and the small intestine were removed by squeezing. One ml of 50 mM phosphate buffer pH 7 was added and the samples were homogenized by slight sonication. They were neutralized by adding a minimum volume of 1 M Tris-base, and centrifuged 30 min at 48000 x g at a temperature of 4°C. The radioactivity of the supernatant and of the sediment was determined by γ-spectrometry. In exp.2 the supernatants were filtered through 0.45 µm filters and analysed by Fast Performance Liquid Chromatography (FPLC) in 50 mM phosphate buffer pH 7 on a TSK G 2000 SW column (8x300 mm) at a flow rate of 0.5 ml/min and a pressure of 5 bar. The chromatography was monitored by UV-spectrometry at 280 nm. Fractions of 0.5 ml were collected and 52 fractions from the start were measured by γ-spectrometry. From each of the remaining animals one faecal pellet was subjected to the same procedure as the intestinal contents.

Results and discussion

One hour post administration the major part of the radioactive Cd was found in the stomach and the intestine (Table 1). Between 16 and 37 of its percentage in each of the contents was in the liquid phase (Table 2). The faeces collected 24 h later contained 64 to 75 % of the administered radioactivity but only 1 % or 0.4% of it was in the liquid phase while 99 % or more were insoluble. Thus the radioactivity in the solid phase was increasing while that in the liquid phase was decreasing.

Table 1: RADIOACTIVITY RECOVERED IN THE INTESTINAL SEGMENTS PLUS CONTENTS (as % of administered amount)

Time of sampling	1 hour post administration		24 hours post administration	
Administered Compound	$^{109}CdCl_2$	^{109}Cd labelled fungus protein	$^{109}CdCl_2$	^{109}Cd labelled fungus protein
Stomach	39	34	0	3
Small intestine	38	33	5	7
Caecum	10	7	11	16
Large intestine	0	1	7	9
Faeces	0	0	75	64

On its passage through the gastrointestinal tract the food becomes acid in the stomach. In the rat the natural pH in the stomach is approximately 3 ,in the small intestine approximately 7. The radioactivity found in the various segments of the gastrointestinal tract plus their contents and in the faeces are demonstrated in Table 1. One hour after the administration the major part of the tracer was found in the stomach and the small intestine. After 24 h the major amount was found in the faeces. In the acid contents of the stomach all of the tracer was soluble irrespective of its form at the time of administration. The chromatographic separation revealed that it was the free Cd ion. The fractions of radioactivity found at a neutral pH in the supernatant of the contents of the stomach, the small intestine and in the faeces are shown in Table 2. One hour post administration 22 % was found in the solution in those animals where the tracer was administered as $CdCl_2$ and 36 % where the tracer was bound to fungus protein. In the contents of the small intestine 16% was soluble where the tracer was administered as a chloride and 37 % where it was administered bound to the fungus extract. In the faecal pellets collected 24 h later 1 % was soluble in the animals having received $CdCl_2$ and 0.4 % in those having received the fungus bound Cd. Chromatographs of the soluble fraction of the neutralized contents of the stomach, small intestine and of faeces are shown in Fig. 1. The UV absorption at 280 nm which was continuously recorded during chromatography is also demonstrated. In the animals where the tracer was administered as $CdCl_2$ the two peaks of radioactivity in the supernatant of the neutralized stomach contents are equal to those in the intestine and in the faeces. In the original fungus extract the chromatograph of the tracer contains two main peaks. The two peaks in the neutralized stomach contents of the animal which had been given the labelled fungus protein are apparently not identical to those of the original labelled protein. The smaller one, however, may have been masked in the graph of the original protein by the left one of the larger peaks. In the graph obtained from the contents of the small intestine the main peak resembles the major one found in the labelled origi-

Table 2:RADIOACTIVITY RECOVERED IN THE SUPERNATANT (as % total radioactivity of sample)

Time of sampling	1 hour post administration		24 hours post administration	
Administered compound	$^{109}CdCl_2$	^{109}Cd labelled fungus protein	$^{109}CdCl_2$	^{109}Cd labelled fungus protein
Stomach	22	36	--	14
Small intestine	16	37	80	23
Faeces	--	--	1	0.4

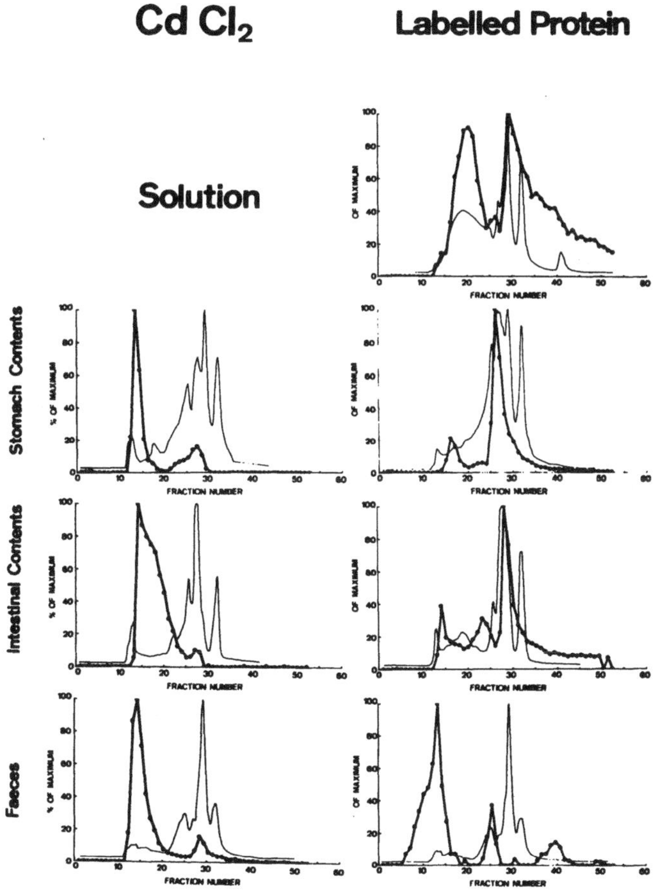

Fig. 1

Chromatographs of radioactive ^{109}Cd (*--*) and UV absorbancy (---) at 280 nm as a percentage of the respective maximum in each of the graphs obtained from solutions of the contents of the stomach, the small intestine and from a faecal pellet of the rat after the administration of labelled $CdCl_2$ (left) and of the labelled extract of the fruit body of the mushroom *Agaricus arvensis* (right).

nal protein while the two minor peaks cannot be identified either in the graph of the original protein or that of the neutralized stomach contents. In the graph obtained from the faecal pellets the most striking feature is the identity of the major peak with that one in the respective graph obtained from the animals which had received the tracer as a chloride.

These findings demonstrate that the binding of Cd in the soluble fraction changes during the passage of the diet through the intestinal tract. In the acid contents of the stomach practically all of it seems to be present in its free ion form irrespective of its original binding. After neutralization soluble Cd binding compounds are formed which were different in the animals where the element has been given as $CdCl_2$ and in those where it has been given bound to the fungus extract. A further minor change which may be due to enzymatic digestion occurs in the small intestine. In the faeces the major part of the element in this fraction appears in the same peak irrespective of the compound administered, but in animals that had received the element bound to the fungus material some of it is still bound to compounds to which it became attached on its way through the intestine.
We know from similar experiments (2) that in the rat only between 0.5 % and 1 % of the administered radioactive $CdCl_2$ is resorbed in the intestine. Most likely this part is derived from the soluble fraction which in itself is decreasing during the intestinal passage of the material while Cd in the solid fraction increases up to 99% of the element in the faeces.

References

1. Friberg, L. 1984. Environ.Health Perspect.Vol. *54,* 1

2. Berg, D. 1984. GSF-Bericht *A 6-84 ISSN 0721-1984* , 61.

3. Shukla, G.S., R.L. Singhal 1984. Can.J.Physiol.Pharmacol. *62,* 1015.

4. Jackl, G.A., W.A. Rambeck, W.E. Kollmer. 1985. Biol.Tr.

Elem.Res. *7*, 74.

5. Revis, N.W., T.R. Osborne. 1984. Environ. Health Perspect. *54*, 83.

6. Rose, H.E., J. Quarterman. 1984. Environ. Res. *35*, 487.

7. Schenkel, H. 1986. Schriften Bundesminister Ernährung, Landwirtschaft, Forsten, Angewandte Wissenschaft, *335*

8. Wagner, G.J., E. Nulty. 1984. J. Toxicol. Environ. Health 13, 979.

9. Diehl, J.F., U. Schlemmer. 1984. Z.Ernährungswiss.*23*, 126.

10. Schellmann, B., M.J. Hilz, O. Opitz. 1980. Z.Lebensm. Unters.Forsch. *171*, 189.

11. Kruse, H., A. Lommel. 1979. Z.Lebensm.Unters.Forsch. *168*, 444.

12. Jackl, G.A., G. Reidel, W.E. Kollmer. 1986. Appl.Radiat. Isot. 38, 431.

TRACE ELEMENT LEVELS and REFERENCE DATA in BODY-FLUIDS and TISSUES

Plenary Lecture given by:

A.M. Reichlmayr - Lais

Chairmen:

U. Weser
P. J. Aggett

Poster Reports:

P. J. Aggett, M. Kirchgessner, J. D. Kruse-Jarres
A.M. Reichlmayr-Lais, U. Weser

LIMITS OF TRACE ELEMENT CONTENTS IN ORGANISM AS PARAMETERS FOR TRACE ELEMENT METABOLISM

A.M. Reichlmayr-Lais, M. Kirchgeßner

Institut für Ernährungsphysiologie, Technische Universität München, D-8050 Freising 12

Abstract

Trace element content in different tissues, especially in blood, plasma, hair and rarely in liver are frequently used as parameters for diagnosing trace element status in practical and scientific work because of the accessibility of sampling and of modern analytical methods. Generally the tissue content of a trace element is influenced by many factors, e.g. homeostasis, homeorhesis, interactions between the elements, age and so on, so that tissue content does not reflect events at cellular sites most sensitive to depletion. But the generalization differs between elements and tissues. For example trace element content in serum does not reflect the status of a trace element very well. Rather anabolic processes lead to a decrease and katabolic processes to an elevation. Thus in gravidity because of anabolism in the case of adequate supply the content of certain elements is decreased in serum and increased in other tissues. Similiar many experiments demonstrated no good correlation between trace element content in hair and status. Altogether in severe deficiency trace element content in serum and also in hair declines clearly whereas a suboptimal status is difficult to diagnose by these parameters.

Because of these uncertainties metabolic and functional parameters are proposed for diagnosing trace element status. Because these parameters are also influenced by many factors it

Trace Element Analytical Chemistry in Medicine and Biology

is a task of scientific experiments to work out defined conditions for practical applications. Altogether it is necessary to find out experimentally adequate and characteristical diagnostical parameters for each element. It is proposed to apply always a package of parameters for one element and to evaluate even the status of other trace elements for differential diagnosis.

Introduction

The modern technique of trace element analysis opens up new possibilities to measure even small concentrations of trace elements in body fluids and body tissues. In addition, many procedures allow simultaneous analysis of several elements in a relatively short time and with relatively small effort. This leads to the possibility that a large amount of data are accumulated which are barely or not at all meaningful when seen critically. We should therefore consider the basic question of the relevance of trace element concentrations in body fluids and tissues. In the following it will be tried to point out some essential limitations, mainly physiological ones, of the meaningfulness of trace element concentrations. Here the serum as being an indicator of the supply will be a main subject.

Concentrations of trace elements in the serum

The status of a trace element is often determined from the concentration in the serum, since this fluid is also easily accessible for practical purposes. Under defined experimental conditions the concentration of a trace element at first seems to be a good parameter, since a measurable reduction of the serum level is measured in the case of significant depletion of an element. In this way it could be observed in an experiment with dairy cows, which had been strongly depleted of zinc, that the zinc content of the serum reacted faster and more strongly to a depletion than the zinc content of milk, blood, bones, and hair (Kirchgeßner and Schwarz 1975). In experiments with rats as well Pallauf and Kirchgeßner (1972) could measure a strong decline in the serum concentration of zinc as early as a few days after feeding of a Zn deficient diet. This is shown in Table 1. This table shows also, however, that the zinc content obviously exhibits a certain rhythmic as the values dropped until the 17th day of depletion, then rose again and later fell again. This rhythmic can cause a wrong estimation of the extent of a depletion, when the values are determined only for one day. This rhythmic can be explained by a mobilization of zinc from the depot as well as by a rhythmical feeding behaviour which can be observed during zinc depletion (eg. Chesters and Quarterman 1970; Williams and Mills 1970) (see section on homeostasis).

The reaction of the serum level of Zn during extreme depletion of Zn could also lead to misinterpretations. As Roth and Kirchgeßner (1979) could demonstrate in rat experiments with an extremely low supply of zinc that was causing considerable clinical symptoms, an even higher zinc content was measured in comparison to higher but nevertheless deficient supplementations of zinc.

The rhythmic changes of the serum Zn level upon depletion as well as the elevated levels in the case of extreme Zn depletion indicate that there is no linear relationship between Zn supply and Zn content in the serum. Therefore it is not possible to correlate the serum contents to a status of supply unambiguously, even in the case of deficient supply.

This is also true for the reactions of the serum level to different supplementations of Zn which are higher than the actual requirements, since there is no unambiguous correlation between repletion and Zn supplementation. In an experiment with dairy cows (Schwarz and Kirchgeßner 1975 a) zinc was depleted followed by repletion with different amounts of zinc. These

Table 1: Zn content of the serum of depleted in comparison to control animals during the course of depletion (Pallauf and Kirchgeßner 1972)

Duration of experiment (days)	Serum zinc level (mg/ml) Depletion	Control
0	2.33	2.33
2	1.63	1.97
4	1.25	2.04
8	1.54	2.25
17	0.90	2.32
25	1.42	2.36
30	1.35	1.75
35	0.88	2.30
42	1.18	1.87

supplementations very quickly led to an increase of the serum zinc content that was related to a certain extent to the amount of zinc supplementation. A supplementation of 22 ppm led to the smallest increase. The highest increase was measured at 54 ppm whereas even higher doses only led to a smaller increase. The highest values were determined as early as after two days of repletion, whereas the zinc content dropped later, obviously to approach a certain value. In this case the decrease was especially pronounced in the case of very high zinc supplementation. The slow rise that was observed in the beginning in the case of very high zinc supplementation is likely to be caused by the regulation of absorption which counteracted a strong increase of zinc level. The fast decrease of serum zinc content after an initial rise can also be explained by the phenomenon of homeostasis and may well signify a rapid deposition of zinc in Zn storage compounds (see section on homeostasis).

Similar correlations are resulting also from the influence of the supply status on the time course of the zinc supply in the serum. In a study on pigs Lantzsch and Berschauer (1983) established that the postprandial plasma level of zinc is inversely related to the supply status. Pigs with a low status of supply of zinc showed a high postprandial increase of the zinc concentration after zinc administration and vice versa. This indicates that there is an increased absorption during a low status of supply. In the reverse case of a high supply status of zinc, the serum zinc level was higher at the time of feeding and decreased rapidly later. Obviously the absorption is strongly inhibited as a consequence of the high status.

The serum zinc level as an indicator of the supply status proved to be problematical especially in the case of gravidity. Numerous results from experiments on animals as well as humans show a decrease of the zinc level in the serum during gravidity (e.g. Kirchgeßner and Schneider 1978, Schneider und Kirchgeßner 1979, Brands et al. 1980). Some experiments that attempted to restore the zinc levels measured before pregnancy by zinc supplementation resulted in abnormalities of the fetuses (Kumar 1976). The decrease of serum zinc concentration during gravidity obviously represents a physiological phenomenon and must not be interpreted as zinc deficiency in every case although zinc deficiency can naturally occur in addition to the physiological decrease. In a study on rats (Schneider and Kirchgeßner 1979) it could namely be demonstrated that the serum zinc level was reduced not only in the case of a deficient supply of zinc but also in the case of sufficient supply. Even a zinc supply that was higher than the requirements led to a decrease of the zinc content of the serum of gravid rats (Kirchgeßner and Schneider 1978).

This decrease of the zinc level can moreover be explained physiologically. This is shown not only by the decrease of the serum zinc level in the case of high zinc supply but also by the increased storage of zinc mainly in the liver which is caused by the anabolic activity during gravidity (Kirchgeßner and Schneider 1978). This relationship is shown in Table 2.

Table 2: Zinc content in the liver and zinc concentration in the serum of gravid and nongravid rats for different zinc supply (Kirchgeßner and Schneider 1978)

Zn concentration in the diet (mg/kg)	Zn content of the liver (mg)		Zn concentration in the serum (mg/ml)	
	gravid	nongravid	gravid	nongravid
20	270 ± 32	183 ± 13	0.74 ± 0.28	1.47 ± 0.12
100	256 ± 27	185 ± 7	1.24 ± 0.34	1.53 ± 0.24

Obviously gravidity induces a shift of the zinc towards strongly retaining organs (anabolism of gravidity) and of course towards organs and products involved in the reproduction (homeorhesis). In general anabolic processes cause an increase in the concentration of a trace element in the tissues whereas the serum level decreases; inversely catabolic processes cause a decrease of the content in the tissues and an increase of the serum level (see Mills 1981).

These examples taken as a whole demonstrate clearly that there are no simple, linear relations between the trace element content of the serum and the status of the same trace element. For the interpretation of a trace element concentration physiological conditions have to be taken into account which are determined by homeostatic and homeorhetic regulation of the trace element metabolism and which are tightly linked to the role of the serum as a distribution pool. Consequently the trace element content of the serum is determined by influx and outflux in relation to requirements and supply, with controlled processes of absorption, storage, mobilization and excretion playing a role. Because of the fundamental importance of homeostasis and homeorhesis these phenomena will be dealt with in detail in a later chapter.

The content of a trace element is influenced not only by these physiologically caused factors but also by further factors under practical conditions. For the element zinc for example, an influence of stress situations, operations, infections, other diseases, drugs and oral contraceptives (see Kirchgeßner et al. 1979) was demonstrated.

Trace element concentration in hair

For a judgement of trace element status frequently hair analyses are suggested, sometimes by commercial companies. On the whole there are no unambiguous results in the literature. Taking zinc as an example, some representative results for the zinc content of hair during depletion shall be demonstrated.

In a model experiment with depleted rats, Pallauf and Kirchgeßner (1973) could not measure an influence of an insufficient zinc supply on the zinc content of hair during a 35 day period of depletion. The authors explained the lack of a measurable effect by saying that the growth of hair had stopped because of the extreme zinc depletion. This is also shown by a lack of growth of the test animals. In this study, even a zinc supplementation in different doses after a previous period of depletion did not lead to a change of the zinc content of hair (Pallauf and Kirchgeßner 1973). On the other hand, a decrease of the zinc content of hair after prolonged depletion was found in studies on cows (Schwarz and Kirchgeßner 1975 b). This was observed since the depletion obviously was not pronounced enough to interfere with the growth of hair. A growth of new hair was observed after shearing and therefore it can be argued that hair deficient in zinc was grown.

Thus the reaction of the zinc in hair to a deficient Zn supply seems to be dependent on the extent of depletion and therefore on the interference with the growth of hair as a consequence of Zn depletion. The only result that is proved unambigously seems to be that a sufficiently long zinc depletion leads to a reduction of the zinc content of hair in different species (e.g. Miller et al. 1965, Groppel and Hennig 1971, Hoekstra et al. 1967). In these experiments the depletion must not be extreme enough to cause by itself interference with the growth of hair.

In order to be able to make a general judgement about the trace element content in hair as a diagnostic parameter, the reaction of the contents to deficient as well as excessive supply should be investigated for each element, since each one of the elements reacts at a different speed and to a different extent to the changes in the status of supply (see Anke 1979). Since, as a rule, the trace element content of hair reacts only very slowly, only long term deficient or excessive states of supply are likely to be reflected.

Apart from nutritive factors, hair colour, type of hair and depth of shearing, performance standard, sex and age (e.g. Miller et al. 1965, Anke 1979) represent further influencing factors. As an example Kauf et al. (1985) could show the influence of age in a study on nursing infants who were tested until the age of a small child. They could detect a postnatal physiological change in the levels of zinc, copper, manganese, and gold which is likely to cover up the dependance of the hair trace indicator on nutritional factors.

Homeostasis

When one wants to judge the value of trace element concentrations as a parameter one has to take into account first that the metabolism of trace elements is controlled by homeostatic regulation (see Kirchgeßner 1987). In this context homeostasis is defined as the attempt of the organism to establish or maintain physiologically necessary and tolerable trace element concentrations in the cells of the body and their compartments in a range of supply that is typical for each trace element. This phenomenon requires regulation of absorption, of intermediary metabolism and of excretion. Here the actual target of homeostasis is in the cell itself whereas the regulation of absorption and excretion may well represent a crude regulation. As a consequence of the aim of homeostasis the absorption rate is increased, storage compounds are mobilized and the excretion is lowered when the supply of an element is deficient. On the other hand an excessive supply can be compensated for by a decrease of absorption, by a deposition of storage compounds as well as by an increase of excretion. In general in vitro as well as in vivo studies show that the absorption of an element is increased in the case of deficient supply.

Weigand and Kirchgeßner (1978) could demonstrate a close correlation between absorption and supply of zinc in an in vivo study, measuring over a broad range of supply. A similar correlation could also be demonstrated in the case of manganese (Weigand et al. 1986). Such correlations may well hold in general for each of the essential trace elements.

An essential homeostatic mechanism in the organism or rather in the cell is the mobilization or deposition of storage compounds. The compensating ability of this mechanism is dependent on the storage capacity and on the ability to mobilize storage compounds. These factors are dependent on the respective element as well as on the individual organs or tissues. For some elements it is known that they are stored preferentially in specific organs like iron in the liver, spleen and bone marrow (see Weinfeld 1970, Lamberti 1970), and copper in the liver (Kirchgeßner and Weser 1965, Graßmann and Kirchgeßner 1973). In these cases the content of the respective elements in specific organs is meaningful as a measure for the status of supply.

Recognizing the presence of these homeostatic regulations which fundamentally influence the utilization of an element it becomes understandable that the absorption rate or the mobilization rate react faster to an insufficient supply than functional compounds. Table 3 shows an example for this phenomenon for the case of iron (Caster and Doster 1979).

Table 3: Effect of different dietary iron intake on hemoglobin concentration of blood, ferritin concentration and iron content of liver (Caster and Doster 1979)

Fe content of the diet	3.5	13	35	117	390	1160
Hemoglobin (%)	6.5	10.8	14.0	14.4	14.2	14.4
Ferritin in liver (ppm)	4.9	5.7	27.9	35.2	54.4	58.2
Total liver iron (ng)	42	76	292	326	442	476

The interplay of homeostatic regulation mechanisms also makes it possible to explain the reaction of the serum to zinc supplementation for different states of supply as well as to different zinc supplementation after previous depletion. That is to say in the case of a higher status of supply or in the case of a higher supplementation a lower amount of zinc is absorbed and a higher amount is transported to the storage organs so that in comparison a lower serum level can be reached in the end than in the case of a lower supply.

Homeorhesis

In addition to homeostasis a further regulatory phenomenon is found during the states of performance of growth, gravidity and lactation which was termed homeorhesis of nutrients by Bauman (1984). This homeorhesis causes a priority of the products and organs of performance to all other organs when the supply is concerned. This could also be shown for some trace elements (see Kirchgeßner et al. 1987). This causes a shift of trace element contents between the different organs and tissues depending on the state of supply.

Kirchgeßner and Spörl (1975 a) could demonstrate a priority of the performance products during gravidity for copper as an example. Gravid rats were provided with different amounts of Cu between insufficient and sufficient supply. When the Cu supply was raised from insufficient to suboptimal consequently an increased Cu incorporation into uterus, fetuses and chorion with amniotic fluid was found that increased no more when the Cu supply was increased further.

During the period of lactation also the transfer of a trace element to the milk has priority over maternal supply. In this way Table 4 illustrates for iron as an example that in a study with rats the Fe level in milk reached a plateau already for a lower Fe supply in relation to the Fe content of maternal liver (Kochanowski and Sherman 1983).

In the case of dairy cows a high priority for the performance product milk could also be demonstrated for zinc as an example (Schwarz and Kirchgeßner 1975 c). The relative zinc excretion with the milk, i.e. the Zn excretion in the milk in relation

Table 4: Fe content in liver and milk of dams on day 20 of lactation dependent on dietary iron content (Kochanowski and Sherman 1983)

Dietary Fe (mg/kg)	35	75	150	300
Liver Fe (mg/gDM)	246 ± 9	481 ± 26	508 ± 58	598 ± 26
Milk Fe (mg/ml)	12.9 ± 0.5	14.5 ± 0.9	14.5 ± 0.9	14.3 ± 0.7

to Zn ingestion, was very high in the case of deficient or suboptimal Zn supply and eventually quickly decreased with rising Zn supply.

This priority of the performance product milk is also guaranteed by the mobilization of maternal trace element reserves. For the element copper Spoerl and Kirchgeßner (1976) could show that the Cu content of whole carcasses of rats after 17 days of lactation was significantly lower than on the day of the litter. Even in the case of a high Cu supply a clear difference could be observed which points out that obviously the reserves accumulated by the anabolism of gravidity are not sufficient for the time of lactation.

The priority of the performance products that is caused by homeorhesis can consequently lead to a shift in the content of a trace element that is dependent on the supply of the respective element. This means that trace element reserves are mobilized in favour of the performance products in the case of suboptimal supply. This also means, however, that trace element contents of performance products react to the status of supply in a delayed way in many cases.

For this reason it is understandable that the Cu content of the milk is reduced only in the case of a pronounced Cu deficiency. According to studies on rats carried out by Kirchgeßner and Spoerl (1975 b) in the suboptimal range of supply, the Cu content of milk could be influenced by the dietary Cu content only in the colostral milk period. On the other hand even for a relatively high Cu supply the Cu content of milk was not increased referring to the fact that the Cu content of milk is obviously strictly controlled.

The Zn content of milk behaves in a similar way as the Cu content. Zn depletion of dairy cows, which led to clinical symptoms, caused a relatively fast decrease of the Zn content of milk, which increased very rapidly upon repletion and decreased somewhat afterwards (Schwarz and Kirchgeßner 1975). This rapid increase caused by repletion argues again in favour of a high priority of the incorporation of Zn into the performance product milk.

Taking into account the duration of lactation the trace element content of the milk seems to be on the whole a relatively good parameter for the diagnosis of a deficiency. For the judgement of the suboptimal range, however, systematic studies are lacking. In the same way studies about possible influencing factors are missing. The trace element content of the milk would be welcome as a parameter for the performance product milk for diagnosis and the derivation of requirements, especially because of the simple sampling procedure.

Anabolism of gravidity

Especially during gravidity a special phenomenon can be observed caused by homeorhetic regulation processes, namely anabolism of gravidity. Anabolism of gravidity, also called superretention, is defined as an incorporation caused by gravidity of a trace element that is higher than the incorporation into the reproduction products. This anabolism of gravidity was examined by Kirchgeßner and Spörl (1975 a) for rats taking copper as an example. Fig. 1 shows the results for the anabolism of gravidity together with the Cu incorporation into the products of reproduction of gravid rats. This study shows furthermore that the anabolism of gravidity is also dependent on the supply. Evidently it cannot be observed in a deficient state of supply of an element.

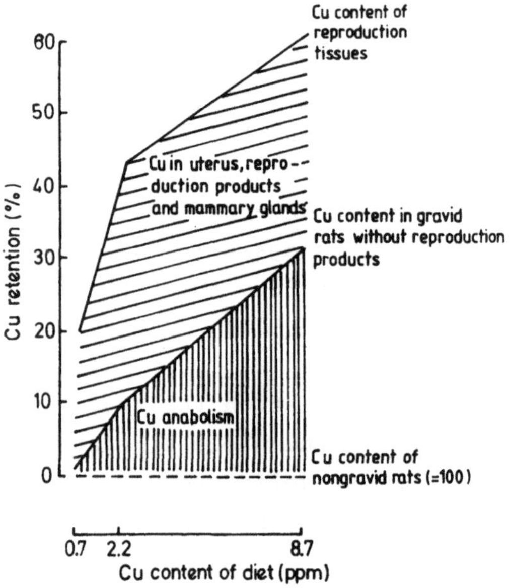

Fig. 1: Cu incorporation of gravid rats in comparison to nongravid rats (Kirchgeßner and Spoerl 1975 a)

For the example of zinc it could be demonstrated that superretention is organ-specific (Schneider and Kirchgeßner 1979). So it could be found out in a study on gravid and nongravid rats that during gravidity the zinc content rose in the liver, however, not in femur and muscle. A somewhat lower content was observed instead in femur and muscle (see Table 5).

Evidently, the anabolism of gravidity causes a shift of a trace element to the deposits. This means that this kind of change in the distribution of an element in the organism, which occurs in the state of gravidity, has to be taken into account for the interpretation of trace element contents.

Table 5: Zn content of liver, femur, and muscle of gravid and nongravid rats (Schneider and Kirchgeßner 1979)

	Liver		Femur		Muscle	
Dietary Zn content	gravid	non-gravid	gravid	non-gravid	gravid	non-gravid
4	207.6[a]	176.1[b]	52.6[a]	60.3[ab]	21.4[a]	22.4[ab]
12	239.6[c]	187.1[ab]	65.3[bd]	78.8[efg]	27.3[cd]	26.8[bcd]
36	275.7[d]	195.4[ab]	72.3[de]	74.9[efg]	21.8[a]	28.6[d]

Interactions

It is shown by a wealth of studies in the field of trace elements that trace elements are linked by numerous metabolic interactions so that the metabolism of an element cannot be viewed in isolation (see Kirchgeßner et al. 1981, 1982). Some of these interactions can be explained by a similar electron configuration so that different elements compete for binding sites (Hill and Matrone 1970). There are, however, numerous interactions between elements which cannot be explained electrochemically. In general the mechanisms of interaction are not well understood. What is more, only consequences of such interactions are known and measurable, mainly in extreme states of supply like deficient or toxic supply of an element. Frequently, such consequences are changes in the trace element content of specific organs and tissues as well as changes in the distribution of a trace element in the organism.

As a suitable example showing the problems of trace element interactions the relations between iron and copper shall be demonstrated. Cu deficiency results in anemia in conjunction with an increase of the Fe content in the liver and spleen, whereas Cu content in liver decreases (e.g. Sourkes et al. 1968, Graßmann and Kirchgeßner 1973). Not only in the case of Cu depletion but in a more general way there are strong interactions between the elements copper and iron, which influence the contents of both elements in the different organs of rats (Graßmann et al., 1983).

Another example is represented by the interaction of nickel with iron and manganese (Kirchgeßner et al. 1984). In a study on rats which were supplemented with an increasing nickel concentration between 0.06 and 600 ppm in the diet, an increased Fe concentration in different body tissues could be measured if the nickel concentration was 10 ppm or higher. The manganese concentration exhibited the opposite effect. It was remarkable that the iron and manganese concentrations adjusted themselves to new levels abruptly when the dietary concentration of nickel had reached a certain level. This points to the fact that interactions between trace elements always take place when there is a change of certain relations between the trace elements. In a further experiment on rats in which Ni and Zn supply varied at the same time it could furthermore be demonstrated that the increase of Fe concentration in the liver for example did no longer occur when higher Zn contents were present in the diet (Kirchgeßner et al. 1985).

This last example especially shows the close interrelation of the metabolism of many trace elements, which make it nearly impossible to interpret the content of a trace element in an organ or tissue. These numerous interactions therefore call for a simultaneous evaluation of the status of several elements in order to judge the status of one trace element.

Metabolic and functional parameters

After having illustrated the basically physiologically determined limitations of a meaningful use of trace element concentrations in tissues and body fluids as diagnostic parameters in research and for practical applications, alternative parameters will be mentioned briefly.

Because of the tendency towards homeostasis in the trace element metabolism mainly metabolic parameters seem to be useful (Reichlmayr-Lais and Kirchgeßner 1985, Kirchgeßner et al. 1985, 1986), because they react very quickly to changes in the supply i.e. the status of supply. Absorption, excretion, and mobilization rates as well as binding capacities of transport compounds are understood as metabolic parameters in this context. Since all the metabolic parameters are influenced in multiple ways (e.g. net requirements, age, time of feeding, interaction with other ingredients of the diet), standardized conditions must be established for the parameters in order to be able to use them under practical conditions. These parameters are of especial importance mainly for trace elements which have been proved to be essential but whose function has not yet been elucidated.

Apart from the metabolic parameters so-called functional parameters are suited for diagnosis in many cases. Such functional compounds are mainly activities of metalloenzymes. Here it has to be taken into account that metalloenzymes do not react on depletion of a certain element in every case. Furthermore tissue-specific differences in the reaction can be observed (see Kirchgeßner et al. 1976). Therefore the respective functional compounds of each element have to be tested for their suitability. In the case of the element zinc alkaline phosphatase seems to be a functional compound which is especially suited, since the enzyme is reduced in bone as well as in mucosa, kidney, and serum upon depletion of zinc (Roth and Kirchgeßner 1974). On the other hand the activity of alkaline phosphatase, like generally the activity of other functional compounds, is also influenced by a number of factors. As an example the activity of alkaline phosphatase is dependent on the dietary protein content. Graßmann et al. (1978) demonstrate a decrease of activity for a low as well as for a high protein supply. Because of the numerous influencing factors, Roth and Kirchgeßner (1979) have worked out a so-called response technique for the rat system as an example. In this method the activity of alkaline phosphatase in the serum is measured before and after a Zn-injection. As is shown in Table 6, in the case of a previous insufficient Zn supply the activity increased after the injection and evi-dently reached a plateau. This response technique should also be tested for other trace elements with suitable metalloen-zymes. In many cases one might possibly get around the problem of nominal versus standard values.

Table 6: Activity of the alkaline phosphatase in serum of rat dependent on Zn supply before and three days after Zn injection (0.8 mg/Zn) (Roth and Kirchgeßner 1979)

Dietary Zn content (ppm)	Alkaline Phosphatase (mU/ml) before Zn injection	Alkaline Phosphatase (mU/ml) after Zn injection	Difference (mU/ml)	Relative activity increase (%)
1,3	23 ± 8	152 ± 36	129	560
4	39 ± 10	162 ± 35	123	315
6	88 ± 41	195 ± 61	107	122
8	116 ± 19	185 ± 72	69	59
10	139 ± 29	193 ± 33	54	39
12	183 ± 44	245 ± 62	62	34
20	175 ± 39	249 ± 66	74	42
100	239 ± 27	246 ± 50	7	3

References

1. Anke, M.. 1965. Arch. Tierernährg. 15, 469.
2. Anke, M., M. Risch. 1979. Haaranalyse und Spurenelementstatus, VEB G. Fischer Verlag, Jena.
3. Bauman, D.E. 1984. In: Herbivore nutrition in the subtropic and tropics (eds. Gilchrist et al.). The Science Press (PTY) Ltd. Craighill, South Africa, p. 505.
4. Brandes, J.M., A. Lightman, J. Itskowitz, O. Zinder. 1980. Biol. Neonate 38, 66.
5. Caster, W.O., J.M. Doster. 1979. Internat. J. Vit. Nutr. Res. 49, 215.
6. Chesters, J.K., J. Quarterman. 1970. Effects of zinc deficiency on food intake and feeding patterns of rats. Brit. J. Nutr. 24, 1061.
7. Graßmann, E., M. Kirchgeßner. 1973. Arch. Tierernährg. 23, 261.
8. Graßmann, E., S. v. Krziwanek, M. Kirchgeßner. 1978. Arch. Tierernährg. 28, 451.
9. Graßmann, E., M. Kirchgeßner, J. J. Kim. 1983. Z. Ernährungswiss. 22, 124.
10. Groppel, B., A. Henning. 1971. Arch. Exper. Vet. Med. 25, 817.
11. Hill, C.H., G. Matrone. 1970. Fed. Proc. 29, 14.
12. Hoekstra, W.G., E.C. Faltin, C.W. Lin, H.F. Roberts, R.H. Grummer. 1967. J. Anim. Sci. 26, 1348.
13. Kauf, E., W. Wiesner, S. Niese. 1985. In: Mengen- und Spurenelemente, Arbeitstagung 1985 (eds. Anke et al.). Karl-Marx-Univ. Leipzig, p. 78.
14. Kirchgeßner, M., U. Weser. 1965. Z. Tierphysiol., Tierernährg. u. Futtermittelkde. 20, 261.
15. Kirchgeßner M., W.A.Schwarz. 1975. Zbl.Vet.Med.A. 22, 572.
16. Kirchgeßner, M., R. Spoerl. 1975 a. Z. Tierphysiol., Tierernährg. u. Futtermittelkde. 36, 75.
17. Kirchgeßner, M., R. Spoerl. 1975 b. Arch. Tierernährg. 25, 505.

18. Kirchgeßner, M., U.A. Schneider. 1978. Arch. Tierernährg. 28, 211.

19. Kirchgeßner, M., F.J. Schwarz, E. Graßmann, H.-P. Roth, A. Schnegg. 1979. In: Spurenelemente-Analytik, Umsatz, Bedarf, Mangel und Toxikologie (eds. Gladtke et al.). Thieme Verlag, Stuttgart, p. 68.

20. Kirchgeßner, M., A.M. Reichlmayr-Lais, F.J. Schwarz. 1981. In: Nutrition in Health and Disease and International Development: XII Internat. Congr. Nutr., Alan R. Liss., Inc., New York, p. 189.

21. Kirchgeßner, M., F.J. Schwarz, A. Schnegg. 1982. In: Clininical, biochemical and nutritional aspects of trace elements (ed. A.S. Prasad), Alan Liss, New York, p. 478.

22. Kirchgeßner, M., R. Maier, A.M. Reichlmayr-Lais. 1984. Z. Tierphysiol.,Tierernährg.u. Futtermittelkde. 52, 217.

23. Kirchgeßner, M., A.M. Reichlmayr-Lais, A.K. Mathur. 1985. Z.Tierphysiol.,Tierernährg.u.Futtermittelkde.53, 214.

24. Kirchgeßner, M., A.M. Reichlmayr-Lais. 1985. In: Trace elements in man and animals-5 (eds. Mills et al.). Commonwealth Agricultural Bureaux, p. 884.

25. Kirchgeßner, M., A.M. Reichlmayr-Lais, E. Weigand. 1986. In: Trace element - Analytical chemistry in medicine and biology. Vol. 4 (eds. P. Brätter, P. Schramel), de Gruyter & Co., Berlin, New York, p. 3.

26. Kirchgeßner, M.. 1987. Übers. Tierernährg. 15, 153.

27. Kochanowski, B.A., A.R. Sherman. 1983. Br. J.Nutr. 49, 51.

28. Kumar, S. 1976. Nutr. Rep. int. 13, 33.

29. Lamberti, G.. 1970. Hippokrates 41, 483.

30. Lantsch, H.-J., Berschauer, F. 1983. Ann. Nutr. Metab. 27, 228.

31. Miller, W.J., G.W. Powell, W.J. Pitts, H.F. Perkins. 1965. J. Dairy Sci. 48, 1091.

32. Mills, C.F.. 1981. Phil. Trans. R. Soc. Lond. B. 294, 199.

33. Pallauf, J., M. Kirchgeßner. 1972. Zbl. Vet. Med. A. 19, 594.

34. Pallauf, J., M. Kirchgeßner. 1973. Zbl. Vet. Med. A. 20, 100.
35. Roth, H.-P., M. Kirchgeßner. 1979. Res. Exp. Med. (Berl.) 174, 283.
36. Schneider, U.A., M. Kirchgeßner. 1979. Nutr. Metab. 23, 241.
37. Schwarz, W.A., M. Kirchgeßner. 1975 a. Z. Tierphysiol., Tierernährg. u. Futtermittelkde. 34, 289.
38. Schwarz, W.A., M. Kirchgeßner. 1975 b. Deutsche Tierärztl. Wochenschr. 82, 141.
39. Schwarz, W.A., M. Kirchgeßner. 1975 c. Arch. Tierernährg. 25, 597.
40. Schwarz, W. A., M. Kirchgeßner. 1975 d. Z. Tierphysiol., Tierernährg. u. Futtermittelkde. 35, 1.
41. Sourkes, T.L., K. Lloyd, H. Birnbaum. 1968. Can. J. Biochem. 46, 267.
42. Spoerl, R., M. Kirchgeßner. 1976. Arch. Tierernährg. 26, 25.
43. Weigand, E., M. Kirchgeßner. 1978. In: Trace elements in man and animals-3 (ed. M. Kirchgeßner). ATW Freising-Weihenstephan, p. 106.
44. Weigand, E., M. Kirchgeßner, U. Helbig. 1986. Biol. Trace Elem. Res. 1, 24.
45. Weinfeld, A.. 1970. In: Iron deficiency. Pathogenesis, clinical aspects, therapy (eds. Hallberg, et al.). Acacemic Press, New York, p. 329.
46. Williams, R.B., C.F. Mills. 1970. The experimental production of zinc deficiency in the rat. Br. J. Nutr. 24, 989.

INVESTIGATION OF NATURAL CHROMIUM LEVELS IN HUMAN BLOOD

Robert R. Greenberg and Rolf Zeisler
Center for Analytical Chemistry, National Bureau of Standards
Gaithersburg, Maryland 20899, USA

Saul Malozowski
National Institute of Child Health and Human Development,
National Institutes of Health, Bethesda, Maryland 20892, USA

Abstract

Chromium is one of the most difficult elements to determine accurately at the naturally occurring, ultratrace levels normally found in uncontaminated samples of whole blood and blood fractions. In view of the importance of this element, both as an essential and a toxic element, a newly developed radiochemical procedure for Cr determination using neutron activation analysis has been applied to samples of whole human blood, blood serum, and packed blood cells collected under clean conditions. Typical Cr concentrations observed in human, whole-blood samples ranged from 0.04-0.3 μg/kg.

Introduction

Chromium was recognized as one of the elements essential for humans more than twenty five years ago (1). However, at elevated concentrations, some forms of Cr are toxic (2) and possibly even carcinogenic (3). In recent years, several analytical techniques have demonstrated the capability to accurately determine chromium at the naturally occurring, ultratrace levels normally found in uncontaminated samples of human serum. A variety of problems, however, have limited

Trace Element Analytical Chemistry in Medicine and Biology

the determination of Cr to this single fraction of human blood. In view of the importance of this element, more detailed investigations of its occurrence and role at natural levels are needed, requiring even more sensitive analytical methods.

A reliable radiochemical procedure for Cr has been developed (4) based upon neutron activation followed by a radiochemical separation. The separation procedure involves the liquid/liquid extraction of Cr(VI) into a 5% solution of tribenzylamine/chloroform, back-extraction into an aqueous solution (2 mol/L) of sodium hydroxide and precipitation as barium chromate. Radiochemical neutron activation analysis (RNAA) is one of the best analytical techniques currently available for trace and ultratrace level determination of Cr in biological materials. It possesses the sensitivity required to determine sub-nanogram amounts of Cr, and the nearly blank-free nature of the technique allows chemical dissolution and separation of Cr from the sample matrix (after irradiation) without regard to reagent blank. In addition, NAA is free from chemical interferences, and most types of matrix effects which can interfere with other, non-nuclear, analytical techniques. However, the nuclear interference from the fast neutron reaction Fe-54(n,α)Cr-51, can be significant. This interference depends upon the fast flux of the irradiation facility, and upon the relative amounts of Cr and Fe in the material. In a typical light water reactor, the fast neutron reaction with one gram of Fe would produce approximately the same amount of Cr-51 as would the thermal neutron reaction on 11 μg of Cr (5). This interference can be minimized, however, by using a highly thermalized facility such as the RT-4 facility of the NBS Research Reactor (6). Even in matrices with Fe concentrations more than 100,000 times greater than that of Cr, such as whole blood, the interference in RT-4 is smaller than the directly produced Cr-51 activity, and thus can be accurately subtracted.

This newly-developed, radiochemical procedure has been used to determine the Cr concentrations in whole blood, serum, and even two packed blood cell samples collected, under clean conditions, from nine individuals.

Experimental

Reagents and Apparatus - The FEP Teflon catheter needles (B-D I.V.CATH) and plastic syringes (B-D PLASTIPAK) used for blood collection were obtained from Becton, Dickinson and Co. Quartz vials used for sample collection and subsequent irradiation were Suprasil AN. All reagents used in chemically processing the samples were ACS reagent grade. The dissolution vessels were machined from solid blocks of TFE Teflon, and had internal volumes of approximately 125 mL. The filtration apparatus was obtained from Millipore, and the liquid reservoir on top of the filter could hold approximately 15 mL of solution.

Sample Collection - The combined Teflon catheter/needle was attached to a syringe and inserted into an arm vein. The needle was immediately withdrawn, leaving the Teflon catheter in place for sampling. Several mL of blood were allowed to flow through the catheter and were discarded, to minimize transfer of contaminants from the needle to the sample. After this rinse, several mL of blood were withdrawn through the catheter and immediately portioned into two preweighed quartz vials, each containing approximately 1 mL of sample. The quartz vials were stoppered with Teflon plugs and individually placed into plastic snap-cap vials. The vials were placed in a plastic box for transport. Similarly, for serum and packed-cell analyses, 10 mL of blood was collected into acid-cleaned, TFE Teflon screwcap vials. The blood was allowed to clot, and the vials were centrifuged. Samples containing approximately 1 mL of serum were decanted into cleaned quartz vials. The blood clots were then pulverized with Teflon spatulas, and packed blood cells were then transferred directly to quartz vials. All samples were lyophilized prior to irradiation.

Irradiations - The samples were irradiated for 8 or 16 hours in the RT-4 facility of the 20 MW NBS Research Reactor. The thermal neutron flux in this pneumatic tube facility is approximately 2.7 x 10^{17} n/m^2s, and the thermal-to-fast neutron ratio is approximately 1500 (6).

Radiochemical Separation Procedure - Full details of the procedure are given elsewhere (4). An outline of the procedure is as follows: addition of 1 mg of Cr carrier; sample dissolution using nitric, sulfuric and perchloric acids; oxidation with potassium permanganate; extraction from a dilute HCl solution into a 5% solution of tribenzylamine (TBA)/chloroform; back-extraction into an aqueous solution of sodium hydroxide, and finally precipitation and filtration (at pH = 6.5) of barium chromate. This procedure is nearly quantitative, and highly reproducible (98.1 ± 0.5%).

Results and Discussion

The determination by NAA of small amounts of Cr in the presence of much larger quantities of Fe is complicated by the (n,alpha) fast neutron interference. This Fe interference was first calculated for the RT-4 facility based upon the measured production of Al-28 from Si-28 via the (n,p) reaction (7). This calculation indicated that 1 gram of Fe would appear to contain 82 ng of Cr when irradiated in this facility. This interference was then measured directly using a small piece of extremely pure Fe metal. The measured Cr-51 was equivalent to 87 ng of Cr per gram of Fe. This results in an average interference (blank) of 0.039 ng of Cr per gram of whole blood for men, and 0.035 μg/kg for women. The average interference for Cr in packed blood cells is equivalent to 0.082 μg/kg.

As a check, two certified reference materials (CRMs) were analyzed along with the samples. Unfortunately, very few materials were available with certified Cr concentrations at the μg/kg level. The CRM with a certified Cr concentration which most closely approximated blood and blood fractions was the NBS Milk Powder Standard Reference Material (SRM 1549), which has a certified Cr concentration of 2.6 ± 0.7 μg/kg on a dry weight basis. Since human whole-blood was found to have Cr concentrations between 0.04 and 0.3 μg/kg on a wet weight basis, and since blood typically contains about 90% moisture, the Milk Powder is a reasonably good match, except for the difference in Fe concentrations. Four samples of the Milk Powder were analyzed for Cr and the results obtained, 2.65 ± 0.17 μg/kg, agree well with the NBS certified value of 2.6 ± 0.7 μg/kg. In addition, three samples of the original NBS Bovine Liver (SRM 1577), were analyzed, and the results obtained, 80 ± 2 μg/kg are in good agreement with the NBS certified value of 88 ± 12 μg/kg.

Chromium concentrations were determined in whole-blood samples from nine individuals, and the results obtained are listed in Table 1. Serum samples have been analyzed for seven of these individuals, and packed cells for two. The uncertainties listed in Table 1 represent 1 sigma counting statistics and are typically about 10% of the Cr concentration. Duplicate samples from a single collection have been run for the whole blood from one person and for the serum from a second. Both duplicate values are consistent with the observed counting statistics. In addition, blood from these

Table 1. Chromium concentrations ($\mu g/kg$) determined in human whole-blood, blood-serum, and packed cells (corrected for Fe interference)

Subject	Sample Date	Whole Blood	Blood Serum	Packed Cells
B1 Male	3/86	0.159±0.023	0.102±0.012	
B1 Male	3/86	0.160±0.016		
B1 Male	9/87	0.226±0.027	0.193±0.021	
B2 Female	3/86	0.273±0.019	0.378±0.024	
B2 Female	3/86	0.445±0.044		
B2 Female	9/87	0.292±0.043	0.454±0.044	
M1 Female	3/86	0.148±0.019	0.117±0.015	
T1 Male	3/86	0.340±0.018		
D1 Male	3/86	0.150±0.016	0.182±0.027	
W1 Male	3/86	0.119±0.025		
J1 Male	9/87	0.280±0.036	0.258±0.038	0.185±0.027
R1 Male	9/87	0.038±0.014	0.077±0.010	
T2 Female	9/87	1.286±0.048	1.743±0.042	0.765±0.058

Notes:
1. Concentrations are in $\mu g/kg$, on a fresh weight basis.
2. Uncertainties are 1 sigma counting statistics.
3. Fe interferences applied for Cr:
   Whole blood (men) - 0.039 $\mu g/kg$
   Whole blood (women) - 0.035 $\mu g/kg$
   Packed cells (both) - 0.082 $\mu g/kg$

two individuals has been sampled on two occasions more than one year apart. No differences in the Cr concentrations, above the analytical uncertainties, can be seen for subject B2 with time. A small, but real, increase in both the whole blood and serum Cr levels from B1 was observed. It should be noted that between the 1986 and 1987 sample collections, individual B1 was taking a daily dietary supplement containing 200 μg of Cr in the form of an amino-acid chelate, while no supplement was taken prior to the 1986 collection. This factor of five increase in Cr intake, compared to the average

American dietary intake of approximately 40 µg/day (8), resulted in a Cr increase of approximately 40% in the blood, and 90% in the serum. Since serum accounts for approximately 50%, by weight, of whole blood in men, there appears to be no change in the Cr concentration in the packed cells of this individual. Thus the large increase in dietary Cr had only a relatively small effect, elevating the serum level for B1 from the second lowest concentration observed in this study, to the median value. Thus, at least for this individual, Cr levels in the blood may be under some type of bioregulatory mechanism.

An interesting observation can be made concerning individual T2. This person had concentrations of Cr in both whole blood and serum which were elevated by factors of 5-10 compared to the other individuals in Table 1. Subsequent questioning revealed that this person has been diagnosed as a borderline diabetic, and has a family history of diabetes (mother). In view of the possible link between Cr and glucose metabolism (9), further studies of the Cr concentrations in whole blood and blood fractions from diabetics may be indicated. However, in view of the difficulty of accurate Cr determinations, at the naturally occurring levels in these materials, carefully validated analytical methods, and rigorous quality assurance procedures should be used for these studies.

Conclusions

The accurate determination of Cr at the naturally occurring levels in biological materials is still a challenge for analytical chemistry. It has been shown that even for a relatively interference-free analytical technique, such as NAA, a chemist has to follow carefully validated procedures to obtain accurate results. The reliability of Cr determinations is well known to be adversely affected by contamination from implements, containers and the environment. Errors in this area can be minimized by careful sample collection, storage and preparation procedures (10, 11). This work has demonstrated that if all sources of systematic error are carefully controlled, RNAA can be used to determine Cr at the sub-ppb level in even the most difficult biological matrices, such as whole blood and packed blood cells. Some of the results of this study point out the need for further investigations of the importance, at ultra-trace levels, of this element in biological tissues and fluids.

Certain commercial equipment, instruments, or materials are identified in this paper in order to adequately specify the experimental procedure. Such identification does not imply recommendation or endorsement by the National Bureau of Standards, nor does it imply that the materials or equipment identified are necessarily the best available for the purpose.

References

1. W. Mertz, Physiol. Rev. 49 (1969) 163-167.

2. U. S. EPA, Quality Criteria for Water, Washington, D. C. (1976).

3. J. C. Arcos, Amer.Laboratory 10 (1978) 29-41.

4. R. R. Greenberg and R. Zeisler, J. Radioanl. Chem. - in press.

5. R. Cornelis, J. Hoste, and J. Versieck, Talanta 29 (1982) 1029-34.

6. D. A. Becker, J. Radioanal. Nucl. Chem 112 (1987) 393-401.

7. R. M. Lindstrom, and R. F. Fleming, Proc. 3rd Int'l. Conf. on Nucl. Meth. in Environ. and Energy Res., Columbia, MO, 1977, 90-98.

8. G. V. Iyengar, Sci. Total Environ. 61 (1987) 235-252.

9. R. A. Anderson, Clin. Physiol. Biochem. 4 (1986) 31-41.

10. J. J. Versieck, J. Res. Nat. Bur. Stand. (U.S.) 91 (1986) 87-92.

11. R. Zeisler, J. Res. Nat. Bur. Stand. (U. S.) 91 (1986) 75-85.

TRACE ELEMENTS IN HUMAN CLINICAL SPECIMENS: EVALUATION OF LITERATURE DATA TO IDENTIFY REFERENCE VALUES.

J.R.W. Woittiez

Netherlands Energy Research Foundation ECN, P.O. Box 1 1755 ZG Petten, (N.H.), The Netherlands

G.V. Iyengar

National Bureau of Standards and U.S. Department of Agriculture Gaithersburg, MD 20899, U.S.A.

Abstract

Reference values are proposed for the concentrations of As, Cd, Co, Cr, Cu, Fe, Hg, Mn, Mo, Pb, Se and Zn in whole blood, blood serum, urine, milk, liver and hair from adult human subjects. For F, I and Ni it was not possible to evaluate reference intervals for all the specimens mentioned above. For several other elements, including Al, B, Br, Cs, Li, Rb, U and V results from selected investigations are listed for information. The major source of information for this evaluation was obtained by requesting a number of analysts, aware of trace element analytical problems, to identify reliable data for clinical specimens for their country. This information was complemented with recent literature data for a number of elements including Al, B, Li, and V. The elements Cu, Fe and Zn appear to show symmetrical distributions in most tissues. Elements like Se and As are well documented, but their contents are influenced by dietary intake. Other elements, such as Cd, Pb and Hg are influenced by environmental factors. Reference values for some tissues could be identified for "difficult" elements like Cr, Mn or Mo. A limited amount of data is available for elements like Al, V, B, Li, and U. Therefore, more investigations are needed for these elements before an evaluation can be attempted. A review article on this subject will be published in the March 1988 issue of Clinical Chemistry.

Trace Element Analytical Chemistry in Medicine and Biology

Introduction

Well-founded reference values or baseline data for trace elements in clinical specimens are needed to interpret the results generated in clinical chemistry laboratories. By definition, reference values reflect the findings in a well defined group of individuals. Further, it is expected that reference values are a reflection of natural levels. It is also recognized that age, sex, living environment, diet etc., influence the concentration levels of certain trace elements.
A comprehensive compilation of literature values was published by Iyengar et al. in 1987 [1]. In 1983, Woittiez exclusively investigated the problem of establishing reference values for 28 elements in human serum [2]. In 1984, Versieck high-lighted the analytical problems of sampling and contamination associated with trace element determinations in human body fluids [3]. In the last few years, there has been further progress in the understanding of the problems of ultra-trace element analytical chemistry of clinical specimens. Therefore, in our opinion the time is ripe to re-evaluate literature data for reference purposes.

Data Base and Statistical Treatment

With a good insight into the merits and demerits of a given analytical technique and available instrumentation, a knowledgeable analyst may be expected to generate reliable analytical results. This concept was followed by requesting a number of analysts, known to be aware of the problems of trace element analytical chemistry, to identify reliable data for selected clinical specimens from their countries. It was emphasized that the analytical quality of the result was crucial. Thus, a total of 1153 mean or median values for 15 elements in 7 human tissues were received. Over 60 investigators reported these results which covered nearly 100 000 individuals. In all, data have been obtained from 55 countries. The database has been compiled in the form of a comprehensive report [4]. It consists of a reported mean result with a standard deviation or a range or median value, the number of individuals sampled, their sex, the country and place of residence as well as the analytical methods used.

Whenever a median or mean value was reported, it has been used in statistical calculations. Values reported as ranges or upper limits are not included in the evaluation. The statistical treatment includes calculation of the overall median with standard deviation, the overall weighted mean with internal and external standard deviation and the geometric mean and standard deviation. For Fe, Cu and Zn these distribution parameters are presented, for the other elements only the median and range are reported.

Results

Elements in the database

The responding analysts gave data on As, Cd, Co, Cr, Cu, F, Fe, Hg, I, Mn, Mo, Ni, Pb, Se and Zn in whole blood, blood serum, urine, milk, liver, hair and erythrocytes. Results for the overall median value, the range and the number of reported sets are given in Tables 1, 2 and 3. The tissue analysed most are hair and whole blood, while Zn, Cu and Pb are the most determined elements. The combined amount of F, Mo, I, Ni and Co results is just over 10% of the total of 1153 reported mean values. Data for erythrocytes and urine are below 10%. The combination of knowledge of analytical difficulties, often linked to low elemental concentrations and sampling [5,6], and the reported number of results per element, may lead to a classification in four groups.

Copper, iron and zinc.

In most tissues and body fluids, Cu, Fe and Zn occur on the mg/kg or mg/L level. Therefore, there are no major analytical problems in the determination of these elements. Many factors, such as stress, are known to influence Zn in serum levels. Zn in milk and Zn in hair reflect nutritional, environmental and genetic factors.

Lead, selenium and iodine

With the exception of lead in blood serum, in most other tissues and body fluids, these elements occur above the 10 µg/kg or 10 µL level.
Pb in serum is less than 1 µg/L. Although there is evidence to suggest

Table 1. Reference values for trace elements in whole blood and blood serum. N = number of sets of results.

	Whole blood (µg/L)			Serum (µg/L)		
	Median	Range	N	Median	Range	N
As	5	2-23	8	3.50	1.7-15.4	5
Cd(*)	1.0	0.3-7.0	53	NV	NV	-
Cr	NV	2.8-45	5	0.19	0.12-2.1	8
Co	NV	NV	-	0.29	0.11-0.45	4
Cu	960	800-1300	16	1100	800-1750	36
F	NV	NV	-	NV	NV	-
I	40	NV	1	60	58-73	3
Fe(**)	445	309-521	13	1060	750-1500	24
Pb	123	8-269	95	NV	NV	-
Mn	13.6	8.0-18.7	8	0.63	0.54-1.76	9
Hg	NV	0.6-59	17	NV	2.2-5.8	5
Mo	1.20	0.8-3.3	3	1.78	0.55-3.00	2
Ni	NV	4.5-28.0	5	NV	2.6;7.5	2
Se	105	58-234	20	96	46-143	16
Zn	6400	4400-8600	19	930	700-120	36

(*) Values for Cd in whole blood are for non-smokers

(**) Values for Fe in whole blood are in mg/L

NV No value to be proposed

Table 2. Reference values for trace elements in milk and liver. N = number of sets of results.

	Milk (µg/L)			Liver (µg/kg)		
	Median	Range	N	Median	Range	N
As	1.25	0.25-24	10	12	5-53	11
Cd	1.3	0.7-4.6	6	1770	300-4100	26
Cr	1.4	0.4-5.1	7	NV	8-72	8
Co	0.5	0.2-3.0	11	30	6-151	7
Cu	290	197-751	28	6500	3200-9900	14
F	12.3	7.0-16.8	9	NV	NV	-
I	68	41-168	8	200	NV	1
Fe(*)	477	202-1710	19	198	46-307	16
Pb	NV	3.6-30	10	500	370-2300	7
Mn	12	3.2-42	15	1200	500-2100	13
Hg	NV	0.2-13.0	12	140	33-490	10
Mo	2.7	1.0-19.5	7	570	360-800	7
Ni	15.3	1.5-39	10	39	9-78	4
Se	18.5	10-62	24	400	190-700	18
Zn	1680	750-4000	25	5630	3230-7000	36

(*) Values for Fe in liver are in mg/kg

NV No value to be proposed

Table 3. Reference values for trace elements in hair and urine.

N = number of results.

	Hair (µg/kg)			Urine (µg/L)		
	Median	Range	N	Median	Range	N
As	260	85-750	17	20	10;30	2
Cd	1150	350-2430	22	0.80	0.5-4.70	11
Cr	460	60-4100	20	0.4	0.24-1.8	7
Co	77	0.4-500	13	NV	1.20-1.35	2
Cu(*)	16	6.8-39	34	38	12-80	5
F	NV	NV	=	NV	NV	=
I	600	270-4200	12	NV	NV	=
Fe(*)	33	13-1710	24	129	1.2-600	4
Pb(*)	11	4.2-52	16	11	6.3-13.0	4
Mn	1200	200-4400	21	0.6	0.5-9.80	4
Hg	3250	500-12200	24	4.30	0.1-2.0	7
Mo	380	100-490	3	33	33;34	2
Ni	44	2.1-1250	7	2.50	2.2;2.7	2
Se	530	200-1400	24	40	7-160	8
Zn(*)	175	124-320	50	449	361-599	6

(*) Values for Cu, Fe, Pb and Zn in hair are in mg/kg

NV No value to be proposed

Table 4. Information values for several trace elements.

Al in whole blood	: 2 - 7.5	µg/L
Al in blood serum	: 1 - 5	µg/L
Al in bone	: 3.6 ± 1.5	mg/kg
Al in liver	: 0.3 - 2.0	mg/kg
Al in hair	: 3 - 10	mg/kg
Al in urine	: 13 - 110	µg/24 hr
B in whole blood	: 31 ± 5.6	µg/kg
B in blood serum	: 22 ± 5	µg/kg
B in erythrocytes	: 2.6 ± 2	µg/kg
Br in blood serum	: 4.5 ± 1	mg/L
Br in hair	: 2 - 10	mg/kg
Br in liver	: 1 - 2	mg/kg
Br in milk	: 1 - 2	mg/L
Br in urine	: 5	mg/24 hr
Cs in blood cerum	: 1	µg/L
Cs in erythrocytes	: 5 - 10	µg/kg
Cs in liver	: 10 - 20	µg/kg
Cs in hair	: 0.1 - 1.0	mg/kg
Cs in urine	: 8 - 20	µg/kg
Li in whole serum	: 0.58 ± 0.06	µg/kg
Li in plasma	: 0.3 ± 0.01	µg/kg
Li in erythrocytes	: 0.12 ± 0.02	µg/kg
Rb in blood serum	: 200	µg/L
Rb in liver	: 4 - 6.5	mg/kg
Rb in hair	: 1 - 2	mg/kg
Rb in urine	: 1 - 3	mg/24 hr
U in liver	: 0.2 ± 0.3	µg/L
U in urine	: 3.6 ± 0.9	ng/L
V in blood serum	: < 0.1	µg/L
V in liver	: 5 - 10	µg/kg
V in hair	: 90 - 160	µg/kg

global variations on Pb in milk, liver and hair, it cannot be ruled out that analytical errors are involved. Selenium is one of the few elements which occurs on the µg/kg level, yet it is relatively easy to control contamination. Selenium levels in whole blood or serum reflect geographical differences and age. The concentration of I in body compartments is influenced by the dietary status of the element. Thus, to interpret results, food intake information should be known.

Arsenic, cadmium, mercury, cobalt and nickel

Generally, these elements occur at the µg/kg or µg/L level in most tissues and body fluids. Therefore, their determination mostly presents a considerable challenge. The levels of As are influenced by regional differences and are subject to short-term effects of intake. Cd is fairly well documented in whole blood and kidney, but moderately in most other specimens. Smoking influences blood Cd-levels. Cd in serum is in the order of 0.1 µg/L. The dietary intake of Hg is reflected in whole blood. Hg concentrations in serum may be as low as 1 µg/L. Accumulation in hair is a way of Hg excretion. Cobalt concentrations in body fluids are well below the µg/L level. As liver is a store for vitamin B12, Co in liver levels may reflect differences in food intake. Values for Ni mainly refer to milk, liver and hair. Values in blood and urine have been investigated in context of occupational exposure.

Chromium, manganese, molybdenum and fluorine

This group of elements is very difficult to determine in biological systems, because of both low concentrations and unresolved methodological problems. The analysis of Cr and Mn is extremely sensitive to contamination. However, the overall median values for these elements in serum are probably close to the real value. Molybdenum is poorly documented, with a few exceptions. Additional literature data on Mo in serum, 0.58; 0.59 and 0.91 µg/L agree with the overall mean. Mo in milk reflects wide variations in Mo contents in food. Fluorine is poorly documented. Most results are for milk. Literature values suggest 160 ± 100 µg/L for whole blood and 0.3 - 1.0 mg/L for urine.

Other Elements

Table 4 presents results on the elements Al, B, Br, Cs, Li, Rb, U and V. These elements have not been included in the data base, as no or scarce information was given by the responding analysts. However, we identified a few results that are backed by analytical quality assurance measures. A striking example is the value obtained for U in urine by Kelly, 3.6 ± 0.9 ngL [7].

Concluding remarks

There has been substantial progress in the last decade in understanding the problems of trace element analysis by the analytical community. Therefore, it has been possible to arrive at reference values in selected clinical specimens. However, many of the available data need confirmation and more investigations are needed for elements such as B, L, Al, V and U.

Literature

[1] Iyengar, G.V., W.E. Kollmer, H.J.M. Bowen, 1978. Elemental Composition if Human Tissues and Body Fluids, Weinheim, Verlag Chemie.

[2] Woittiez, J.R.W., 1983. Elemental Analysis of Human Serum and Serum Protein Fractions by Thermal Neutron Activation.PhD-Thesis, University of Amsterdam.

[3] Versieck, J., 1984. Accuracy in Trace Element Analysis, CRC Crit. Rev. in Clin. Lab. Sci., 22, 97.

[4] Iygengar, G.V., 1985. Concentration of 15 trace elements in some selected adult human tissues and body fluids of clinical interest from several countries. Results from a pilot study for the establishment of reference values. Jülich Report Jul-1974.

[5] Versieck, J., R. Cornelis. 1980. Normal levels of trace elements in human blood plasma or serum, Anal. Chim. Acta, 217.

[6] Heydorn, K., 1980. Aspects of Precision and Accuracy in Neutron Activation Analysis, RISO National Laboratory, RISO-R-419, 1980.

[7] Kelly, W.R., J.D. Fasset, S.A. Hotes, 1987. Determination of picogram quantities of uranium in human urine by thermal ionisation mass spectrometry, Health Physics, 52, 331.

PB, CD, AND MN IN FETAL, INFANT, AND ADULT RIB BONES FROM URBAN AND INDUSTRIAL AREAS OF GERMANY

M. Hedrich[1], R.L. Bergmann[2], U. Rösick[1], P. Brätter[1], and K.E. Bergmann[2]

1) Hahn-Meitner Institut Berlin GmbH, Trace Elements in Health and Nutrition, Glienicker Str. 100, D-1000 Berlin 39

2) Bundesgesundheitsamt Berlin, Institute of Social Medicine and Epidemiology, General-Pape Str. 62, D-1000 Berlin 42

Summary

Zeeman compensated graphite furnace AAS was used to determine the concentrations of lead , cadmium , and manganese in wet digested rib bones of 83 fetuses and children from Berlin up to two years of age as well as in rib bones of 55 adults from Weiden /Bavaria. Referred to defatted dry bone material the following median values were determined: 0.9 µg Pb/g, 0.02 µg Cd/g, and 1.1 µg Mn/g in fetal bones from Berlin and 6.4 µg Pb/g, 0.06 µg Cd/g, and 0.2 µg Mn/g in adult bones from Weiden. A comparison with published data did not reveal an elvated lead burden in reference to Weiden inhabitants caused by the local lead glass industry. The behavior of manganese in the fetal skeleton is an interesting point. Up to an age of ten months a drastic depletion of 14% Mn per month was observed. Mineral composition was measured by ICP optical emission spectrometry. Comparing the P contents of the fetal and the adult rib bones, no difference could be detected (102 ± 4 and 103 ± 5 mg P/g, respectively), while Ca concentrations with an average value of 215 ± 9 mg/g in fetal bones were about 7% lower than in adult bones (230 ± 13 mg Ca/g).

Introduction

As a preparatory to an epidemiological study concerning intake, fixation, and remobilisation of bone seeking essential and toxic trace elements in the fetal and infant skeleton, Pb, Cd, and Mn contents were determined in bone samples of stillborn and deceased children from Berlin. Special attention was drawn to Mn which has been supposed to play an important part in metabolism of the fetal skeleton [1-3].

Trace Element Analytical Chemistry in Medicine and Biology

Apart from this question the attempt was undertaken to find out by bone analyses, if the inhabitants of Weiden/Bavaria have been or still are subject to elevated heavy metal burden due to the local lead glass industry.

Materials and Methods

Two groups of patients were subject to our examinations. Fetal bone samples were taken autoptically from the sixth rib during the years 1982 through 1984 in different hospitals of Berlin. The 83 cases examined (47 male and 36 female) included 26 stillbirths between the 23rd and 42nd week of pregnancy, 23 newborns who deceased during the first month of their life, and 34 children who did not reach one year of age (among them 16 who died because of sudden infant death syndrom). Their biological ages, calculated by the period of gestation + lifetime - 40 weeks, ranged from -4 to 23 months, averaging 1 ± 5 months. From every member of the adult group a piece of the sixth rib bone was taken during autopsy between 1984 and 1986 in the Institute of Pathology of the urban hospital of Weiden/Bavaria. Death of the 55 cases under study (38 male and 17 female) was primarily caused by coronary circulation failure, embolism of the lungs, and accidents. Their ages ranged from 20 to 90 years with an average of 60 ± 18 years.

Samples were taken with a normal scalpel, freed of surrounding tissue and cut into discs about 2 mm thick with a diamond saw using 70% ethanol as cooling agent. The bone discs were extracted with ether in a Soxhleth apparatus for six hours and subsequently dried resulting in defatted dry bone material. Portions of 50 to 150 mg of each dried sample were dissolved in 1 ml of concentrated nitric acid and wet digested under pressure over a period of nine hours at 160°C in quartz vessels enclosed in teflon coated bombs. The obtained solutions were diluted with doubly distilled water to a volume of 5 or 10 ml.

The elements Pb, Cd, and Mn were determined directly in the digested solutions by Zeeman compensated graphite furnace AAS (Perkin-Elmer 5000/Zeeman spectrometer) with careful attention paid to the conditions of the STPF concept. Calibration was performed by standard addition method. H-5 animal bone, a standard reference material of the IAEA, served as a quality control and a self-synthesized monetite ($CaHPO_4$) with very low concentrations of Cd and Pb was used to supervise contamination of our digestion procedure [4,5]. Contents of Ca and P were measured by ICP optical emission spectrometry (Applied Research Laboratories 3520) from a 50fold dilution of the digested solutions.

For statistical evaluation a Kolmogorov-Smirnov test was run for log-normality with critical levels of significance as given by Lilliefors [6]. The results are presented as cumulative frequency distributions on a probability scale with a logarithmically transformed abcissa (figure 1).

Results and Discussion

Mineral Composition

Rib bones from fetuses and adults show almost the same mineral composition except for a 7% difference in Ca concentrations (table 1). The Ca/P ratio is lower in fetal bones which is probably due to a larger portion of octacalciumphosphate in the mineral phase that changes into hydroxyapatite at increasing age.

Table 1: Mineral Composition of Examined Rib Bones

Source	Ca [mg/g]	P [mg/g]	Ca/P (molar ratio)
Berlin (fetal)	215 ± 9.4	102 ± 3.7	1.62 ± 0.02
Weiden (adult)	230 ± 12.9	103 ± 5.3	1.72 ± 0.03

a) referred to defatted dry matter

Lead

Figure 1 and table 2 show our own results as well as those of Liese [7] and Ermer [8], who examined rib bones of adults from two different groups of Berlin patients (65 and 110 cases, respectively) with a similar age distribution (35 - 76 years and 38 - 79 years). These patients lost a piece of their fifth or sixth rib bone during thoracical surgery in 1977 or 1978. Both groups were not known to be occupationally exposed to lead.

The linear functions followed by the cumulative frequencies in the probability plot reveal log-normal distributions of individual values within the groups. This is confirmed by a Kolmogorov-Smirnov test. Large differences in concentrations between fetal and adult rib bones are noticeable. When interpreting this it has to be considered that the lead burden of inhabitants of large german cities has decreased drastically during the last decade. Wagner et al. [9] found that the blood lead level of Berlin school students in 1985 had dropped to about 50% of its original value in 1976. Also Drasch et al. [10] were able to show that the lead contents in pelvic bone of urban males had decreased between 1974 and 1984 by the factor of 1.9 ± 0.2 for all age groups from 20 - 70 years. If this reduction is assumed to proceed linear with time, an extrapolation of the geometrical means for rib bones from Berlin adults from 1977 onto 1982 would result in 7.9 µg Pb/g, leading to a ratio of $[Pb]_{fetal\ rib\ bone} / [Pb]_{adult\ rib\ bone} = 0.12$ valid for Berlin. A similar ratio (0.15) is obtained in recalculating the lead values published by Barry [11] for rib bones of british fetuses and adults before 1979.

Table 2: Pb, Cd, and Mn in Rib Bones of Fetuses and Adults from Berlin and Weiden/Bavaria

Source	N	Range [µg/g]	Median [µg/g]	Geom. Mean [µg/g]	SDF[a)]	Arithm. Mean [µg/g]	Dependence on Sex	Dependence on Age
				Pb				
Berlin	83	0.3 - 3.4	0.87	0.90	1.57	1.0 ± 0.5	no	no
Weiden	55	2.6 - 19.1	6.5	6.88	1.65	7.8 ± 4.1	yes	yes
Berlin [7]	64	5.1 - 23.5	11.2	11.4	1.42	12.1 ± 4.2		
Berlin [8]	110	5.8 - 31.0	11.1	11.5	1.41	12.2 ± 4.6	yes	yes
				Cd				
Berlin	83	0.003 - 0.206	0.020	0.021	2.48	0.031 ± 0.033	no	yes
Weiden	55	0.010 - 0.272	0.059	0.063	2.11	0.081 ± 0.058	no	no
Berlin [7]	65	0.007 - 0.089	0.032	0.033	1.93	0.040 ± 0.023		
				Mn				
Berlin	82	0.09 - 5.5	1.03	0.81	2.71	1.2 ± 1.1	no	yes
Weiden	55	0.06 - 0.71	0.16	0.18	1.71	0.21 ± 0.13	no	no

a) standard deviation factor = antilog (standard deviation of the mean derived from log-transformed data)

If the inhabitants of Weiden are or are not subject to a heavy metal burden caused by the local lead glass industry, cannot finally be concluded by the data collected so far. Extrapolating the values of Ermer [8] and Liese [7] from 1978 onto 1985, a lead concentration in rib bone of 6.8 µg/g would be the result for Berlin inhabitants. In comparison to this, 6.9 µg Pb/g is the geometrical mean value of the Weiden group. Considering further that there is no difference in the total lead burden between today's urban and rural population any more [10], one could conclude that the Weiden group under examination was not exposed to an elevated lead pollution. For a final decision on this question, however, examination of a suitable, that means unexposed, control group from a neighboring region still has to be undertaken.

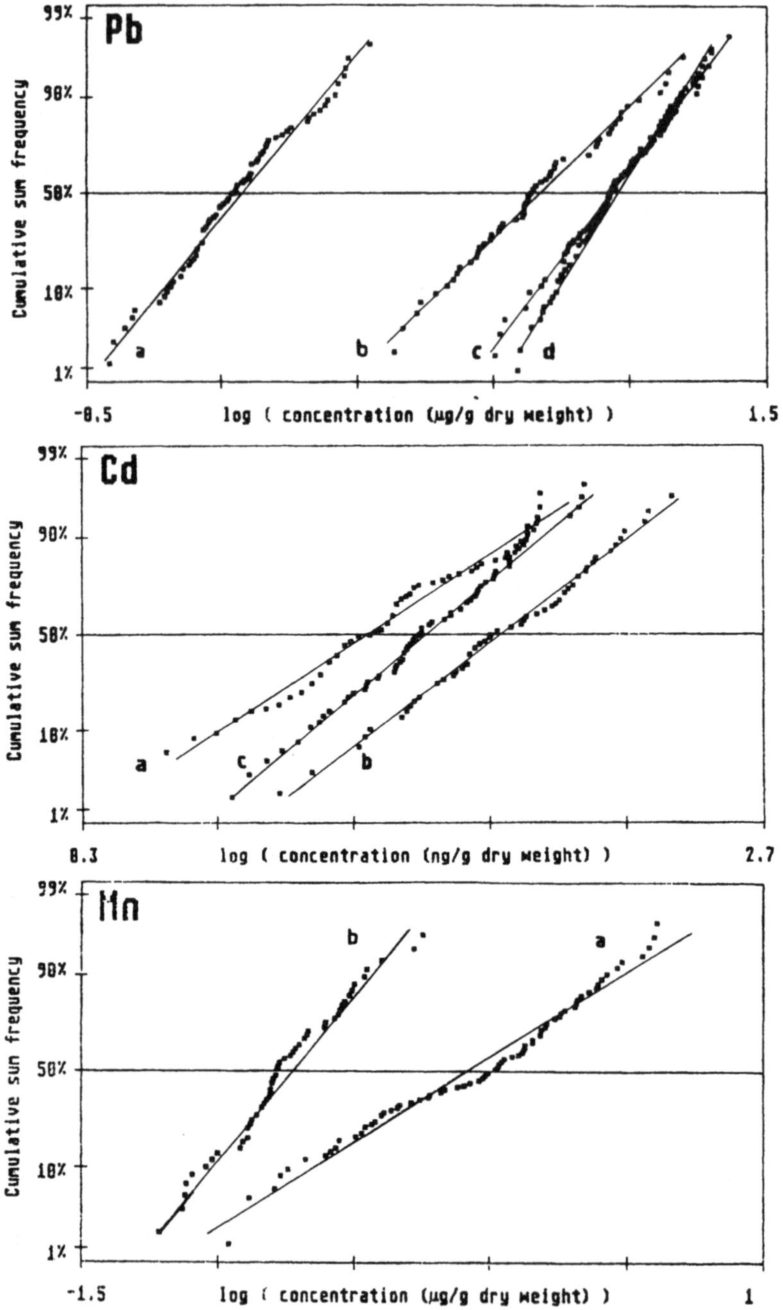

Figure 1: Pb, Cd, and Mn in rib bones of fetuses and adults. (a) Berlin (fetal); (b) Weiden/Bavaria (adult); (c) Berlin (adult) [7]; (d) Berlin (adult) [8]

Cadmium

Considering the element cadmium we also observe log-normal frequency distributions within the examined groups showing similar scatterings that are larger than those of the lead values (figure 1 and table 2). Similar to lead, fetal bones from Berlin show a lower cadmium level than Berlin adults [7] although differences are less distinct than in the case of lead. Compared to Berlin adults, average cadmium values for Weiden adults are twice as high. Whether this difference actually shows an elevated body burden could be better indicated in determining the cadmium contents of the kidneys instead of the bones of this populational group.

Bryce-Smith et al. [12] determined an average of 5.2 µg Cd/g wet weight in rib bones of 26 british stillborn children from Birmingham. These bone samples were collected before 1977 and their cadmium level, which exceeds the average cadmium value of Berlin fetuses (after 1982) by 750 times, suggests tremendous local differences in Europe.

Manganese

Looking at manganese, log-normal frequency distribution does not fit as well as for lead and cadmium and fetal manganese values exhibit a larger scattering than those of adults (see figure 1 and table 2). In contrast to the other two elements there are substantially higher manganese concentrations in Berlin fetal rib bones than there are in Weiden adults. An explanation for this might have to do with the essential importance of this element for the proper development of fetal bone and cartilage.

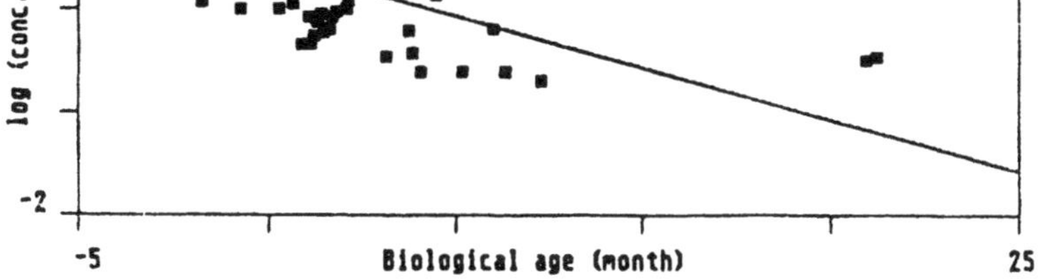

Figure 2: Decrease of Mn concentrations in fetal rib bone at increasing age

As figure 2 shows, drastic changes in manganese concentrations of the bones take place during the perinatal period. At first glance the depletion takes place exponentially. The drawn line represents a monthly decrease of 14% manganese in bone. The cause for this behaviour is unknown and subject to speculations. Either the requirement for manganese is not sufficiently met by supply and reserves or the needed amount in the bones decreases with increasing skeletal growth and less is transported into this tissue as the infant grows older. Supporting the second hypothesis is that, compared with the fetal skeleton, the much smaller manganese content of adult bone is sufficient for normal skeletal metabolism and does not indicate a state of deficiency. No age dependent change of concentrations in the adult group could be observed.

References

[1] L.S. Hurley. 1981. Phil. Trans. Soc. London B Biol. Sci. 294, 145.

[2] G. Saner, T. Dagoglu, T. Özden. 1985. Am. J. Clin. Nutr. 41, 1042.

[3] L.G. Strause, J. Hegenauer, P. Saltman, R. Cone, D. Resnick. 1986. J. Nutr. 116, 135.

[4] M. Hedrich, U. Rösick, P. Brätter, R.L. Bergmann, K.E. Bergmann. 1987. In: 4. Colloquium Atomspektrometrische Spurenelementanalytik (B. Welz, ed.) Bodenseewerk Perkin-Elmer, Überlingen, p. 461.

[5] M. Hedrich, U. Rösick, P. Brätter, R.L. Bergmann, K.E. Bergmann. 1987. In: Trace Elements in Environmental History (G. Grupe and B. Herrmann, eds.) Springer Verlag Berlin, p. 17.

[6] H.W. Lilliefors. 1967. J. Amer. Statis. Assoc. 62, 399.

[7] T. Liese. 1982. Dissertation Freie Universität Berlin.

[8] A. Ermer. 1986. Dissertation Freie Universität Berlin.

[9] H.M. Wagner, N. Englerth, Ch. Krause. 1987. In: Schwermetalle in der Umwelt (E. Lahmann and K. Jander, eds.). G. Fischer Verlag Stuttgart, p. 113.

[10] G.A. Drasch, J. Böhm, C. Baur. 1987. Sci. Total Environm. 64, 303.

[11] P.S.I. Barry. 1981. Brit. J. Industr. Med. 38, 61.

[12] D. Bryce-Smith, R.R. Deshpande, J. Hughes, H.A. Waldron. 1977. Lancet 1, 1159.

A STUDY OF TRACE ELEMENTS IN HUMAN PLACENTA, BLOOD AND HAIR FROM THE YUGOSLAV CENTRAL ADRIATIC

M. Horvat, P. Stegnar, A.R. Byrne, M. Dermelj
"J. Stefan" Institute, 61111 Ljubljana, Yugoslavia

Z. Branica
Gynaecology Clinic, 58000 Split, Yugoslavia

Abstract

From a group of 34 women mainly resident in the Islands of the Yugoslav Central Adriatic Sea, scalp and pubic hair, and maternal and umbilical cord blood were sampled at parturition. The pregnant women were selected on the basis of a questionnaire about seafood consumption in order to identify a group with increased mercury intakes, and to study the transport of inorganic and methylmercury across the placental barrier. In addition to Hg and MeHg, analyses of As, Cd, Sb, Cu, Se and Zn in placental and blood samples were performed.

Results indicate that concentrations of methylmercury are increased in umbilical cord blood and inorganic mercury increased in placenta, showing active transport of MeHg across the placental barrier and removal of inorganic Hg before it can be transferred to the foetus. Lower maternal blood and placental Zn levels in comparison to literature data were found. A significant correlation for As between maternal and umbilical cord blood was found, with higher levels of As in maternal blood. No significant correlations with Se were observed, nor any notable differences in the concentrations of the other trace elements measured.Since speciation of mercury at low concentration levels (ng/g) in such tissues is demanding, methodological procedures are also described.

Trace Element Analytical Chemistry in Medicine and Biology

Introduction

As prenatal life is the most sensitive human developmental stage, there is considerable interest in studying toxic effects and deficiencies of trace elements. Determination of trace element contents and their distribution in maternal and neonatal tissues may provide useful information on "normal" and "abnormal" levels during gestation and birth. As reviewed by Koos and Longo (1) in many methylmercury poisoning epidemics, the newborn child was also affected, as methylmercury readily crosses the placental barrier and is accumulated in the foetus. The human placenta plays an important protective role in preventing inorganic mercury intoxication (2).
Some studies suggest that Zn and Cu also have an important role during pregnancy (3,4) and therefore they also received attention in our study. Relatively little data are available on the transport of non-essential trace elements from mother to foetus. On the basis of mercury levels found so far in the Yugoslav population (5,6) within the framework of a long-term programme for pollution monitoring and research in the Mediterranean Sea (MED POL Phase II), where concentration levels of MeHg in hair ranged from 0.09-4.75 µg/g, this population cannot be considered to be at risk. Our further studies are directed toward pregnant women and the foetus, which have been identified as groups at special risk. Special attention in the first phase of the present study was drawn towards the optimization of analytical methods. Some preliminary results for Hg-total, MeHg and As in scalp and pubic hair and Hg-total, MeHg, As, Se, Zn, Cu in maternal and umbilical cord and placenta are discussed.

Materials and methods

All studied samples together with information considered useful for the interpretation of data (age, location, occupation, number of seafood meals per week) were taken at the time of birth in the Maternity Hospital in Split. Hair samples were cleaned and stored according to the IAEA recommendation (7). Only the first cm of occipital hair was taken for analysis. Maternal blood was taken by venepuncture. The whole umbilical cord and maternal blood were heparinized and stored in precleaned polyethylene containers in a deep freeze until analyzed. Placentae samples were cut from two parts; from the outer part which belongs to the chorion membrane and the inner part belonging to the amnion membrane. Samples were placed in precleaned polyethylene containers and stored in a deep freeze. Samples were analyzed by methods developed at our laboratory, and together with their primary references, are listed in Table I.

Table 1: Methods for Determination of Studied Analytes

Analyte	Method	References
Hg-total	NAA, pyrolysis, trapping	8, 12
	CVAAS acid decomposition	9
MeHg	GC-ECD, Volatisation	10, 11
Se	NAA pyrolysis	12
	NAA solvent extraction, 4-NPD	13
As	NAA; solvent extraction of iodide	14
Cu,Zn,Cd	NAA carbamate extraction	15

NAA - neutron activation analysis
GC-ECD - gas chromatography-electron capture detector
CVAAS - cold vapour atomic absorption spectrometry

Results and discussion

The accuracy and precision of the analytical results obtained for Hg, As, Se, Cd, Zn, Cu, Sb were checked by the analysis of standard reference materials. Comparison of our results with certified values showed good agreement and hence the reliability of the procedures for routine work. In the absence of any certified reference material for MeHg, however, accuracy of analytical methodology has to be assured by interlaboratory comparisons, standard addition experiments and comparison of the results obtained by different isolation techniques, which are described in detail elsewhere (10, 11). On the basis of a comparison of the results obtained by diffent approaches to methylmercury determination, we have chosen the volatilization GC techniqe, which is very sensitive, selective (for all sample types in this study) and especially suitable for routine work.

Mean values and ranges of concentration of total Hg and MeHg in all examined samples are presented in Table 2.

Table 2: Mean Values , Standard Deviation and Ranges of Concentration of Total and Methyl Mercury

sample		Hg-total	Methylmercury	% methylation
scalp hair (µg/g), n = 34		1.48 ± 0.71 (0.42 - 3.28)	1.15 ± 0.66 (0.20 - 3.36)	72 ± 17 (40 - 100)
pubic hair (µg/g), n = 29		1.39 ± 0.80 (0.55 - 3.04)	1.11 ± 0.69 (0.41 - 2.90)	79 ± 18 (33 - 100)
maternal blood (ng/ml), n = 34		3.7 ± 1.9) (1.2 - 9.6)	3.1 ± 2.0 (0.3 - 7.9)	73 ± 29 (21 - 100)
umbilical cord blood (ng/ml), n = 34		7.7 ± 5.1 (1.2 - 21.1)	6.7 ± 4.8 (0.8 - 21.3)	80 ± 19 (43 - 100)
placenta (ng/g), n = 34	-AMNION	12.9 ± 7.7 (2.8 - 37.3)	7.0 ± 5.1 (1.4 - 25.6)	56 ± 26 (10 - 100)
	-CHORION	13.4 ± 7.0 (4.3 - 30.7)	6.7 ± 4.7 (1.6 - 21.9)	50 ± 23 (14 - 100)

mean ± stand.dev.
min. - max.

Table 3: Correlation Coefficients of Hg and MeHg in Scalp and Pubic Hair Versus Blood and Placentae Samples

			Scalp hair r	Scalp hair p	Pubic hair r	Pubic hair p
pubic hair		Hg-T	0.86	< 0.001		
		MeHg	0.74	< 0.001		
maternal blood		Hg-T	0.66	< 0.001	0.76	< 0.001
		MeHg	0.53	< 0.001	0.66	< 0.001
umb. cord blood		Hg-T	0.68	< 0.001	0.80	< 0.001
		MeHg	0.62	< 0.001	0.54	0.004
placenta	amnion	Hg-T	0.51	0.005	0.74	< 0.001
		MeHg	0.42	0.017	0.59	0.002
	chorion	Hg-T	0.61	< 0.001	0.73	< 0.001
		MeHg	0.62	< 0.001	0.78	< 0.001

r = correlation coefficient
p = significance

It is evident that concentrations of Hg and MeHg in the groups of women examined, known to be exposed through seafood consumption, are at acceptable levels and therefore they cannot be considered to be at risk. The number of seafood meals per week is evidently not a precise enough criterion for estimating the exact weekly intake of mercury. In spite of this, a significant positive correlation between the number of seafood meals per week and mercury in scalp ($r=0.44$, $P=0.01$) and pubic hair ($r=0.47$, $P=0.009$) was found. No correlation with other criteria (age, occupation, location) was observed. The good correlation between mercury concentrations in scalp and pubic hair (Table 3) lead us to the conclusion that pubic hair could also be a useful indicator to estimate Hg exposure through seafood consumption. Since scalp hair may be exposed to external sources and exhibits differences across the scalp, pubic hair could be a better indicator. This is also confirmed by the better correlation coefficients between Hg and MeHg pubic hair levels than in all other samples examined (Table 3). The Hg and MeHg levels in scalp and pubic hair are higher than those in maternal blood by a factor varying between 250-300, which is in agreement with literature data. The maternal blood: umbilical cord blood relatioship is expressed by the equation y(maternal blood)=0.27(umbil. cord blood)x +1.61. At more elevated levels of Hg in maternal blood (7ng/ml), the blood ratio is about 3, which is in accordance with results obtained by other autors (16).

The relatively high concentrations of MeHg in umbilical cord blood and the fact that the percentage of MeHg in placenta is lower than in all other samples (hair, blood) confirm efficient placental transfer of MeHg and suggest removal of inorganic Hg before it can be transferred to the foetus. Diffences between chorion and amnion found by other authors (17) were not confirmed in the present study, probably due to the relatively low concentrations and/or because not enough precautions were taken to remove adhering blood during sampling. The Se concentrations were not correlated with Hg in placentae or any other sample examined.

The applicability of the Hg content of hair as an integrator of exposure over a period of time has been proved in many studies. The positive correlation between Hg and MeHg concentrations in hair and placenta, which was found in this work, suggests that more attention should be given to the significance of Hg and MeHg levels in placenta as estimates of exposure of the infant during pregnancy. Further discussion of this problem must await the results of additional studies, which are now in progress in our laboratory.

There are very few data in the literature about the ability of As to cross the human placenta. The origin of As in this study (Table 4), is probably due to seafood consumption, where it is mostly present as organic As(arsenobetaine). A strong correlation between As levels in maternal and cord blood (r=0.86, P=0.000) was found; their ratio is 0.75 indicating higher As levels in maternal blood. These results suggest more study on placental transfer of organic As should be done.

The results obtained for Cu in whole blood (Table 4) are not directly comparable with other studies, since it is known that plasma Cu is a better indicator of Cu status of a subject than whole blood (3); but placental Cu levels lay within the range of normal pregnancy (19).

Table 4: Mean Values, Standard Deviations and Ranges of Concentration of Se, As, Cu and Zn in Studied Samples

		Se (ng/g; ng/ml)	As (ng/g; ng/ml)	Cu (µg/g; µg/ml)	Zn (µg/g; ug/ml)
scalp hair			24.9 ± 13.6;n=22 (9.4 ÷ 59.2)		
maternal blood		111 ± 28;n=15 (72 - 170)	11.6 ± 8.8;n=26 (0.9 - 28.9)	0.96 ± 0.34;n=26 (0.25 - 2.04)	3.92 ± 1.25 (1.55 - 7.06)
umb.cord blood		124 ± 24;n=15 (81 - 161)	8.0 ± 7.5;n=26 (0.9 - 31.0)	0.44 ± 0.21;n=26 (0.24 - 11.6)	1.96 ± 1.22 (1.01 - 6.27)
placenta	amnion	196 ± 61;n=34 (83 - 339)	6.5 ± 5.8 (0.8 - 22.5)	0.58 ± 0.15;n=25 (0.19 - 0.79)	5.80 ± 2.64 (1.30 - 10.3)
	chorion	196 ± 63;n=33 (137 - 317)	5.7 ± 5.1 (1.0 - 20.0)	0.58 ± 0.22;n=20 (0.22 - 0.96)	6.19 ± 1.88 (2.1 - 8.6)

mean ± standard deviation
ranges
n - number of samples

From the results for Zn presented in Table 4, it is evident that all the maternal blood levels are relatively low compared to the most frequent values found in the literature (6-7 µg/ml) (19), while umbilical cord blood is comparable with literature data (20). Consistent with low blood Zn levels, lower concentrations were also found in placentae in

comparison with literature data (18). The results for Zn obtained in the present study suggest attention should be given to possible effects of Zn deficiency in the studied population.
Concentrations of Cd and Sb were very low and below the limits of detection (10 and 1 ng/g, respectively) in some samples, and are not discussed further here.

Acknowledgements

We gratefully acknowledge the assistance of Nurse L.Amulić of the Gynaecological Hospital in Split for providing the samples used in this study. We also thank D.Konda, A.Prosenc, J.Smrke and A.Vakselj for the exellent technical assistance, and Dr. A.Sabadin for statistical evaluation of data.

Literature

1. Koos B.J., K.D.Longo. 1976. Am.J.Obstet.Gynecol.126. 390.
2. Suzuki T., J.Yonemoto, H.Satoh, A.Naganuma, N.Imura, TKigawa. 1984. J.Appl.Toxicol 4, 249.
3. Underwood E.J.: Trace Elements in Human and Animal Nutrition. 1977. A.P.Inc., New York.
4. Rösic E., U.Rösic, P.Bratter, G.Kynast. 1983. Trace Element - Analytical Chemistry in Medicine and Biology. Walter de Gruyter, Berlin. New York. Vol.2. 463 - 482.
5. Dermelj M., M.Horvat, A.R.Byrne, P.Stegnar. 1986. Chemosphere 16 (4), 887.
6. WHO, Health Effects of Methylmercury in the Mediterranean Area, Report on a joint WHO/FAO/UNEP meeting, ICP/CEH 054. 1987. Athens.
7. IAEA Report RL-50. 1978. Vienna.
8. Kosta L., A.R.Byrne. 1969. Talanta 16, 127.
9. Horvat M., T.Zvonarić, P.Stegnar. 1986. Vestn. Slov. Kem. Drus. 33 (4), 475.
10. Zelenko V., L.Kosta. 1973. Talanta 20, 115.
11. Horvat M., K.May, M.Stoeppler, A.R.Byrne: submitted for publication.
12. Byrne A.R., L.Kosta. 1974. Talanta 21, 21.
13. Dermelj M., G.Hancman, A.Gosar, A.R.Byrne, L.Kosta. 1985. Vestn. Slov. Kem. Drus. 32, 127.

14. Byrne A.R., A.Vakselj, 1974. Croat. Chem. Acta 46 (3), 225.

15. Dermelj M., A.Vakselj, V.Ravnik, B.Smodiš. 1979. Radiochem. Radioanal. Letters 41 (2), 149.

16. Galster W.A. 1976. Environ. Health Perspect. 15, 135.

17. Garett N.E., R.J.B.Garret, J.W.Archdeacon. 1979. Toxicol. Appl. 22, 649.

18. Damsgard E., K.Heydorn, N.Horn. 1983. Trace Element - Analytical Chemistry in Medicine and Biology. Walter de Gruyter, Berlin. New York. Vol. 2, 499-516.

19. Iyengar G.V. 1987. Biological Trace Element Research 12, 263.

20. Åbdulla M., L.Löfberg et al. 1983. Trace element - Analytical Chemistry in Medicine and Biology. Walter de Gruyter, Berlin. New York. Vol. 2, 517-531.

COMMERCIAL DIAGNOSTIC HAIR ANALYSIS: ACCURACY OR FANTASY - A CASE STUDY

A.O. Lee Jones, R.M. Jacobs

U.S. Food and Drug Administration (FDA), Washington, D.C. 20204 and San Francisco, CA. 94102

N.B. Ranney, E.L. Summers

U.S. Federal Trade Commission (FTC), Washington, D.C. 20850

Abstract

As part of an investigation into the reliability of hair analysis for trace elements and potential fraudulent misrepresentation of the results in assessing nutritional status, a commercial testing laboratory (CTL) was sent 6 samples of human hair collected from a single individual. Results obtained by the CTL for the 6 identical test samples were highly variable (CTL, 11-162 %, C.V. vs. FDA, 2-13 %, C.V.). Only 11 % of the results agreed with FDA analyses within ± 10 %. Based on the investigation, an indictment and court trial by FTC of the CTL's owners on charges of fraudulent practices resulted in court decisions that barred CTL from representing in any advertisement aimed at direct consumers that hair analysis accurately reflected the nutritional status of an individual and that recommendations for vitamin, mineral, or other dietary supplements can be based on the analysis of one's hair.

Introduction

The efficacy of trace element analysis of human hair as a diagnostic tool for elemental nutritional status of individuals

Trace Element Analytical Chemistry in Medicine and Biology

has yet to be demonstrated. Nevertheless, the number of commercial laboratories performing these analyses either as physician-requested tests or as direct consumer-purchased tests is large. Hair analysis is being advertised and sold to the consumer by mail, through food supplement stores, advertisements in "health" magazines, beauty parlors, and other places of commerce (1, 2, 3).

This "diagnostic test" is usually based upon the results of a single, simultaneous or sequential multielement analysis of a small hair sample. The report generated from this analysis is sent directly to the consumer or health professional, and frequently contains an assessment of the individual's health status. Mineral and vitamin supplements or a detoxification regimen may be recommended in the report to alleviate or correct any claimed diseased states, deficiency states, or toxicity states. However, accurate trace element determinations by simultaneous multielement analytical techniques are technically demanding, and require skill and good judgement on the part of the analyst.

In view of the popularity of elemental analysis of hair, the suspect analytical procedures practiced by the industry, and the lack of efficacy of hair analysis as a "diagnostic test" for nutritional status, a study was designed to evaluate the analytical performance and claims of a single, local (Washington, D.C. area) laboratory that offered hair analysis.

Materials and Methods

All the hair that was used to evaluate the CTL performance originated from a single person (male, age 21, brown hair color). A stainless steel surgical scissor was employed to cut this hair into segments (< 2 cm). The cut hair was mixed and rinsed thoroughly with deionized distilled water (DDW), dried overnight at 37°C, and stored in a self-sealing plastic bag. Care was taken to protect the hair from contamination. Six individual one gram hair test samples were removed from the original sample for the CTL analyses and placed in individual self-sealing plastic bags. These hair test samples were identified numerically and transferred to FTC custody. At 2 week intervals, the FTC sent 2 hair test samples to the CTL for elemental analysis. Seven test samples from the same original sample were analyzed by FDA in sets of 3 and 4 test portions, two years apart.

The FDA test samples were washed sequentially twice with n-hexane, twice with ethanol, and 5 times with DDW, then freeze-dried. The amount of hair test portions used for FDA's analysis varied from approximately 0.1 g to 1.0 g. Three test portions were analyzed 2 years prior to the other 4 test portions. Known quantities of Cd, Cu, Fe, Mg, Mn, P, Pb, and Zn were added to one of the earlier test portions for a recovery determination. Analysis of a National Bureau of Standards Standard Reference Material (NBS-SRM 1577 Bovine Liver) was employed to validate the FDA methodology.

FDA test portions were digested in a mixture of concentrated nitric, 70 % perchloric and concentrated sulfuric acids until evolution of sulfuric acid fumes. Care was taken not to char the digestion mixture at any time. The digests were quantitatively transferred into 10 mL volumetric flasks and diluted to volume with DDW. The multielement measurements were carried out by inductively coupled plasma atomic emission spectrometry (ICP-AES, Model 975 Jarrell-Ash, Waltham, MA.) for the following elements: Ca, Mg, Na, K, P, Cr, Mn, Mo, Fe, Cu, Zn, Se, Cd, Pb, and Ni.

After FDA had completed all its analyses, FTC delivered the CTL results in the form of patient reports to the FDA for evaluation. FDA's analytical data and proficiency evaluation of CTL were submitted to the FTC as evidence. An indictment and trial of CTL were pursued for violation of the FTC Act on fraudulent practices in the advertisement and sale of chemical hair analysis to the public.

Results

The CTL reported values for 18 elements (Ca, Mg, Na, K, P, Si, Cr, Mn, Mo, Fe, Cu, Zn, Se, Al, Cd, Pb, Hg, Ni) determined by "direct current plasma method" as per their advertisement. The CTL results were highly variable, as shown in Table 1. The

Table 1. Comparison of Mean Elemental Values between CTL and FDA

Elements	Mean Elemental Values, ug/g[a]					
	CTL	C.V. (%)	FDA	C.V. (%)	Recovery[b] (%)
Ca	563 ± 264	47	454 ± 58	13	-
Mg	37.9 ± 10	27	31.2 ± 1	4	97
Na	113 ± 37	33	<20		-
K	110 ± 53	48	<25		-
P	92.8 ± 45	48	153 ± 3	2	103
Si	6.86 ± 5.5	80	NR[c]		-
Cr	0.873 ± 0.34	39	< 0.5		-
Mn	5.40 ± 8.3	154	0.386 ± 0.01	4	102
Mo	0.389 ± 0.23	59	< 0.4		-
Fe	28.7 ± 18	63	13.6 ± 0.9	6	109
Cu	21.3 ± 13	62	14.3 ± 0.8	5	99
Zn	190 ± 22	11	179 ± 6	4	95
Se	19.2 ± 5.8	30	< 1.5		-
Al	4.15 ± 6.7	162	NR[c]		-
Cd	0.609 ± 0.45	74	1.15 ± 0.1	8	93
Pb	3.18 ± 2.0	64	16.0 ± 0.3	2	97
Hg	1.32 ± 1.2	87	ND[d]		-
Ni	0.475 ± 0.30	64	< 0.46		-

[a]CTL results for each element are the mean value ± S.D. of 6 separate analyses. FDA elemental values are the mean ± S.D. of 6 values. Nine elements (Ca, Mg, P, Mn, Fe, Cu, Zn, Cd, Pb) were present in quantities that could be accurately determined by FDA. Other elements (Na, K, Si, Cr, Mo, Se, Ni) were less than the reliable quantifiable limits (10 X detection limit) based on a test portion of 1.0 g in 10 mL, final concentration. Coefficient of variation (C.V.) represent a measure of precision.

[b]Recovery of fortified elements (Mg, P, Mn, Fe, Cu, Zn, Cd, Pb) at one times the quantitated elemental concentration determined in the hair (except Mg, at 5 times) was performed by FDA.

[c]Not reported by FDA due to possible contamination from glassware employed.

[d]Not determined.

precision was poor based on the coefficient of variation that varied between 11 to 162 percent. The CTL reported values for 9 elements (Na, K, Si, Cr, Mo, Se, Al, Hg, Ni) that were not reported by FDA. The elemental values for 6 of these elements (Na, K, Cr, Mo, Se, Ni) were above FDA's reliable quantifiable limits, and would have been detected and quantified by FDA, if they had been present at levels reported by the CTL.

FDA analyses revealed that the original hair had a high degree of homogeneity. For the specified elements, the precision based on the C.V. ranged from 2 to 13 percent. Recovery of known amounts of 8 elements (Mg, P, Mn, Fe, Cu, Zn, Cd, Pb) added to one test portion varied between 93 to 109 percent (Table 1). The elemental analysis of the NBS-SRM 1577 Bovine Liver revealed that the FDA analytical procedure employed for human hair was accurate, and within the 95 % confidence limits for Na, K, Mn, Fe, Cu, Zn, and Cd.

Based on the means of FDA elemental values, the CTL came within ± 10 % of the FDA value only 11 % of the time (Table 2). Two values for Ca, 1 for Mg, 3 for Cu, and 3 for Zn fell within this limit. This limit ± 10 % used as an evaluation range was based on the limits employed for serum mineral chemistries proposed by the U.S. Center for Disease Control (± 4% to ± 10%) These limits are currently used for proficiency testing during inspections for licensing U.S. clinical laboratories that perform serum mineral determinations (4).

The CTL classified the elemental results of each sample into 3 ranges ("Low", "Normal", "High") for "Nutrient Minerals" and 2 ranges for "Toxic Minerals" ("Normal", "Toxic"). Variability in analytical results led to variable classifications for the same elements. No two reports were identical.

Classification of hair values as "Low", "High", or "Toxic" resulted in recommendations for either dietary supplements or a regimen for removal of toxic metals, or both as shown in Table 3. In addition, all the reports recommended the consumption of a variety of vitamin and non-vitamin substances to "enhance optimal health".

Two major decisions resulted from the trial prohibiting the

owners of the CTL from: (1) receiving orders from or reporting to consumers directly (i.e., performance or marketing of hair analysis), and (2) representing in any advertisement aimed at direct consumers, either directly or by implication, that the hair analysis accurately reflected the nutrient mineral content of the body and that recommendations for vitamin, mineral, or dietary supplements can be based on the analysis of one's hair (5).

Table 2. Accuracy of CTL Compared to FDA Elemental Values[a]

	FDA X	CTL Elemental Values (ppm)						
Element	(ppm)	420R	416R	187	186R	290R	323R	Score[b]
Ca	454	765	216	587	435	952	422	2/6
Mg	31.2	54.0	43.5	35.9	28.3	25.6	40.2	1/6
Na	<20	74.9	130	142	142	58.0	133	0/6
K	<25	66.1	77.0	191	134	138	54.0	0/6
P	153	76.9	88.1	62.0	53.0	177	99.5	0/6
Si	NR[c]	5.83	4.50	6.79	3.52	17.7	2.79	
Cr	<0.5	0.869	0.229	1.04	1.18	0.861	1.06	0/5
Mn	0.386	2.91	3.12	1.45	1.16	1.48	22.3	0/6
Mo	<0.4	0.568	0.354	0.598	0.136	0.093	0.584	0/3
Fe	13.6	61.7	8.21	16.7	29.2	28.1	28.5	0/6
Cu	14.3	15.3	12.3	15.1	47.5	22.2	15.6	3/6
Zn	179	155	181	192	215	185	209	3/6
Se	<1.5	14.7	15.0	16.8	22.1	17.2	29.7	0/6
Al	NR[c]	2.87	1.58	0.297	0.409	17.7	2.04	
Cd	1.15	0.546	0.091	0.779	1.39	0.266	0.583	0/6
Pb	16.0	3.22	1.29	6.45	1.56	1.87	4.66	0/6
Hg	ND[d]	0.325	0.249	3.27	0.857	1.22	2.02	
Ni	<0.46	0.607	0.126	0.409	0.860	0.706	0.143	0/3
							Overall	9/83

[a] CTL results were of 6 separate analyses. FDA elemental values were reported as a mean of 6 values. Nine elements (Ca, Mg, P Mn, Fe, Cu, Zn, Cd, Pb) were present in quantities that could be accurately determined by ICP-AES. Other elements (Na, K, Cr, Mo, Se, Ni) reported by FDA were based on the reliable quantifiable limits (RQL) (test portions of 1.0 g in 10 mL).

[b] Accuracy Score was based on the fraction of values within ± 10% of the FDA mean value and greater than 10% of the RQL value. CTL values less than RQL were not used.

[c] Not reported.

[d] Not determined.

Table 3. Conclusions and Recommendations of CTL[a]

Subsample (Patient No.)	Conclusions: Deficiencies	Conclusions: Excesses	Recommended Supplements	Recommend to remove toxic metals[b]
420R	K, Zn	Ca, Fe	K orotate (120 mg), Zn orotate (100 mg)	no
416R	Si, Cr, Fe	Na	Cr (1 mg), Fe (50 mg), Si (1000 mg)	no
187	Fe	Na, K, Hg	Fe (50 mg)	yes
186R	Si	Cu, Na, Cd	Si (1000 mg)	yes
290R	Mo	Ca, P, Al	Mo (50 ug)	yes
323R	K, Si	Na, Mn, Se	K orotate (120 mg), Si (1000 mg)	no

[a] Supplements were recommended for daily consumption. In addition to the elemental supplements, CTL recommended the following supplements to "enhance optimal nutrition": Vitamin (Vit) A (20,000 units); Vit B50 complex (2 tablets); Vit B15 (150 mg for 20 days,thereafter 50 mg); Vit D (400 units); Vit E (400 units); Lecithin (3600 mg); Mucozyme (3 tablets); Vit F (3 tablets).

[b] Substances recommended by the CTL to remove toxic metals: Ascorbic Acid (8000 mg); PABA (100 mg); L-Methionine (2 g); Sodium Alginate (1350 mg); Bran (3 g); Garlic (540 mg); Dry Aloe (200 mg).

Discussion

From an analytical perspective, many elements can be identified and accurately determined in hair. The degree to which each element can be determined depends on a number of factors: (1) the quantity present in hair, (2) the preparative procedures used prior to measurement, and (3) the analytical technique utilized (6, 7, 8, 9, 10, 11, 12, 13). Whereas many elements are present in sufficient quantity and may be determined by simultaneous or sequential multielement analysis of a single test portion, many elements may require either special preparative techniques or different analytical techniques in order to be accurately determined.

Unlike the commercial hair analysis survey reported by Barrett (3), the elemental profile of the human hair sample was well-characterized in this study. FDA employed a similar analytical technique (ICP-AES) to that advertised by the CTL (DCP-AES). FDA procedures were successfully validated. Moreover, because analytical results for those elements reported by FDA were reproducible over a range of test portion sizes, it was concluded that the original sample was adequately homogeneous. However, the elemental analysis of 6 test samples reported by CTL from the same human hair sample resulted in poor precision for most of the elements reported.

Inaccuracies in the CTL's results contributed to multiple and and variable conclusions similar to observations in Barrett's survey (3). Ultimately, the CTL recommended a host of supplements to "enhance optimal health", to aid in the utilization of minerals, to cure deficiency, and to treat toxicity. These supplements were a collection of vitamin, mineral and non-food substances. Many were inappropriately labeled (e.g., "vitamin B50 complex", "vitamin F"). The CTL listed as supplements to enhance optimal health or remove excess metals, vitamin A, vitamin E, and vitamin C at 4, 13, and 133 times greater than the U.S. Recommended Daily Allowances (RDA) levels (14) (Table 3). For Cr and Fe, the supplements recommended provide quantities greater than those estimated as "a safe and adequate daily dietary intake" for adults (15) or the U.S. RDA, respectively. The use of the combination of substances for "removal of toxic metal" is a practice totally devoid of scientific merit, especially so, considering the poor reliability of the CTL's results (Table 3).

Apart from analytical validity, severe problems in interpreting the mineral composition of hair may completely negate the value of hair analysis for the individual consumer as a sole indicator of nutritional or toxicological status (1, 8, 13, 16, 17, 18). A variety of hair treatments (e.g. bleaching, dyeing, permanent waving), as well as natural hair color, alters trace element concentrations (1, 19). Age and sex must be considered in comparing hair trace element concentrations (20, 21). Little information is available concerning the extent to which hair concentrations of specified elements correlates with concentrations in other organs, tissues, subcellular fractions, or metabolic pools (1, 16, 22, 23, 24, 25). Furthermore, the "normal" range of elemental hair concentration with physiological and health-related significance is not known (1). As stated by Hambidge (1) and Kelvay (17), very little is known about the utility of hair analysis in clinical medicine. Low values for essential elements or high values for potentially toxic elements have not always been shown to indicate either deficiency or toxicity, respectively. Beneficial effects of either group or individual therapy based on hair analysis have not been established (17).

Considerable concern remains regarding the usefulness of hair analysis in predicting the nutritional status of individuals. It is necessary for health professionals and the public to be aware of the limited value of hair analysis and of its potential to confuse and mislead. However, as shown in this study, the efficacy of hair analysis as a diagnostic tool may be overshadowed by the concerns of the analytical reliability of elemental analyses performed by the commercial testing laboratories. Stringent quality assurance procedures are absolutely necessary for those laboratories performing such analyses.

References

1. Hambidge, K.M. 1982. Am.J.Clin.Nutr. 36, 943.

2. Fenner, L. 1983. FDA Consumer. April, p. 16.

3. Barrett, S. 1985. JAMA. 254, 1041.

4. Code of Federal Regulations. 1974. Public Health. Clinical Laboratories, Title 42, Part 74.

5. FTC vs. Furman, 1985-1 CCh Trade Case (CCH) 66486 (ED Va 1985).

6. Hambidge, K.M., M.L. Franklin, J.A. Jacobs. 1972. Am.J.Clin.Nutr. 25, 384.

7. Hilderbrand, D.C., D.H. White. 1974. Clin.Chem. 20, 148.

8. Lazar, P. 1974. JAMA. 229, 1908.

9. Flynn, A. 1977. J.Appl.Nutr. 29, 51.

10. Salmela, S., E. Vuori, J.O. Kilpio. 1981. Anal.Chim. Acta. 125, 131.

11. Fletcher, D.J. 1982. Postgrad.Med. 72, 79.

12. Kollmer, W.E. 1983. Sci.Total.Environ. 27, 251.

13. Rivlin, R.S. 1983. Am.J.Med. 75, 489.

14. Code of Federal Regulations. 1987. Food and Drugs, Title 21, Parts 100 to 169, p.20.

15. Recommended Dietary Allowances. 1980. In: Natl.Acad.Sci. (Nutr. Research Council, ed.). Washington, D.C., p. 178.

16. Wibowo, A.A., B. Brunekreef, E. Lebret, H. Pieters. 1980. Intl.Arch.Occup.Environ.Health. 46, 275.

17. Kelvay, L.M., B.R. Bistrian, C.R. Fleming, C.G. Newmann. 1987. Am.J.Clin.Nutr. 46, 233.

18. Medeiros, D.M., A. Mazhar, E.W. Burnett. 1987. Nutr.Res. 7, 1109.

19. Eads, E., C.E. Lambdin. 1973. Environ.Res. 6, 247.

20. Petering, H.G., D.W. Yeager, S.O. Witherup. 1971. Arch. Environ.Health. 23, 202.

21. Petering, H.G., D.W. Yeager, S.O. Witherup. 1973. Arch. Environ.Health. 27, 327.

22. McBean, L.D., M. Mahloudji, J.G. Reinhold, J.A. Halstead. 1971. J.Clin.Nutr. 24, 506.

23. Kollmer, W.F., D. Berg. 1979. J.Radioanal.Chem. 52, 189.

24. Iyengar, G.V., W.E. Kollmer. 1986. Intl.J.Trace.Elements. Med. 3, 25.

25. Iyengar, G.V. 1986. Food.Lab.Newsletter. 7, 13.

REGIONAL DISTRIBUTION OF TRACE ELEMENTS IN THE HUMAN BRAIN

E. Andrási, M. Mohos

Institute of Inorganic and Analytical Chemistry, L. Eötvös University, Budapest, Muzeum krt. 4/B, 1088

L. Bezur

Institute of General and Analytical Chemistry, Technical University, Budapest, Gellért tér 4, 1111

Zs. Molnár

Reactor Laboratory, Technical University, Budapest, Müegyetem rkp. 9, 1111

Abstract

The aim of our recent work was the determination of trace element concentrations in human brain parts of persons having no primary lesions in their nervous system. The selected brain parts represented the sensory and motor centres, the limbic system as well as the white tissue. To remove organic substances different sample preparation methods were used. To analyse the brain parts prepared from both hemispheres solid sample /NAA, DC-AES/ and solution analytical /ICP-AES and AAS/ methods were applied. The techniques used had been tested by the analysis of the NBS Bovine Liver SRM. On the basis of the investigations the performance and limitations of the independent methods were compared. As a result the regional distribution of trace element concentrations in the human brain and the "healthy" mean values characteristic of this age-group 65-75 were determined.

Trace Element Analytical Chemistry in Medicine and Biology

Introduction

We believe that all disorders of function must have an simple explanation and that element analysis must be a method complementary to biochemical studies. In the case of the brain there are specific sites characterized by their involvement in specific functions /1/. We determined the regional distribution of trace elements in "normal" /healthy/ human brain in order to establish in our further experiments the connection between the primary lesions in the nervous system and the content of trace elements.

Sample Treatment

Tissue samples were dissected from 10 defined regions from corresponding right and left sides of 9 brains of deceased persons aged from 65 to 75 years. There was no sign of disease of central nervous system in any persons. Essential trace elements /Ba, Cd, Co, Cr, Cu, Fe, Mn, Ni, Sn, Sr, Zn/ and some other elements /Al, B, Pb, Ti/ were also examined.
To standardize the sample treatment which considerably influences the results of the analysis we endeavoured to control the sample collection, sample handling and sample storage. Brain parts of various antecendents /fresh, liophilised or conserved in formalin/ were examined /2/.
Because of the different water content of the brain parts, it was necessary to determine factors of the mass proportion of fresh, dried and ashed samples /3/.
To remove organic substances, two basic sample treatment methods were used: dry ashing /450 ^{O}C, 36 h, quartz vessel/ and wet digestion in PTEE bomb. For wet digestion bidestilled Carlo Erba 65% nitric acid was used in bomb /150 ^{O}C, 2 h, sample 100 mg/. We examined the decomposition efficiency making and alteration in the time and temperature of digestion /Table 1/.

Table 1.

ICP-AES Results as a Function of Time and Temperature of Digestion /µg/g dry weight/

Brain part	Time /h/	and Temperature /°C/ of Digestion	Element Cu	Mn	Zn
Occipitalis area	2	150	26,4	2,40	111
/right side/	2	220	26,5	2,25	109
Corpus callosum	2	150	16,4	1,24	41,1
	4	150	15,4	1,31	39,8
Cerebellum	2	220	27,9	2,44	69
/right side/	4	220	23	2,14	75

Applied Techniques

To analyse the brain parts solid sample and solution analytical methods were used. The samples dried at 105 °C were analysed directly by NAA method. DC-AES method was used to determine the trace element concentrations of the ashed samples. Solutions obtained by wet digestion were measured by ICP-AES and AAS methods. The conditions for measurements are described in 3.

Results and Discussion

Digestions carried out in PTFE bomb are advantageous because of the small risk of loss and contamination as well as the complete decomposition /Table 1/. Ashing loss and contamination were not detected /Table 2/.

Table 2.

Concentrations measured by ICP-AES in the case of Ashing /A/ and Parr Bomb Digestion /B/ as well as determined by NAA /C/ /µg/g dry weight/

	Cerebellum /left side/			Corpus callosum		
	A	B	C	A	B	C
Cu	29,3	27,9	29,4	24,7	27,6	24,0
Mn	2,3	2,4	2,7	1,4	2,0	1,8

NBS Bovine Liver was determined to prove our methods /Table 3/. The results for three elements showed good agreement with the certified values.

Table 3.

Analytical Results for NBS Bovine Liver 1577 /µg/g dry weight/

Element	Certified Values	ICP-AES	AAS	NAA	DC-AES
Cu	193±10	205	193	195	172
Mn	10,3±1,0	11,5	11,5	10,5	8,1
Zn	130±13	149	123	152	

Brain parts of which we possessed sufficient quantities were analysed by the above mentioned independent method. The concentrations determined in different laboratories can be seen in Table 4.

The regional distribution of trace element concentration in human brain and the "normal" mean values characteristic of age-group 65-75 were determined. Concentrations of Cu, Mn, Zn in different brain parts are summarized in Table 5.

Table 4.
Concentrations of a Few Elements in a Right Side Thalamus /μg/g dry weight/

Element	ICP-AES	AAS Flame	AAS GF	NAA	DC-AES
Cu	14,5	13,4	15,0	16,0	17,3
Mn	1,9	2,5	1,7	1,8	1,9
Zn	45,2	40,2		36	

Table 5.
Mean Concentration Values of Some Elements in Various Brain Parts from Both Hemispheres of 9 Persons Measured by ICP-AES Method /μg/g dry weight/

	Cu	Mn	Zn
1. Occipitalis area	27,9±5,3	1,84±0,65	77,0±15,7
2. Caudatum	34,3±9,6	2,54±0,79	81,6±8,5
3. Putamen	36,0±4,6	2,65±0,48	66,7±18,5
4. Pallidum	34,4±7,9	3,35±1,13	71,7±10,3
5. Thalamus	16,1±5,1	1,84±0,62	75,6±10,2
6. Pulvinar thalami	18,3±6,0	1,76±0,52	50,1±13,7
7. Hippocampus	29,0±9,2	1,64±0,34	80,1±12,0
8. Cerebellum	28,8±4,5	2,40±0,38	82,6±9,7
9. Corpus Callosum	21,5±8,5	1,46±0,41	59,4±18,8
10. Cerebellar Cortex	20,2±6,9	1,89±0,79	56,5+15,5

From these results it can be seen that the NAA and the ICP-AES methods are suitable for the determination of the trace element content in human brain parts. The NAA method is ideal for biological sample analysis. Several elements could be determined using more than one isotope and this helps

in checking the interference exemption. Since no chemical process is involved this minimizes the possibility of contamination or incomplete separation. The ICP-AES technique is almost insensitive to matrix composition which results in a high detection power and wide linear range. For a few elements /e.g. Co, Cr, Ni, Pb/ the using of GFAAS method is suggested because of the small quantity of these elements.
In spectrographic analysis the matrix effect due to thermal evaporation of the sample is high, calibration must be done with a standard addition technique.
In the present work the establishment of normal levels and the relation between the concentration values of elements determined by independent methods was our principal aim.
It can be seen that the concentrations measured in different laboratories are in a good agreement with one another.

References

1. Schneider, M. 1971. Einführung in die Physiologie des Menschen, Springer, Berlin

2. Andrási, E., T. Tenczer. 1986. MSZV Előadásai, 30, 485-490

3. Andrási, E., Zs. Gyimesi, J. Bérczi, L. Bezur. 1987. In: Proc., 30th Hungarian Annual Conference on Spectral Analysis, Debrecen, p. 47.

4. Bérczi, J., D. Bódizs, G. Keömley, Zs. Molnár, 1982. Egyetemi jegy-zet, BME Tanreaktora, Budapest

5. Bezur, L., L. Ernyei, E. Pungor. 1987. In: Proc., Winter Conference on Plasma and Laser Spectrochemistry, Lyon, p. 241.

LITHIUM IN BIOLOGICAL AND DIETARY MATERIALS

G.V. Iyengar

National Bureau of Standards, Gaithersburg, MD 20899, U.S.A.
and Department of Agriculture, Beltsville, MD 20705, U.S.A.

W.B. Clarke
McMaster University, Hamilton, Ontario, L8S4K1, Canada.

R.G. Downing
National Bureau of Standards, Gaithersburg, MD 20899, U.S.A.

J.T. Tanner
Food and Drug Administration, Washington D.C. 20204, U.S.A.

Abstract

Progress in biological trace element research has been heavily dependent on the availability of reliable analytical methods to carryout accurate determinations, as witnessed by the slow to no-progress scenario with respect to several trace elements at ultratrace levels of concentrations in biological systems. Boron and Lithium are no exceptions in this context, and the lack of progress in this case is further accentuated by the fact that no known essential functions in animal metabolism have yet been demonstrated for these two elements. Before the beneficial aspects can be conclusively answered, it is of primary concern to search for reliable analytical methodology. The technique based on the combination of neutron activation analysis and mass-spectrometry (NAA-MS) for the simultaneous determinations of B and Li developed by Clarke et al (1) has the required sensitivity ($<10^{-8}$ g g^{-1}) to deal with the natural concentration levels of these two elements in biomaterials. The method has been successfully used to quantify B and Li in several biological and dietary reference materials (1). The results for selected dietary standards are shown in Table 1.

Trace Element Analytical Chemistry in Medicine and Biology

Table 1

Boron and Lithium in Dietary RMs (dry basis, Mean ± 1 S.D..) (1)

Standards		**Lithium** (ng/g)	**Boron** (µg/g)
Rice flour	(NBS SRM 1568a)	4.7 ± 0.2	0.65 ± 0.02
Non-fat milk powder	(NBS SRM 1549)	9.6 ± 1.8	2.13 ± 0.07
Wheat flour	(NBS SRM 1567a)	56.4 ± 0.6	0.58 ± 0.02
Bovine liver	(NBS SRM 1577a)	176.0 ± 5.0	0.59 ± 0.08
Citrus leaves	(NBS SRM 1572)	209.0 ± 6.0	63.10 ± 0.60
Brewer's yeast	(NBS SRM 1569)	678.0 ± 9.0	5.64 ± 0.13

This technique has also been applied to determine B and Li in mixed total human diets obtained from an ongoing international nutrition project, for which foods representative of the U.S. diet, comprising of 201 items from the United States Food and Drug Administration's Total Diet Study, are being collected, total diet composites prepared and analyzed (2). The U.S. diets represent the intake of 25-30 year old adult males, and the two collections analyzed are from the Southeast (Georgia, Texas, West Virginia) and North Central (Ohio, Iowa, Kansas) regions. These results are shown in Table 2. Also shown in Table 2 are concentrations of B and Li in a mixed diet reference material (IAEA-H9) and a Turkish diet sample.

Table 2

Determination of Boron and Lithium in Human Mixed Total Diets by NAA-MS (Dry Basis, Mean ± 1 S.D.)

	Lithium (ng/g)	**Boron** (µg/g)
USDIET-1	59.8 ± 0.9	3.59 ± 0.33
USDIET-2a	63.4 ± 0.7	2.53 ± 0.02
USDIET-2b	64.5 ± 0.6	2.62 ± 0.02
USDIET-2c	62.1 ± 0.7	2.41 ± 0.02
IAEA-H9	35.0 ± 0.3	3.47 ± 0.25
ARAS-1	102.0 ± 3.0	4.58 ± 0.09

Initial estimates show that the average intake of an adult American amounts to 1.5 mg of B and 31 µg of Li (Table 3). No difference was observed for Li concentration in diets between the two regions. The Li intake calculated from the present study indicates a lower intake in comparison with earlier published results (3,4), and is comparable with the recent findings reported for Canada (5). The B values between the 2 U.S. regions vary considerably, being 3.6 and 2.5 µg/g in Southeast and Northcentral, respectively. The average intake, namely 1.5 mg agrees with the intake reported for Canada (5) and Finland (6). Results reported for U.K. show high intake as well as wide variation in the intake (5). Boron being rich in plant foods, such variations may be anticipated in different diets or in diets from different regions of a country. Additional dietary collections from other regions in U.S. are being analyzed.

Table 3

Comparison of Daily Dietary Intakes of Lithium and Boron in Canada, Finland, U.K. and the U.S.A. (Mean ± 1 S.D.)

Country	Lithium (µg)	Boron (µg)
Canada	22 ± 2.1 (6)	1330 ± 130 (6)
Finland	-	1700 (4)
United Kingdom	107 ± 53 (5)	2819 ± 1554 (5)
United States	31 ± 1.2	1521 ± 377

References

1. Clarke, W.B., Koekebakker, M., Barr, R.D., Downing, R.G. and Fleming, R.F. 1987. Appl. Radiat. Isot. 38, 735.

2. Iyengar, G.V., Tanner, J.T., Wolf, W.R. and Zeisler, R. 1987. Sci. Total Environ. 61, 235.

3. Reference Man. 1975. ICRP-23, Pergamon Press, New York.

4. Hamilton E.I. 1979. The Chemical Elements and Man. John Wiley, New York.

5. Clarke, W.B., Gibson, R.S., J. Food Comp. Analysis. (in press).

6. Acta Agriculturae Scandinavica. 1980. Supplement 22.

THE DISTRIBUTION OF FLUORINE OVER HAMSTER TOOTH GERMS

W.J.M. Lenglet, C.J.J. Stoof, R.D. Vis
Natuurkundig Laboratorium der Vrije Universiteit, Amsterdam, The Netherlands

Abstract

The influence of fluoride administration on tooth development in vivo has been measured by means of the precise determination of the location where F is incorporated in the germs. For the analysis, a proton microprobe was used to induce the nuclear reaction $^{19}F(p,p'\gamma)^{19}F$. In order to improve the peak to background ratio, γ-rays were detected anti-coincident with the cyclotron burst, made possible by the half life of 87 nsec of the 197 keV nuclear level in ^{19}F.

Introduction

After the initial reports on fluorosis, a fluoride induced change in enamel structure, more studies on morphological changes in the secretory ameloblasts have been published (1-3). In spite of these reports, no satisfactory description is given for the mechanism by which F^- affects the enamel deposited during the exclusively secretory phase of ameloblasts. Lyaruu (4) et al. demonstrated that F^- at all doses used (between 5 and 20 ppm in organ culture) induced hypermineralization of enamel secreted in vivo just prior to administration of F^-. It was shown that F^- decontrols enamel formation leading to hypermineralization of the prefluoride enamel. In order to determine the location of fluor in the growing tooth germs, the proton microprobe of the Vrije Universiteit (5) was

Trace Element Analytical Chemistry in Medicine and Biology

used in two different operation modes. With high spatial resolution the distributions of the matrix elements Ca and P were determined using micro-PIXE and with a lower resolution, imposed by the much lower cross sections for nuclear reaction as compared with inner shell ionization, the distribution of F. During the F-scans simultaneously the Ca and P distributions were measured again in order to keep track of the accurate position on the specimen.

Experimental

Maxillary first molar tooth germs from neonatel hamsters (4-5 days; bodyweight 4-5 g) were used. Dissection of upper jaws was performed 24 hrs after intraperitoneally administration of 20 mg/kg bodyweight NaF. Germs were fixed for 1 hr in 4% paraformaldehyde in a 0,1 M phosphate buffer, rinsed and dehydrated through a graded series of ethyl alcohol. After clearing with propylene-oxide, samples were embedded in Spurr's low viscosity resin. In this way, free fluoride ions were removed in order to collect only data from mineral bound F. 40-80 μm thick sections were cut and used for the microprobe experiments. Fig. 1 gives a schematic representation of such a section. Indicated is the location where line scans were made.

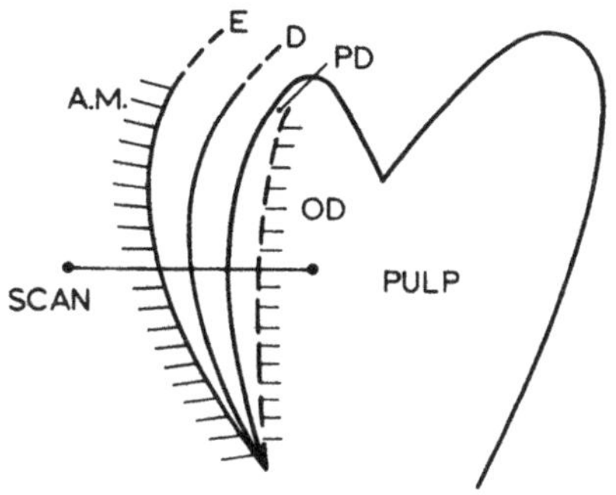

Fig. 1. A schematic representation of a tooth germ; the figure shows a typical scan along the various layers (AM = ameloblasts, E = enamel, D = dentin, PD = predentin, OD = odontoblasts.

For the experiments a 3.0 MeV protonbeam was used with a 8 μm lateral resolution for Ca and P measurement and a 20 μm resolution for F. The $^{19}F(p,p'\gamma)^{19}F$ nuclear reaction was used for the detection of fluorine. An electronic anti-conincident circuit was used to shut the detector during the cyclotronbursts and open it in a time window of about 100 ns in between bursts (6). Due to the half life of the 197 keV nuclear level in ^{19}F this procedure results in a considerable peak to background ratio increase, since all prompt events induced by the beam are not seen by the detector. Fig. 2 gives spectra of the F-measurements with and without the use of this electronics. The improved signal to noise ratio is obvious.

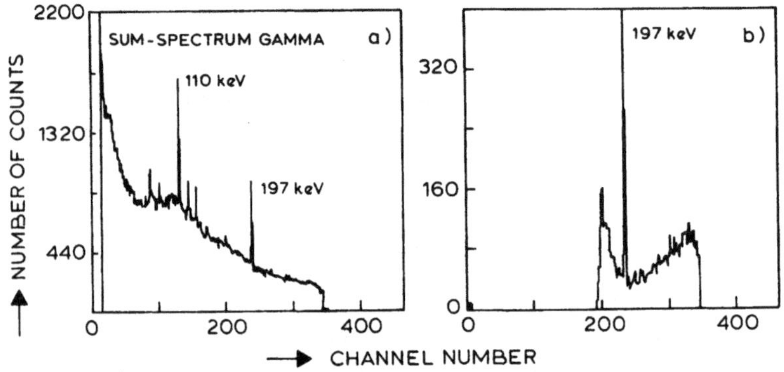

Fig. 2. γ-spectra of F-measurements
a) without anti-coincident electronics
b) with prompt events suppression.

For the detection of the γ-rays an intrinsic planar Ge detector was used equipped with a pulsed optical feedback preamplifier. For the detection of the X-rays of Ca and P a Si(Li) detector with standard electronics was used. The obtained γ- and X-ray spectra (32 along a line scan) were analysed using suitable peak fitting codes; the peak contents were represented per position along the track as a histogram.

Results and Discussion

Fig. 3 gives results of three scans as indicated in fig. 1.

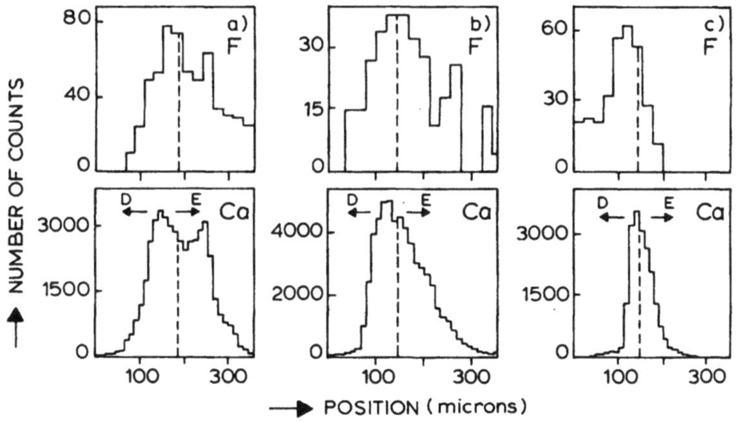

Fig. 3. Distributions of Ca and F across the fluoride treated tooth germs. The positions of the enamel layer (E) and the dentin (D) are indicated.

The major purpose of our fixation procedure was removing free and loosely bound fluoride ions from the specimen. The work of Weatherell et al. (7) showed that in control rats the concentration of fluoride in forming enamel decreased as the tissue matured. During or just before the enamel mineralization the fluoride concentrations were relatively high ($\sim$150 ppm). No precise time-scale regarding the enamel development of rat tooth is mentioned in the paper of Weatherell. In our experiments no fluorine has been detected in the control animals. Hammarström (8) suggested that part of the fluoride is either free, or bound to the organic matrix and perhaps subsequently lost as part of the matrix is withdrawn from the enamel during mineralization. The fixation procedure enhances the effect of fluorine removal, which could account for our negative results for fluorine in case of the control hamsters.

The fact that fluorine is detected in both mineralized layers (fig. 3, detin and enamel) seems to support the original hypothesis that in case of fluoride administration fluoride binds to the mineral part of the calcified tissue. The profound degree of hypermineralization in one of the specimens (see fig. 3.a) is clearly reflected in an elevated concentration of F, Ca and P (not shown). In the fluoride induced maturation stage of enamel development, the organic matrix is almost completely removed, which makes assumed preferential matrix binding improbable.

The germs which were studied were all in the secretory and maturation phase of enamel development. The bulk composition of mature enamel is exactly known: it consists of calcium hydroxyapatite. The detected P/Ca ratio of the tooth germs can be compared with the theoretical value for $Ca_5(PO_4)_3OH$; the P/Ca ratio gives an indication of the stage of development of the tooth germs. The Ca concentration of mature enamel is about 37%, while the P concentration is 18.3%. The experimentally determined P/Ca ratios for both a control and a fluoride administered sample are presented in fig. 4. Also here, indications may be found for a certain degree of hypermineralization.

In conclusion, the presence of rather low quantities of F^- during the development of tooth germs seriously seems to affect the structure and the chemical composition of the germs. Fluoride tends to concentrate on the dentin/enamel junction leading to hypermineralization at least on that location.

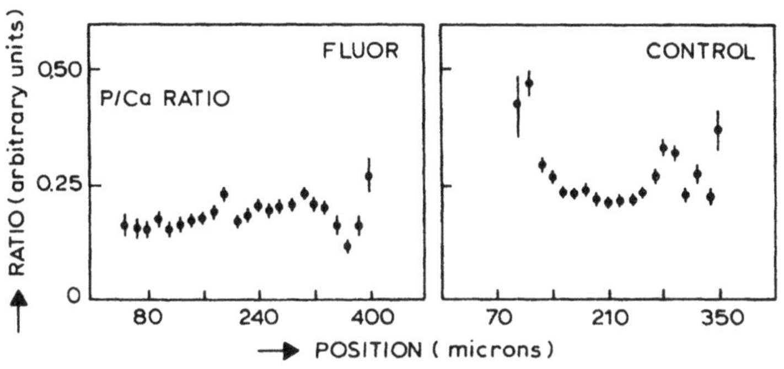

Fig. 4. The location dependent P/Ca ratio for a fluoride administered and control sample respectively.

References

1. Kruger, B.J. 1968.
Arch. Oral Biol. 13, 969

2. Walton, R.E., D.R. Eisenmann. 1974.
Arch. Oral Biol. 19, 171

3. Shinoda, H. 1983.
In: Mechanisms of Tooth Enamel Formation (Syga S ed).
Quintessence, Tokyo p. 273

4. Lyaruu, D.M., M. de Jong, A.L.J.J. Bronckers, J.H.M. Wöltgens. 1986.
Arch. Oral Biol. 31, 109

5. Bos, A.J.J., C.C.A.H. v.d. Stap, J.C. den Ouden, R.D. Vis, H. Verheul. 1986.
J. of Trace and Microprobe Techniques 4, 57

6. Lenglet, W.J.M. 1988.
Thesis, Vrije Universiteit Amsterdam, 79.

7. Weatherell, J.A., D. Deutsch, C. Robinson, A.S. Hallsworth. 1975.
Nature 256-5514, 230

8. Hammerström, L. 1971.
Scan. J. Den. Res. 79, 369

SPECIAL ELEMENTS : ALUMINIUM and PLATINUM

Plenary Lecture given by:

F. Alt

Chairman:

B. Welz

Poster Reports:

W. Kollmer, E. Steinnes, B. Welz

PLATINUM METALS IN THE ENVIRONMENT - A CHALLENGE TO TRACE ANALYSIS

F. Alt

Institut für Spektrochemie und angewandte Spektroskopie,
Bunsen-Kirchhoff-Str. 11, D-4600 Dortmund 1, Federal Republic of Germany

Abstract

The role of platinum as environmental element is discussed. Relevant digestion, separation and determination methods for platinum in biotic and environmental materials are encountered with special regard to the determination of natural platinum concentrations.

Introduction

Platinum metals play a significant part in the fields of industry, in the diverse electro-, measurement- and control-techniques, in the manufacture of laboratory equipment, in the construction of chemical apparatus and in medical fields. Especially the catalytic properties, the resistance against certain working materials and against corrosion, as well as the anti-cancerogenic effect of cis-platinum in the treatment of specific types of tumors are exploited. More elaborate utilization of platinum as working material is restricted mainly by the rare occurrence - the average natural concentrations of platinum within the crust of the earth being only 0.005 g/t. Up to now, the platinum elements were of insigni-

Trace Element Analytical Chemistry in Medicine and Biology

ficant environmental relevance, as in most cases, operations with platinum metals were carried out in closed systems. In the industry, only very slight emissions of platinum occur due to effective filter devices and recycling procedures. It is known that platinum salts possess great allergy-promoting potentials and that in addition, they may cause a disease called platinosis. Very little or nothing at all is known about toxic, bio-catalytic or even essential effects of small amounts of platinum.

By the introduction of catalytic cleaning of motorcar exhaust gases, a new pollution source of especially platinum emissions was introduced leading to uncontrolled emissions into the environment. Tests made in the USA, where motorcar catalyzers were introduced more than ten years ago, disclosed that, depending on the mode of driving, up to 2 µg platinum and more were emitted per one driven kilometer (1). In 1986, Hodge and Stallard (2) were able to prove as much as 0.7 µg/g platinum and 0.3 µg/g palladium in roadside dusts. These amounts of platinum metals, although their local distribution is mostly in the close vicinity of dense-traffic roads, are no longer negligible, especially in view of the general acceptance that natural concentrations of platinum in the environment lie in the lower pg/g-regions and below.
Environmental concentration of platinum have in this case already increased by several orders of magnitude. For an objective risk estimation, data would have to be available which give information about the ubiquitous concentrations of platinum metals, about the kind and the amount of emitted platinum metals, about the plant and animal consumption of platinum metals and last not least about human consumption of platinum metals, as well as about further related problems. Analytical chemistry is indeed a decisive aid here in solving the whole problematic situation. It is faced with exacting requirements as far as informations about the lowest limit of detection (femtogram level), the detectable concentration

levels ranging over 10 orders of magnitude, the question of the kind of platinum metal compounds (key-word speciation) and the kind of materials (water, soil, biotic material, catalyzer material, filter and others) are concerned. Desirable would be the development of a universal analytical procedure that would offer reliable and at the same time economic solutions for the majority of the problems named. But unfortunately, there is only narrow scope for such an undertaking, which would have to include as an added difficulty, different strategies for the high and for the low concentration regions.

In the following, the analysis of platinum shall be dealt with, although this involves a lot which could be said in respect to the two other important catalyzer metals, palladium and rhodium. Further emphasis shall be given to the pg-ng/g-region, that is to say, to procedures which are possibly suitable for the detection of the natural concentration of platinum. For concentration levels >ng/g, there are reliable procedures including various sample preparation techniques and determination methods available.

In Table 1, an overview over known values of naturally occurring concentrations of platinum is given. The data show clearly that there are not many reliable values available for this determination level. Either, it is highly probable that the values attained are wrong or they involve a great factor of unreliability. Even the otherwise so reliable technique of RNAA in this case yields data, which differ by three orders of magnitude during operation with the same material. Thus, a great lack of reliable data exists and an attempt is made in the following to show up some analytical strategies which may be suitable to close this gap.

Material	Pt [ng/g]	Method	Reference
NBS Orchard leaves	89.2 ±15.4	RNAA	3
SRM 1571	0.2 ± 0.2	RNAA	4
	<1	RNAA	5
NBS Bovine liver	0.070 ± 0.33	RNAA	4
SRM 1577	<3	RNAA	5
Bowen's Kale	189 ± 8.5	RNAA	3
Red algae	0.08 - 0.19	Ion exchange	6
Brown algae	0.22 - 0.32	Ion exchange	
Ocean sediments	0.7 -21.9	ETAAS	
Ocean waters (Pacific ocean)	173 -228 pg/L		
Citrus leaves	0.04 - 0.10 ±0.04	RNAA	4
Human liver	0.054 - 0.240	RNAA	4
Plant on ultra-basic soils	100 - 800	RNAA	7
Blood serum	<0.1	RFA	8
Blood plasma	<40	MS	
Dental enamel	<90	MS,NAA	
Roadside dust (contaminated)	40 - 700	Ion exchange	2

Table 1: Natural platinum concentrations in some biotic and environmental materials

Direct Determination of Platinum in Solid Samples

The attempt of direct determination without sample preparation fails in respect of almost all relevant concentration regions. This is due to incompensable or incorrectable matrix interferences, which often can only be avoided by employing elaborate sample preparation techniques. This applies to the neutron activation analysis, the proton induced X-ray fluorescence as well as to the solid state atomic absorption spectrometry, to name but a few. As an exception one would like to name here the determination of platinum in samples of human serum taken from patients who have received medical treatment with cis-platinum. After diluting, the determination is carried out via Zeeman-AAS and the addition method for calibration method. After fivefold dilution, the limit of detection lies at 20 ng/mL, which is fully sufficient for the control of samples from the clinical area, because during treatment, the platinum concentrations are normally found at the µg/mL level. Interferences by non-atomic absorption during measurement with ET-AAS may be compensated via background correction with the Zeeman technique.

Even if one succeeded in eliminating all matrix interferences during the direct determination of solids, the platinum determination of pg/g-concentrations would fail, because of the insufficient sensitivity of the physical measuring methods in question. A chemical sample preparation including decomposition, separation and preconcentration of the platinum sample is therefore unavoidable. But here, the aim should be to attain an analyte solution containing only the platinum to be determined. Only with such analyte solution the power of detection of the chosen analytical method can be fully exploited.

Relevant Digestion Methods for the pg-ng/g-Region

Digestion methods for the pg-ng/g-region should fulfil the following operation conditions:

1. The sample must be completely decomposed. Solid decomposition residues may contain undissolved platinum or else they may recombine with platinum from the analyte solution by adsorption or reduction. Not oxidized dissolved organic residues cause interferences during subsequent chemical separation and preconcentration procedures.
2. Decomposition vessels should be made of quartz glass. This is the only material of sufficiently high purity grade, and it can easily be vapor-cleaned with nitric acid and reconditioned. Decompositions with hydrofluoric acid cannot be carried out in vessels made of quartz. Here, one should resort to PTFE or glassy carbon serving as vessel material. No experience has been gathered as far as the behaviour of pg-amounts of platinum reacting to this vessel materials is concerned; this needs to be investigated first.
3. The sample decomposition should be carried out in a closed system. Thereby, the surface of the decomposition vessel can be kept small and furthermore, only a minimum of high-purity grade decomposition agents (nitric acid, hydrochloric acid, hydrofluoric acid) is required for the procedure. Also reduced by this is the danger of contamination.

At present, there are only few techniques known, which fulfil all the above named conditions (Table 2). The technique of high pressure ashing (9) belongs into this group. The attainable pressure and temperature condition it involves (130 bar and 300^{0}C) ensures the complete mineralization for most of the biotic materials concerned. Serving as material for the decomposition vessels is quartz glass. For decompositions with hydrofluoric acid, these vessels can be replaced by vessels made of glassy carbon. During one analytical run, up to 5 g dried biotic material may be decomposed within a period of three to four hours.

Technique	Reagents	max. sample amount	suitable for pg and ng-levels	suitable for µg and mg-levels
dry digestion	air, $Mg(NO_3)_2$	5 - 10 g	no	no
wet digestion	HNO_3, HCl, $HClO_4$, H_2F_2	5 - 10 g	no	yes
pressure digestion PTFE vessels	HNO_3, HCl	1 g	no	yes
high pressure digestion				
quartz vessels	HNO_3, HCl	5 x 1 g*	yes	yes
glassy carbon vessels	HNO_3, HCl, H_2F_2	6 x 100 mg*	yes?	yes
combustion with oxygen	O_2	300-500 mg	yes	yes

* Five, respectively six vessels can be inserted in the high pressure asher for one decomposition run.

Table 2: Digestion methods for the determination of platinum in biotic and environmental materials

The procedure of combustion in an oxygen flux within a dynamic system (10) also fulfills the above named operating conditions to a great degree. Here, we have an almost closed system made of quartz glass, in which sample burning is performed. However, only biotic materials with low silicate content are suitable, as treatment with hydrofluoric acid is not possible in this system. During one analytical run lasting about one hour, maximally 300 to 500 mg dried sample material may be decomposed. A newly developed device makes possible 8 to 10 decompositions in one hour.

When larger sample amounts have to be decomposed, an open decomposition is often preferred, during which under certain circumstances, biological materials may be dry-ashed. For this, various ashing aids are necessary. After decomposition, a relatively large volume of acid for the dissolution of the ashing residues is required. Losses of platinum have been observed in such procedures (11). Open wet digestion with acids have been found suitable for the determination of platinum concentrations in the upper ng/g-range and above, only. Here too, one can expect platinum losses or errors of results due to contamination in the pg-region. Thus, no open decomposition is feasible for the problematic situation with which we are faced here.

Relevant Platinum Separation and Preconcentration Modes for the pg-Region

Operational instructions covering the separation of platinum from samples and sample solutions are often described in the literature, and the number of new contributions is steadily growing. Employed are all methods known in analytical chemistry, including methods of liquid-liquid distribution, of sorption, of co-precipitation as well as of chromatography such as ionexchange methods and ionchromatographic methods, electrolytic preconcentration methods - to name just the most important ones.

At this point, the well known fire-assay technique shall be mentioned, which is even employed today, e.g., in the analysis of ores. However, it cannot be employed for platinum separation in the pg-region. All other methods are in principle applicable, after appropriate modification and adaption to the problem in question. The following prerequisites should be observed:

1. The separation yield should amount to >90%;
2. The chemical reagents used should be reduced to a minimum-
3. The analyte solution attained after separation should be free from matrix components, which may cause interferes during the subsequent analysis;
4. The isolated platinum should be contained in a minimal solution volume or it should be collected on a smallest possible sample carrier.

Practical work has shown that it is not always possible to obtain an analyte solution by only one single separation step which would render it suitable for subsequent, interference-free analysis. An example for this is posed by the extraction mode; here, only the combination of two sequential separation reactions leads to success.

The electrolytic precipitation of platinum as a method of separation and preconcentration was intensively investigated by R. Eller (12). He describes an electrolytic flow cell in which the platinum precipitates from the sample solution by cathodic reduction (Fig. 1). The shape of the cathode was chosen to also serve as sample carrier in ETAAS or in TRXFA. Great problems were encountered, when it came to the choice of the material for the electrodes. Pyrographite and glassy carbon suffered corrosive attacks, due to the high current

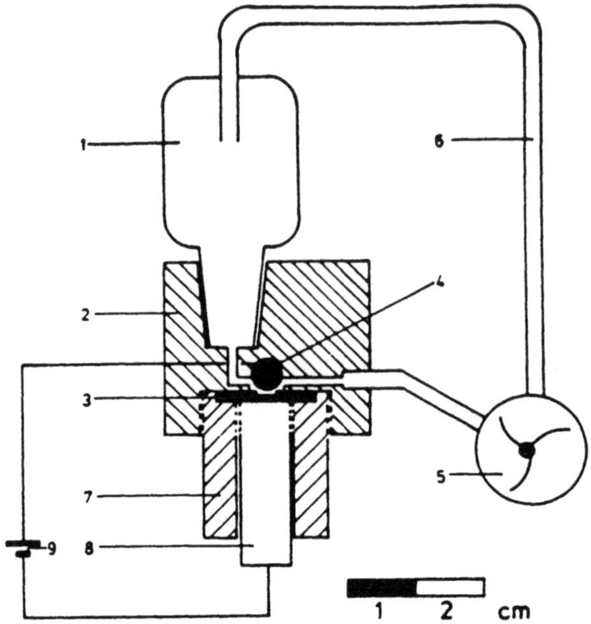

Fig. 1: Flow through electrolysis cell with platform cathode (12)

1. Stock container. 2. PTFE holder for electrodes. 3. Platform cathode (pyrographite). 4. Anode (glassy carbon). 5. PTFE - pump. 6. PTFE - tubes. 7. PTFE holder for the cathode. 8. Brass contact for the cathode. 9. voltage.

densities that are necessary for the precipitation of platinum from solutions with very low concentrations. Accordingly, bad finding rates and bad reproducibilities resulted. Electrodes made of a different material, e.g. rhodium, did not bring about any decisive improvement. Tests with aluminium electrodes are in progress, but a final positive result cannot yet be envisaged.

When employing chromatographic methods, there are two approaches, along which analyte solutions can be attained. Hodge and coworkers (6) isolated the anionic species of some platinum metals with the aid of a suitable anion exchanger; in addition, they carried out a subsequent cleaning procedure with a single-bed anion exchanger. Employing the graphite furnace AAS as a determination step in the procedure, they were able to determine pg-amounts of platinum in sea waters, in sediments and in organisms. The separation yield in respect of the individual analytical steps amounted from 35 to 90%, which is far from ideal.

During the second approach, the platinum is adsorbed to a small section of the exchanger in a batch procedure. After centrifugation the anion exchanger is digested and the platinum is finally determined in the decomposition solution (13). A second cleaning step must possibly be included here. In this way, the elution step of platinum from the anion exchanger can be avoided. This step was probably responsible for the bad results obtained by the procedure employed by Hodge et al (6).

The ionchromatographic determination of platinum with direct uv or amperometric detection in the sub-ppm-region proved successful according to Rocklin (14). This procedure offers the possibility for the determination of platinum species.
All above described procedures involve strategies which promise to be successful in the determination of platinum in

the pg-region. However, intensive work in view of further optimization is still necessary.

Relevant Determination Methods for Platinum in the pg-Region

The foregoing chapter, in which separation and preconcentration methods for the analysis of platinum are treated, already transgressed onto the subject of determination methods, because in most cases, they are related in their conception to one another, anyhow. Table 3 will give an overview over limits of detection which can be attained with the individual methods stated.

Spectrophotometry	0.5 µg/mL
Chemoluminescence	1 ng/mL
Kinetic methods	0.5-2 ng/mL
ICP-OES	0.5 ng/g
Flame-AAS	1 µg/mL
ETAAS	2 ng/mL; 20 pg
DPP	10 ng/mL
PIXE	0.5 ng/g
XRFA	5 µg/g
TRXRF	2 ng/mL; 20 pg
INAA	1 ng/g
RNAA	0.01-0.1 ng/g
SIMS	0.5 ng/g
GDMS	0.5 pg
TAMS	10 pg/g
Laser-ET-AFS	sub ng/mL
Laser-flame-AFS	ng/mL
Ionchromatography	sub ng/g

Table 3: Determination methods and detection limits for platinum

The most powerful methods, the tandem mass spectrometry and the radiochemical neutron activation analysis (RNAA), involve very elaborate techniques and are generally accessible to very few analysts only. It is in particular the RNAA, which is taken as a reference method for the compilation of reference values and for the control of accuracy and reliability of other procedures. For this purpose, the activity of the daughter nuclide 199 Au is measured, after interfering activities are eliminated, during the course of which the activities of the nuclides 197 Au, 198 Au and 199 Au of gold from the sample must be detected quantitatively and corrected (4). It is at this stage, where a great source of error can be assumed, possibly explaining the great differences which occurred during the determination of natural platinum concentrations in several materials (Table 1). Measuring platinum-nuclides imposes similar difficulties, especially for the fact that they possess only a short half-value time of maximally a few days.

Laser atomic fluorescence spectroscopy involves very elaborate instrumentation, with detection limits lying in the lower pg-regions. On the other hand, at present, the glow discharge mass spectrometry (GDMS) promises to become a most powerful detection method. Under favourable operating conditions, it is likely that fg-amounts of platinum can be detected with it. This would mean a most significant step forward in the determination of natural environmental concentration of platinum. No experience has been gathered so far with this technique, so one can assume that some time will pass until real samples can be tested.

For platinum determination, one of the most frequently employed method is the graphite furnace AAS (ETAAS) with a detection limit of a few ng/mL, or 20 to 50 pg absolute, respectively. It is therefore suitable for the determination of omnipresent concentrations of platinum only after appropriate preconcentration, and only in such cases, if these are not lower than 10 to 20 pg/g. The restriction of this method and others is imposed by the maximally possible amount of sample from which platinum can be separated quantitatively. This amount can be estimated to only a few grammes.

Finally, the X-ray fluorescence analysis with total reflection excitation (TRXRF) shall be mentioned. Its power of detection is similar to that of the graphite furnace AAS, but in addition, it facilitates multielement analysis. Quartz targets serve as sample carriers, but also electrodes of pyrographite can be employed, on which traces of platinum can be isolated electrolytically.

Strategies for the Determination of Platinum

In the final part, some details shall be given concerning the strategies, which we developed in order to determine platinum in contaminated and uncontaminated biotic and environmental materials. The respective concentration region extends over 6 orders of magnitude, and it is obvious that such wide range cannot be determined with only one procedure. The requirement was therefore that the individual steps of the procedure should fulfil all operational conditions as stated in the foregoing chapters and an additional prerequisite was the simple operation with routinely employed techniques, that would be available possibly in every analytical laboratory. Wherever these conditions could not be met, compromises had to be made.

For sample decomposition, we principly decided on the technique of high pressure ashing, because it is suitable for almost all of the materials in question as well as for the entire concentration range quoted. Furthermore, it is suitable for routine analysis and sources of error occurring during sample decomposition are reduced to a minimum. A disadvantage is the elaborate technical equipment required, but this is compensated by the named advantages. As a compromise, an open wet digestion may be employed in the µg - mg/g region.

For separating the platinum from the decomposition solution, there are a number of methods available which, after appropriate modification, are suitable for all relevant concentration regions. We decided in this case on separation modes via extraction and sorption.

Concerning the choice of an appropriate determination method, we had to resort to the ETAAS. Although it does not offer the best power of detection, it can be employed in routine and it is available in every laboratory. In some cases, TRXRF was alternatively employed. Great advances are expected in view of the GDMS, but this technique was not available for our current investigations.

The course of the procedure (15) for the determination of platinum in the ng- and µg-region is presented on a flow diagram (Fig. 2). The dried, well homogenized sample is weighed into the decomposition vessel, treated with digestion acids and decomposed in the high-pressure asher. A typical temperature programme is shown in the flow diagram. After cooling, the solution is diluted with three molar hydrochloric acid. The platinum is then extracted in the form of its dithizone complex with isobutyl methyl ketone. The determination of the platinum concentration is made directly in the organic phase with ETAAS. Calibration of the procedure is made with appropriately extracted platinum standard solution. By varying the weighed-in amount, the aliquot part of the analyte solution and the volume of the organic phase, the

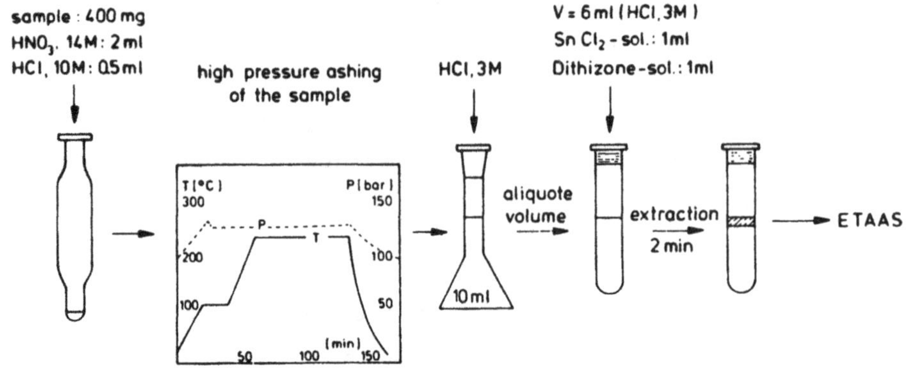

Fig. 2: Flow diagram of the platinum determination procedure (15).

procedure can be modified to be suitable for the different concentration ranges, and it can even be extended into the mg/g-region. Interferences by matrix effects of different sample kinds were not observed. The application of the procedure for the determination of platinum in contaminated plant and soil samples proved to be very successful.

The ETAAS was applied also for the detection of pg-amounts of platinum. This required a concentration of platinum after separation to a volume that could be completely pipetted into the graphite tube, that is to say, a volume of 20 to 50 µL. All attempts to concentrate the organic phase after the just described dithizone extraction failed. Co-extracted iron from the sample and tin, which is required for the reduction of platinum IV to platinum II, as well as excessive dithizone reagent caused very marked signal depressions and unspecific background absorption which proved to be no longer correctable or compensatable. We therefore decided on a multistep procedure. After high pressure decomposition of the sample, the platinum is adsorbed on activated carbon in form of its hexamethylene-dithiocarbamidate complex. The activated carbon is then centrifuged, washed with hydrochloric acid and subjected to a second high pressure decomposition step. From the resulting analyte solution, the platinum is now extracted with trioctylamine and isobutyl methyl ketone. The organic phase is then transferred into a special vessel and is vacuum-reduced to a volume of approx. 20 µL. The remaining volume containing the whole sample platinum is then injected into the graphite tube, where it can be measured practically without interferences. For calibration, it is quite sufficient, if the platinum standard solutions are submitted to the last extraction step and to the vacuum-reduction step. The yield of platinum separation was found to be greater than 90% for the whole procedure. The concentrated organic phase may furthermore be subjected to other suitable determination methods such as GDMS and TRXRF.

With a procedure as outlined above, we hope to possess an analytical tool, especially in connection with GDMS, to facilitate the determination of ubiquitous concentrations of platinum.

Acknowlwedgement: This work was supported by the 'Bundesministerium für Forschung und Technologie' and by the 'Ministerium für Wissenschaft und Forschung des Landes Nordrhein-Westfalen'.

References

1. Rosner, G., R. Hertel. 1986. Staub Reinhaltung der Luft 46, 281-285.

2. Hodge, V.F., M.O. Stallard. 1986. Environ.Sci.Technol. 20, 1058-1060.

3. Nadkarni, R.A., G.H. Morrison. 1977. J.Radioanal.Chem 38, 43-449.

4. Zeisler, R., R.R. Greenberg. 1982. J.Radioanal.Chem. 75, 27-37.

5. Tjioe, P.S., K.J. Volkers, J.J. Kroon, J.J.M. Goeij, S.K. The. 1984. Int.J.Environ.Anal.Chem. 17, 13-24.

6. Hodge V.F., M. Stallard, M. Koide, E.D. Goldberg. 1986. Anal. Chem. 58, 616-620.

7. Valente, I.M., M.J. Minski, P.J. Peterson. 1982. J.Radioanal.Chem. 71, 115-127.

8. Iyengar G.V., W.E. Kollmer, H.J.M. Bowen. 1978. The Elemental Composition of Human Tissues and Body Fluids, Verlag Chemie, Weinheim • New York.

9. Knapp, G. 1984. Fresenius'Z.Anal.Chem. 317, 213-219.

10. Knapp, G., S.E. Raptis, G. Kaiser, G. Tölg, P. Schramel, B. Schreiber. 1981. Fresenius'Z.Anal.Chem. 308, 97-103.

11. McGahan, M.C., M. Tyczkowska. 1987. Spectrochim.Acta 42B, 665-668.

12. Eller, R. 1986. Dissertation, Mainz.

13. Branch, C.H., D. Hutchinson. 1986. J.Anal.At.Spectrom. 1, 433.

14. Rocklin, R.D. 1984. Anal.Chem. 56, 1959-1962.

15. Alt, F., U. Jerono, J. Messerschmidt, G. Tölg. 1988. Submitted to publication.

DETERMINATION OF BASELINE PLATINUM LEVELS IN BIOLOGICAL MATERIALS

Rolf Zeisler and Robert R. Greenberg
Center for Analytical Chemistry, National Bureau of Standards, Gaithersburg, MD 20899

Abstract

A radiochemical neutron activation analysis (RNAA) procedure has been developed that is sufficiently sensitive for the determination of baseline platinum concentrations. This procedure is based on the element-specific separation of gold for interference-free counting, thus including the activation products of platinum, ^{199}Au, and of gold, ^{198}Au, for the determination of platinum and gold. The RNAA procedure has been applied to human liver tissues and biological Standard Reference Materials (SRMs). As an important addition to the investigated materials, the analysis of SRM 2670 "Toxic Metals in Freeze-Dried Urine" for normal and elevated concentration levels is discussed. In this urine pool sample, the natural platinum concentration is 8 ng/L. In human liver samples from 1980 the platinum concentrations ranged from 5 to 57 ng/kg (wet weight).

Introduction

The likelihood of platinum accumulation in the biosphere is increasing with the expanding use of this noble metal in industrial and medical applications. To assess the conceivable role of the element and its compounds as a potential hazard to human and environmental health as well as to better understand its biological effects, investigators need

Trace Element Analytical Chemistry in Medicine and Biology

extensive data on its presence at natural levels. Nevertheless, analytical methods to obtain those measurements have not been readily available. Since the naturally occurring concentrations in biological materials are ordinarily at or below the μg/kg level, the determination of platinum in biological tissues and fluids is a difficult task. As a consequence, baseline levels in a wide range of biological materials are still unknown.

Most of the current analytical work has focussed on the determination of elevated levels of platinum in biological tissues and fluids when investigating the properties, metabolism, pharmacokinetics, toxicity, etc. of platinum compounds that are administered for their antitumor activities. Here, a variety of common analytical techniques is applicable, and analytical chemists are to be concerned more with the reliability and accuracy of the results than with the questions of detectability. The common concentration levels in these investigations are 0.1 to 10 mg/kg. However, natural-matrix certified reference materials that can be used for validation are just emerging and little is known about the accuracy of the techniques used.

Among the established techniques for the determination of these elevated concentrations is neutron activation analysis which has played a major role. Either in its instrumental (INAA) or radiochemical (RNAA) form it has demonstrated its applicability to various research problems (1,2). RNAA procedures were shown to be necessary to achieve lower detection limits for the determination of platinum at natural concentrations.

We have developed a radiochemical neutron activation analysis (RNAA) procedure that is sufficiently sensitive for the determination of baseline platinum concentrations (3). This procedure encompasses the element specific separation of gold products of platinum, ^{199}Au, and of gold, ^{198}Au, for the determination of platinum and gold. The RNAA procedure has been applied to human liver tissues and biological Standard Reference Materials (SRMs). As an important addition to the investigated materials, this contribution discusses in detail the analysis of SRM 2670 "Toxic Metals in Freeze-Dried Urine" for normal and elevated concentration levels.

Experimental Section

The RNAA scheme was based upon a radiochemical separation of the activation products from essentially all other nuclides formed in the biological matrix. The method used the inherent sensitivity of the activation reaction $^{198}Pt(n,\gamma)$ ^{199}Pt and counting of the daughter nuclide ^{199}Au. The remaining interference of the secondary reaction $^{197}Au(n,\gamma)$ $^{198}Au(n,\gamma)^{199}Au$ from gold in the sample was quantitatively assessed and corrected for. Optimization of the irradiation times minimizes this interference.

For the determination of natural concentrations, 20 g of fresh tissue or 20 mL of fluid were used. The materials were prepared for irradiation in a class 100 clean room. The analysis of SRM 2670 used the freeze-dried stock material which was dissolved in 5 mL of ultrapure water (4) to assure a representative sample. This solution was transferred to the irradiation containers. These containers were made from linear polyethylene (LPE) and had been cleaned in aqua regia. The samples were freeze-dried in their containers and the containers were heat-sealed and double bagged in LPE. Standards were prepared from solutions of the pure metals dried on filter paper and sealed similarly in LPE. Samples and standards were packaged together in irradiation rabbits. The samples were irradiated in the RT-3 pneumatic tube facility of the National Bureau of Standards Research Reactor (5) at a fluence rate of $5.6 \cdot 10^{17}$ $n \cdot m^{-2} \cdot s^{-1}$. The natural-level urine was irradiated for 3600 s, biological tissues were irradiated between 900 s and 3600 s depending on their gold concentration, and elevated-level urine samples were irradiated for 36 s. At the midpoint of the irradiation time, the rabbits were retrieved from the reactor, inverted and sent back for continued irradiation. This achieves a nearly homogeneous irradiation of the rather large samples and standards.

Approximately one day after irradiation, the outer sample container was removed and the inner bag containing the sample was opened. The opened bags were allowed to remain for two hours in glass beakers with 50 mL water, 25 mL concentrated nitric and 25 mL concentrated hydrochloric acids, and 85 mg of gold carrier. The bags were washed with water and the washing water was added to the appropriate beaker. A second wash followed with 100 mL of 8 mol/L sulfuric acid for three hours, adding this washing solution also to the sample beakers. An additional 100 mL of water was added to each sample and the solutions were gently heated overnight (80 °C). The next morning the heating rate was increased and the remaining water, nitric acid and hydrochloric acid were allowed to evaporate. The temperature was then slowly increased until the solution was just below the boiling point of sulfuric acid. The solutions were covered with watch glasses and were heated at this temperature until the metallic gold had precipitated. The separated gold samples were sequentially washed in water, nitric acid, water, hydrofluoric acid, water, hydrochloric acid, and finally water. The samples were then heated to 500 °C to ensure that no sulfuric acid remained. They were then cooled, weighed for yield determination, sealed in polyethylene bags and finally counted in a Compton suppressor gamma spectrometer (6). It should be noted that the amines and/or amino acids in urine greatly enhance the formation of the gold precipitate.

Therefore, unirradiated urine or aspartame was added to the solution of biological tissues. This led to more efficient separations of gold in these matrices.

Results and Discussion

Certified biological reference materials are needed to address the question of accurate platinum determinations and to assess the reliability of the various analytical techniques. However, in materials with low natural concentrations, as well as in some materials with elevated concentrations, the necessary determinations for certification by reliable and independent techniques (7) have not yet been accomplished. Table 1 lists the platinum and gold results obtained by this RNAA procedure on Standard Reference Materials. The table gives weighted mean values for platinum and gold and their standard deviation for each set of determinations. Where available, the table includes literature values (8) and other results (9).

Table 1. Platinum and Gold Content of Biological Standard Reference Materials (Concentrations in ng/kg)

SRM No.	Material	Pt	Au
1571	Orchard Leaves	190±50 [3][a]	1430±80 [4]
	Ref. (8):	89000-1200000	970-1800
1572	Citrus Leaves	70±20 [2]	107±4 [2]
1577	Bovine Liver	70±20 [3]	56±7 [3]
2670	Urine, low level	8.3±0.4 [2]	8.0±0.3 [2]
	Urine, elevated	124900±1100 [4]	244400±1200 [4]
	AAS Ref. (9):	106000±4000 [3]	

[a]Weighted mean concentration and standard deviation of the mean with [n] degrees of freedom.

Comparable results for platinum are available only on the SRM 2670 at the elevated concentration. The agreement between the atomic absorption determinations and RNAA would be satisfactory for certification. Nevertheless, possible bias in either technique needs to be evaluated.

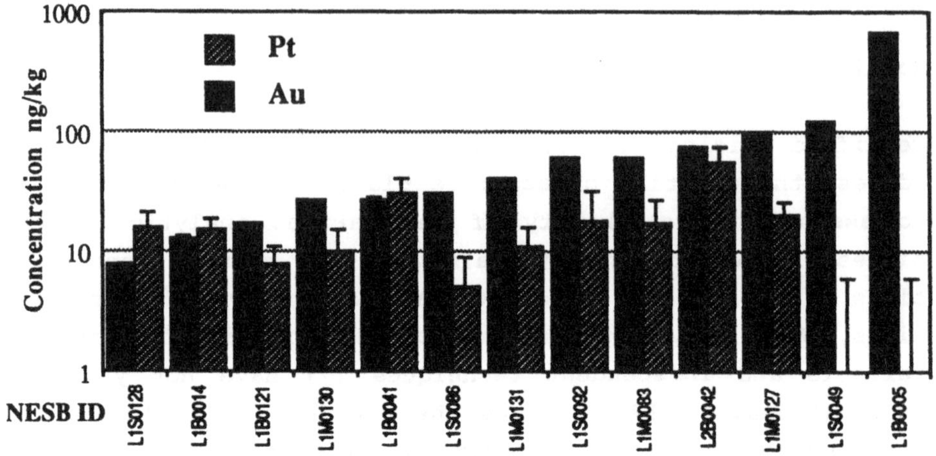

Figure 1. Platinum and gold concentrations in human livers from the 1980 NESB collection. The samples have been ranked towards increasing gold concentration, which have been obtained in all investigated tissues.

The application of the improved RNAA methodology to specimens from the National Environmental Specimen Bank (NESB) (10) yielded results on eleven human livers from the 1980 collection. The results are summarized in Figure 1. The individual data have been ranked according to increasing gold concentrations. The individual specimens were selected after the assay of approximately 30 elements which included INAA determinations of gold. In this way the irradiation could be optimized depending on the gold content of the livers. The baseline levels found in the 1980 human livers are the only data available on past platinum concentrations. Of course, the specimen bank would enable an extensive survey of past samples. These data should then be compared with the current 1988 concentrations to evaluate whether a significant change might have occurred.

Acknowledgement

This work was supported in part by the U.S. Environmental Protection Agency, Office of Research and Development, Office of Health Research.

References

1. Zeisler, R., Lux, F., Seidenberger, H., Schoenenberger, H., and Beck, W., 1979. In: Nuclear Activation Techniques in the Life Sciences 1978. IAEA, Vienna, p. 467.

2. Taskaev, E., Karaivanova, M., Grigorov, T., 1988. J. Radioanal. Nucl. Chem., 120, 75.

3. Zeisler, R. and Greenberg, R. R. 1982. J. Radioanal. Chem., 75, 27.

4. Kuehner, E. C., Alvarez, R., Paulsen, P. J., Murphy, T. J., 1972. Anal. Chem., 44, 2050.

5. Becker, D. A., 1986. In: Technical Activities 1986, Center for Analytical Chemistry. NBSIR 86-3486, U. S. Nat. Bur. Stds., Gaithersburg, MD, p.41.

6. Vogt, J. R., Zeisler, R., 1983. In: NBS Reactor: Summary of Activities. NBS Techn. Note 1178, U. S. Nat. Bur. Stds., Gaithersburg, MD, p.154.

7. ISO Guide 31, 1981. Contents of Certificates of Reference Materials. Ref. No. ISO GUIDE 31-1981 (E).

8. Gladney, E. S., O'Malley, B. T., Roelandts, I., Gills, T.E., 1988. Standard Reference Materials: Compilation of Element Concentration Data for NBS Clinical, Biological, Geological, and Environmental Standard Reference Materials. U. S. Nat. Bur. Stds. Spec. Publ. 260-111, Gaithersburg, MD.

9. Rains, T. C., 1985. Report of Analysis, U. S. Nat. Bur. Stds., Gaithersburg, MD.

10. Wise, S. A. and Zeisler, R., 1984. Environ. Sci. Technol., 18, 302A.

TRACE ELEMENT ANALYSIS OF SKELETAL MUSCLE IN PATIENTS WITH TESTICULAR TUMOUR BEFORE AND AFTER CISPLATINUM TREATMENT

J. Boman, E. Selin

Department of Physics, Chalmers University of Technology, S-412 96 Göteborg, Sweden.

C. Drott, K. Lundholm, B. Unsgaard

Department of Surgery 1 and General Oncology, Sahlgren's University Hospital, S-413 45 Göteborg, Sweden.

Abstract

Trace element analysis was made before and after Cisplatinum treatment on biopsies from skeletal muscle in eight patients suffering from testicular tumour. Energy Dispersive X-Ray Fluorescence (EDXRF) was used as analytical method. Trace element concentrations in blood serum and in urine were followed during the treatment for three of the patients. The concentrations of copper, zinc and selenium in skeletal muscle increased by a factor of two after treatment, whereas the bromine concentration decreased by almost a factor of three. At the onset of the Cisplatinum treatment there was an increase in the excreted bromine in urine and a parallel decrease in the serum concentrations for the patients whose urine and serum levels were followed. Very little is known about the biological function of bromine in humans.

Introduction.

Cisplatinum is a chemotherapeutic agent that has been widely used since the mid 1970:s. It has a remarkable effect on germ cell tumours and is also used against an increasing number of other malignancies, e g gynecological, urinary bladder, head and neck tumours. The dose limiting toxic effects are myelosuppression, ototoxicity, peripheral neuropathy and nephropathy. The nephrotoxic effect is primarily renal tubular damage. Apart from a rise of serum creatinine this may cause renal wasting of certain ions. Cisplatinum induced renal magnesium wasting may lead to hypo-

Trace Element Analytical Chemistry in Medicine and Biology

magnesia. This may induce muscular twitching tetany and general convulsions (1,2,3). It is plausible that other electrolyte disturbances may occur due to losses through the urine. The main ions in the body, e g potassium, sodium, calcium and chlorine are regularly analysed in the clinical routine, and deficiences are compensated by increased intake. Ions of less concentration are seldom analysed but may be wasted during Cisplatinum treatment. These trace elements are important for many metabolic pathways, and depletion might lead to derangements of cellular function and give additional side effects of treatment.
The aim of the present study was to assess skeletal muscle trace elemental concentrations in a group of men suffering from tumour of the testis before and after Cisplatinum treatment, respectively. For a few of the patients trace elements in serum and urine were also measured.

Material and Methods

Eight patients with non-seminomatous germ cell tumours of the testis were studied. Their mean age was 25 years (range 19-44 years). The chemotherapy consisted of Cisplatinum, 20 mg/m^2, intravenously on days 1-5, Vinblastine, 0.15 mg/kg, on days 1-2 and Bleomycin, 30 mg, weekly. Each cycle of treatment had a duration of three weeks. Seven patients received three cycles and one patient four cycles of this regimen. In conjunction with the Cisplatinum administration three litres of a balanced salt solution and 40 mg of Furosemid were given to all patients in order to decrease renal toxicity. Four of the patients were randomized to receive total parenteral nutrition (TPN) during the five days of Cisplatinum administration. TPN included a daily amount of electrolytes and trace elements (calcium 5 mmol, magnesium 9 mmol, iron 50 µmol, manganese 40 µmol, zinc 20 µmol, copper 5 µmol, fluorine 50 µmol and iodine 20 µmol). The remaining four patients received no nutritional or trace element support.

Before and after the treatment a muscle biopsy was taken from the vastus lateralis muscle using a trocar technique. For three of the patients serum samples were taken at regular intervals and 24 hour collections of urine were obtained during the days of Cisplatinum administration. All samples were stored in acid rinsed plastic tubes to avoid trace element contamination. The tubes were immediately frozen and kept at $-70^{o}C$ until analysis. The solid (muscle) samples and the liquid ones (serum, urine) were then subjected to somewhat different procedures. A standard reference solution of 50 µl with known concentrations of Sc and Y was added to 50 µl of the liquid samples, which were placed on thin plastic foils, mounted onto conventional photo frames. The samples were then dried at low temperature (approximately 4 ^{o}C) until they were completely dry.
For the solid samples it was not possible to add any kind of internal standard. Instead about 5 mg of each biopsy was dried at low temperature and pressed in a hydralic press into flat, thin samples. These were then placed directly onto the same kind of thin foils as used for the liquid samples.

Analysis by X-ray Fluorescence

The trace element determinations were made with Energy Dispersive X-Ray Fluorescence (EDXRF) analysis, which is a non-destructive, multielement method (4,5,6,7). The spectrometer used in the present work is based on three axial geometry: The beam paths of [1)]the primary radiation from the X-ray tube, [2)]the exciting radiation from the secondary (molybdenum) target and [3)]the characteristic radiation from the elements in the sample form three mutually orthogonal angles. This geometry has the advantage of reducing the background radiation relative the characteristic signals. The X-radiation from the sample - a dried biological specimen placed on a thin plastic foil as described above - is detected by a Si(Li)-detector. The information in the X-ray spectra is then by use of physical data converted into relative

element concentrations.
For the liquid samples (serum and urine) absolute concentrations will be obtained from comparison with the known concentrations of Sc and Y in the added standard solution. For the muscle samples absolute concentrations will be obtained when one makes use of the information available in the X-ray spectra in the peaks of scattered radiation from the secondary target(6). As trace element determination by EDXRF is independent of the chemical form of the elements the errors in the analyses are mainly due to uncertainties in the physical parameters and to counting statistics. We estimate that the errors range from a few to about thirty per cent for well defined X-ray peaks. For small peaks with few counts the errors may be larger.

Results

There were no apparent differences in the measured element concentrations in muscle between the TPN patients and the patients who did not receive TPN. We have therefore treated them as one group. The results of the measurements for the eight patients before and after Cisplatinum treatment, respectively, are shown in Table 1A and B. The spread in the individual values is - as seen from the table - sometimes rather large, and we therefore present the median as well as the mean values for comparison.
The most significant change in the mean element concentration as a result of Cisplatinum treatment occurs for bromine ($p < 0.003$), which is reduced almost three times. Increases occur for copper ($p < 0.05$), zinc ($p < 0.02$) and selenium ($p < 0.04$). It is interesting to note that the concentrations of potassium and rubidium are practically the same before and after Cisplatinum treatment.
Serum samples were collected and analysed from three of the TPN-patients during the course of the treatment, namely in week 0 (before treatment), and week 1, 4 and 7 (when Cisplatinum was given every day). During these weeks 24 hour collections of urine

Table 1 A and B: Element concentrations in Vastus Lateralis in Eight Patients Suffering from Testicular Tumour Before(1A) and After(1B) Cisplatinum Treatment.

	Element	Mean Concentration (ppm dry weight) ± Standard Error	Median Concentration (ppm dry weight)
A.	Cl	7.600 ± 3.100	4.000
	K	11.500 ± 1.500	10.200
	Ca	240 ± 60	160
	Fe	500 ± 170*	420
	Cu	140 ± 70*	85
	Zn	240 ± 40*	220
	Se	0.7 ± 0.1	0.7
	Br	9.6 ± 2.0	8.6
	Rb	23 ± 2	23
B.	Cl	6.900 ± 2.000	4.600
	K	13.600 ± 2.700	12.500
	Ca	300 ± 60	290
	Fe	430 ± 80	380
	Cu	330 ± 60	330
	Zn	470 ± 80	470
	Se	1.2 ± 0.2	1.1
	Br	3.5 ± 0.7	2.9
	Rb	24 ± 4	25

*Based on seven values due to possible contamination of one sample.

was sampled and analysed for excreted elements each day. An additional serum sample was also analysed for all patients at approximately two years after treatment, but no significant deviations from normal values were found. The serum element concentrations for the three patients were not significally different from reference values, nor did they change appreciably during treatment except for the bromine concentration, which decreased from an abnormally high value (mean value: 11.4 ± 3.6 ppm) to a value within the range of reference values. In parallel with the decrease of bromine in serum - and in muscle - the excretion of bromine in urine increased drastically when the Cis-platinum treatment started. For one of the patients the excreted bromine amounted to about 50 mg per day during the first week of treatment.

Acknowledgement

The authors are grateful for technical assistance from Ing. P. Standzenieks and for financial support from the Swedish Natural Science Foundation.

References

1. Schilsky, R.L., T. Andersson. 1979. Ann.Int.Med. 90, 929-931.
2. Macaulay, V. M., R.H.J. Begent, M.E. Phillips, E.S. Newlands. 1982. Cancer Chemother Pharmacol. 179-181.
3. Vogelsang, N. J., J.L. Torkelson, B.J. Kennedy. 1985. Cancer 56, 2765-2770.
4. Alenius, G., H. Riedl, A. Rindby, E. Selin, P. Standzenieks. 1977. Physica Scripta 15, 220-224.
5. Rindby, A., E. Selin, P. Standzenieks. 1982. In: Trace Element - Analytical Chemistry in Medicine and Biology, Vol 2 (P. Brätter, P. Schramel, eds.). Walter de Gruyter, Berlin·New York. pp 721-727.
6. Selin, E., P. Standzenieks, S. Larsson, A. Rindby. 1983. Report in Swedish to the Swedish Work Environment Fund, 80-0680, Nr709.
7. Larsson, S., I. Karlberg, H-I Petersson, E. Selin. 1986. Trace elements in medicine 3, 5-10.

Analytical and Pharmacokinetic Studies of Pt-Based Antitumor Agents in Biological Fluids

S. Caroli, F. Petrucci, F. La Torre, A. Alimonti, A. Cifani
Istituto Superiore di Sanità, Viale Regina Elena 299, 00161-Rome, Italy

C. Dominici, M.A. Castello
Department of Pediatrics, University "La Sapienza", Viale Regina Elena 324, 00161-Rome, Italy

Abstract

The combination of ICP-AES with HPLC was found to be particularly suitable for the determination of the various chemical forms under which Pt occurs in biological fluids when chemotherapy with Pt-based drugs is carried out. The study was centered mainly on Cisplatin and Carboplatin, for which preliminary in vitro and in vivo results are discussed.

Introduction

With the discovery of the cytotoxic activity of Pt coordination compounds (1), successful clinical application of Cisplatin (cis-diamminedichloroplatinum II) soon followed. This antitumoral drug is, to date, still the most widely used in spite of the fact that it was the first of its kind to be employed. However, in recent years, the high toxicity of Cisplatin has promoted intense research to identify Pt-based substances which have less serious side effects but similar therapeutic efficacy (2, 3). At present the most promising second-generation Pt drug would appear to be Carboplatin (cis-diammine-1,1-cyclobutanedicarboxylate-platinum II) for which preclinical studies have demonstrated a greater therapeutic index compared with the parent compound (4).

Trace Element Analytical Chemistry in Medicine and Biology

These two drugs show one and the same trend in that both are irreversibly bound to intra- and extracellular protein components as well as to DNA (5, 6, 7). In particular, the formation of adducts with extracellular protein moieties, mainly plasma proteins, prevents a large amount of the element from interacting with DNA, a critical target for cytotoxicity. In fact, only the so-called free drug can passively diffuse into malignant cells where substitution of the anionic groups allows the element to covalently bind to DNA, thus inhibiting replication (8). Various in vitro investigations have confirmed that the bond with plasma proteins virtually nullifies the cytotoxic activity of such drugs (9, 10, 11). Consequently, in terms of pharmacokinetic and pharmacodynamic evaluations, the determination of total Pt plasma levels is not a reliable indication of the relationship linking tissue exposure with therapeutical response and toxicity.

On the other hand, it is only on the basis of the concentration of free Pt that a dependable, possibly overestimated, indication of parent drug plasma levels can be had (12). Studies carried out so far have solved the problem of simultaneous assays of plasma aliquots of free and total Pt through centrifugal ultrafiltration of the fluid with microconcentrators with cut-off values of between 10,000 and 50,000 daltons (13, 14). With this procedure, therefore, also the amount of complexed Pt can be determined as the difference between the total concentration of the element and that pertaining to free Pt. It should be clearly borne in mind however, that this approach is not suited for more specific analyses, such as the distribution of Pt in the plasma among various protein species or the investigation of the nature of low molecular weight compounds in which Pt might be present in the ultrafiltrate fraction, these last being considered as the entirely bioavailable aliquot of the element. Thus the aim of this work is to develop a more adequate analytical approach capable of simultaneously separating and quantifying the different protein species to which Pt is bound as well as the free portion of the drug. This was attempted by exploiting the wide investigative potential of Inductively-coupled Plasma Atomic Emission Spectroscopy (ICP-AES) combined with High Performance Liquid Chromatography (HPLC).

Experimental

The study investigated two pharmaceutical products which represent well the ample capabilities of the Pt-drug family, namely Cisplatin and Carboplatin, available commercially as Platamine (Farmitalia Carlo Erba) and Paraplatin (Bristol). Fresh human plasma samples at volumes of 10 and 100 ml, respectively, were incubated with 20 mg of Platamine (corresponding to 56.2 mg l^{-1} Cisplatin) and with 13.8 mg of Paraplatin (corresponding to 69.0 mg l^{-1} Carboplatin). In fact, the composition of the two products is very different; Platamine consists of 12.6 % NaCl, 84.6 % mannitol and 2.8 % Cisplatin while Paraplatin is made up of 50 % mannitol and 50 % Carboplatin. In this way it was possible to attain the same initial concentrations of Pt as the element itself. Incubation times were 0, 2, 4 and 24 h for Cisplatin whereas for Carboplatin also 48, 72 and 120 h were chosen. These additional sampling times were necessary owing to the different binding kinetics of the latter substance, as was ascertained through preliminary assay. An <u>in vivo</u> study was undertaken on a 6-months-old boy affected with hepatoblastoma and treated with Carboplatin at doses of 600 mg m^{-2} body surface area over a period of 48 hours. Blood samples were taken during the infusion at the 24th hour as well as at 0.5, 5, 24 and 72 h after completion of the treatment. In both studies plasma samples were immediately frozen and stored at -30 °C until analysis. Protein species were separated by means of an HPLC apparatus as specified in Table 1.

Table 1 - HPLC Apparatus Specification

Pump	Waters model 510 (USA)
Injector	U 6 K
Column	Pharmacia, Superose 6 preparative grade, HR 16/50
Detector UV	Waters model 450, variable wavelength (USA)
Recorder	Perkin-Elmer model LC-100 Integrator (USA)

The use of a preparative column was necessary because up to 2 ml plasma had to be injected in order to be able to nebulize a sufficient amount of the eluate into the torch (only ca. 3 % of this is actually used in the conventional pneumatic spray system of an ICP spectrometer). Information on the characteristics of the ICP-AES detector used in this research is given in Table 2. It should be emphasized that the chromatograph is directly connected to the spectrometer so that the eluted fractions are sent on-line to the detector. Efficiency of protein separation is more than satisfactory in spite of the fact that a preparative instead of analytical column had to be used. An example of a chromatogram obtained in this manner is given in Fig. 1 at different input amounts of plasma samples.

Table 2 - ICP-AES Apparatus Specification

Spectrometer	Jobin-Yvon 32-38 PI VHR (ISA Instruments, France)
RF Generator	DURR-JY 3248, frequency 56 MHz, maximum power output 2.2 kW
Induction Coil	6 coils, o.d. 32 mm, height 30 mm
Torch	INSA, demountable
Nebulizer	Concentric-pneumatic type with Scott chamber
Monochromator	HR 1000 M, focal length 1 m, Czerny-Turner mounting equipped with a 3600 grooves mm^{-1} holographic plane grating, entrance and exit slit widths: 40 μm
Recorder	Perkin-Elmer model LC-100 integrator (USA)
Computer	Apple II, Version Pascal 2-5

The working conditions chosen to perform the analyses are listed in Table 3. Calibration of the column was accomplished using the reference substances listed in the said table. The relationship logMW = 0.75t + 10.66 could thus be deduced (MW = molecular weight in daltons, t = time of retention in min, r = 0.995). This equation was calculated without taking into account the data pertaining to DNP-L-alanine and dextran blue, the molecular weights of which were respectively lower and higher than the exclusion limits of the column. Even in this case however, when all the data points are included, linearity is still satisfactory (r = 0.93) while the other parameters of the equation are only slightly altered. The transient signals originating in the ICP-AES system by the amounts of Pt present in the eluted fractions were recorded and measured through the peak heights. Preliminary calibrations were achieved by injecting into the torch calibration standards with decreasing quantities of Pt as Carboplatin (20, 10, 5, 2, 1, 0.4 and 0 /ug) in physiological solution. The calibration equation thus arrived at is M = 0.049h - 0.158 (with M = mass of the element in g and h = peak height in mm, r = 0.999).

Table 3 - Operative Parameters of the HPLC-combined ICP-AES Analytical System

A) HPLC	
Solvent	Phosphate buffer at pH 6.8 containing 0.07 M KH_2PO_4, 0.1 M NaCl and $6 \cdot 10^{-4}$ M NaN_3
Flow	1.0 ml min^{-1}
UV wavelength	280.0 nm
Calibrants	Solutions of aldolase, catalase, chymotripsinogen, cytochrome-c, dextran blue, DNP-L-alanine, ferritin, myoglobin, ovalbumin
B) ICP-AES	
Nominal power	1.7 kW
Argon flows	Plasma gas, 16 l min^{-1}; coating gas, 0.4 l min^{-1}; aerosol, 0.5 l min^{-1}
Pt wavelength	(II) 214.4 nm

Finally, the nature and purity of the various protein fractions were further investigated by immunoelectrophoresis in order to confirm their identity (see Table 4).

Table 4 - Working Conditions for Immunoelectrophoretic Analysis

Method	Immunoelectrophoresis according to Grabar and Williams
Buffer	Veronal pH 8.6
Antiserum	Anti-whole human serum; anti-human gamma-globulin
Running conditions	First dimension: 2.5 h at 5 mA/slide; second dimension: overnight in humidity chamber

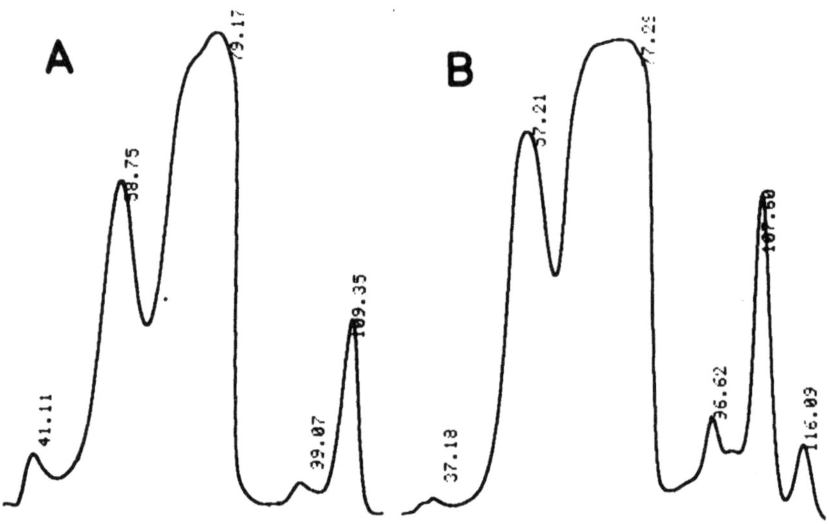

Fig. 1 - HPLC separation of plasma proteins through elution on preparative column, with UV detection. A, 0.5 ml plasma; B, 1.0 ml plasma.

Results and Discussion

The on-line coupling of the two techniques allowed useful information to be gained on the distribution of the two drugs among different protein fractions as well as to quantitate the amount of combined and free Pt. The analytical approach adopted was characterized by a linear dynamic range for element analysis of between 0.2-20 μg, a detection limit as low as 38 ng (2-sigma criterion) and precision of $\pm$ 2.5 % as calculated on 10 consecutive measurements performed on a sample containing 2 μg of Pt as Carboplatin. Recovery experiments were also more than adequate affording amounts as high as 99 % of those injected for collection at the column outlet for both drugs in physiological media. It is worth noting (and to a certain extent even surprising) that Cisplatin and Carboplatin elute with different retention times (112 and 103 min, respectively) notwithstanding the fact that their molecular weights are well below the separation capability of the column employed. This behavior is shown in Fig. 2 and can probably be ascribed to polar interactions occurring with different binding strength between the Pt-containing molecules and the adsorbing support. The correspondence between the UV chromatographic profile recorded for the eluted proteins and the pattern of ICP-AES detection of Pt contained in these fractions can be seen in graphs in Fig. 3. On the basis of these findings it can be assumed that Pt is bound to albumin and globulin species (as confirmed by immunoelectrophoretic assays) as well as to still unidentified protein species of relatively low molecular weight. In fact, plasma samples incubated with Cisplatin for 2 h at 37 °C and then subjected to ultrafiltration with Centricon-10 (Amicon Corp., USA),

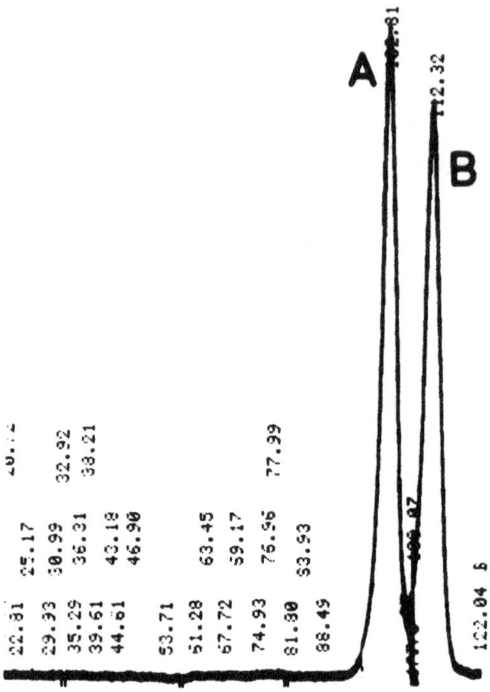

Fig. 2 - Elution of Carboplatin (A) and Cisplatin (B) as detected through ICP-AES detection.

cut-off value at 10,000 daltons, clearly show that there is a not negligible amount of the element linked to compounds of molecular weight below the said threshold. This assumption is further corroborated by the recordings illustrated in Fig. 4. Insofar as the binding kinetics of the Pt drugs is concerned, Table 5 lists the various amounts of bound and free Pt at the pre-established incubation times. For Cisplatin the free aliquot is virtually absent 24 h after starting the experiment while in the case of Carboplatin it is still clearly detectable 120 h later, the half lives being respectively 81 min (time interval 0 - 4 h) and 30 h (time interval 0 - 120 h). Diversity of behavior can be traced back to the different nature of ligands present in the drug molecules. It can be assumed that the relative ease with which Cl groups are released by Cisplatin makes

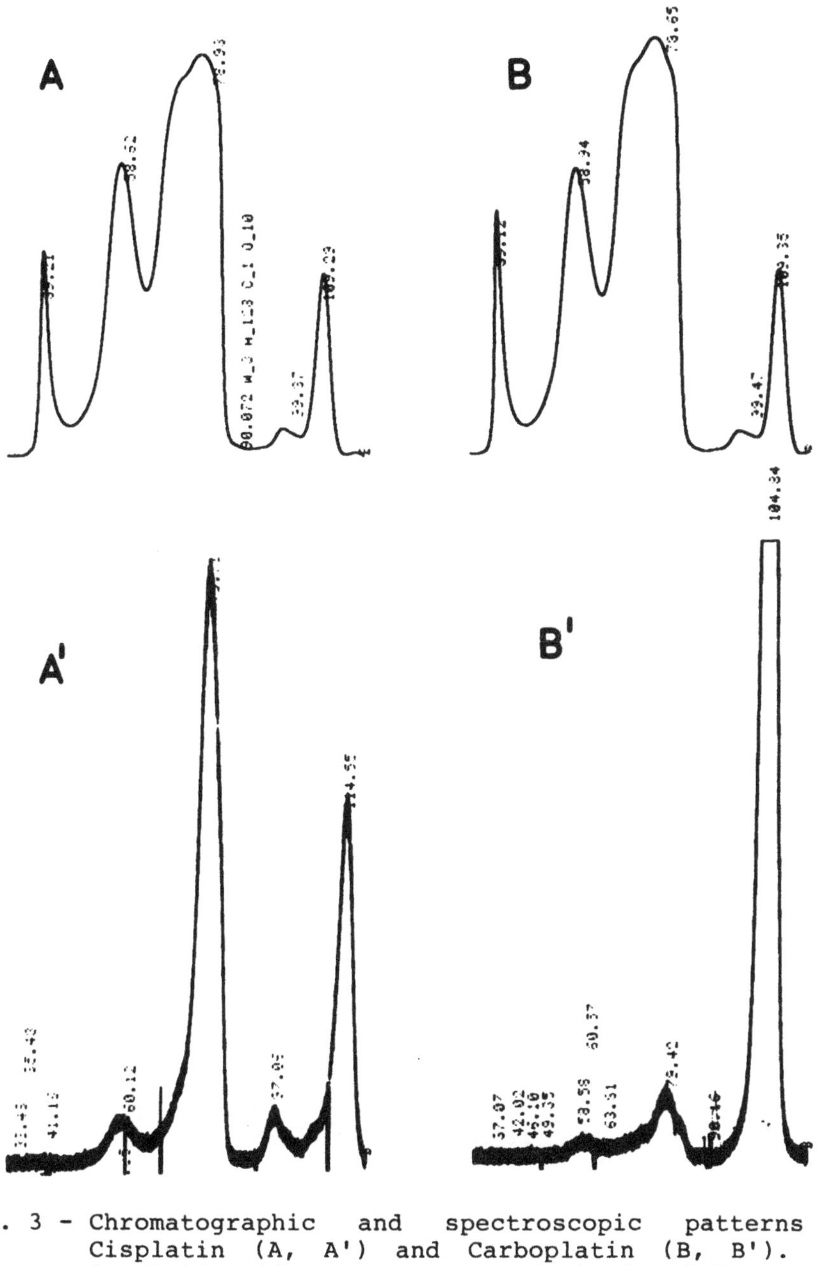

Fig. 3 - Chromatographic and spectroscopic patterns of Cisplatin (A, A') and Carboplatin (B, B'). The incubation times were respectively 2 and 6 h.

Table 5 - In vitro Study: Percentage of Free and Bound Drugs

Incubation Times (h)	Cisplatin Free Pt (%)	Cisplatin Bound Pt (%)	Carboplatin Free Pt (%)	Carboplatin Bound Pt (%)
0	91	9	100	0
2	24	76	80	20
4	10	90	71	29
24	0	100	50	50
48			22	78
72			16	84
120			9	91

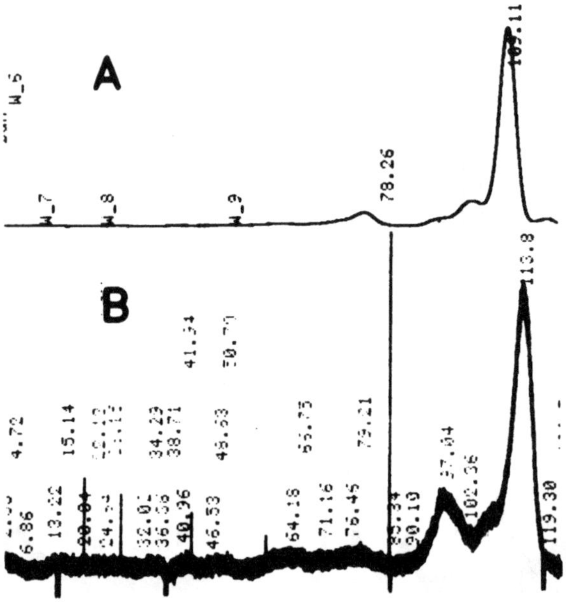

Fig. 4 - Evidence for Pt-protein species of molecular weight less than 10,000. A, HPLC recording; B, ICP-AES recording.

it more prone to form bonds with receptor groups of proteins (15). Predictions of a greater availability of free Pt in the case of Carboplatin are confirmed by the data obtained in the *in vivo* study. In this last situation, however, the plasma levels of free drug are the result of input due to infusion and output due to either diffusion in the extravascular compartment or urinary excretion. Based on previous studies (16, 17), it is known that the urinary excretion of Carboplatin is higher than Cisplatin, and therefore depending on the administered dose, more Carboplatin is excreted and less retained in the body compared to Cisplatin. The greater bioavailability of Carboplatin as predicted by the *in vitro* model, allows a good tissue exposure to the drug, as is demonstrated by the results summarized in Table 6.

Table 6 - *In vivo* Study: Platinum Plasma Levels During and Following Infusion of Carboplatin

Sampling Time	Total Pt (/ug/ml)	Free Pt (%)	Bound Pt (%)
24th h of the infusion	2.90	77	23
30th min following infusion	1.80	53	47
5th h following infusion	1.00	17	83
24th h following infusion	0.95	0	100
72th h following infusion	0.46	0	100

For Carboplatin the percentage of free drug during infusion and towards the end ranges from 77 to 53 %, compared to 5.4 % for Cisplatin as previously reported (18).

The experimental evidence gained in this study sheds further light on what should be meant by the term "free Pt", which in practice, after completion of the infusion, mainly consists of Pt bound to compounds of molecular weight below 10,000 daltons. This is of particular importance as the cytotoxicity of these species has not yet been investigated. In conclusion, the combined analytical system adopted in this study seems to afford an expedient and easy-to-handle approach for investigating some of the basic aspects of the action mechanism of Pt-based drugs.

References

1. Rosenberg, B., L. Van Camp, J.E. Trosko, V.H. Mansour. 1969. Nature 222, 385-386.

2. Burchenal, J.H., K. Kalaher, K. Dew, L. Lokys. 1979. Cancer Treat. Rep. 63, 1493-1498.

3. Wolpert-DeFilippes M.K. 1980. In: Cisplatin. Current Status and New Developments, (A.W. Prestayko, S.T. Crooke and S.K. Carter, eds.). Academic Press, New York, pp 183-191.

4. Rose W.C., J.E. Schurig. 1985. Cancer Treat. Rev. 12 (Supp. A), 1-19.

5. Repta A.J., D.F. Long. 1980. In: Cisplatin. Current Status and New Developments, (A.W. Prestayko, S.T. Crooke and S.K. Carter, eds.). Academic Press, New York, pp 285-304.

6. Gaver R.C., A.M. George, G. Deeb. 1987. Cancer Chemother. Pharmacol. 20, 271-276.

7. Pinto A.L., S.J. Lippard. 1985. Biochim. Biophys. Acta 780, 167-180.

8. Zwelling L.A., K.W. Kohn. 1979. Cancer Treat. Rep. 63, 1439-1444.

9. Gormley P.E., J.M. Bull, A.F. Le Roy, R. Cysyk. 1979. Clin. Pharmacol. Ther. 25, 351-357.

10. Holdener E.E., C.H. Park, R.J. Belt, R.L. Stephens, B. Hoogstraten. 1983. Eur. J. Cancer Clin. Oncol. 19, 515-518.

11. Takahashi K., T. Seki, K. Nishikawa, S. Minamide, M. Iwabuchi, M. Ono, S. Nagamine, H. Horinishi. 1985. Jpn. J. Cancer Res. 76, 68-74.

12. Himmelstein K.J., T.F. Patton, R.J. Belt, S. Taylor, A.J. Repta, L.A. Sternson. 1981. Clin. Pharmacol. Ther. 29, 658-664.

13. Bannister S.J., L.A. Sternson, A.J. Repta, G.W. James. 1977. Clin. Chem. 23, 2258-2262.

14. Dominici C., A. Alimonti, S. Caroli, F. Petrucci, M.A. Castello. 1986. Clin. Chim. Acta 158, 207-215.

15. De Waal W.A.J., F.J.M.J. Maessen, J.C. Kraak. 1987. J. Chromatog. 407, 253-272.

16. Belt, R.J., K.J. Himmelstein, T.F. Patton, S.J. Bannister, L.A. Sternson, A.J. Repta. 1979. Cancer Treat. Rep. 63, 1515-1521.

17. Koeller, J.M., D.L. Trump, K.D. Tutsch, R.H. Earhart, T.E. Davis, D.C. Tormey. 1986. Cancer 57, 222-225.

18. Dominici, C., F. Petrucci, S. Caroli, A. Alimonti, A. Clerico, M.A. Castello: J. Clin. Oncol. (Submitted to).

ALUMINIUM KINETICS DURING HEMODIALYSIS:

EFFECTS OF THE ALUMINIUM CHELATOR DESFERRIOXAMINE

A. Kaenzig, R. Knutti, Ch. Schlatter

Institute of Toxicology, Swiss Federal Institute of Technology,
8603 Schwerzenbach, Switzerland

U. Vogelsang, K. Zaruba

City Hospital Waid, 8037 Zurich, Switzerland

Abstract

The aluminium (Al) chelator desferrioxamine (DFO) enhances the Al elimination from plasma into dialysate during hemodialysis (HD) due to an increase of the plasma Al concentration gradient, presumably as a result of tissue Al mobilization, and due to the favourable kinetic properties of the low molecular weight DFO-Al chelate been formed.

Introduction

Al accumulation may be the cause of vitamin D resistant osteodystrophy, non-irondeficient microcytic anemia and dialysis encephalopathy in patients on chronic intermittent HD [1]. Nowadays dialysate with Al concentration below 10 μg/l and the restricted use of Al-containing phosphate binders are routine measures preventing excessive tissue Al loading.

A therapeutic strategy against Al toxicity symptoms must be able to mobilize Al body stores from different tissues, i.e. bone, brain, liver and skeletal muscle. The chelating agent desferrioxamine has been successfully utilized in the treatment of dialysis patient suffering from Al-related bone disease [4], neurological disorder [6] and anemia [5].

During HD plasma Al is removed across the dialyzer membrane mainly depending on the free concentration gradient between the dialysable (ultrafiltrable) plasma

Trace Element Analytical Chemistry in Medicine and Biology

Al fraction and the Al concentration in the dialysate [2]. In situation of Al concentration in plasma of about 100 µg/l and in dialysate of < 5 µg/l we reported a net Al transfer from plasma into dialysate of 50-500 µg during a 4 hours HD [3]. Following these basic Al transfer studies we examined the different effects of DFO on the Al kinetics during HD.

Patients and Methods

Al kinetics were studied in 4 chronic renal failure patients on a schedule of 4 hours of maintainance HD 3 times a week. Acetate dialysate, polysulfone Hemoflow F60 (Fresenius AG) dialyzers and a single pass dialysate proportioning system (Fresenius MTS 2008) with water purified by reversed osmosis were used [3]. The patients were treated with i.v. administration of 20 or 40 mg of DFO/kg body weight immediately following the HD sessions.

The following samples were taken as described earlier [3]; (1) during the HD sessions before the first DFO dose, and (2) during the HD sessions 44 hours after the first DFO dose:

- Blood ultrafiltrate (2 x 10 ml) was taken both, at the start and at the end of the HD.
- 1-2 ml of blood was taken after 15 and 240 min directly at the inflow and outflow site of the dialyzer into precleaned plastic syringes (Monovette[R], Sarstedt AG) containing 2 mg of Al-free Na_2EDTA. Additional blood samples were taken after 60 and 180 min during the HD 44 hours after the first DFO administration. In each case a second blood sample was taken for the determination of the packed cell volume.
- Dialysate (2 x 10 ml) was collected after 5, 15, 60, 120, 180 and 240 min both, at the inflow and outflow site of the dialyzer into precleaned plastic tubes.

Further sample processing and the Al determination by electrothermal atomic absorption spectrophotometry was carried out as previously described [7].

Plasma Al concentrations at the end of the HD session and at the dialyzer outflow site were corrected for hemoconcentration, because of the loss of plasma fluid by ultrafiltration as measured by an increase of the packed cell volume [3,8].

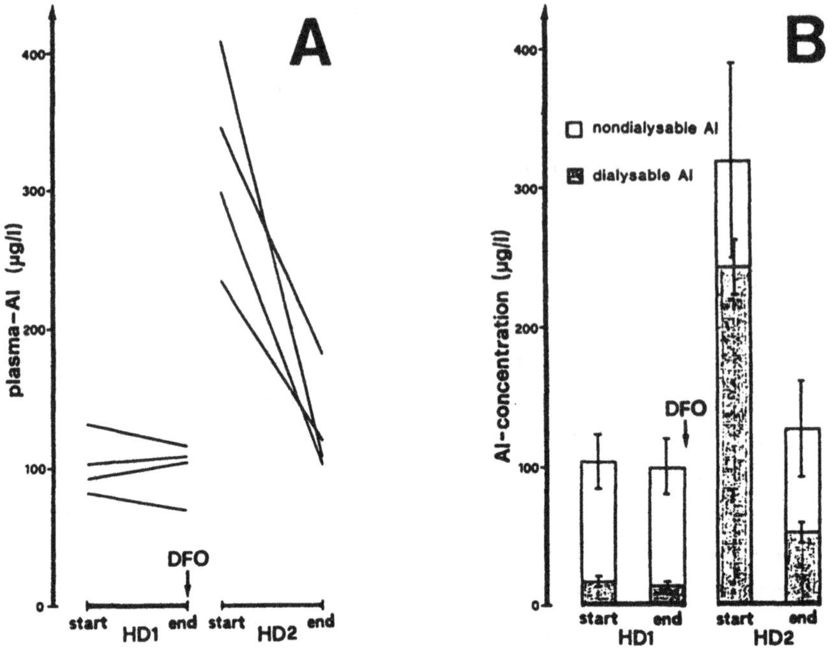

Figure 1A: Plasma Al concentration at the start and at the end of the 4 hours HD before and after DFO treatment.

Figure 1B: Dialysable and nondialysable plasma Al concentration (mean±SD) before and after DFO treatment.

Results and Discussion

There was a distinct increase of the individual predialytic plasma Al levels as mentioned by several authors before, presumably due to the mobilization of accumulated tissue Al (**Figure 1A**). Mean plasma Al increased from 110±14 µg/l before DFO, to 321±77 µg/l (mean±SD) 44 hours after the first DFO administration. Simultaneously to the elevation of the total plasma Al concentration we observed a marked increase of the corresponding ultrafiltrable and therefore dialysable plasma Al fraction as depicted in **Figure 1B**. Mean ultrafiltrate Al fraction increased from 16.1±3.3 % to 76.1±6.3 % (mean±SD); this effect can be explained by the formation of the low molecular weight DFO/Al complex (about 600 dalton). These results are in good agreement with data from literature [9,10].

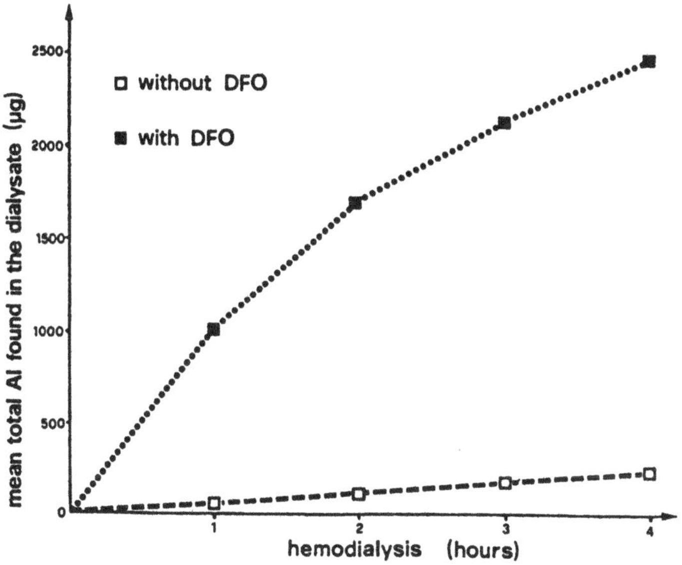

Figure 2: Mean cumulative Al eliminated in dialysate in the course of the 4 hours of HD before DFO treatment and 44 hours after the first DFO administration.

The increase of the total plasma Al concentration and especially the large dialysable fraction under DFO therapy resulted in a tenfold elevation of the mean cumulative Al elimination from 250±140 µg to 2480±830 µg during 4 hours of HD (**Figure 2**). The Al removal during HD remained elevated in all patients throughout the 2-8 weeks of DFO therapy [11].
The good agreement between the total Al amount removed from plasma and simultaneously eliminated in the dialysate as described previously without DFO treatment [3] could be confirmed in this part of the study (**Figure 3**).

The plasma Al clearance (mean±SD) during HD without DFO was only 11.4±4.8 ml/min even with the highly permeable polysulfone membrane, due to the pronounced binding of plasma Al to high molecular weight plasma proteins, e.g. transferrin. Under DFO therapy mean plasma Al clearance was 47±18 ml/min as a result of the increased ultrafiltrable and therefore dialysable plasma Al fraction. These data are consistent with plasma Al clearance values of highly permeable membranes found in literature [12,13]. Conflicting to the findings of

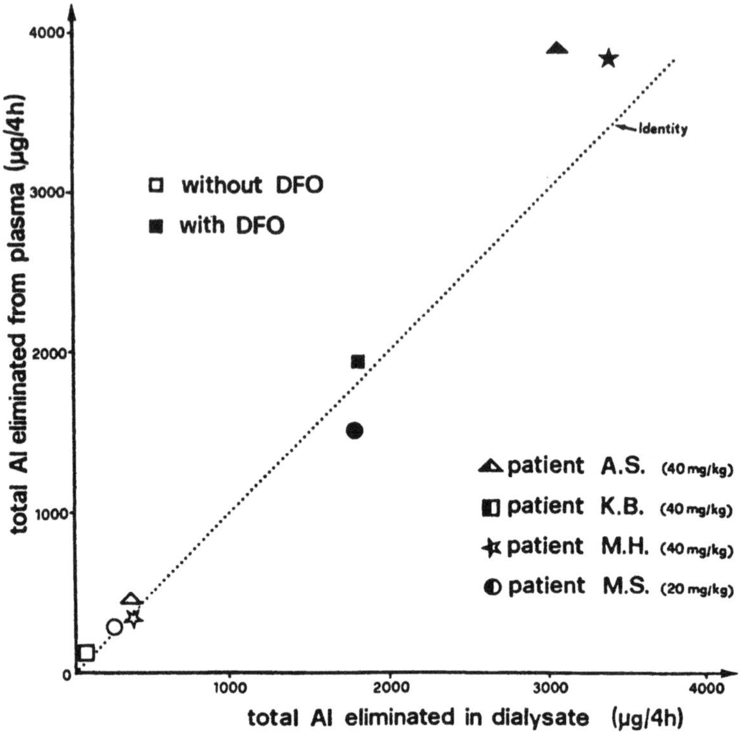

Figure 3: Relationship between plasma Al elimination calculated from arterio-venous plasma Al differences and calculated from dialysate Al concentration differences across the dialyzer, during the 4 hours HD before and after the first DFO dose.

Chang & Barre [13] and Bonal et al. [14] our data showed a higher plasma Al clearance in the presence of DFO during HD also with conventional cuprophane membrane [11].

In the course of the 4 hours HD there was always a continuous decrease of the plasma Al clearance from 62±11 ml/min after 15 minutes, to 30±9 ml/min after 240 minutes of HD (mean±SD), in consequence of the diminishing dialysable plasma Al fraction. However, the calculated Al dialysance values, which consider only the free plasma/dialysate Al concentration gradient remained approximately constant [11]; indicating, that there was no marked permeability reduction of the polysulfone membrane, e.g. by protein adsorption.

Conclusion

This study demonstrates, that the administration of DFO to Al loaded dialysis patients has two major positive effects referring to the Al kinetics during HD:

I. An increase of the total plasma Al, presumably due to the mobilization of accumulated Al in the tissue.

II. An increase of the ultrafiltrable and therefore dialysable plasma Al portion, explainable by the formation of the low molecular weight A/DFO chelate.

References

1. Wills M.R. and J. Savory. 1983. Lancet 2, 29-34.

2. Graf H., H.K. Stummvoll, V. Meissinger, J. Kovarik, A. Wolf, W.F. Pinggera. 1981. Kid. Int. 19, 587-592.

3. Kaenzig A., R. Knutti, Ch. Schlatter, U. Vogelsang, K. Zaruba. 1987. In: Trace Element Analytical Chemistry in Medicine and Biology, Vol. 4 (P. Brätter and P. Schramel, eds.). Walter de Gruyter, Berlin New York. pp. 463-471.

4. Ackrill P., J.P. Day, F.M. Garstang. 1982. Proc EDTA-ERA 19, 203-207

5. Swartz R.D., J. Dombrouski, M. Burnatowska-Hledin, G.H. Mayor. 1985. Proc. EDTA-ERA 22, 101-105.

6. Milne F.J., B. Sharf, P. Bell, A.M. Meyers. 1983. Clin. Nephrol. 20, 202-207.

7. Kaenzig A., R. Knutti, Ch. Schlatter, U. Vogelsang, K. Zaruba. 1987. In: Fortschritte in der atomspektrometrischen Spurenanalytik, Vol. 4 (B. Welz, ed.). pp. 527-538.

8. Cutler R.E., S.C. Forland, P.G. St. John Hammond, J.R. Evans. 1987. Ann. Rev. Pharmacol. Toxicol. 27, 169-191

9. Milliner D.S., G. Hercz, J.H. Miller, J.H. Shinaberger, A.R. Nissenson, J.W. Coburn. 1986. Kid. Int. 29 (Suppl. 18), S100-S103

10. Nambu M., K. Kumano, S. Kusakari, K. Sakurai, T. Sakai. 1986. Jpn. J. Artif. Organs 15, 1219-1222.

11. Kaenzig A.. Ph.D Thesis, ETH Zurich. in preparation.

12. Malberti F., M. Surian, G. Colussi, C. Minoia, A. Salvadeo. 1986. In: Al and other trace elements in renal disease (A. Taylor, ed.). Bailliere Tindall, England. pp. 203-208.

13. Chang T.M.S., P. Barre. 1983. Lancet 2, 1051-1053

14. Bonal J., J. Montoliu, J. Löpez Pedret, E. Bergadä, L. Andrew, L. Revert. 1984. Proc. EDTA-ERA 21, 366-370.

INFLUENCE OF RENAL IMPAIRMENT ON ALUMINIUM KINETICS

Robert A. Yokel, Patrick J. McNamara
College of Pharmacy and Graduate Center for Toxicology, University of Kentucky, Lexington, KY, 40536-0082

Abstract

Partial nephrectomy (Nx) decreased the $t_{\frac{1}{2}}$ of iv aluminium (Al) elimination in rabbits and altered Al distribution, favoring Al disposition into bone. The altered kinetics and enhanced toxicity of Al in Nx rabbits may be due to a greater percentage of smaller molecular weight Al species produced by uremia.

Introduction

Aluminium accumulation has been implicated in the production of toxicity to the central nervous, skeletal and hematopoetic systems. Renally-impaired/anephric patients seem to be more susceptible to Al accumulation and toxicity, presumably due to loss of function of the major organ of Al elimination, the kidneys. However, the influence of reduced renal function on Al kinetics has not been well defined. To assess the influence of renal function on the distribution and elimination of Al, intravenous Al was repeatedly administered to unoperated renally-intact (I) and Nx rabbits. The kinetics of Al following selected iv Al injections was determined. Tissue Al distribution was assessed five weeks after completion of Al injections.

Trace Element Analytical Chemistry in Medicine and Biology

Methods

Each of 71 rabbits, 2.1 - 3.6 kg, was assigned to one of six groups, I-Na, I-Al-50, I-Al-100, Nx-Na, Nx-Al-50, and Nx-Al-100, where Na = sodium and Al = aluminium and 50 and 100 = µmoles of Al/kg (as the lactate) given iv five times weekly for four weeks. Nx rabbits were prepared by the remnant kidney model (1). Blood samples were withdrawn before and ~ 5, 10, 20, 30, 45 and 60 min and 2, 4, 8, 12 and 24 hr after the 1st, 5th and 20th Al injection. Sera were obtained, diluted 11 times with 0.2% HNO_3 and compared to Al standards in 11x diluted rabbit serum matrix by electrothermal atomic absorption spectroscopy (ETAAS) (2). The rabbits were sacrificed five weeks after completion of Al injections. Selected tissues were harvested and analyzed for Al by ETAAS (3). Creatinine clearance (Clcr) was determined in all rabbits. All rabbits were weighed biweekly.

Results

Renal function of Nx rabbits was approximately 23% of I rabbits (Clcr of 0.65 vs. 2.7 ml/min/kg). Creatinine clearance of Nx-Na rabbits improved during and after Na injections. In contrast, Clcr of Nx-I-50 rabbits during and after Al injections averaged 66% of pre-injection Clcr. Renal function of I-Al-100 rabbits during and after injections averaged 58% and 73% of pre-injection Clcr. Injection of 50 µmole Al had less effect on I rabbits. Mortality due to Al and/or Nx was 0, 0, 8, 29, 20 and 100% in the I-Na, I-Al-50, I-Al-100, Nx-Na, Nx-Al-50 and Nx-Al-100 groups. I-Na and I-Al-50 rabbits gained 24% body weight during the nine week study. I-Al-100 and Nx-Na rabbits gained 5 and 10% whereas Nx-Al-50 rabbits lost 3% body weight. Serum Al prior to Al injections averaged 95-140ng/ml (Table 1). Sera obtained 24 hr after the 4th and 19th Al

Table 1. Serum Aluminium Concentration - ng/ml

Group	I-Na	I-Al-50	I-Al-100	Nx-Na	Nx-Al-50
Prior to any Al injection	-	95±30[a]	110± 54	-	140± 64
24 hrs after the 4th daily Al injection	-	147± 49	370±122	-	357±202
24 hrs after the 19th of 20 Al inj	-	427±149	805±329	-	704±177
5 wks after the last Na or Al inj	72±24	187± 32	251±124	96±71	251±134

[a]Mean ± SD of n=12, except I-Al-50(n=9), Nx-Al-50(n=14, except 5 weeks after last Al injection n=12).

injection and 5 weeks after the last iv Al injection demonstrated an elevation of Al (Table 1) suggesting a persistence of Al in the blood, and perhaps elsewhere, long after termination of Al exposure. Serum Al after iv Al injections decreased in an apparent biphasic manner. The elimination $t_{\frac{1}{2}}$ was shorter in Nx than in I rabbits, (approximating 10 and 22 hours respectively, although $t_{\frac{1}{2}}$ could not be accurately estimated because serum Al was already quite low 4 hr after Al injection and samples could not be obtained beyond 24 hr after Al injection). The shorter $t_{\frac{1}{2}}$ of Al in Nx rabbits, in spite of decreased renal function, appears to be due to the greater % of ultra-filterable Al species (< 30,000 daltons) in serum from Nx rabbits than I rabbits (Figure 1). Tissue Al concentrations five weeks after Al injections (Table 2) demonstrated a significant Al accumulation in adrenal, bone, heart, kidney, liver, lung, spleen, stomach and some CNS regions. The only significant difference between I-Al-50 and Nx-Al-50 rabbits

Table 2. Tissue Aluminium Concentration - μg/g Dry Tissue [Organ Aluminium Content as Percentage of Body Burden]

	Renally-Intact Rabbits			Partially-Nephrectomized Rabbits	
Group	(I-Na)	(I-Al-50)	(I-Al-100)	(Nx-Na)	(Nx-Al-50)
Tissue					
Brain-frontal gray	0.8±0.5[a.)]	1.2±0.2	1.8±0.9[b.)]	0.8±0.9	1.9±0.9[b.)]
Brain-white	0.7±1.1	1.0±1.0	0.7±0.6	0.3±0.4	0.7±0.9
Brain-hippocampus	1.0±0.9	1.3±0.5	1.8±0.5	0.2±0.2	1.8±0.9[b.)]
Brain-cerebellum	0.6±0.8	1.3±0.7	1.0±0.6	0.5±0.5	1.0±0.3[b.)]
Spinal cord	0.8±0.7	0.8±0.5	1.3±1.1	0.9±0.9	1.2±0.8
Sum of CNS[c.)]		[.006]	[.004]		[.01]
Adrenal	0.5±0.3	13±5[b.)] [.01]	24±5[b.)] [.01]	0.2±0.2	10±6[b.)] [.01]
Bone	28±10	55±25[b.)] [26]	79±26[b.)] [23]	29±17	91±25[b.)d.)] [58]
Heart	1.9±1.1	6.9±6.2[b.)] [.05]	9.8±1.6[b.)] [.04]	1.2±0.8	9.9±2.2[b.)] [.08]
Kidney-cortex	0.4±0.4	46±29[b.)] [0.3]	113±72[b.)] [0.4]	0.8±0.6	44±41[b.)] [0.3]
Kidney-medulla	0.5±0.4	18±16[b.)] [.06]	26±5[b.)] [.04]	0.8±0.8	41±29[b.)] [.09]
Liver	0.9±0.5	424±270[b.)] [64]	908±401[b.)] [62]	0.6±0.6	275±69[b.)] [37]
Lung	0.7±0.5	19±6[b.)] [0.2]	111±46[b.)] [0.5]	1.4±1.2	20±11[b.)] [.02]
Muscle	0.4±0.4	0.4±0.4 [1.5]	1.7±1.3 [0.7]	0.9±0.5	1.1±0.7 [1.4]
Spleen	0.6±0.6	3882±2101[b.)] [7.2]	14419±5357[b.)] [12]	0.7±0.8	846±742[b.)] [1.4]
Stomach	1.2±1.3	10±5[b.)] [0.8]	26±9[b.)] [0.9]	1.8±1.1	33±17[b.)] [2.2]

a.) Mean ± SD of n = 12 except I-Al-50 rabbits, where n = 9.

b.) Different from rabbits with same renal status receiving 0 Al, by t-test at $p < 0.05$ after significant ANOVA at $p < 0.05$

c.) Estimate of total central nervous system Al content based on average Al concentration of 5 regions assayed times CNS weight.

d.) Different from I rabbits receiving the same Al injection dose by t-test at $p < 0.05$.

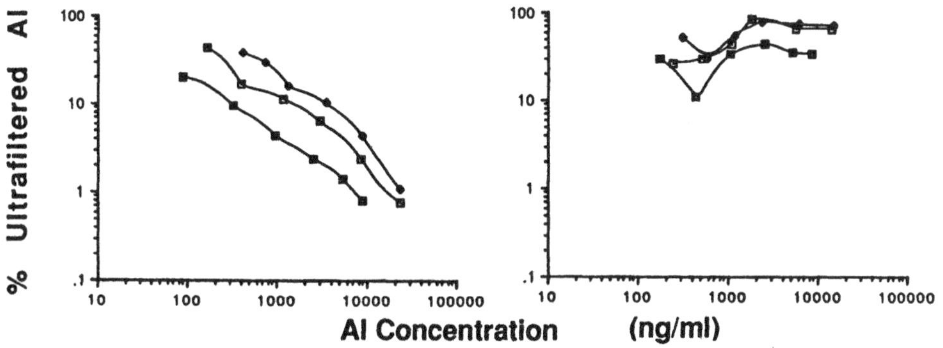

Figure 1. The percentage of total Al that was ultrafilterable in I (left panel) and Nx (right panel) rabbit serum incubated with Al _in vitro_. Results are from 3 rabbits in each group. The placement of the lines was determined by graphic interpolation. From (4).

was an increase in bone Al in the latter group. Calculation of the total Al in the organs sampled (Al concentration x estimated organ weight) minus the Al content of Na treated rabbits showed that 20 - 23% of the total Al administered was still in the I-Al-50, I-Al-100 and Nx-Al-50 rabbits. Bone accounted for 58% of the total estimated Al in Nx-50 rabbits compared to 23-26% in I-Al-100 and I-Al-50 rabbits.

Discussion

Al rapidly leaves the blood by distribution into tissues and by renal elimination. Although Nx decreases Al clearance, the reduction in Al clearance is not as great as expected from the decrease in Clcr because uremia increases the percentage of Al that is ultrafilterable. Nevertheless, serum Al was higher in Nx-Al-50 than I-Al-50 rabbits. The total body burden of Al in Nx rabbits was not greatly affected by Nx, compared to I rabbits, although more Al was in the bone of Nx rabbits. Al appeared to be nephrotoxic as evidenced by decreased Clcr. Greater toxicity was observed in Nx than I rabbits receiving the same Al exposure, as evidenced by increased mortality and greater inhibition of body weight gain. Only bone Al was significantly greater in Nx-Al-50 rabbits than I-Al-50 rabbits. This suggests that one or more factors other than total Al concentration must contribute to the increased toxicity of Al in uremia. Uremia evidently alters the chemical species of Al, increasing the percentage of smaller molecular weight (filterable) species, which may have different distributional properties than larger Al species. This is suggested by the increased deposition of Al into bone of Nx rabbits. These smaller species may be more toxic. Other concurrent changes associated with uremia (for example, accumulation of metabolic products and pH changes) may potentiate Al toxicity. A total lack of renal function, as in end stage renal failure, would be expected to result in significantly different Al kinetics than observed in the present study.

Acknowledgement

Supported by NIH grants RO1 ES02676 and KO4 ES00174.

References

1. Heimberg, K.W., C. Matthews, E. Ritz, J. Augustin and W. Hasselbach. 1976. Europ. J. Biochem. 61, 2070.

2. Oster, O. 1981. Clin. Chim. Acta. 114, 53.

3. Yokel, R.A. and J.M. Melograna. 1983. Biol. Tr. Elem. Res. 5, 225.

4. Yokel, R.A. and P.J. McNamara. 1987. Fed. Proc. 46:4, 1147.

THE INFLUENCE OF SUCRALFATE ON SERUM ALUMINIUM LEVELS IN HEALTHY VOLUNTEERS

M. Benedik, J. Drinovec

University Medical Center, Department of Nephrology, Center for Dialysis, 61000 Ljubljana, Zaloška 7, Yugoslavia

E. Kozak, V. Hudnik

Boris Kidrič Institute of Chemistry, 61115 Ljubljana, Hajdrihova 19, Yugoslavia

Abstract

Serum aluminium levels were measured in 8 healthy volunteers (4 men, 4 women) with mean age 38.3 ± 4.7 years before and at the end of a 6-day treatment of sucralfate in granular form. During the study, no other medication was used. Serum aluminium levels were determined by electrothermal atomic absorption spectrometry(ETAAS). Before and at the end of sucralfate treatment serum aluminium levels were less than 10 µg/l, i.e., below the detection limits of our laboratory. It can therefore be concluded that the use of sucralfate in subjects with normal renal function does not significantly influence serum aluminium levels.

Introduction

Aluminium is one of the most common elements in the earth's crust. The total body content of aluminium in a healthy subject is small - only about 30 mg. In persons with normal renal function, aluminium which is reabsorbed after ingestion is cleared by the kidneys (1). Aluminium - containing drugs

Trace Element Analytical Chemistry in Medicine and Biology

are an important source of aluminium (2). In patients with renal failure, dialysis encephalopathy, aluminium osteopathy, non-iron deficient microcytic hypochromic anemia, and porphyria cutanea tarda, has been described as a consequence of chronic aluminium exposure (3).
Sucralfate, a complex of aluminium hydroxide and sulfated sucrose is currently used for the treatment of peptic ulcer disease. It forms a protective barrier on the ulcer base and protects the ulcer from the effects of pepsin, acid and bile salts (4,5). Brogden at al. proved that, after ingestion of ^{14}C labelled sucralfate, only 0.5 to 2.2 % of a dose was excreted in the urine over a 4-day period. In animal studies, 85 to 95 % of a dose was located in the gastrointestinal tract (5). Kinoshita et al. demonstrated that plasma aluminium levels in patients with normal renal function, who were on long-term sucralfate therapy, were not significantly different from plasma aluminium levels in control subject (6).

Subjects and methods

9 healthy volunteers were included in the study after informed consent. One volunteer dropped out from the study, due to epigastralgia during sucralfate ingestion. 4 men, 4 women, with mean age 38.3 ± 4.7 years completed the study. Before the beginning of the study, venous blood sample was taken to determine serum creatinine level. For its analysis, standard method of our laboratory was used. Sucralfate in granular form was given to every volunteer during the 6 days, 4 g daily, divided into four doses. The first venous blood sample (6 ml) was taken before the first dose of sucralfate in the morning on an empty stomach. The second venous blood sample was taken 12 hours after the last dose of sucralfate in the morning on an empty stomach. During the study, neither antacids or any other drugs were taken. Venous blood samples

were taken by disposable needles into polyethylene vessels. Immediately after collection the serum was separated by centrifugation for 10 minutes at 3000 rpm. Before they were analysed, serum samples were stored in polyethylene vessels at -20°C. Serum aluminium levels were determined by electrothermal atomic absorption spectrometry (ETAAS). Measurements were made with a Perkin Elmer 300 S spectrometer equipped with HGA 72 graphite tube furnace. Measurement conditions are provided in the Table.

Table. Measurement Conditions

Wavelength	309.3 nm
Spectral bandwith	0.7 nm
Drying	150°C, 30 s
Ashing	1400°C, 50 s
Atomization	2700°C, 7 s
Purge gas	argon 50 ml/min
Sample volume	10 µl

Results

Mean serum creatinine level before the beginning of the study was 84±15 µmol/l (mean±SD). Mean serum aluminium level in 8 healthy volunteers before the beginning of sucralfate therapy was less than 10 µg/l. At the end of a 6-day therapy with sucralfate, serum aluminium level was also less than 10 µg/l, i.e., below the detection limits of our laboratory.

Discussion and conclusion

Much is known about the accummulation of aluminium in patients with chronic renal failure (3). Kaehny et al. showed that aluminium may be absorbed from the gastrointestinal tract in great quantities after administration of aluminium hydroxide (7). Knoll et al. proved that in healthy volunteers serum aluminium levels increased as early as 1 hour after the ingestion of 2.4 g of aluminium hydroxide and remained increased for 6 hours after ingestion. The excretion of aluminium in urine was increased for 4 days (8). Kellinghaus and Graefe compared the influence of 3 different, single-dose phosphate binders (including aluminium) on serum aluminium levels in patients with renal failure not yet on dialysis with healthy subject. The resorption of aluminium from these phosphate binders was nearly identical in uremic patients and healthy controls (9).

The ingestion of aluminium contained in sucralfate is quite great. As Lione points out, sucralfate treatment, 4 g daily, corresponds to 828 mg of aluminium intake per day, because 1 g of sucralfate contains 207 mg of aluminium (2). Leung et al. comparatively studied the influence of aluminium hydroxide and sucralfate on serum aluminium levels in 12 patients with end stage renal failure. 6 patients received continuous ambulatory peritoneal dialysis (CAPD) and 6 patients received hospital hemodialysis 12 hours per week. Serum aluminium levels were not significantly different when either aluminium hydroxide or sucralfate were used. When aluminium hydroxide was not given for 3 weeks, serum aluminium levels were significantly lower (10). Kinoshita et al. compared plasma aluminium levels in 17 patients with ulcer disease on chronic therapy with sucralfate for 8 to 10 weeks. Daily dose was 4 g per patient. Control group consisted of 6 subjects who did not take any aluminium-containing antacids or sucralfate for one year. Plasma aluminium levels in patients on chronic sucralfate therapy were not significantly different from plasma aluminium levels in control group (8.41 ± 6.0 µg/l versus 6.5 ± 7.87 µg/l). In this study urinary aluminium excretion was not measured (6).

In our study serum aluminium levels before and after the 6-day administration of sucralfate were the same, i.e., below 10 µg/l. It is possible that insignificant differences of serum aluminium levels existed, but this was not proved by our laboratory.

It can be concluded from our study that a short sucralfate therapy in persons with normal renal function has no significant influence on serum aluminium levels.

References

1. Ott, S.M. 1985. Amer.J.Kidney.Dis. 6, 297-301.

2. Lione,A. 1985. Gen.Pharmac. 16, 223-228.

3. Savory, J., R.L. Bertholf, M.R. Wills. 1985. Clin.End. Metab. 14, 681-702.

4. Anonymous. 1984. Med.Lett. 26, 43-44.

5. Brogden, R.N., R.C. Heel, T.M. Speight, G.S. Avery. 1984. Drugs. 27, 194-209.

6. Kinoshita, H., K. Kumaki, H. Nakano, K. Tsuyama, R. Nagashima, M. Okada, and B. McGraw. 1982. Res.Commun. Chem.Pathol.Pharmacol. 35, 515-517.

7. Kaehny, W.D., A.P. Hegg, A.C. Alfrey. 1977. New.Eng.J. Med. 296, 1389-1390.

8. Knoll, O., H. Kellinghaus, H.P. Bertram, H. Zumkley, U. Graefe. 1984. Contr.Nephrol. 38, 24-31.

9. Kellinghaus, H. and U. Graefe. 1985. Nieren-und Hochdruckkrankheiten, 14, S. 36-40.

10. Leung, A.C.T., I.S. Henderson, D.J. Halls, J.W. Dobbie. 1983. Brit.Med.J. 286, 1379-1381.

THE INTESTINAL SPECIATION OF GALLIUM: POSSIBLE MODELS TO DESCRIBE THE BIOAVAILABILITY OF ALUMINIUM.

Farrar G, Morton A.P., and Blair J.A.

Biology Division, Dept Pharmaceutical Sciences, Aston University, Aston Triangle, Birmingham, B4 7ET, UK.

ABSTRACT

The binding of aluminium to certain ligands in the intestine may alter the bioavailability and subsequently affect the neurotoxic nature of aluminium. The ligands maltol, citrate and fluoride, common additions to food and water supplies have been shown to affect the bioavailability of gallium, a Group 3 element chemically similar to aluminium. Maltol and citrate enhanced gallium absorption from the small intestine in fasted rats only. Fluoride however, inhibited the uptake of gallium in both fed and fasted rats. Nutritional status is therefore an important factor when considering the bioavailability of aluminium. A large molecular weight (M.W.) species has been identified in the soluble phase of gut washings after the oral administration of gallium. Although present in both fed and fasted rats the large M.W. species is predominant in the fed state. It is possible that the formation of this large M.W. species restricts the absorption of gallium in the fed state.

INTRODUCTION

The neurotoxicity of aluminium is well established in the aetiology of dialysis encephalopathy (1) and dialysis osteomalacia (2). It has also been suggested as a possible causative agent in the development of senile dementia of the Alzheimer type (3) and Parkinsonian dementia of Guam (4). Since water- borne exposure to aluminium in dialysate fluids has largely been eliminated by haemofiltration methods (5) it is thought that the major exposure route to aluminium is via the gastrointestinal tract.

Chemical reactions in the gastrointestinal tract between the hydrolysed Al^{3+} cation and possible complexing agents may be of importance. If complexes with Al^{3+} form which promote uptake of aluminium by the epithelial cells of the small intestine there may be increased neurotoxic implications. Conversely aluminium complexes which form, but are not absorbed to any appreciable extent may be useful as a binding agent preventing uptake in people who are exposed to unusually large quantities of Al^{3+}.

Our laboratory has begun preliminary investigations into three possible such ligands, citrate, maltol and fluoride. Using the chemically similar Group 3 element gallium (radioisotope ^{67}Ga, half life 78 hours) as a marker for aluminium (6) *in vivo* tissue distribution and *in vitro* everted sac experiments were performed in the rat to assess the effects of the above ligands on the bioavailability of gallium. Gel- filtration and ion exchange chromatography were also used to investigate the species present in the soluble phase of gut washings after oral administration of gallium.

Trace Element Analytical Chemistry in Medicine and Biology

MATERIALS AND METHODS

IN VIVO INTESTINAL ABSORPTION EXPERIMENTS

Wistar rats (180g) were either fed or fasted and given an oral dose of the appropriate solution (see table 1 for exact dosing regimes). 4 hours after administration the rat was sacrificed by stunning and cervical dislocation. All major organs, the femur and a portion of skeletal muscle were removed for determination of ^{67}Ga.

EVERTED SAC EXPERIMENTS

Everted jejunal sacs 7cm in length were prepared using the method of Wilson and Wiseman (7) and filled with approximately 0.5ml Krebs Henseleit bicarbonate buffer (KHBB). All gut sacs were incubated for 20 minutes (at 37°C and bubbled with 95% O_2/5% CO_2) in 10mls of KHBB containing $Ga(NO_3)_3$ (2μM), 5-10μCi ^{67}Ga and the appropriate concentration of citrate, maltol and fluoride (see results table 2, for exact concentrations).

GEL FITRATION CHROMATOGRAPHY (G.F.C.)

Sephadex G15 was eluted with 25mM Tris-HCl and 100mM NaCl, pH 7.4 at a flow rate of 3ml min^{-1} on a 60 x 1.5cm glass column.

The small intestine of animals that had been orally dosed with gallium was excised and washed with 20ml isotonic saline. The washings were centrifuged at 40K rpm for 1 hour and applied to the G15 column for speciation analysis.

RESULTS

In all *in vivo* intestinal absorption experiments significantly more gallium was identified in the tissues in the fasted state when compared to the fed (table 1).

Chromatographs of the soluble phase of gut washings after oral administration of gallium have identified both a large and low molecular weight species, peaks one and two respectively (figures 1a and 1b). Peak one was significantly greater in fed rats than fasted rats (fed rats, average peak one percentage - 42% ±1.59, fasted rats, average peak one percentage - 26% ±2.70, fed peak one > fasted peak one, $P< 0.1\%$, df=9). Peak two is thought to be associated with the hydrated gallium ion which alone eluted at fraction 18.

The food additive maltol was found to enhance the uptake of gallium after oral administration in fasted rats only (table 1). Similar observations were made with the acidity regulator citrate (table 1). Fluoride however, was shown to inhibit the uptake of gallium in both fed and fasted rats (table 1).

In jejunal everted sacs, the presence of maltol and citrate in the mucosal incubating buffer promoted the serosal transfer of gallium (table 2). The transport of gallium however, was reduced in the presence of fluoride (table 2).

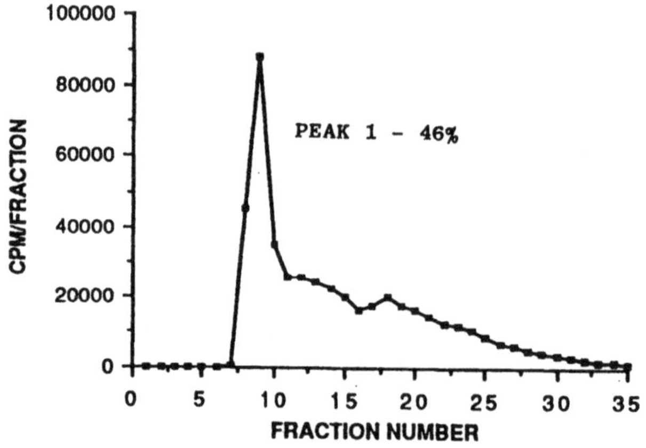

FIGURE 1A. G15 chromatograph of soluble gut washings from FED rat.

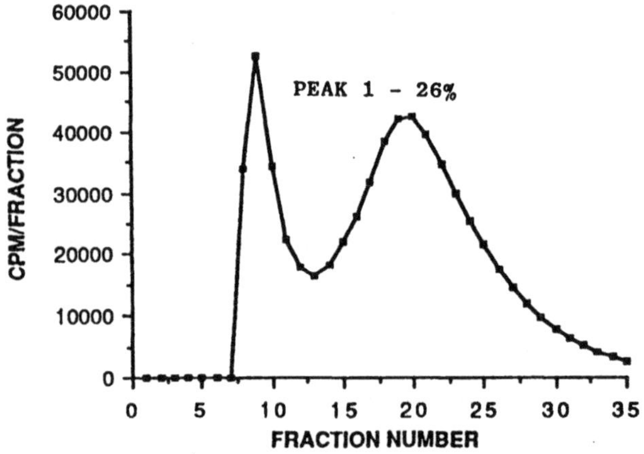

FIGURE 1B. G15 chromatograph of soluble gut washings from FASTED rat.

TABLE ONE. THE EFFECT OF MALTOL, CITRATE AND FLUORIDE ON THE *IN VIVO* INTESTINAL ABSORPTION OF GALLIUM IN FED AND FASTED RATS.

DOSE (0.3mls)	FED	FASTED
Control, gallium only (10^{-3}M)	0.0766 ±0.0066	0.342 ±0.046
Test, gallium/maltol (10^{-3}M/10^{-2}M)	0.0681 ±0.0113	0.664 ±0.0749**
Control, gallium only (10^{-6}M)	0.0311 ±0.0026	0.158 ±0.013
Test, gallium/citrate (10^{-6}M/5 x 10^{-4}M)	0.0281 ±0.0027	0.408 ±0.068**
Test, gallium/fluoride (10^{-6}M/2.5 x 10^{-4}M)	0.0187 ±0.0021**	0.0395 ±0.0026***

Notes.
1) *P<5%, **P<1%, ***P<0.1% (unpaired students t-test), indicating significant differences from the control dose (gallium only) of the group. N=5(±SEM) in all cases.
2) The above values represent the sum of ^{67}Ga identified in liver, kidney, spleen, heart, brain, femur and muscle.

TABLE TWO. THE EFFECT OF MALTOL, CITRATE AND FLUORIDE ON THE MUCOSAL UPTAKE AND SEROSAL TRANSFER OF GALLIUM IN MID-JEJUNAL EVERTED SACS.

INCUBATING MEDIUM (KHBB containing)	MUCOSAL UPTAKE	SEROSAL TRANSFER
Gallium only- 500ng/ml (2μM)	0.561 ±0.034	0.0070 ±0.0013
Gallium/maltol 2μM/20μM	0.47 ±0.078	0.011 ±0.00073*
Gallium/citrate 2μM/1mM	0.43 ±0.063	0.0138±0.00019***
Gallium/fluoride 2μM/0.5mM	0.66 ±0.063	0.0034 ±0.00038*

Notes.
1) Results expressed as ng Ga/mg wet weight tissue, N=5 (±SEM).
2) *P<5%, **P<1%, ***P<0.1% (unpaired students t-test) indicating significant differences from the control (gallium only) experiment.

DISCUSSION

Nutritional status (i.e. fed or fasted) appears to be a important contributor to the bioavailability of gallium. Significantly more gallium was observed in the tissues of all fasted rats when compared to the fed (table 1). Gel filtration chromatography has identified both large and small molecular weight species binding to gallium in the soluble phase of gut washings in both fed and fasted rats. The large molecular weight species identified in both fed and fasted rats was significantly increased in fed rats (figure 1a). The smaller weight peak co-chromatographed with the hydrated gallium ion and was present to a greater extent in the washings of the fasted rat (figure 1b). When the large molecular weight species was redosed to both fed and fasted rats gallium was not detected in any of the tissues examined. The formation of this as yet unidentified species in combination with a reduction in the concentration of the free hydrated gallium ion may be responsible for the reduced absorption of gallium in the fed state.

All 3 species under investigation (maltol, citrate and fluoride) were observed to effect gallium uptake in fasted animals and jejunal everted sacs (tables 1 & 2). It is likely that the gallium-ligand complex has formed under these circumstances and the ligand exerts its effect upon the bioavailability of gallium. The enhanced uptake of gallium in the presence of maltol and citrate is probably indicative of the formation of neutral lipid soluble complexes which can travel across the epithelial cell layer intracellularly rather than through the limited paracellular route (8,9). Ion exchange chromatography (I.E.C.)has shown the gallium-maltol species to be present only in the fasted state. In the fed state it is likely to be decomposed. I.E.C. has also shown $Fe^{3+}$to displace gallium from the gallium-maltol complex leaving gallium in the free hydrated form. In the fed state free gallium is thought to be associated with insoluble phosphates and hydroxides (8) which in combination with the large molecular species identified significantly reduce the transport of gallium. It is likely that gallium and citrate follow a similar pattern of speciation in the lumen of the small intestine.

Fluoride is known to form tight complexes with aluminium (8), and unlike maltol and citrate influences the bioavailability of gallium in both fed and fasted rats. It is likely that fluoride resists competition from other ligands in food for gallium and hence is effective in both nutritional states.

The jejunal everted sac results were similar to those of the fasted rat. Analysis by gel filtration and ion exchange chromatography indicated that the species present in the mucosal incubating fluids were similar to the species present in the intestinal lumen of the fasted rat.

The results outline the importance of nutritional status and food and water additives in influencing the bioavailability and possible neurotoxicity of aluminium. Any further understanding into the mechanisms of intestinal aluminium absorption need to carefully consider the complex interactions between dietary ligands and aluminium.

REFERENCES

1. Wills M. R. & Savory J. Lancet ii, 29 (1983)

2. Malluche H. H. & Faugere M. Am. J. Kid. Dis. 6, (5), 336 (1985)

3. Perl D. Env. Health. Persp. 63, 149 (1985)

4. Garruto R. M. et. al., Proc. Nat. Acad. Sci. USA. 81, 1875 (1984)

5. Ward M. K. et al., Lancet i, 841 (1978)

6. Farrar G, Morton A.P, Blair J. A. In 'Proceedings of Conference, Heavy Metals in the Environment, New Orleans, Vol 2 pp14-16. CEP consultants, Edinburgh (1987)

7. Wilson T. H. & Wiseman G. J. Phys. (London) 123, 116 (1954)

8. Bruce-Martin R. Clin. Chem. 32, (10), 1797 (1986)

9. Finnegan M. M. et. al., Inorg. Chem. 26, 2171 (1987)

NUTRITION and FOODSTUFFS

Plenary Lecture given by:

P. J. VanSoest

Chairmen:

P. Brätter
N. I. Ward

Poster Reports:

I. Lombeck, R. Schelenz

ANALYSIS AND CLASSIFICATION OF DIETARY FIBRE

Peter J. VanSoest

Cornell University, 324 Morrison Hall
Ithaca, NY 14853, USA

L.H.P. Jones

Institute for Animal and Grassland Production
Hurley, near Maidenhead, Berks
England SL6 5LR

Abstract

Dietary fibre has come to the fore in human nutrition because of the epidemiological association of its protection against various human diseases. Formerly assumed to be energetically inert, fermentation of fibre can provide significant energy and can modify metabolism. The analysis for dietary fibre is rendered difficult because of the diversity of components falling within the purview of the definition. The physicochemical properties of dietary fibre are most important in affecting fecal losses. The microbial component arising from fermentation of fibrous carbohydrates is most important. Since microbes require minerals, their growth becomes a mechanism of loss. On the other hand, the production of short chain volatile acids (SCFA) may promote colonic absorption through the efficient absorption of these substances in the colon. Since dietary fibres vary in their capacity to ferment and grow bacteria, the various fibre sources vary in their capacities to affect utilization of various mineral ions.

Introduction

The analyses of feeds and foods for fibre content has

Trace Element Analytical Chemistry in Medicine and Biology

traditionally involved crude fibre, standardized ca 1860 but not invented by Henneberg (1). No one knows who actually invented the method but it may go back to paper-making efforts in the middle ages (2). The oldest extant analysis for fibre was on Indian corn by John Gorham of Harvard in 1820 (2). Crude fibre was introduced into nutrition and animal feeding before an adequate biochemical understanding of carbohydrate digestion and fermentation developed. Thus it has presented an anomaly with its continued use up to the present, in that its chemistry does not accord with application. Fibre is not indigestible by most animal species and affords energy through gut fermentation and absorbtion of SCFA. The fraction that is fermented can vary from up to 10 times the crude fibre value. Hence the many efforts to replace crude fibre with more rational analyses.

The earliest efforts tended to focus upon idealistic chemical fractions, viz cellulose, lignin etc. and various elaborate schemes were proposed for fractionation and analysis by Paloheimo, Harwood, and Gaillard (2). These efforts were not successful in displacing the simple crude fibre procedure because of the larger labour and time cost, and the traditional entrenchment of crude fibre. The development of the detergent procedures removed some of these obstacles (2). The notion of using enzymes to prepare dietary fibres goes back to at least Williams and Olmstead (3) who used a sequence of enzymatic digestions to analyze the diet and faecal residues from human subjects. These authors were sufficiently sophisticated to also analyze for the SCFA in the faeces, and suggested the fermentive digestion of cellulosic carbohydrates in man. Their method was clumsy and tedious and was not seriously pursued as an analytical technique. Little was contributed from the human side to fibre studies until the advent of Southgate's papers (4,5) on available and unavailable carbohydrates.

These studies plus that of the development of the detergent analyses (6) emphasized the failure of crude fibre to recover unavailable residues present in relevant diets, and it was their intent in estimating the available energy in food and feed that was driving purpose for these analyses of fibre. Slightly later in the 1970's other objectives were pursued which included human disease patterns (7) and the significance of gut fermentation and the undigested residues (8,9). These factors promoted a comprehensive look at the components in undigested residues and the development of a definition of dietary fibre: the polymers present in plant food that are resistant to mammalian digestive enzymes. For man and monogastric animals this proved to contain more than the insoluble matrix fibre (NDF) and included "soluble fibres" such as pectins and gums.

From an idealistic point of view the methods that were gravimetrically based came under criticism because of the probable loss of the soluble fraction, which might be recovered by precipitation of 4 volumes of ethanol. Alternatively the enzymatically undigested residue could be hydrolyzed and analyzed for component sugars as an endpoint. Originally Southgate measured these spectrophotometrically (4) but later GLC and HPLC have come into vogue (10). However, regardless of the sophistication of sugar analysis, the integrity of results depends on the reliability of the isolation of enzymatically indigestible matter. The sugar-based methods have the problem that the non polysaccharide fractions viz lignin, cutin, Maillard products, tannins and unavailable fractions are not accounted for.

Attempts to standardize and collaborate enzymatic procedures include the efforts edited by James and Theander (10). These included the procedures of Southgate, Theander, and the NDF, the Swedish Asp procedure, later promoted by the AOAC in another collaboration (11). Another British effort has been that of Cummings et. al., (12). Some of these procedures notably those of Theander, and Cummings and Englyst use HPLC or GLC to assay sugar components (12).

The partial success of the AOAC procedure has derived from the advent of Termamyl^{R}, which is a heat-stable amylase with a low calcium requirement. This enzyme will operate in boiling neutral detergent. At lower temperatures (< 60°C) Termamyl may retain some hemicellulasic activity, which is inactivated at higher temperatures (13). Termamyl hydrolyzes starch to oligosaccharides, that are insoluble in 80% ethanol. These plus protein would remain in the residue. The AOAC procedure utilizes an amyloglucosidase and a protease to remove the respective contaminants. Alternatively the residual matter may be dialyzed to remove interfering matter (14).

The successful removal of starch is not always achieved. So-called resistant starches are peculiar to certain foods: in particular, peas, bananas, and potatoes (15). Removal of starch requires an initial step of gelatinization to render starches accessible by the enzymes. Boiling for an hour or autoclaving for shorter periods have been popular. Alternatives are to gelatinize in the cold with a chaotropic reagent, such as 8M urea, 4 M ammonium thiocyanate etc. The disadvantage of hot treatments is that proteins are denatured, hemicellulosic and pectinic polysaccharides are degraded and indigestible synthetic Maillard products are accrued to the lignin residue. Most of the values of Klason lignin from the Southgate or Asp procedures show elevation relative to the direct acid-detergent analysis (10). This increase is due to the synthesis of lignin artifact in the sample treatments. Here one must distinguish methodological artifact from that present in food as eaten. This is one of the more difficult problems of food analysis.

Failure to remove starch may arise from excessive heat treatment, or by the failure to achieve gelatinization of the more crystalline starches. Starches may also retograde (recrystallization) into difficult-to-digest forms (15). A further problem is that foods containing inhibitory tannins may offer special problems in enzymatic digestion. It may be very hard to distinguish Maillard products from leather formed in the tanning reaction.

Definition of Dietary Fibre

Dietary fibre has been defined as the complex macromolecular substances in food plants that are resistant to mammalian digestive enzymes (16,17). The definition has tended to focus on the polysaccharide components that are the dominant portion of the complex. However, signficant amounts of resistant mineral, lipid, protein and polyphenolic components occur and modify the matrix polysaccharides with which they are associated. A description of the major components and the types or parts of food in which they occur is given in Table 1. Because of the adopted definition, soluble polysaccharides including pectin and gums have been included in the dietary fibre complex since these substances are also resistant to digestion. The definition is a compromise, since these soluble materials are not fibrous in a physical sense. However, they can elicit responses in the digestive tract similar to the true insoluble matrix fibres. Bulk laxatives containing large amounts of polysaccharide gums also fall into this class.

Criticism of this definition of dietary fibre includes emphasis upon carbohydrates and the overlooking of non carbohydrate components that may have major effects. These include lignin, cutin, Maillard polymers (2) tannins and tannin-protein complexes (18) that are also resistant to digestion. The strict definition of unavailable carbohydrate

Table 1. Classification of Dietary Fibres According to Type and Source

Matrix: fibre I: insoluble

Type	Source	Function
cellulose		primary plant cell wall
hemicellulose	cereal	secondary wall
crude lignin		
true lignin	brans	woody tissue (secondary wall rigidifying factor)
cuticle (cutin)	unpeeled vegetables and fruits	skin, peel
protein (extensin)		secondary wall component
Soluble carbohydrates related to the dietary fibre complex		
pectin	fruits and vegetables	cellular membranes
beta glucan	oats, rye, barley	storage
psyllium	psyllium seed	storage
gum, arabic, guar	tropical legumes	exudates
microbial and	food additives	
algal gums	food additives	
methyl or hydroxy-methylcellulose	food additive	Synthetic from wood cellulose
Other noncarbohydrate fractions in fibre*		
Maillard products	insoluble matrix and some soluble	polymers from sugar and amino acids, produced in baking, frying, extruding
Tannins and Tannin-protein complexes	protective antimetabolic substances	Secondary plant substances

*Phytate is not considered as a dietary fibre fraction.

also overlooks resistant and heat-damaged starches, some of which reach the large bowel (19) and thus can have a fibre effect, and that all of these "unavailable" carbohydrates may support substantial fermentation and contribute dietary metabolizable energy through absorbed SCFA.

The attitude regarding starch contamination of the dietary fibre residue is open to different views. Resistant starches as they occur in food as eaten, may very well escape the digestion processes to escape to the colon or caecum, and then ferment with other fibrous carbohydrates. Does not this fit within the definition of dietary fibre? On the other hand, the puristic chemist will insist upon ideal composition of the dietary fibre residue, and will go to undue lengths to purify his product from starch contamination. Unfortunately there is no clear endpoint as to what is "resistant". Some information shows that intestinal escape of starch in monogastric animals is related to dietary fibre intake (19,20) i.e. rate of passage, also known to be important in the ruminant relative to starch escape from the rumen.

The "purity" argument also has problems with non-carbohydrate compounds, particularly protein associated with plant cell wall and other insoluble more slowly digestible proteins and heat-damaged proteins that may be slowly and incompletely digested. The purists have tended to correct by Kjeldahl analyses for N x 6.25 and thus have departed from the biological definition. Most plant cell walls contain proteins and perhaps other nitrogenous compounds that may have a relatively resistant core, particularly if lignins and tannins are present (2).

The Cornell 1977 Fibre Study

A human feeding study was conducted at Cornell on 24 young men, who were fed on low and high fibre-containing diets for

80 days. Diets were controlled and fed in a metabolic ward. Total faecal collection was made, and chromium and polyethylene glycol were used for transit measurements. Four sources of fibre were fed in the form of breads supplemented to a low fibre basal diet consisting mainly of meat, eggs and mashed potatoes. The low fibre control diet included white bread without added fibre and provided about 4 g (of dietary fibre) per day per subject. Fiber-containing diets were equalized and averaged about 20 g per day per subject. Maximum intakes were up to 40g/day. Details of experimental design and other protocol have been published elsewhere (21, 22, 23). The four breads were to include coarse wheat bran, finely ground wheat bran, a wood-derived cellulose, and cabbage. Fibre sources were fed at equalized levels relative to total dietary fibre. The choice of fibre diets allowed the comparison of highly fermentable cabbage, moderately fermentable bran and relatively unfermentable wood cellulose. Comparison of coarse and finely ground bran assayed the effect of grinding upon particle size. Comparison of cabbage and bran allows a comparison of cereal versus vegetable type of fibre. Vegetable fibres contain more pectin and virtually no lignin, while bran contains no pectin and is high in hemicellulose. Passage (transit) was measured by administering appropriate markers on a weekly basis.

The responses of the individuals fed the various fibre sources confirm the expectation that the fibre sources are unique and elicit varied effects upon fermentive digestion, passage and stool composition (Table 2). The digestion of the fibre sources was highest for cabbage and lowest for wood cellulose. Highly significant individual differences were observed among subjects in ability to digest fibre; this was negatively associated with passage rate (24). The cellulose fermentation appeared to have a depressing effect on fibre digestion in the subsequent diet (24). The mechanism of this depression cannot be attributed to passage since the wood cellulose was less effective than coarse bran in decreasing retention of food residues. Some subjects fed wood cellulose over a prolonged period were able initially to ferment more than 60% of this source but totally lost this ability upon extended feeding (25). This was verified by examination for extended cellulolytic bacteria in the faeces (26). The ability of the fibre sources to increase passage rate and reduce retention time in the digestive tract also varied widely (Table 3). Coarse bran elicited the shortest retention times while those for cabbage did not differ from the low fibre control. Fine grinding of bran markedly decreased the passage effect consistent with the presumption of reduction of volume from collapse of the cellular structure, a characteristic shared by the finely ground wood cellulose. The results with fine bran have been confirmed by other workers (27, 28).

The results of the study shows that dietary fibre sources are of unequal value as rated by hydration, stool volume and composition, transit and microbial responses. The generally poorer microbial responses of fine bran and wood cellulose may be the result of chemostatic effects where increased passage influences microbial efficiency (24). This efficiency rises with faster turnover, because faster transit causes the voiding of a younger microbial population with less death and recycling of nutrients. Slower turnovers result in more VFA and less microbes. It has been suggested that water-holding capacity promotes faecal response (29), which Stephen and Cummings (30) attacked by pointing out that a high degree of fermentation may ruin this capacity. However, hydration capacity and cation exchange can be positive factors in promoting fermentibility and that if one considers the replacement of fermented fibre with microbial growth, the original hypothesis may still stand (31). A comparison of the physical properties of some fibre sources is shown in Table 4.

Table 2. Composition and faecal responses of the fibre sources used in the Cornell Study (21, 22, 23)

Component	Cabbage powder	Coarse bran	Fine bran	Wood cellulose
Bulk volume (ml/g)	12.3[a]	8.2[b]	5.9[c]	13.9[a]
Particle size (microns)	484[b]	744[a]	173[c]	143[c]
Digestiblity				
cellulose (%)	81[a]	41[b]	53[b]	20[c]
hemicellulose (%)	93[a]	58[b]	56[b]	50[b]

[abcd] Numbers with differing superscripts are significantly different ($p < .05$).

Table 3. Stool Charactaristics and Transit (mean retention of Cr) Time in the Cornell Study (21, 22, 23)

	Basal	Cabbage	Coarse bran	Fine bran	Wood cellulose
Stool wt (g/day)	79[a]	87[a]	144[c]	111[b]	127[b]
Percent Water	75.2[b]	78.1[c]	77.6[c]	73.5[a]	75.1[b]
Residual Fibre (g/day)	1.1[a]	3.1[b]	8.9[c]	8.3[c]	14.2[d]
Faecal Nitrogen (g/day)	1.0[a]	1.1[a]	1.9[b]	1.1[a]	1.0[a]
Transit time (hr) Cr	62[a]	64[a]	41[c]	57[a]	48[b]

[abcd]Numbers with differing superscipts are signficantly different ($p < .05$).

Table 4. Physical Properties and Fermentability of Some Sources of Dietary Fibre (31, 32, 33)

Source	Cation exchange meq/g	Hydration[a] ml/g Dietary fibre	Hydration[a] ml/g Faecal fibre	Unfermentable residue[b] %
Cabbage	1920	2.8	2.3	9
Coarse bran	870	1.7	1.9	54
Fine bran	870	1.1	ND[c]	53
Wood cellulose	50	.6	.64	80[d]
Psyllium	30	.9	3.1	72
Pectin	2270	11.0	--	0
Bacteria	>2000	--	4.0	30

[a]Hydration capacity of isolated fibres at 89 milliosmolar/l (2 atm). Measured values are somewhat variable (32). The more conservative values have been chosen.
[b]Fermentation for 24 hr using human faecal inoculum.
[c]Not determined. The value of 1.1 has been used in calculations for Figure 2.
[d]The fermentibility of wood cellulose is highly variable depending upon donor of faecal inoculum. Digestion balances in the feeding study varied from 20 to 70% for wood cellulose recovered in faeces.

Effects of Dietary Fibre Source Upon Mineral Balances

Polyvalent ions are held tightly on the exchangeable surface of fibres. For example, when an unbuffered $CaCl_2$ solution is added to a exchangeable fibre source the pH can drop 1-2 units because of the release of adsorbed hydrogen. The strength of this ionic chelation raises questions about the availability of adsorbed minerals and the possible negative influence of fibre upon mineral absorption. Our balance study showed that the availabilities of Mg, Zn, Cu, and Mn were not adversely affected by high exchange quality fibre from wheat bran and cabbage (Table 5).

Table 5. Mineral balance (amount/day) from the Cornell Study (12 male students)

	Diets				
	Coarse bran	Fine bran	Wood Cellulose	Cabbage	Low fibre control
Fe (mg)	7.9**	8.2**	10.4	9.9*	11.3
Ca (g)	.36	.33	.45*	.26*	.35
Mg (g)	.23**	.24**	.06*	.21*	.14
Zn (mg)	1.9	3.1	-1.0**	6.8*	4.4
Cu (mg)	-.05	-.09	-.35	+.45+	.06
Mn (mg)	-2.2	-1.4	-4.4**	-1.6	-1.5

+ $P < .1$; * $P < .05$; ** $P < .01$

The addition of some fibres, particularly cabbage, improved certain mineral balances. On the other hand, lignified wood cellulose, low in exchange, reduced the mineral balances of Zn, Cu, Mg and Mn. The behavior of Ca and Fe appeared to be inverse to those of the aforementioned elements, and would agree as to the negative effects of fibre upon Ca and Fe. However, considering the amounts of fibre fed (up to 40 g/day) these small negative balances do not elicit much alarm that such higher fibre diets will induce serious deficiences.

Other studies also have shown diverse behaviour of dietary fibres upon mineral excretion; some positive, and some negative (35, 36, 37, 38). The earlier work of Reinhold et. al., (39) showing the negative effects of wheat bran upon zinc balance, involved red wheat that is higher in polyphenolics than white wheat, which probably has no adverse effect upon Zn. The curious effect of the respective dietary fibres in our study was that balances of Zn, Mg, Cn and Mn are inverse to cation exchange capacity, but more related to fermentability (Table 4). Balances of Ca and Fe are seemingly related to the exchange capacity.

Mineral Elements Associated With Fibre and Microbes

Dietary fibres have the ability to bind and hold metal ions on their surfaces in much the same way that soil, organic matter or clay minerals are able to hold cations in soil. This cation exchange serves as a bank exchanging K, Ca, Na, Mg for hydrogen when the pH drops and recharging as new cations become available when saliva and ingesta are mixed. However, some of this exchange also exists on the microbial cell walls (Table 4). When one considers that bacteria also require trace elements and that fermentable fibre promotes their growth, a microbial mechanism for faecal mineral loss is probable. The bacteria also produce much SCFA which are absorbed efficiently by the colonic mucosa (8,9). The absorption of the SCFA is accompanied by mineral ion exchanges, both into the colonic lumen and across to the blood (40). Two mechanisms are suggested: one for mineral loss by bacterial feeding, but another for absorbtion of available quantities of colonic mineral ions that may be swept across the colonic membrane by SCFA. The permeability of the colon has been underestimated (41).

The buffering capacity of foods derives in part from the physical effects they elicit in the gut and upon their intrinsic composition affecting the cation exchange. Since the fermentation of carbohydrates in the caecum or colon leads inevitably to production of large amounts of organic acids, their removal by absorption and the recycling of mineral ions are essential processes in the maintenance of normal pH in the gut. The contribution of the more soluble and mobile feed components to the buffering is comparatively ephemeral, since their residence time is limited i.e., they are either fermented, absorbed or washed out. Thus, it is the slower digesting solid matter--the fibre-that contributes more to the maintenance of the fermentive environment (42).

Exchangeable groups in fibres include carboxyl, amino, nonhydrogen bonded hydroxyls, and phenolic hydroxyls, all of which have some affinity toward binding of metal ions. Thus the surface properties of fibre, hydration and cation exchange, are intercorrelated (r about 0.7) (43). Both microbes and plant cell walls are negatively-charged (44) and attach to fibrous particles through their exchangeable surface by means of divalent cation liganding (45,46) (probably magnesium). Rumen bacteria similarly attach to the suface of cellulosic fibres, a means of protecting their substrate, and avoidance of secreting extra-cellular enzymes (2).

We have developed several systems for measuring cation exchange in plant cell walls. Direct measurement of H^+ exchange with acid is apt to degrade sensitive carbohydrate structures and produce artifacts. The first values by us involved the use of lithium binding (43). Lithium is weakly bound and variability is encountered in the washing procedures to remove unbound lithium. Calcium and barium proved unsatisfactory because of sulfate interference. Later values were obtained with Cu^{++} (47) via a modification of the procedure of Keijbets and Pilnik (48). This method is limited to measurement at pH 3.5 because of the instability of cupric ion at higher pH. More recently the stronger binding rare earth ions (praseodymium and neodymium) have been applied in a new procedure (49). Values obtained by Cu and Pr are shown in Figure 1, and at pH 3.5 are approximately equal, while values at pH 7.0 obtainable only with Pr are about double those at pH 3.5. Copper is attracted by amino groups for which rare earth has lesser affinity. Both Cu and Pr have high affinities for phenolic groups and the exchange values have high correlations with lignin content (Table 6).

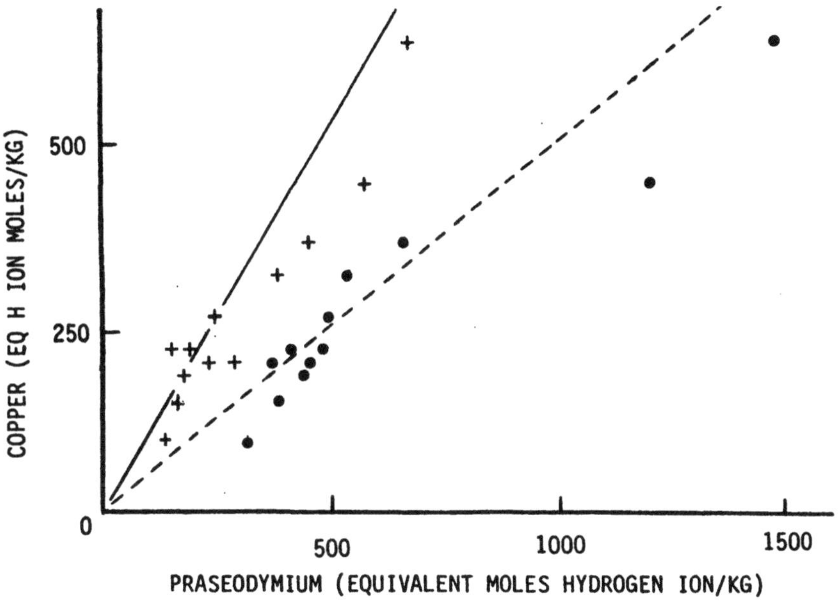

Figure 1. Comparison of three different measurements of cation exchange: Copper II at pH 3.5 and praseodynium III at pH 3.5 (+) and 7.0, (●) (50).

Table 6. Correlation coefficients (r) of CEC determined with copper (II) and praseodymium (III) at specified pH with composition and fermentation kinetics of batch-isolated netural-detergent fibres (50).

Parameter	Cu (II) pH 3.5	Pr (III) pH 3.5	Pr (III) pH 7.0
Lignin	0.763***	0.693***	0.843***
Hemicellulose (%)	0.486	0.563*	0.481
Cellulose (%)	0.114	0.032	0.155
Nitrogen (%)	0.699**	0.500*	0.579*
Cu at pH 3.5		0.963***	0.949***
Pr at pH 3.5			0.941***

Regression equations are significant at: * = P < 0.10; ** = P < 0.05; *** = P < 0.01.

Exchange is not the only way mineral elements are held in plant cell walls. Silicon for example exists as opal in graminaceous cell wall (51) and can form a skeletal system that offers structural and biological resistance (2). Rice and oats are the most important food plants with this characteristic. The silica could also offer the opportunity of mineral binding (52). Other minerals that may be held structurally in plant cell walls include Fe, and rare earths (53). Some of these elements may be soil born and become incorporated in the structure of the plant. Iron in this form is unavailable and does not exchange with other forms iron. Plants will deposit potentially toxic metal ions in the cell wall matrix where they are benign and probably not available to digestion or fermentation (51).

References

1, Tyler, C. 1975. Albrecht Thaer's hay equivalents: fact or fiction? Nutr. Abstr. and Rev. 45:1.

2, Soest, P.J.van 1982. Nutritional Ecology of the Ruminant. pp. 374. Cornell Univ. Press, Ithaca, NY

3, Williams, R.D. and W.H. Olmsted. 1936. The effect of cellulose, hemicellulose, and lignin on the weight of the stool: A contribution to the study of laxation in man. J. Nutr. 11:433.

4, Southgate, D.A.T., 1969. Determination of carbohydrates in foods, I. Available carbohydrate, J. Sci. Fd. Agric. 20:326-330.

5, Southgate, D.A.T. 1969. Determination of carbohydrates in foods, II. Unavailable carbohydrates, J. Sci. Fd. Agric. 20:331-336.

6, Soest, P.J.van 1967. Development of a comprehensive system of feed analyses and its application to forages. J. Anim. Sci. 26:119-128.

7, Burkitt, D.P. 1973. Some diseases characteristic of modern civilization. Brit. Med. J. 1, 274-279.

8, McNeil, N.I., J.H. Cummings and W.P.T. James. 1978. Short chain fatty acid absorption by the human large intestine. Gut 19, 819-824.

9, Stevens, C.E. 1977. Comparative physiology of the digestive system. In Duke's Physiology of Domestic Animals. ed M.J. Swenson. pp. 216-232. Ithaca and London, Comstock Publ. Assoc.

10, James, W.P.T. and O. Theander. 1981. The Analysis of Dietary Fiber in Food. Eds. W.P.T James and O. Theander. Marcel Dekker, Inc. New York, NY.

11, Prosky, L., A.S.P. Nils-George, I. Furda, J.W. Devries, T.F. Schweizer and B.F. Harland. 1984. Determination of total dietary fiber in foods, food products and total diets: Interlaboratory Study. J. Assoc. Off. Anal. Chem. 67:1044-1052.

12, Cummings, J.H., H.N. Englyst and R. Wood. 1985. Determination of dietary fibre in cereals and cereal products - collaborative trials. Part I: Initial Trial. J. Assoc. Publ. Analysts 23:1-35.

13, Mascarenhas-Ferreira, A., J. Kerstens and C.H. Gast. 1983. The Study of several modifications of the neutral detergent fibre procedure. An. Feed Sci. and Tech. 9:19-28.

14, Jeraci, J.L., P.J. Van Soest, J.B. Robertson and B.J. Lewis. 1988. New urea enzymatic dialysis procedure for total dietary fibre. Paper presented at the NE Regional Sect. AOAC, Nashua, NH. June 22, 1988.

15, Englyst, H., H. Trowell, D.A.T. Southgate and J.H. Cummings. 1987. Dietary fiber and resistant starch. Am J. Clin. Nutr. 46:873-874.

16, Trowell, H. 1975. Dietary changes in modern times. In: Burkitt DP, Trowell, H. eds. Refined carbohydrate foods and disease. London: Academic Press.

17, Soest, P.J.van 1985. Definition of fibre in animal feeds. In: Recent Advances in Animal Nutrition - 1985 Haresign and Cole; Butterworths Pub.

18, Wursch, P. 1979. Influence of tannin-rich carob pod fibre on cholesterol metabolism in the rat. J. Nutr. 109(4):685-692.

19, Stephen, A.M., A.C. Haddad and S.F. Phillips. 1983. Passage of carbohydrate into the colon. Direct measurements in humans. Gastroenterology 85:589-95.

20, McBurney, M.I. and L.U. Thompson. In press. Dietary fiber and energy balance: integratrion of the human ileostomy and in vitro fermentation models. An. Sci. Feed Tech.

21, Heller, S.N., L.R. Hackler, J.M. Rivers, P.J. Van Soest, D.A. Roe and B.A. Lewis. 1980. Dietary fiber: the effect of particle size of wheat bran on colonic function in young adult men. Am. J. Clin. Nutr. 33:1734-1744.

22, Ehle, F.R., J.B. Robertson, and P.J. Van Soest. 1982. Influence of dietary fibers on fermentation in the human large intestine. J. Nutr. 112:158-166.

23, Wrick, K.L., J.B. Robertson, P.J. Van Soest, B.A. Lewis, J.M. Rivers, D.A. Roe and L.R. Hackler. 1983. The influence of dietary fibre source on human intestinal transit and stool output. J. Nutr. 113:1464-1469.

24, Soest, P.J.van 1981. Some factors influencing the ecology of gut fermentation in man. In Gastrointestinal Cancer: Endogenous Factors. Banbury Report 7, pp. 61-69. Cold Spring Harbor, New York.

25, Soest, P.J.van 1984. Some physical characteristics of dietary fibers and their influence on the microbial ecology of the human colon. Proc. Nutr. Soc. 43:25-33.

26, Jeraci, J.L. 1984. Use of the high fibre chemostat to study interactions among gut microflora. PhD Thesis, Cornell University.

27, Kirwan, W.O., A.N. Smith, A.A. McConnell, W.D. Mitchell and M.A. Eastwood. 1974. Action of different bran preparations on colonic function. Brit. Med. J. 4:187-189.

28, Dokkum, W.van, N.A. Pikaar and J.T.N.M. Thissen. 1983. Physiological effects of fibre-rich types of bread 2. Dietary fibre from bread: digestibility by the intestinal microflora and water-holding capacity in the colon of human subjects. Brit J. Nutr. 50:61-74.

29, McConnell, A.A., M.A. Eastwood and W.D. Mitchell. 1974. Physical characteristics of vegetable foodstuffs that could influence bowel function. J. Sci. Fd. Agric. 25:1457-1464.

30, Stephen, A.M. and J.H. Cummings. 1979. Water-holding by dietary fibre in vitro and its relationship to faecal output in man. Gut 20, 722-729.

31, McBurney, M.I., P.J. Horvath, J.L. Jeraci and P.J. Van Soest. 1985. Effect in vitro fermentation using human fecal inoculum on the water-holding capacity of dietary fibre. Brit. J. of Nutr. 53:17-24.

32, Horvath, P.J. 1984. The measurement of dietary fibre and the effects of fermentation. PhD Thesis, Cornell University.

33, Soest, P.J.van, P.J. Horvath, M.I. McBurney, J.L. Jeraci, and M.S. Allen. 1983. Some in vitro and in vivo properties of dietary fibers from noncereal sources. In: Unconventional Sources of Dietary Fiber, ed. by Furda. I. Amer. Chem. Soc., Washington, D.C. p. 135-141.

34, Soest, P.J.van and L.H.P. Jones. unpublished data.

35, Aar, P.J. van der, G.C. Fahey, Jr., S.C. Ricke, S.E. Allen and L.L. Berger. 1983. Effects of dietary fibers on mineral status of chicks. J. Nutr. 113:653-661.

36, Cummings, J.H., D.A.T. Southgate, W.J. Branch, H.S. Wiggins, H. Houston, D.J.A. Jenkins. 1979. Digestion of pectin in the human gut and its effect on calcium absorption and large bowel function Br. J. Nutr. 41(3):477-485.

37, Sandstead, H.H., J.M. Munoz, R.A. Jacob, L.M. Klevay, S.J. Reck, G.M. Logan, Jr., F. Dintzis, G.E. Inglett, and W.C. Shuey. 1978. Am. Jour. Clin. Nutr. 31:180-184.

38, Ward, A.T. and R.D. Reichert. 1986. Comparison of the effect of cell wall and hull fiber from canola and soybean on the bioavailability for rats of minerals, protein and lipid. J. Nutr. 116:233-241.

39, Reinhold, J.C., B. Faradji, P. Abadi and F. Ismail-Beigi. 1976. Decreased absorption of calcium, magnesium, zinc and phosphorus by humans due to increased fiber and phosphorus consumption as wheat bread. J. Nutr. 106:493-503.

40, Engelhart, W.v. and G. Rechkemmer. 1983. Absorption of inorganic ions and short-chain fatty acids in the colon of mammals. In: Intestinal transport. Fundamental and comparative aspects. eds. M. Gilles-Baillien, R. Gilles. Springer Verlag, Berlin, Heidelberg. pp.26-45.

41, Shiau, S.Y. and G.W. Chang. 1986. Effects of certain dietary fibers on apparent permeability of the rat intestine. J. Nutr. 116:223-232.

42, McBurney, M.I., P.J. Van Soest and J.L. Jeraci. 1987.

Colonic carcinogenesis: the microbial feast or famine mechanism. Nutr. Cancer 10:23-28.

43, Soest, P.J.van and J.B. Robertson. 1976. Chemical and physical properties of dietary fibre. Proc. of the Miles Symposium. Nutr. Society of Canada. Halifax, N.S. pp.13-25.

44, Marquis, R.E., K. Mayzel and E.L. Carstensen. 1976. Cation exchange in cell walls of gram-positive bacteria. Can J. Microbiol. 22:975-982.

45, Marshall, K.C. 1980. Bacterial adhesion in natural environments. In: Microbial Adhesion to Surfaces. Ed. R.C.W. Berkely et. al., Society of Chemical Industry. London. pp.187-196.

46, Stotzky, G. 1980. Surface interactions between clay minerals and microbes, viruses and soluble organics, and the probable importance of these interactions to the ecology of microbes in soil. In: Microbial Adhesion to Surfaces. Ed. R.C.W. Berkeley. Society of Chemical Industry, London. pp.231-248.

47, McBurney, M.I., P.J. Van Soest and L.E. Chase. 1983. Cation-exchange capacity and buffering capacity of neutral- detergent fibres. J. Sci. Food Agric. 34, 910-916.

48, Keijbets, M.J.H. and W. Pilnik. 1974. Some problems in the analysis of pectin in potato tuber issue. Potato Res. 17:169-177.

49, Allen, M.S., M.I. McBurney and P.J. Van Soest. 1985. Cation-exchange capacity of plant cell walls at neutral pH. J. Sci. Food Agric. 36:1065-1072.

50, McBurney, M.I., M.S. Allen and P.J. Van Soest. 1986. Praseodymium and copper cation-exchange capacities of neutral-detergent fibres relative to composition and fermentation kinetics. J. Sci. Food Agric. 37:666-672.

51, Jones, L.H.P. and K.A. Handreck. 1967. Silica in soils, plants and animals. Advan. Agron. 19:107.

52, Jones, L.H.P. 1978. Mineral components of plant cell walls. Am. Jour. of Clin. Nutr. 31:94-98.

53, Robinson, W.O., H. Bastron, and K.J. Murata. 1958. Biogeochemistry of the rare earth elements with particular reference to hickory trees. Geochimica Cosmochimica Acta. 14:55-61.

SELENIUM STATUS IN HEALTHY BELGIAN ADULTS: ASSESSMENT OF ADEQUACY BY VARIOUS BIOCHEMICAL INDICES AND RESPONSE TO SUPPLEMENTATION.

J. Nève[+], F. Vertongen[++], A. Peretz[+++] and S. Van Erum[++]

Free University of Brussels, [+]Toxicology and Bioanalytical Chemistry Unit, Campus Plaine, 205-1, B-1050 Brussels; [++]Department of Medical Chemistry, Saint-Pierre Hospital, rue Haute, 322, B-1000 Brussels; and [+++]Department of Rheumatology, Saint-Pierre Hospital, rue Haute, 322, B-1000 Brussels, Belgium.

Abstract

Selenium (Se) status in Belgium was established in a large adult group (n= 145) in years 1985-6 by determination of plasma Se (mean ± SD: 84 ± 13 µg/L); erythrocyte Se (0.40 ± 0.09 µg/g Hb); urine Se (18.4, range 4-54 µg/g creat.); and erythrocyte Glutathione Peroxidase (GPx) activity (7.4 ± 2.0 U/g Hb). There was no influence of age and sex between 20 and 79 years. Plasma and erythrocyte Se levels were significantly lower than those from years 1980-1. Platelet as well as plasma GPx were significantly correlated with plasma Se in a smaller group. Two supplementation trials with different Se forms, 100-200 µg Se/d for 60 d, resulted in significant responses in plasma and urine Se, as well as in platelet GPx, but plasma and erythrocyte GPx were not modified. The responses in platelet GPx, that occured for plasma Se levels in the range 110-120 µg Se/L, suggest that Se status of investigated groups is insufficient for an optimal function of GPx.

Introduction

When compared to other parts of the world, inhabitants of West-European countries such as Belgium, France or Germany, have intermediate plasma Se levels, in the range 50-100 µg Se/L. Their dietary intake is near the low limit of the estimated safe and adequate dietary intake for the element of 50-200 µg Se/d. These two parameters, that are generally correlated, can however led to an erroneous assessment of Se status as they are subject to large geographic and time-to-time variations (1-3). Moreover, homeostatic regulation of Se body levels remains poorly known, although adaptative mechanisms to low Se intake have been characterized in humans (4). Actually, few attempts have been made

Trace Element Analytical Chemistry in Medicine and Biology

in selected population groups to define "optimal Se levels", i. e. those offering maximal protection for health by the essential trace element.

A promising approach to this purpose is the measurement of the only known biologically active Se: the activity of the selenodependent enzyme glutathione peroxidase (GPx). Several authors demonstrated that erythrocyte GPx is positively correlated with blood Se in subjects with deficient or low Se status, but not in subjects with higher Se levels (2,5-7): the relation shows a plateau above a threshold value that corresponds to a level where Se requirement for the enzyme is met (2,7). As erythrocyte GPx activity only offers a poor sensitivity to changing Se status due to the relatively long life span of red blood cells in the circulation, studies were recently done using platelet GPx. Se-deficient humans, either from low Se areas, or with reduced Se status due to abnormal intake (8-10), showed a rapid response in platelet GPx during supplementation with various Se forms, as well as a plateau of activity after 15 to 30 days supplementation, indicating that Se requirement is satisfied. In these studies, no response in erythrocyte GPx could be observed after that period of time, while plasma GPx produced controversial data (8-10).

The present study assembles several results on the determination of Se levels and GPx activities in various biological compartments from adult population groups in Belgium, and critically examines the relationship between the different indices. Then, assessment of Se status is completed by reporting the results from two supplementation trials including the responses in functional indices of Se status (11,12).

Subjects and Methods

The large population group investigated consisted in 145 healthy Belgian adults, 20-79 years old, living in the area of Brussels. Blood was collected by venipuncture before breakfast. Plasma was separated by aspiration after centrifugation and isolated red cells were washed twice with isotonic saline. Urine was simultaneously collected.

Supplementation studies were performed in two adult groups either with DL-selenomethionine (Calbiochem-Behring), 100 µg/d for 60 days (n=10); or with Se-enriched yeast capsules (Celnium 50, Expansion Aromatique Française, Sanofi, France), 100 µg Se/d for 30 days, then 200 µg Se/d for the next 30 days (n=10). Groups receiving a placebo were simultaneously studied (n=6 and n=10).Blood and urine samples were collected before the study and after 5, 15, 30, 35, 45 and 60 days supplementation (11,12).

Plasma Se was determined by a direct graphite furnace AAS procedure with Cu-Mg as a modifier and Zeeman background correction (13). Erythrocyte Se, expressed per g haemoglobin (Hb), and urine Se, expressed per g creatinin, were determined by graphite furnace AAS after wet ashing and extraction with an aromatic o-diamine (14).Accuracy of both procedures was established by use of serum and urine reference materials (Nycomed) and participation in interlaboratory comparaison trials. GPx activities were measured by the modified method of Paglia and Valentine (22) using H_2O_2 as substrate. Platelets extracts were prepared according to Levander (8). Results were expressed in µmol NADPH oxidized per min per g Hb (U/g Hb) in erythrocytes, in U/L in plasma, and in mU/10^{10} cells or in U/mg protein in platelets.Other details are given elsewhere (11-15).

Differences between groups were assessed by the t test or by analysis of variance. Multiple analysis of variance was used for time and treatment significance in the supplementation trials (12).

Results

Table 1 shows the usual values for Se indices determined in years 1985-1986 in the different groups of Belgian adults examined, 20-79 years old.

Indice	Usual level			n obser vations	n indivi duals	♂/♀
	Mean	SD	range			
Plasma Se, µg/L	84	13	55-125	145	145	69/76
Erythr. Se, µg/g Hb	0.40	0.09	0.20-0.65	145	145	69/76
Urine Se, µg/g creat.	18.4	-	4-54	86	86	47/39
Plasma GPx, U/L	112	21	65-170	142	36	17/19
Erythr. GPx, U/g Hb	7.4	2.0	3.0-13.0	145	145	69/76
Platel. GPx, mU/10^{10}c.	18.8	6.6	6.5-29.5	62	16	7/9
U/mg prot.	125	31	65-210	80	20	10/10

Plasma and erythrocyte Se as well as erythrocyte GPx activities were normally distributed contrary to urine Se. Age and sex had no influence on these indices between 20 and 79 years of age. Low plasma and erythrocyte Se levels below 20 and above 80 were previously documented (16). Mean values reported for plasma and erythrocyte Se were lower ($p < 0.001$) than those determined in years 1980-1981: 97 ± 14 µg Se/L plasma and 0.49 ± 0.07 µg Se/g Hb, but erythrocyte GPx activity was not different (16).No significant correlation could be established between the three Se levels or between plasma Se and erythrocyte GPx activity in the large group investigated.

Plasma and platelet GPx were determined in years 1985-1986 in two smaller groups of subjects, but values were examined several times during two months.

Significant correlations could be established between plasma Se and plasma GPx ($r=0.49$, $p<0.001$ and $r=0.39$, $p<0.001$) as well as between plasma Se and platelet GPx ($r=0.31$, $p<0.01$ and $r=0.29$, $p<0.01$).

The effects of Se supplementation on these indices were described in detail elsewhere (11,12).Plasma and urine Se rapidly responded to supplementation, contrary to erythrocyte Se . Despite of parallel increases, no clear relation could be established between plasma and urine Se levels. Plasma and erythrocyte GPx did not respond to supplementation, unless after two months. As a consequence, no correlation could be demonstrated in supplemented subjects between plasma Se and plasma GPx. Platelet GPx moderately but significantly increased during supplementation, showing a plateau of activity after 15 days in subjects receiving DL-selenomethionine, 100 µg/d, but continued to increase in subjects receiving Se-yeast, 200 µg/d during the second month (11,12). The significant increase in activity was observed for plasma Se levels in the range 110-120 µg Se/L. Such as indicated in Figures 1 and 2, strong correlations could be calculated with plasma Se during the two supplementation trials.

Discussion

Usual plasma Se levels reported in this study are in the same range than those for surrounding countries (1-3). They approximatively correspond to a dietary intake in the range 40-50 µg Se/d (1-3). These values are however lower than those previously established by us or by other authors (16), confirming the time-to-time variability in plasma Se levels, and suggesting that Se status is progressively decreasing in Belgium.

The absence of correlation between Se levels is not unusual when considering subjects with a rather similar Se intake. We previously established for Belgium inhabitants that the relation between blood Se and erythrocyte GPx only reached significance in subjects with reduced Se status due to prolonged total parenteral nutrition or gastro-intestinal disorders. A threshold value of 60 µg Se/L plasma was calculated above which no correlation exists with erythrocyte GPx activity (2,5,16). This observation suggested that Se status above this level was sufficient for erythrocyte GPx activity.

More sensitive functional indices of Se status were then assessed by the determination of plasma and platelet GPx activities and investigation of their responses to supplementation. Significant correlations could already be established before intervention between plasma Se and plasma or platelet GPx,

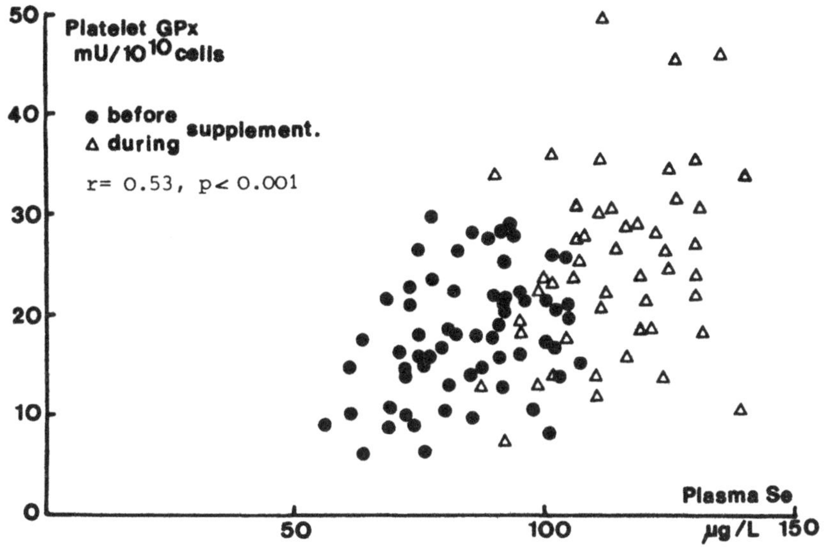

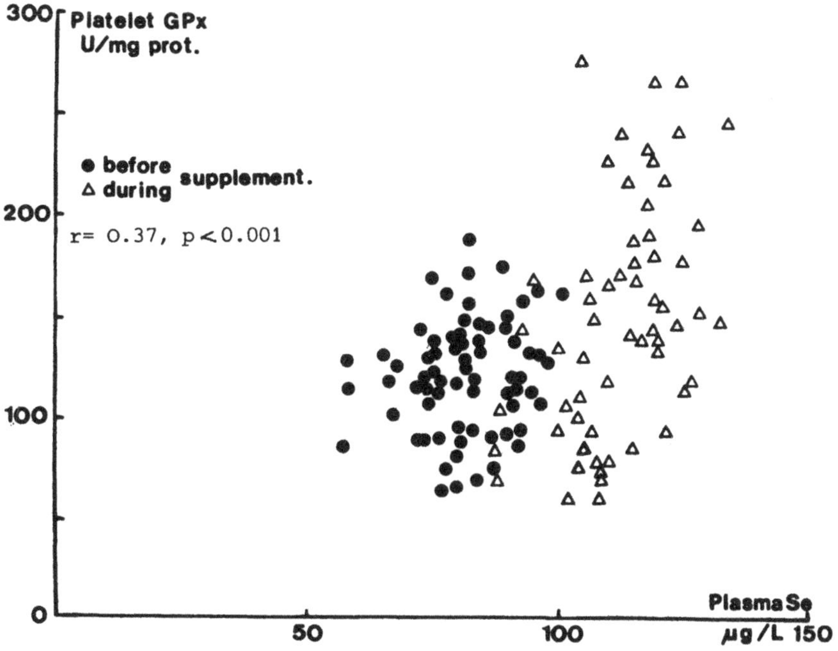

Figures 1 and 2: correlations between plasma Se and platelet GPx activity in two groups of Belgian subjects before and during Se supplementation. (top: trial with DL-selenomethionine; bottom: trial with Se-enriched yeast).

suggesting that Se status was influencing specific enzyme activity in these biological compartments. Such as reported by other authors in subjects with rather low Se status (plasma Se levels in the range: 50-70 µg Se/L) (8-10), we were also able to disclose in the two investigated groups,with plasma Se levels of 70-100 µg Se/L , a significant response in platelet GPx activity by Se supplementation. Levels of 110-120 µg Se/L were associated with higher platelet GPx activities, occuring 15 to 30 days after the beginning of supplementation. Such responses suggest that Se status was inadequate for an optimal function of platelet GPx. The absence of plateau in our second study performed with a higher Se intake during the second month of supplementation could indicate that the optimal intake level was not reached. Such a conclusion is however prematurate before making intervention studies during longer periods of time. Factors such as the chemical form of additional Se deserve also to be more carefully considered.

Acknowledgements: this work was supported by the Fonds National de la Recherche Scientifique et Médicale de Belgique and by the Ministères de l'Emploi, du Travail et du Budget de Belgique.

References

1.Nève,J.,Henry,M.,Peretz,A.,Mareschi,J.P. 1987. Cah.Nutr.Diet.12, 145.
2.Nève,J.,Vertongen,F.,Molle,L. 1985. Clin.Endocrinol.Metab.14, 629.
3.Thorling,E.B.,Overvad,K.,Geboers,J. 1986.Ann.Clin.Res. 18,3.
4.Robinson,J.R.,Robinson,M.F.,Levander,O.A.,Thomson,C.D. 1985. Am.J.Clin. Nutr. 41,1023.
5.Nève,J.,Vertongen,F.,Molle,L. 1984. In: Industrial Uses of Se and Te. (Se Te Develop. Assoc. ed), Darien, Co,537.
6.Levander, O.A. 1986. J.Am.Coll.Nutr. 5,37.
7.Nève,J. 1988. In: Selenium in Medicine and Biology (J.Nève, A. Favier,eds) W. de Gruyter, Berlin, in press.
8.Levander,O.A.,Alfthan,G.,Arvilomni,L, et al.1983.Am.J.Clin.Nutr.37,887.
9.Tomson,C.D.,Ong,L.G.,Robinson,M.F.1985. Am.J.Clin.Nutr. 41,1015.
10.Menzel,H.,Steiner,G.,Lombeck,F.,Ohnesorge,F.K.1983.Eur.J.Pediatr. 140,224.
11.Nève,J.,Vertongen,F.,Capel,P. 1988. Am.J.Clin.Nutr. in press.
12.Nève,J.,Van Erum,S.,Dramaix,M. et al. 1988. In: Selenium in Medicine and Biology, ibid ref 7.
13.Nève,J.,Chamart,S.,Molle,L. 1987. In: Trace Element Analytical Chemistry in Medicine and Biology (P.Brätter, P.Schramel,eds) W. de Gruyter, Berlin, 3,139.
14.Nève,J.,Hanocq,M.,Molle,L.,Lefebvre,G.1983. Analyst, 107,934.
15.Paglia,D.E., Valentine,W.N. 1979.J.Lab.Clin.Med. 70,158.
16.Nève,J.,Vertongen,F.,Peretz,A.,Carpentier,Y.A.1988, submitted.

EFFECTS OF INCREASING DIETARY SELENIUM INTAKE IN HUMANS: PRELIMINARY RESULTS OF A LONGITUDINAL STUDY.

G.C.Guidi, G.Bellisola, R.Schiavon

Institute of Clinical Chemistry, University of Verona, Italy.

S.Galassini

Institute of Biological Chemistry, University of Verona, Italy.

N.Q.Liu*, G.Moschini**

I.N.F.N. National Laboratories, Legnaro, Padua, Italy.
* Application Division of Nuclear Physics, I.H.E.P., Beijing, China.
**Department of Physics, University of Padua, Italy.

Introduction

Selenium plays important biological functions in humans, as pointed out by epidemiological studies on Keshan disease(1), on risk factors for cancer(2) and cardiovascular diseases(3). The Se-dependent enzyme Glutathione-peroxidase (GSH-Px: E.C.1.11.1.9) acts as a crucial protective effector against damages due to organic hydroperoxides. Recently a role has been suggested for GSH-Px in modulating arachidonic acid metabolism in platelets(4). For the preceeding reasons, interest about Se role is continuously rising, chiefly to clarify the relationship between intake and biological effects; this can be represented by a dose-effect curve, where area of pharmacological actions is set between optimum intake and intoxication with a range spanning over several orders of magnitude(5).

To acquire more knowledge about some biological effects after increasing oral daily Se intake, we performed a trial with Selenite supplementation in 5 adult male volunteers (physicians, all from our staff, having had a deep discussion about possible risk of toxicity).

Materials and methods

Volunteers were orally supplemeanted with Na selenite solution (Se 2 g/l); on first week they intaked 100 µg Se/day then progressively increased 100 µg Se/day for each of the following 6 weeks; last dose (700 µg/day) was mantained for further 3 weeks. During the assumption period biochemical parameters to evaluate hematologic, liver and kidney functions were performed by routine methods. Plasma and 24 h urine were collected weekly for Se determination by proton-induced X ray emission (PIXE) using a preconcentration technique previously described(6). Red blood cells (RBCs) were separated in 3 fractions (Top, Medium, Bottom)(7) and platelets (PLTs) were obtained from platelet rich plasma(4). GSH-Px activities in RBC and PLT lysates were performed as described(8) and expressed as IU/g hemoglobin or protein respectively.

Results

RBC GSH-Px (Fig.1): following progressive Se intake the activity slightly decreased, then raised quite sharply most evidently for Top RBCs, representing the youngest cell fraction; that was less evident for Medium and Bottom fractions. On the whole no significant variations occurred from basal to final values.

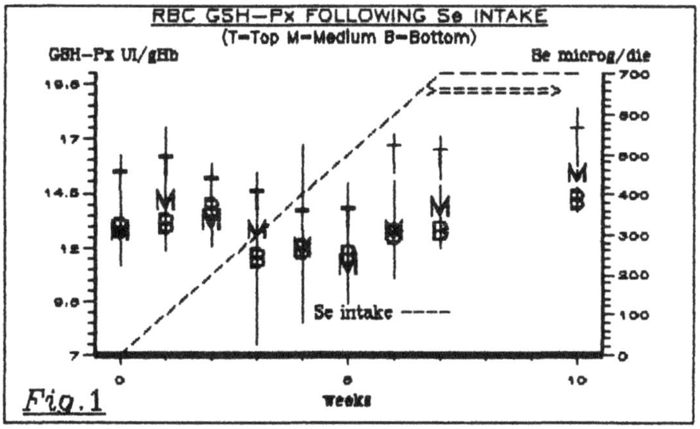

Fig.1

PLT GSH-Px (Fig.2): after a lag of about 15-20 days the activity increased from the third week (300 µg/day) of Se assumption and then seemed to plateaue with a slight decrease on sixth week (600 µg/day). At the end of supplementation the PLT activity (mean ± s.e.m.) was practically doubled (186.5 ± 5.5 IU/g prot. vs.94.9 ± 5.9 ;$p<0.01$) .

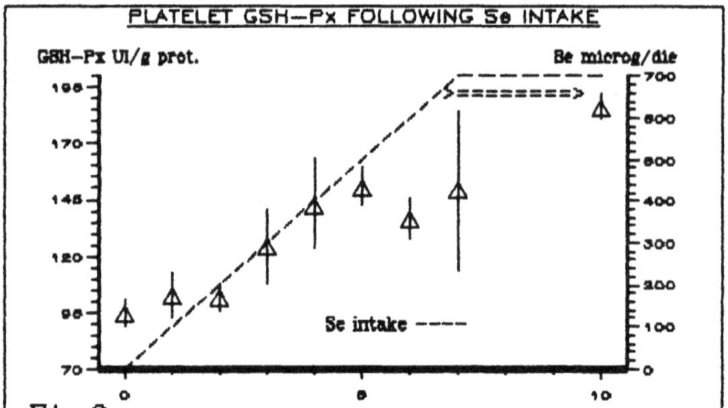

Fig.2

Plasma and urine Se (Fig.3): The mean initial plasma Se value was 62 ± 13 µg/l, then rose up to 128 ± 6 µg/l at the end. The greatest increase was observed within the first four weeks, then the increase did not exceed 10%. Se urinary excretion raised from initial 22 ± 8 µg/day up to 229 ± 54 µg/day as final value showing a strict linear correlation with dose (r = 0.897 ; p<0.01). *Se clearance studies* demonstrate a progressive increase from 0.29 up to 1.73 ml/min (Fig.4). The ratio between endogenous creatinine clearance and Se clearance decreased from about 400 to about 120.

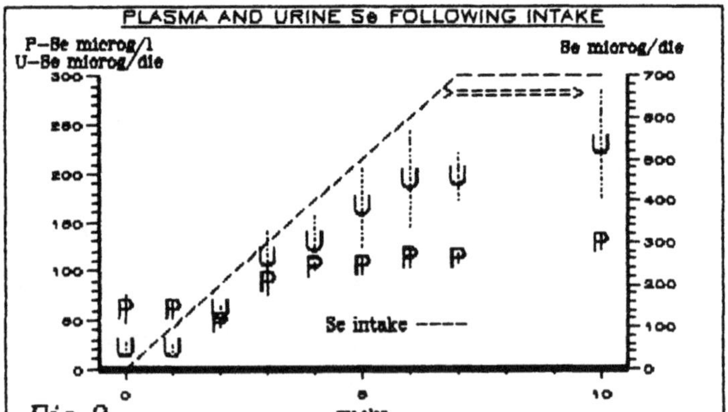

Fig.3

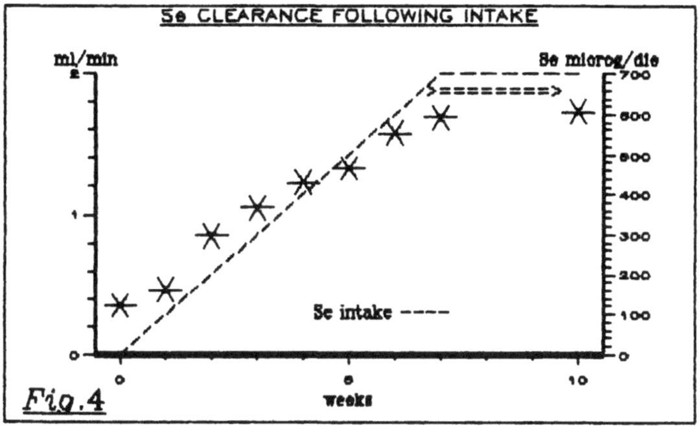

Fig.4

Discussion

The regimen of daily Se intake employed was intentionally get to dosages described as subtoxic and beyond the limit of safe and adequate intake; but some others employed doses higher than ours(9). Moreover epidemiological studies on chronic dietary Se overintake for years showed little toxic concomitant effects(10). Survey of principal biochemical parameters in our subjects pointed out absence of evident toxic manifestations.

The results obtained allow the following comments:

a) RBC GSH-Px is slowly influenced by increase in Se intake. Since RBC half-life span corresponds to the whole period of our study (70 days), we had expected a significant increase in activity, especially in Top fractions. On the contrary RBC GSH-Px is a poor index of Se status during selenite supplementation, as already described(11).

b) On the other hand PLT GSH-Px represents a sensitive and

faster index of Se status. If we consider that PLT life span from intramedullary megakariocyte stage is about 14 days, at third week of our trial we measured GSH-Px activity from a population completely developed under Se supplementation. Decrease of PLT GSH-Px activity observed at sixth week might be related to plateau of plasma Se occurred two weeks before.

c) Mean Se urinary excretion parallels with the progressive increase of Se intake and corresponds to a fraction lying between 32 and 38% of all doses, but the initial 100 µg/day where the excreted fraction was 22% of the administered, suggesting that renal regulation is a critical mechanism in retaining Se when the intake is low, not being yet the chief excretion way during overload. That seems confirmed by plasma Se levels showing limited variations starting from the fourth week. The low Se clearance values suggest a mechanism of active tubular reabsorption.

Conclusion

Despite Se daily oral supplementation exceeding 200 µg/day is regarded as subtoxic, there are some evidences that biological positive effects might be obtained with higher doses; of particular interest may be the described regulatory action on PLT arachidonate metabolism in the presence of raised GSH-Px activity.

References

1. Chen, X., Yang, G., Chen, J., Chen, X., Wen, Z., Ge, K. 1980. Biol.Trace Element. Res. 2, 91.

2. Salonen, J.T., Salonen, R., Lappeteläinen, R., Mäenpää P.H., Alfthan, G., Puoka, P. 1985. Br. Med. J. 290, 417.

3. Salonen, J.T., Salonen, R., Pentillä, I., Herranen, J., Jauhiainen, M., Kantola, M., Lappeteläinen, R., Mäenpää P.K., Alfthan, G., Puska, P. 1985. Am. J. Cardiol. 56, 226.

4. Guidi, G.C., Schiavon, R., Biasioli, A., Perona, G. 1984.J. Lab. Clin. Med. 104, 574.

5. Norheim, G. 1986. Acta Pharmacol. Toxicol. 59 (supp.7),177.

6. Buso, G.P., Colautti, P., Moschini, G., Xusheng, H., Stievano, B.M.1984. Nucl. Instr. Meth. Physics Res. 231, 177.

7. Murphy, J.R. 1973. J. Lab. Clin. Med. 82, 334.

8. Günzler, W.A., Kremers, H., Flohé, L. 1974. Z. Klin. Chem. Klin. Biochem. 12, 1974.

9. Thomson, C.D., Robinson, M.F., Campbell, D.R., Rea, H.M. 1982.Am. J. Clin. Nutr. 36, 24.

10.Yang, G., Wang, S., Zhan, R., Sun, S. 1983. Am. J. Clin. Nutr. 37,872.

11.Gebre Medhin, M., Gustavson, K.H., Gamstorp, I., Plantin, L.O. 1985. Acta Paediatr. Scand. 74, 886.

SELENIUM IN FOOD AND NUTRITION IN FRANCE

M. Simonoff, G. Simonoff
C.E.N.B.G.-U.A.451 du C.N.R.S.-Chimie Nucléaire
Le Haut Vigneau, 33170 BORDEAUX-GRADIGNAN (France)

C. Conri, B. Cornaille
Service de Médecine Interne, Hôpital du Haut Lévêque
33605 BORDEAUX-PESSAC (France)

Abstract

The selenium levels in various samples (meats, fish, vegetables, fruits and other foodstuffs) representive the whole food intake consumed in the Bordeaux region and in France have been measured by a proton-induced-X ray emission (PIXE) method. Using the INSEE data we have calculated the selenium included in average daily dietary intake according to the different regions, the social and professional classes. Moreover selenium content of several therapeutic nutritional regimens was also determined (diabetes, hypercholesterolemia, gout, hypocalorics, lithiasis...)

Keywords

Selenium, Foods, France, Particle Induced X-Ray Emission (PIXE), Nutrition, Selenium dietary intake

Trace Element Analytical Chemistry in Medicine and Biology

I. Introduction

A nutritional deficit in selenium may be a risk factor for cardiovascular diseases (1). The anticarcinogenic properties of selenium, in addition, have been demonstrated in numerous animal experiments, recently reviewed by Shamberger (2), and in several epidemiological studies (3,4, 5). A study involving 750 hospital patients in Bordeaux (6, 7) showed that the serum selenium level was significantly lower in most pathological states (cancer, diabetes, cardiovascular disease, alcoholism), compared to healthy controls. Whether these decreased levels were cause or effect has not been established.

The most obvious medical use of selenium at the present time is to prevent the development of a deficit during total parenteral nutrition (8, 9). It would seem, however that supplementation could be of more widespread use to correct nutritional deficits in man, in particular deficiency arising in breast-fed infants because of inadequate maternal intakes of Se since the level of Se in human milk is directly related to maternal intake (10). A somewhat more speculative use would be as a prophylactic anticarcinogenic agent (11).

To the best of our knowledge, there are no published measurements of the mean daily selenium intake in France, or of the selenium content of those foods which are generally consumed. In the present study we have measured the selenium levels in various samples of meats, fish, vegetables, fruits and other foodstuffs by a proton-induced X-ray emission (PIXE) method.

Using the INSEE (Institut National de la Statistique et des Etudes Economiques) data we calculated the average daily intake according to the different French regions, the social and professional classes. Moreover the selenium content of several nutritional regimens was also determined.

II. **Materials and methods**

1. Food samples

The choice of samples was based on their popularity according to information received from various government agencies and from the directors of supermarkets.

The samples were cooked or left raw, according to French custom, and placed in a 100 mL Pyrex flask, then weighed directly (wet weight). The sample weights were generally in the range 1-2 g, except for vegetables where ~ 5g samples were taken because of the low selenium levels encountered.

2. Chemical pre-concentration of selenium

The method is described in detail elsewhere (12). Selenium is determined in the samples by PIXE after wet ashing and coprecipitation on tellurium. The metallic residue is dissolved in nitric acid. A known volume of a 25 mg/L yttrium solution is added as an internal standard. Recovery for this preconcentration step is measured with the use of carrier-free ^{75}Se.

3. PIXE irradiation

A beam of 3.2 MeV protons from the 4MV Van de Graaff accelerator at the "Centre d'Etudes Nucléaires" in Bordeaux is used to generate X rays on the subsequent targets. Photon spectroscopy was carried out by means of a 100 mm^2 intrinsic germanium X ray detector.

The limit of detection for selenium under the present experimental conditions is of the order of a nanogram per target.

4. INSEE data

These data (13) give the dietary intake per year for each product in France (regions, social and professional classes).

5. Diets

The mean daily intake of selenium was calculated for different diets :

- Normal civilian and military hospitals, school cafetarias.
- Therapeutic for specific disorders.

The selenium intake was calculated as the product of the measured levels in the samples and the amounts of the foodstuffs for each of the different diets.

III. **Results and discussion**

1) Mean results

The results of the measurements in various foodstuffs have been presented elsewere (14) and can be summarized as follows (Table I).

Product	Mean selenium content ng/g
Meat	80
Fish	200
Cereals	80
Milk	⩽13

Table I

2) Data per regions and professions

Using the INSEE tables (13) giving the dietary intake for each product in France along with the region and the social and professional class of the people we have calculated the mean dietary intake. It ranges from 40 to 60 µg/day. The details are given in tables II and III.

Selenium (μg/day) for different professions	
Farmers	59.5
Agricultural workers	52.2
Liberal professions	43.9
High grade employees	39.8
Employees	42
Working class	45

Table II

French Regions	Se μg/day
South West	52
Paris	40
North	48
East	49
Mediterranean	48
National Mean	45.9

Table III

Our values lie in the lower range of data from european countries except for some parts of Finland. No major differences are noted according to the professional classes. The lowest contribution of selenium in the dietary intake corresponds to the region of Paris, resulting mostly from a lower consumption of bread, eggs, cereals and fats.

3) Regimens

Selenium contents of several nutritional regimens were also determined (15) in normal and therapeutic diets.

a) Normal diets

* Diets from the Haut-Leveque Hospital (Bordeaux)

Mean 57 μg/day
Range (49.5-65)μg/day

* Diets from the Military Hospital

Mean 72.5μg/day
Range (63.8-81.1)μg/day

* Open-air schools

Mean 43 μg/day
Range (34-57)μg/day

* Vegetarian diets 42 μg/day

b) Therapeutic diets

* Calcium calculus 24 μg/day (Restricted : milk, cheese, tripe, shellfish, fish, dried fruits)
* Oxalate calculus 31 μg/day (Dairy products, meats, eggs, fish restricted but not excluded)
* Urate calculus 48 μg/day (excludes ; delicatessen, game, meat, mushrooms, offal, but the satisfactory Se intake is obtained from dairy products, meats, lean fish and cereals)
* Diabetic 30 μg/day (with 120 g of carbohydrates)
* Hypocaloric 44 μg/day
* Hypercholesterolemia 66 μg/day

Selenium intake in these different regimens lies within the national mean range except for recommended oxalic and calcic lithiasis diets.

Acknowlegments

The authors thank Mrs Regnier, Mrs Barrot and Pr. Cuisinier-Raynal for their contribution with the regimens.

References

1.Salonen, J.T., Alfthan, G., Huttunen, J.K., Pikkarainen, J., Puska. P. 1982. Lancet. 2, 175-179.

2-Shamberger, R.J. 1986. Trace Elements in Medicine. 3, n°3, 105-111.

3-Schrauzer, G.N., White, D.A., Schneider, C.J. 1977. Bioinorg. Chem. 7, 23-24.

4-Schrauzer, G.N., White, D.A., Schneider, C.J. 1977. Bioinorg. Chem. 7, 35-56.

5-Salonen, J.T., Alfthan, G., Huttunen, J.K., Puska, P. 1984. Am. J. Epidemiol. 3, 342-347.

-Salonen, J.T., Salonen, R., Lappete Lainen, Maenpaa, P.H., Alfthan, G., Puska, P. 1985. Br. Med. J. 290, 417-420.

6-Simonoff, M., Conri, C., Fleury, B., Ducloux, G., Berdeu, B., Llabador, Y.,Moretto, P. First Meeting of the International Society for Trace Element Research in Humans, Palm Springs, California, 1986.

7-Conri, C., Simonoff, M., Fleury, B., Ducloux, G. First Meeting of the International Society for Trace Element Research in Humans,Palm Springs, California, 1986.

8-Neve, J., Molle, L., Timmermans, J. 1985. Pharmakon. 62, 95-105.

9-Kein, C.L., Ganther, H.E. 1983. Am. J. Clin. Nutr. 37,319-328

10-Kumpulainen,J.,Salmenpera, L., Slimes, M.A. Koivistoinen, P. and Perkeentupa, J. 1985 Am. J. Clin. Nut. 42 , 829

11-Combs, G.F., Jr and Clark, L.C. 1985. Nutr. Rev. 43, 325

12-Simonoff, M., Hamon, C., Llabador, Y.,Moretto, P., G. Simonoff. 1988. Nucl. Inst. Meth. (in press)

13-Beyer, M.N., Mercier M.A. Collection de l'INSEE - Ménages M117/1986

14-Simonoff, M., Hamon, C., Moretto, P., Llabador, Y. 1988. Journal of Food Science (in Press)

15-Cornaille, B. 1988. Thèse médecine Bordeaux

DETERMINATION OF MINOR AND TRACE ELEMENT IN HUMAN DIET BY ATOMIC ABSORPTION SPECTROSCOPY

Ş.Mumcu, N.K.Aras
Department of Chemistry Middle East Technical University
06531 Ankara, Turkey

Introduction

Knowledge on the minor and trace element content in diet is of interest on first aid in nutrition planning and evaluation. Therefore there is a need for investigation of intake, absorption, excreation and metabolizm of elements (1).

The trace element studies on diets and biological materials in Turkey are not very extensive and few results have been published in literature (2-4). We are involved with scientist from ten other countries in a coordinated research program organized by IAEA to determine trace elements in diet and their daily dietary intake (5). For this purpose we have selected different population groups, collected their diet samples by duplicate portion technique and analyzed for their trace element content by Atomic Absorption Spectrometry and Neutron Activation Analysis. Here we present results obtained by AAS representing the diet of a Middle upper income group.

Experimental

In order to select sampling site, we have considered some parameters such as socio-economical, cultural and regional variations. For this study we have selected a population group of urban middle upper class. They were members of the Chemistry Department, Middle East Technical University, (METU), or

Trace Element Analytical Chemistry in Medicine and Biology

their close relatives. Since Ankara is a large metropolitan city, it recieves vegetables and fruits from every part of the country. Therefore the food taken is not a typical Ankara region but could be representative of large cities of Turkey. The purpose of this study was to find the trace element content of diet of normal healthy population. Therefore we followed these requirements for selection of subjects:

1- They are all between 18-60 years old and healthy individuals under no regulated diet and taking no drugs of any kind.

2- They do not have more than a meal outside the home in a day, and they are not supposed to change their eating habits during the collection period.

3- The women subjects are not pregnant.

There are number of different methods for collection of diet samples. One of them is preparation of meals in the laboratory by considering the eating habits of the selected region. The second procedure is duplicate portion technique (6,7). We have chosen the second technique in our study. We have selected one member from each of the families and the sample collection method was explained to them. We gave each of the 4 female and 6 male subjects a 5-liter and a 1-liter precleaned polyethylene container for food collection.

Duplicate portion technique method is based on the collection of duplicate portion of everything test subjects eat and drink during the collection period. This includes all the meals coffees, teas all other beverages and alcoholic drink. The unedible parts, skin of fruits, seeds, bones etc. were not collected. Then the plastic containers were placed to a deep freezer. Samples were collected for three consecutive days for each subject during May 27, 1986 to August 29, 1986 representing typical summer food consumed by the middle high income urban families.

Homogenization is one of the most important steps for the trace element analysis of the total diet samples. Since it consists of various food items such as fats, vegetables, it must be homogenized with great care before analysis. The samples were homogenized with a homogenizer having titanium blades. Homogenization was done two minutes at 1500 rpm and then five minutes at 3000 rpm. This is found to be adequate for this type of diet. Homogenized samples were freeze dried and powdered in an all teflon cylindrical mill. Samples were weighed before and after homogenization and freeze drying in order to find their water content.

In this study we tried several dry- and wet-ashing procedures with different acids an acid combinations in order to find the most suitable digestion procedures for the diet samples and IAEA-H9 mixed human diet standard which were used for quality assurance purposes. Approximately 300 mg sample were placed in a precleaned teflon lined acid digestion bombs and 6 ml of 1:1:1 mixture of concentrated HNO_3, H_2SO_4, and H_2O_2 were added. It was digested overnight at 120°C in an oven. Six parallel solutions were prepared from each diet sample and from each parallel solution six or more readings were taken during the analysis.

In 1985 IAEA developed a mixed humen diet RM's, IAEA-H9. This is important because no mixed diet reference material was commercially available from any of the RM manufacturers. We participated in an intercomparison study for IAEA-H9 and analyzed for Na, K, Ca, Mg, Fe, Zn and Cu by AAS. Our results were in good agreement with those of certified and literature values (8). We used this reference material during the analysis of our diet samples for quality assesment of our analytical data.

In this study, Perkin Elmer Model 305-B Atomic Absorption Spectrometer is used for the determinations of Na, K, Ca, Mg, Fe, Zn and Mn in mixed diet samples. For Na, K, Ca, and Mg continious nebulization was applied as aspiration mode because of higher dilution factors of diet samples for these elements.

For Zn, Fe and Mn discrete nebulization was applied. Concentration of Na, K, Ca and Mg were calculated by direct calibration and those of Fe, Mn and Zn by standard addition methods.

Results and Discussion

The average water content of METU samples was 69.4 ± 6.3%. The dry weights per day per person was 507 ± 148 g/day. The dry weights of these samples are less than many European countries (6), perhaps due to the less meat and meat products consumed in Turkey.

The average concentrations of elements found in METU samples are given in Table 1 and compared with the average elemental concentrations of diets consumed in USA (9) and Norway (7).

Table 1. Comparison of the Elemental Concentrations in Diets Based on Dry Weight Basis in METU Samples and Other Countries

Element	METU	USA(9)	NORWAY (7)
Na (g/kg)	11.1 ± 2.0	6.5 ± 0.1	9.0 ± 3.7
K (g/kg)	5.4 ± 2.0	6.2 ± 0.2	7.7 ± 2.8
Ca (g/kg)	1.34 ± 0.39	1.64 ± 0.05	3.25 ± 1.9
Mg (mg/kg)	613 ± 111	608 ± 10	-
Fe (mg/kg)	33.8 ± 7.1	30.0 ± 1.1	29.0 ± 10
Zn (mg/kg)	25.1 ± 6.8	29.6 ± 0.5	19.0 ± 6.0
Mn (mg/kg)	16.0 ± 7.6	5.60 ± 0.03	-

As it is seen, the average concentrations of K, Ca, Mg, Fe and Zn are comparable with other countries. The average Mn concentration of METU sample is about three times greater than the USA values. This might be due to the high tea consumption in Turkey. As known Mn content of tea is very high, 20 mg/kg (11). The Na concentrations in METU samples are higher than the other countries. This is due to the high consumption of table salt in Turkish diets.

The average dietary intakes of elements per day per person in METU samples and few other countries are given in Table 2.

Table 2. Comparison of Daily Dietary Intakes of Elements by METU Subjects and in Other Countries (mg/day/person)

Element	METU	USA (10)	NORWAY (7)	GERMANY (6)	UK (11)	RDA (a)
Na	5480	2560	2970	4500	4440	1100 - 3300[b]
K	2670	2410	2540	2400	2510	1875 - 565[b]
Ca	652	700	1070	380	956	800
Mg	303	238	-	308	249	300
Fe	16.3	13.2	10.1	13.9	11.3	10 - 18
Zn	12.6	12.9	7.1	11.8	9.1	15
Mn	8.4	2.4	-	2.7	4.6	2.5 - 5.0

a. Recommended Dietary Allowance for Adults (12)

b. Estimated Safe and Adequate Dietary Intakes (12)

Recommended Dietary Allowances (RDA) for adults suggested by the Food and Nutrition Board (12) and estimated safe and adequate Dietary intakes (ESDDI) for Na and K are also given in Table 2. As seen, the results are comparable except Na and Mn which are higher because of the reasons mentioned above. The daily Ca intakes of METU subjects are somewhat lower than the other countries. This might be due to the low consumption of dairy products in Turkish diet which provide the greatest amount (60%) of dietary calcium.

Acknowledgement

This work is supported in part by the International Atomic Energy, IAEA, under Research Contract No 3575/R3/RB, TUR/5/011 and by Middle East Technical University under contract AFP 87-01-03-02. We thank Drs.R.Parr, O.Y.Ataman, M.Volkan, and A.Ersöz for their interest and help during the many stages of this work.

References

1. Mertz, W.1981, Science 213, 1332-1338.

2. Aras, N.K., I.Gökmen, 1986. Collection of Data on the Trace Elements in Diet, Human Tissues and Body Fluids in Turkey.

3. Aykut, M., A.Baysal, 1984. Turkish Bull. Hyg.Exp.Bio 41, 53.

4. Beyhan, Y., A.Baysal, 1978. Hacettepe Bull.Med. 11, 66.

5. Parr, R.M., International Collaborative Research Programme on Minor and Trace Elements, 1985. IAEA, Vienna.

6. Schelenz, R., 1977. J.Radioanal.Chem. 37, 539-546.

7. Bibow K, G.Riis, B.Salbu, 1984. Naringsforskning 28, 84-88.

8. Parr, R.M., Report on the Intercomparison run IAEA-H9 for the Determination of Inorganic Constituents of Mixed Diet Material. 1987, Vienna, IAEA.

9. Iyengar, G.V., JT.Tanner, W.R.Wolf, R.Zeisler, 1987. Sci. Tot.Envir. 61, 235-252.

10. Pennington, J.A.T., B.E.Young, B.D.Wilson, 1986. J.Am.Diet Assoc. 86, 876-891.

11. Hazel, T., 1985. Wld.Rev.Nutr.Diet, 46, 1-123.

12. Recommended Dietary Allowances, 1980. National Academy of Sciences, Washington D.C.

DISTRIBUTION OF TRACE ELEMENTS IN FOOD DETERMINED BY PIXE ANALYSIS

M. Yukawa, K. Kitao
National Institute of Radiological Sciences
9-1, Anagawa-4-chome, Chiba-shi 260 Japan

HPLC-combined PIXE and AAS were used for analyzing metal-binding substances in food.

Introduction

It is important in the field of nutritional and medical sciences to know spatial distribution and chemical forms of trace elements in food in relation to their retention and metabolism in human body. In the case of whole body injestion such as small fishes, nutritional intakes of Ca, Fe, Zn, Se and Rb were higher than the case of the flesh only. In this study, High Performance Liquid Chromatography (HPLC) equipped with a gel permeation column was applied to separate components in the water-soluble fractions of fresh tomato, tomato juice and yogurt in order to obtain information about chemical forms of elements. Particle Induced X-ray Emission (PIXE) analysis and Atomic Absorption Spectrometry (AAS) were used to analyze trace elements in the components.

Materials and methods

[Sample Preparation]
Fresh tomato and yogurt were homogenized using a Teflon homogenizer at $0^{\circ}C$. Homogenized fresh tomato and yogurt, and tomato juice were centrifuged at 1800×g for 30 min. The resultant supernatants were filtrated with 0.45μm membrane filter before use for HPLC.

[Gel Permeation Chromatography]

A 100μl of the filtrated supernatants of the samples was applied to a gel permeation column (Hitachi GELPACK GL-W-540) and eluted with 10 mM Tris-HCl buffer solution (pH 7.5) containing 0.15 M KCl at a flow rate of 1.0 ml/min. Molecular absorbance from 200 nm to 360 nm during elution was monitored with three dimensional chromato-system (Hitachi 3-D Chromatograph) and 1 ml portion of eluted solution was collected by a fraction collector to analyze elements in each fraction.

[PIXE and AAS analyses]

The fractionated supernatants by HPLC were freeze-dried and fixed on a sample holder with epoxy resin for elemental analysis by PIXE method. AAS method was also used for determining element contents in the fractionated supernatants. The contents of elements in the whole supernatants and pellets after centrifugation of samples were determined by AAS method.

Results

Fig.1 shows the contents of elements determined by AAS in supernatant and pellet fractions after centrifugation of tomato juice, fresh tomato

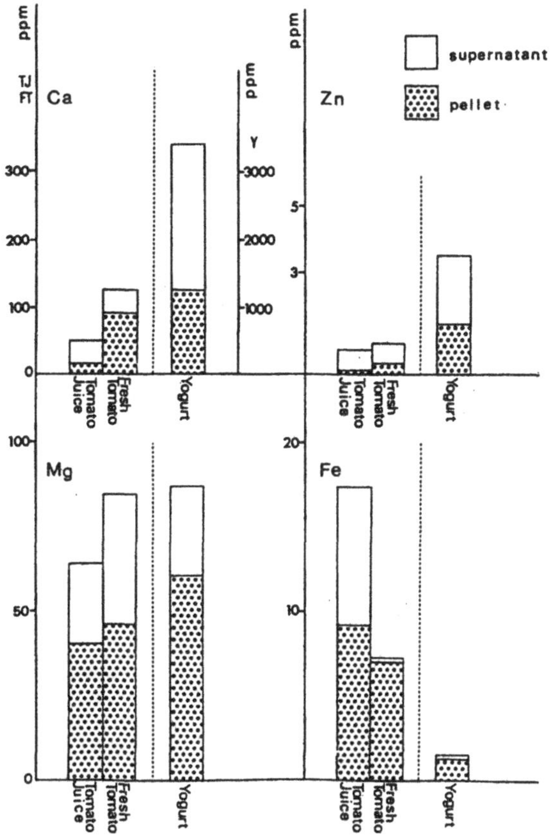

Fig.1 Content of metals in supernatant and pellet after centrifugation of tomato juice, fresh tomato and yogurt

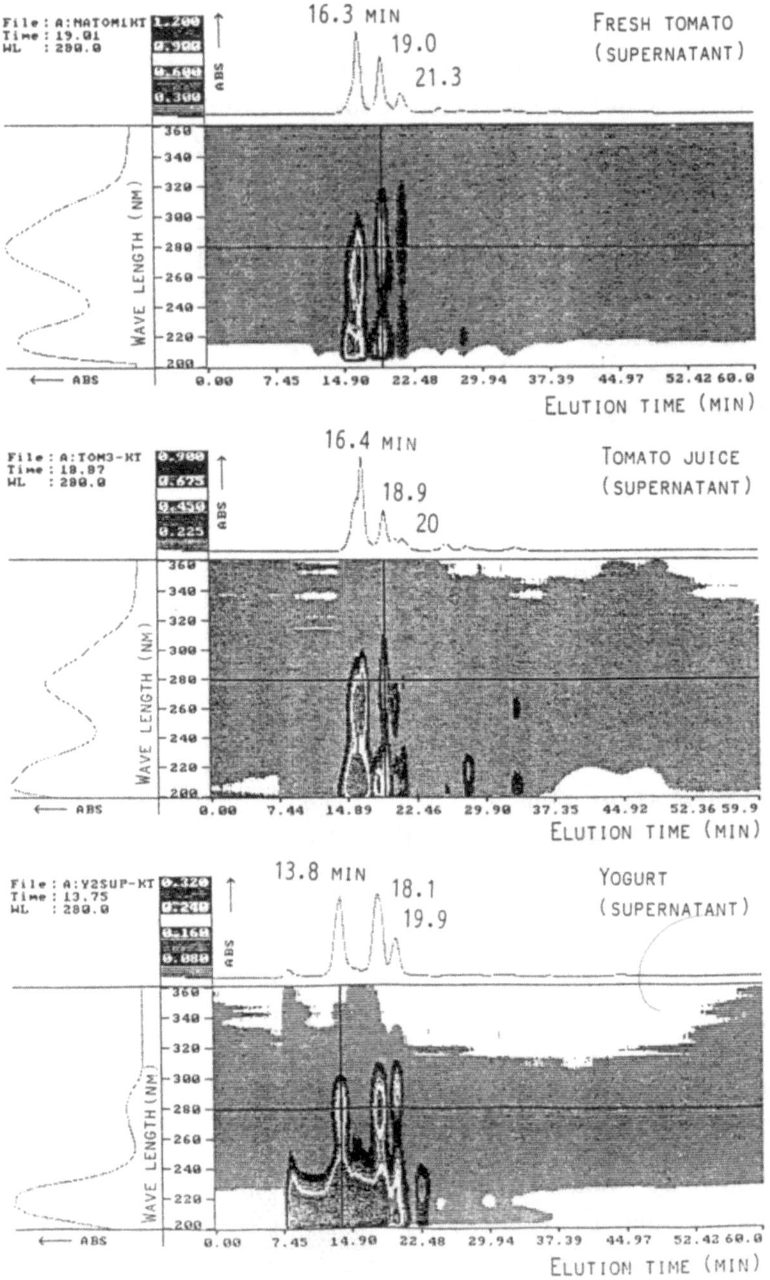

Fig.2 Gel permeation chromatographs of fresh tomato, tomato juice and yogurt supernatants by HPLC.

and yogurt. Total contents of Ca and Zn were significantly higher in yogurt than those in tomato juice and fresh tomato. Total Mg content showed almost the same level in all samples. Total Fe content in yogurt was very low as compared to other samples. The contents of Ca and Zn in the supernatants were more than 50 % of the total except Ca content in fresh tomato. The content of Mg in the pellets was slightly higher than in the supernatants. Fe contents in the supernatants in fresh tomato and yogurt were markedly low, but tomato juice supernatant contained about a half of the total Fe and total Fe content was more than twice of those of other two samples. HPLC chromatographs of the supernatants are shown in Fig.2. The centers of these graphs are contour expression of molecular absorbance and upsides show the chromatograph along the cursor in the contour map (at 280 nm, in this case) in parallel with X-axis. The left side graphs are absorption spectra along another cursor in

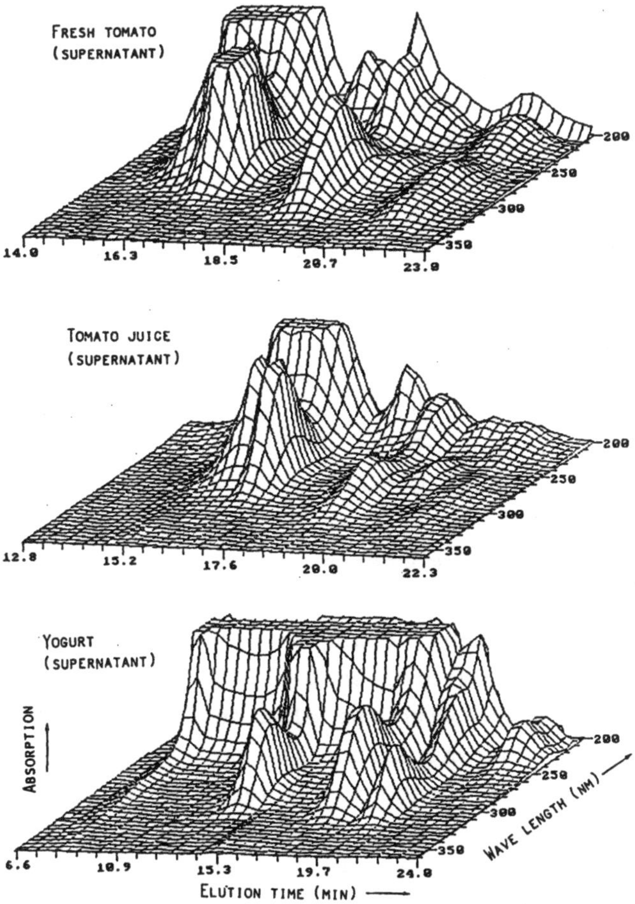

Fig.3 Three dimensional maps of the main component fractions separated by HPLC.

parallel with Y-axis. Tomato juice supernatant showed a very similar elution pattern to fresh tomato. Yogurt supernatant contained higher molecular weight components as compared to tomato, and some substances which have a peak at 280 nm were clearly shown in the contour figure, indicating that yogurt contains higher content of proteins than tomato. Molecular weights in the fraction time at 14 min and 16 min were estimated as about 32000 and 3000 dalton, respectively. Fig.3 demonstrates expanded three - dimensional chromatographs of the main component fractions shown in the contour maps of the HPLC, in which elution time, absorbance and spectral structure of the components were illustrated. In Fig.4, the concentration changes of elements determined by PIXE are shown with the elution time. Concentrations of Zn and Fe in tomato juice were the highest in the fraction of elution times at 16 and 17 min respectively, and concentration of Ca in

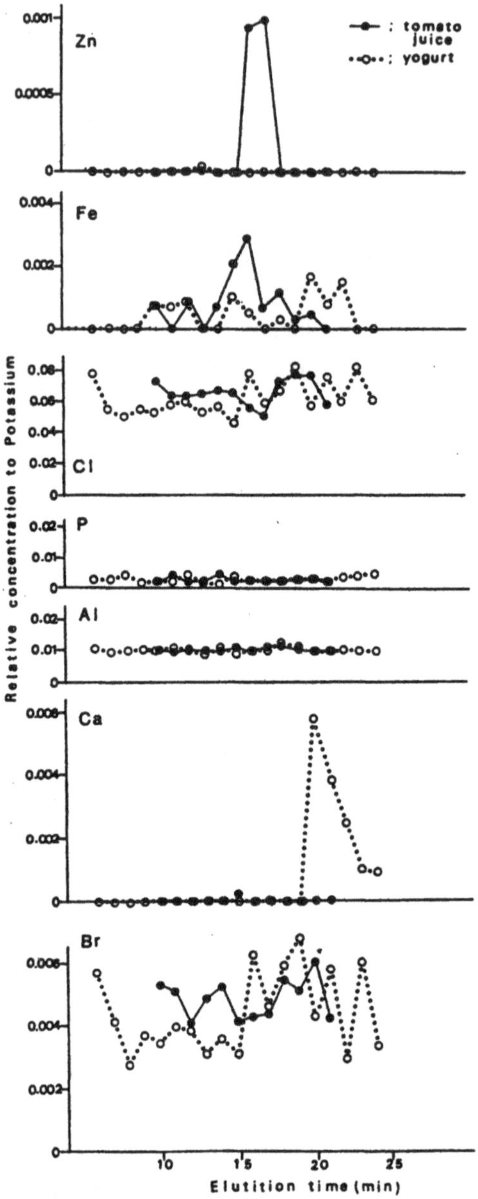

Fig.4 Distribution of elements in fractionated supernatants of tomato juice and yogurt determined by PIXE analysis.

yogurt was the highest in the fraction of elution time at 20 min. These results coincided with the results obtained by AAS analysis of fractionated supernatants. In addition, AAS analysis indicated that Mg had concentration peak near fraction number 20.

Discussion

In the tomato juice and fresh tomato supernatants, the fraction of elution time at 16 min contained Fe and the fraction of elution time at 16.5 min contained Zn. Both fractions had an absorption peak near 262.5nm. The fractions having an absorption peak near 280 nm were also observed in these supernatants and one of these fractions (elution time at 20 min) contained Mg, though there is no metal in another fraction (elution time at 19 min). Yogurt supernatant had two fractions having a strong absorption at 280 nm, one of which (elution time at 18 min) contained a small amount of Zn, but no metal was observd in another fraction of elution time at 14 min. The fraction having an absorption peak near 292 nm was obtained elution time at 20 min and this fraction contained Ca and Mg. Fe was eluted at 16 min in the same manner as the case of tomato, but this fraction did not have any clear absorption peak. These results indicate that the yogurt supernatant contains much higher content of proteins than tomato and that small amount of Zn might bind to these proteins. In this study, characteristics of metal-binding substances were investigated from molecular size and spectral aspects mainly in water-soluble components in tomato juice, fresh tomato and yogurt. However, identification of molecular species of these metal-binding substances is not carried out and water-insoluble components are not analyzed yet. These problems are now under investigation.

MULTI-ELEMENTAL ANALYSIS OF COMMERCIAL BABY MILK FOODS BY ICPSMS AND INAA

Steven Durrant, Nicola Walker, and Neil I. Ward

Department of Chemistry, University of Surrey, Guildford, Surrey GU2 5XH

Abstract

This paper reports on the evaluation of 4 commercial baby/ infant milk 'foods'. Fifteen elements were determined by ICPSMS and eleven by INAA. Comparison is made between these two techniques and where known, with manufacturers' values. The significance of these values for baby/infant health is discussed.

Introduction

Milk generally provides the sole source of nourishment during infancy and its elemental content is therefore of fundamental importance to child development and health. Certain essential trace elements (Ca, Co, Cr, Cu, F, Fe, I, Mg, Mn, Mo, Ni, Se, Si, Sn, V, and Zn) maintain the structural and functional integrity of the developing baby and marginal deficiency or excess can induce significant physiological changes. For example, zinc deficiency has been related to retarded growth from infancy to adolescence and maternal iodine deficiency may cause cretinism (1).

Despite the recent revitalisation in the practice of breast feeding (2) it is still of concern that there are chemical, biological, and physiological differences between human breast milk and its substitutes. The relationship between formulae feeding and increased infant morbidity and mortality may not only be due to unsanitary handling or environmental pollutants but also insufficient nutritional quality (3). This latter consideration is especially true of the trace elements.

Various studies have compared the availability of trace metals in human breast milk and its substitutes (4,5). In general there is a high degree of uncertainty as to the specific elemental levels necessary in such milk substitutes to facilitate the requirements for 'normal' child health. This is due in part to the limited number of accurate multi-element studies completed on milk formulae or powders. Reported literature values of the elemental composition of powdered milk range from levels of 17800 $\mu g.g^{-1}$ for the major elements (such as K) to less than 0.5 $ng.g^{-1}$ for certain toxic elements (such as Hg). Typically atomic absorption spectrometry (AAS) has been employed using a mono-elemental determination procedure requiring acid sample digestion (6). In this study four milk 'foods' were taken for analysis viz: Milupa Milumil babymilk (Milupa); Cow and Gate Premium baby milk food (Premium); Wyeth Progress modified milk-drink (Wyeth); Gold Cap Sma baby-milk food (Sma).

INAA was employed since it is well established in the field of multi-element analysis. As reported elsewhere (7), ICPSMS is a relatively new but very promising method which also offers excellent limits of detection (typically $ng.ml^{-1}$ or less in solution).

Experimental

Inductively coupled plasma source mass spectrometry

This was undertaken using slurries and open nitric acid

digests* in the range 0.4 to 10 $mg.ml^{-1}$ dissolved solids content. These samples were run against standard solutions of 1 $\mu g.ml^{-1}$ in 1% v/v HNO_3.

Instrumental neutron activation analysis

All samples were irradiated in the Imperial College Reactor Centre CONSORT Mk 11 reactor at the maximum power of 100 kW. A short irradiation time (t_i=5 min., t_d=1 min., t_c=8 min.) was possible with the use of the In-Core-Irradiation-System (ICIS) (the thermal flux was 2.3 x 10^{12} neutrons $cm^{-2}.s^{-1}$). The samples were counted by a Ge(Li) detector linked to a Nuclear Data ND 6600 MCA System.

Results

The results are displayed in Table 1. Manufacturers' (where known), ICPSMS and INAA values are given for the four powders. The ICPSMS data reflect fairly good agreement with manufacturers' values. The INAA data are consistent with the other determinations save for Al and Cu (both elevated). The former is probably due to the $^{32}P(n,\alpha)$ ^{28}Al reaction for which there has not been a correction. This is supported by the relatively high P levels. The high Cu values are due to the influence of ^{24}Mg (1014 keV) gamma-ray peak and increased Compton background induced by ^{24}Na and ^{32}P reducing the ^{66}Cu (1039 keV) gamma-ray peak efficiency.

An indication of the accuracy of the results obtained by INAA is obtained by the analysis of the SRM IAEA-All milk powder, and a selection of elemental concentrations is shown in Table

*Authors' Warning: The use of concentrated nitric acid digestion of organic matter in digestion bombs is NOT recommended for the analysis of milk samples (especially those containing appreciable 'fats'). Such a procedure produces an extremely violent explosion.

Table 1. Comparison of Elemental Content of Baby Milk Powders: Manufacturers[a], ICPSMS[b] and INAA[c] values

Elemental Content*												
	Milupa			Premium			Wyeth**			Sma		
Element	Com.[a]	ICPSMS[b]	INAA[c]	Com.	ICPSMS	INAA	Com.	ICPSMS	INAA	Com.	ICPSMS	INAA
Na	1700	1369 ±44	1527 ±81	1400	1214 ±4	1213 ±47	350 (2042)	1905 ±30	2358 ±367	1180	969.1 ±40.0	1165 ±102
Mg		471.3 ±15.3	514.4 ±34.0	400	406.6 ±4.3	450.4 ±29.7	94 (548.4)	756 ±7.0	861.4 ±87.6	420	492.3 ±20.1	489.6 ±14.1
Al		13.6 ±0.6	15.8 ±2.7		7.7 ±0.6	9.8 ±1.4		8.75 ±0.87	28.6 ±5.2		6.00 ±0.57	11.5 ±0.7
P	3900	3386 ±105	-	2100	1942 ±21	-	940 (5484)	6043 ±51	-	2590	2115 ±99	-
S		#(i)	2835 ±484		#(i)	2975 ±725		#(i)	3042 ±292		#(i)	831 ±388
Cl		5236 ±793	3939 ±206	3150	3874 ±67	3828 ±247	750 (4376)	4674 ±255	5237 ±780	3140	3077 ±338	3268 ±53
K		6393 ±82	8670 ±317	5250	5592 ±85	6924 ±694	1050 (6126)	7916 ±51	11319 ±688	4400	5149 ±191	7017 ±856
Ca	5100	5704 ±121	5260 ±404	4200	4813 ±112	4721 ±115	1150 (6709)	8173 ±285	9575 ±1299	3480	3731 ±143	3736 ±32
V		<0.02	0.02 ±0.01		<0.02	0.05 ±0.02		<0.02	0.04 ±0.02		<0.02	<0.002
Mn	0.9	0.79 ±0.09	0.42 ±0.04	0.55	1.10 ±0.17	1.16 ±0.08	0.17 (0.99)	2.59 ±0.03	3.10 ±1.66	1.26	1.25 ±0.19	1.58 ±0.14

Table 1 (continued)

	Elemental Content*											
	Milupa			Premium			Wyeth**			Sma		
Element	Com.[a]	ICPSMS[b]	INAA[c]	Com.	ICPSMS	INAA	Com.	ICPSMS	INAA	Com.	ICPSMS	INAA
Fe	31	37.6 ±8.7	-	40	36.0 ±6.5	-	12 (70.0)	75.3 ±13.0	-	53	41.5 ±16.0	-
Cu	1.9	1.71 ±0.01	2.45 ±0.33	3	3.07 ±0.28	6.20 ±0.91	0.58 (3.38)	4.72 ±0.32	6.60 ±1.13	3.9	5.13 ±0.03	4.01 ±0.92
Zn	29	28.4 ±0.4	(29.1[d] ±0.5)	30	30.7 ±0.5	(32.0 ±0.2)	4.3 (25.1)	29.4 ±2.6	(36.6 ±0.4)	39	43.2 ±3.0	(44.1 ±0.3)
Mo		1.01 ±0.39	-		0.23 ±0.07	-		<0.07	-		<0.07	-
Cd		0.17 ±0.08	-		0.45 ±0.39	-		0.33 ±0.15	-		0.37 ±0.11	-
I	0.15	#(ii)	0.12 ±0.07	0.55	#(ii)	0.43 ±0.17	0.069 (0.40)	#(ii)	0.73 ±0.33	0.54	#(ii)	0.42 ±0.02
Pb		1.02 ±0.27	-		0.41 ±0.19	-		0.47 ±0.15	-		0.16 ±0.03	- -

* Elemental content $\mu g.g^{-1}$ dry weight

** $\mu g.ml^{-1}$ reconstituted. Bracketed values are calculated assuming a (scoop) mass of 4.8g per 28 ml and expressed in $\mu g.g^{-1}$

#(i) Not determined as O_2^+ interferences occur at mass 32.

#(ii) Memory is severe for this element. Range indicated was 0.21 to 0.45 $\mu g.g^{-1}$.

a Manufacturer's recommended value

b Inductively coupled plasma source mass spectrometry

c Instrumental neutron activation analysis

d Flame Atomic Absorption Spectrometry

\- Not measured

Table 2. Elemental Content* of IAEA-All Milk Powder

Element	Recom.# (Lit.)	This work INAA **
Na	4420 ± 332	3969 ± 152
Mg	1100 ± 83	1179 ± 165
Al	(1.3)	2.45 ± 5.0
Cl	(9080)	9343 ± 404
K	17200 ± 998	17671 ± 2039
Ca	12900 ± 800	12849 ± 650
V	(0.10)	0.21 ± 0.01
Mn	(0.377)	<0.02
Cu	(0.26-0.84)	<1.20
Mo	0.11 ±0.01	0.11 ± 0.02
I	(1.5)	0.28 ± 0.01

* $\mu g.g^{-1}$ dry weight

** t_i = 5 min., t_d = 1 min., t_c = 8 min.

\# Recommended value: IAEA or (Literature value)

Table 3. Calculated Daily Elemental Doses+

	Element	Milupa	Premium	Wyeth	Sma	Rec.#
Major* (mg)	Ca	857**	554	644	464	360
	Mg	79	52.8	52.6	56	50
	P	655	277	526	347	240
Essential* (µg)	Zn	4872	3960	2408	5226	3000
	Fe	5208	5280	6720	7102	10000
	I	25	73	38.6	72	40
Toxic** (µg)	Al	2285	1016	840	804	
	Cd	28.6	59.4	31.7	49.6	
	Pb	171.4	54.1	45.1	21.4	

\+ Based on manufacturers' feed information for a six month old infant

* Based on manufacturers' elemental levels

** Based on ICPSMS elemental data

\# Recommended dietary allowances (Zn, Fe, I). Guidelines only (Ca, Mg, P). (Taken from reference 8).

2. Similar results were obtained by analysis of SRM's by ICPSMS.

Implications of Elemental Levels

Based on the manufacturers' elemental concentrations, or if these are not available ICPSMS values, the daily dose of nine elements (Ca, Mg, P, Zn, Fe, I, Al, Cd, and Pb) are calculated (Table 3). The manufacturers' recommended number of feeds were used for this calculation. The daily dose is adequate for the limited number of elements for which there are recommended values (8). The exceptions are Zn for Wyeth, Fe for all powders and I for Milupa and Wyeth. Calculations based entirely on the ICPSMS or INAA data lead to the same conclusions.

The zinc content of milk formula is essential especially during the early stages of development. Substantial zinc deficiency can result in foetal and child growth retardation/ inhibition. Although several cases of inadequate zinc status of infant milk formulae have been reported, such that zinc supplements have been required, the values of zinc measured in this study provide an adequate daily dose. Wyeth Progress although marginally inadequate in zinc is not meant to be used as the sole source of infant nutrition.

The normal term infant has substantial "stores" of iron and this element is found in breast milk typically at only 0.3-0.6 $\mu g.ml^{-1}$ (9). Whether iron levels in milk formulae are currently too low therefore requires careful consideration. Unfortunately space precludes further discussion save to say that, as noted above, low iodine levels also have important implications for infant health. The values of toxic elements such as Al may also be of concern.

All manufacturers' note on their products that breast milk is the best nourishment for babies or 'breast is best'. Our lack of sufficient data and the above indications that some milk foods may be lacking in some important elements can only reinforce this sentiment.

References

1. Underwood, E.J., 1977. Trace Elements in Human and Animal Nutrition. 4th edn., Academic Press, New York.

2. World Health Organisation. 1981. International Code of Marketing of Breast Milk Substitutes. WHO 1, 13.

3. Porter, J.W.G. 1978. Proc. Nutr. Soc. 37, 225.

4. McMillan, J.A., S.A. Landaw, F.A. Oski. 1976. Pediatrics. 58, 686.

5. Stastny, D., R. Vogel, M.F. Picciano. 1982. Fed. Proc. 41, 350.

6. Yin, Z. 1985. Fenxi Cestil Tongbas. 4(5), 22-24.

7. Gray, A.L. 1985. Spectrochim. Acta. 40B (10-12), 1525-1537.

8. National Academy of Sciences Committee on Dietary Allowances. 1980. Recommended Dietary Allowances. 9th edn., National Academy of Sciences, Washington.

9. Hambidge, K.M. 1983. In: Trace Element Analytical Chemistry in Medicine and Biology. Vol. 2, (P. Bratter, and P. Schramel, eds.). Walter de Gruyter. Berlin. pp.3-18.

TELLURIUM IN FOODSTUFFS AND ITS INTESTINAL ABSORPTION IN THE RABBIT*

T. Kron, Ch. Hansen, E. Werner

Gesellschaft für Strahlen- und Umweltforschung, Institut für Biophysikalische Strahlenforschung, Paul Ehrlich-Str. 20, D-6000 Frankfurt/M., FRG

Abstract

The concentration of tellurium in various foodstuffs has been measured with graphite furnace atomic absorption spectrometry. It was mostly found to be below 7 ng/g wet weight. A daily dietary intake for man under normal conditions of about 1 µg tellurium can be estimated from this.
The percentage intestinal absorption of Na_2TeO_4 in a rabbit was determined by applying the double isotope technique with radioactive tracers and was found to be about 40 %.

Introduction

Tellurium is a rare element with a natural abundance in the earth's crust of about 1 µg/kg. It reaches the environment of man due to its increasing usage in industry. Tellurium is used as a semiconductor in the electronic industry in solar cells and in gamma ray detector systems as well as an accelerator in the rubber industry. However, the main quantities of this element are used in metallurgic industry. In 1984 it was estimated that about one third of the tellurium used to im-

* dedicated to Prof. Dr. W. Pohlit on his 60th birthday

Trace Element Analytical Chemistry in Medicine and Biology

prove the quality of steel, gets lost in slag or by emission (1). Yet data on the natural concentration of tellurium in foodstuffs and on the transfer into animals and humans is only available to a limited extent.

Therefore, the aim of this study was to determine the content of tellurium in foodstuffs in order to derive an estimation for the daily dietary intake of tellurium in man under normal conditions.
The intestinal absorption from ingested tellurium was investigated with rabbits applying the double isotope technique.

Materials and methods

Foodstuffs were purchased at local markets. The samples were cut into small pieces and dried at temperatures below 100°C to prevent losses through boiling. Aliquots of approximately 100 mg dry weight were transferred into 10 ml teflon containers. The organic matter was wet ashed in the closed containers under pressure with concentrated HNO_3 for 4 h at 135°C (2). During this procedure unpredictable amounts of hexavalent tellurate were reduced to the tetravalent stage. Therefore, a distinction between different chemical forms of tellurium was not possible. After evaporating the nitric acid the reduction process was completed by heating the sample with 600 µl 5m HCl for at least one hour at 95°C. The remaining solution was transferred from the teflon containers to polypropylene tubes. The tetravalent tellurium only has been extracted from the hydrochloric solution with 150 µl methyl isobutyl ketone (MIBK) (3).
Aliquots of 20 µl of the organic phase were transferred into the graphite tube of the atomic absorption spectrometer (Perkin Elmer PE 2380 with HGA 500). The absorption was measured at the wavelength of 214.3 nm with deuterium background com-

pensation in the peak heights mode.

The limit of detection (2 SD of the background) for tellurium in MIBK was 4 ng/ml, the reproducibility for a solution containing 50 ng/ml tellurium was intra assay 7 % (n=12) and inter assay 12 % (n=12).

A first series of measurements was run with sample amounts of 100 mg, which leads to a limit of detection of 7 ng/g of the organic matter. In a second series the amount of foodstuff was enlarged and the ashing process repeated, until the remaining solution was clear. Applying this preparation the detection limit of tellurium in foodstuffs could be 10 fold increased.

The recovery was tested with radioactive ^{123m}Te added to 12 different samples prior to ashing. The gamma radiation was measured with a sodium iodide detector. For all cases the recovery rate was above 94 %.

Na_2TeO_4 was administered to an anaesthetized rabbit intravenously and perorally. The injected solution contained 180 µg tellurium labelled with radioactive ^{123m}Te and the fraction administered via a stomach tube 2500 µg labelled with another gamma ray emitting isotope, ^{121m}Te.

Blood samples of 3 ml were collected in the first 12 h after injection every hour, then one sample in 6h. The blood was centrifuged and the plasma separated. After 50 h the animal was killed. Urine was taken from the bladder.

The activity of both tellurium isotopes has been detected with a sodium iodide detector in blood plasma, erythrocytes and urine. The fractional intestinal absorption F_i has been calculated from the ratio between p.o. and i.v. applicated tracer:

$$F_i = c_{po} A_{iv} / c_{iv} A_{po}$$

(c concentration of the tracer in plasma or urine 50 h after injection, A administered dose, index po,iv application form)

Table 1: Tellurium Concentration in Foodstuffs
(given in ng/g wet weight)

	this work	H.A.Schroeder (4)
Pork loin	<7	<100
Pork fillet	<7	
Pork liver	<7	<100
Pork kidney	<7	
Bread (rye)	<7	<100 (whole wheat)
Flour (white)	<7	<100 - 2930 (n=9)
Flour (wholemeal)	<7	
Baking powder	<7	9710
Instant coffee	<7	<100 - 1690
Tea (teabag)	11	
Mixed herbs (dried)	12	
Parmesan (grated)	13	
Parsley	<7	370
Lettuce	<7	
Potatoes	3.2	210 - 220
White onion	0.8	5500 - 14310
Garlic	0.9	30970 - 72860
Mushrooms	1.0	360
Tomatoes	<0.7	640 - 1380
Cucumber	4.5	220 - 340
Apple	3.0	

Results and discussion

Table 1 shows the total tellurium concentration in various foodstuffs. From the data in the first column a daily dietary intake (ddi) of man under normal conditions in the range of 1 µg tellurium can be estimated.

This value is considerably lower than the ddi of 112 µg given by H.A.Schroeder and A.P.Nason in 1967 (4). They also found higher tellurium concentrations in foodstuffs as shown in the second column of table 1.

The fraction of ingested tellurium, which is absorbed in the intestine can be derived from the kinetics of the element in the organism after p.o. and i.v. application. Figure 1 shows the activity found per ml plasma in dependence of the time after injection of radioactive labelled sodium tellurate to a

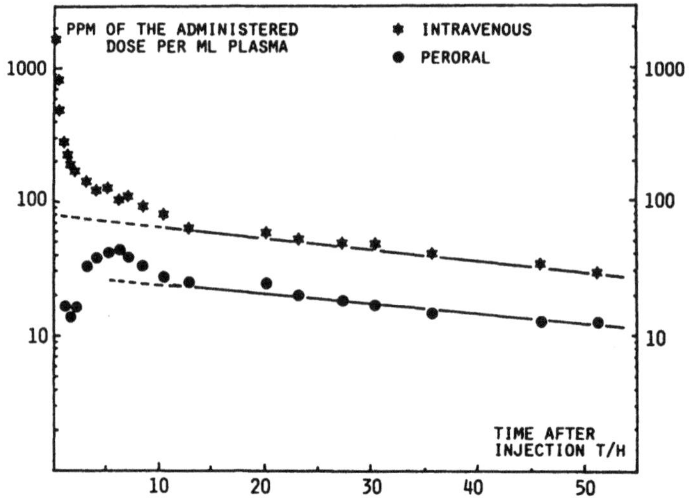

Figure 1: Tellurium Kinetics in the Blood Plasma of a Rabbit (injection of 180 µg (t = 0 min) and peroral gavage (t = 70 min) of 2500 µg Te as Na_2TeO_4)

rabbit. The values are given in a semilogarithmic scale as fraction of the administered dose.

The upper curve shows the kinetics after injection, and the lower one after gavage of Na_2TeO_4 to the same rabbit with an interval of 70 minutes. Six hours after the peroral application the intestinal absorption was terminated and both curves are in parallel with a half life of about 35 hours. This indicates that the metabolic behaviour of i.v and p.o. administered tellurate is similar. This hypothesis is also supported by other results of this study:

- The tellurium concentration in urine was high enough to be measured by AAS without ashing. Therefore, it was possible to distinguish between tetravalent and hexavalent tellurium. There was no change in the oxidation state after absorption into the organism, neither after intravenous nor after peroral application.
- The red blood cells of the rabbit were washed 5 times and

the tellurium activity was measured. After both application forms about 10 % of the tellurium in the whole blood was found in this fraction.

The assumption that p.o. and i.v. given tracers behave in the same way in the metabolism is essential to the application of the double isotope technique. The percentage intestinal absorption in the rabbit was found to be 39 and 41 % from the ratio between p.o. and i.v. administered tracer in plasma and urine respectively. The accuracy of the measurement is ± 10 % of the given value.
This finding leads to another argument for a ddi of about 1 µg. The tellurium concentration was measured with atomic absorption spectrometry in both blood plasma and urine of 8 healthy volunteers of both sexes and found to be below 0.7 ng/ml for every case. Therefore, the amount of tellurium which is excreted per day via urine is below 1 µg.
As the renal excretion is the main pathway of tellurium after injection to animals (5,6), the daily amount of this element absorbed into the human organism is probably also below 1 µg.

REFERENCES:

1. Einbrodt, H.J., S.Michels. 1984. In: Metalle in der Umwelt (E. Merian, ed.). Verlag Chemie, Weinheim. pp.561-569

2. Kotz, L., G.Kaiser, P.Tschöpel, G.Tölg. 1972. Z.anal.Chem. 260, 207.

3. Havezov, I., N.Jordanov. 1974. Talanta 21, 1013.

4. Schroeder, H.A., J.Buckman J.J.Balassa. 1967. J.chron.Dis. 20, 147. and Nason, A.P., H.A.Schroeder. 1967. J.chron. Dis. 20, 671

5. Hollins, J.G.. 1969. Health Physics 17, 497.

6. Wright, P.L, M.C,Bell. 1966. Am.J.Physiol.210, 6.

CADMIUM-BIOAVAILABILITY AND THE INFLUENCE OF FEED COMPONENTS IN CHICKENS

W.A. Rambeck, C. Bruckner, S. Meier, H. Zucker, W.E. Kollmer*

Institut für Physiologie, Physiologische Chemie und Ernährungsphysiologie, Tierärztliche Fakultät, Ludwig-Maximilians-Universität, München, Germany

*Institut für Strahlenbiologie, Gesellschaft für Strahlen- und Umweltforschung mbH, München, Germany.

Abstract

In chicken feeding experiments the effects of iron, vitamin C, selenium and pectine on Cd-bioavailability were studied. While the retention in the kidney was not influenced by the chemical forms $CdCl_2$, Cd-phytate, Cd organically bound in wheat or by the pectine added to the feed, the level of Se and vitamin C as well as the level and the chemical valency of Fe in the diet exerted a considerable influence.

Introduction

Effects of nutritional factors on the influence of dietary cadmium are known for a long time. These components may influence the absorption and the tissue retention of the heavy metal and therefore affect cadmium toxicity. In chicken we recently observed strong effects of calcium (1), zinc (2), phytate (3) and of fiber protein and fat (4) on Cd-retention. The purpose of the study reported here was to assess the

Trace Element Analytical Chemistry in Medicine and Biology

effects of iron, selenium, pectin and vitamin C on Cd-bioavailability in chickens. Additionally, effects of Cd-binding forms on the retention of the heavy metal in kidney were studied.

Material and methods

Male broiler chicks (Lohmann B 975, one day old) received after a one week preperiod a diet, containing different concentrations of Cd either as $CdCl_2$, Cd_3phytate or as Cd-contaminated wheat. After that time, the animals were killed and the Cd content in the kidney and partially also in the liver was determined by flameless atomic-absorption spectroscopy. Cd_3phytate was prepared in analogy to the preparation for Zn_3phytate described by Morris and Ellis (5). For the analyses of Cd the organs were homogenized and a sample of the homogenate of 70 to 140 mg was digested under pressure by HNO_3, analysed by flameless atomic absorption spectroscopy (AAS) as described by Hettich et al. (6). The analytical quality was checked by repeated determinations of Cd in the US National Bureau of Standards (NBS) bovine liver standard reference material 1577a which was treated in the same way as the samples of chick liver and kidney. The certified content of the NBS standard is 440±60 µg Cd/g. Fifteen samples of this standard yielded 452±26 µg/Cd/g. Nine determinations in one sample of the standard yielded 437±22 µg/Cd/g.

Results and discussion

Comparing the bioavailability of different Cd-binding forms it can be seen from Figure 1 that the Cd-retention in the kidney is the same, when $CdCl_2$, Cd_3phytate or Cd-contaminated wheat is added to the chicken diet at a concentration of 1 ppm.

Fig.1 Cd-Content in Chick-Kidney after feeding $CdCl_2$ ▭, Cd_3phytate ▨ or Cd-contaminated Wheat ▩

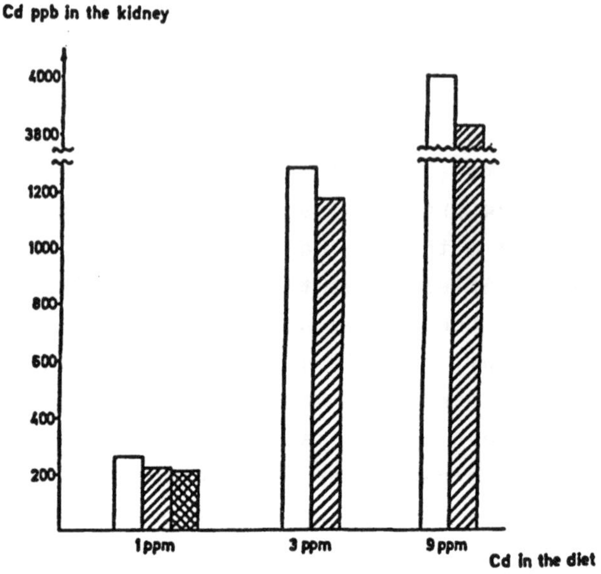

Table 1. Influence of Fe^{2+} and of Vitamin C in the Diet on Cd-Retention in the Kidney at a Level of 3 ppm $CdCl_2$ in the Diet (n = 12)

Fe^{2+} concentration in the diet (ppm)	Vitamin C (%)	Cd-concentration in the kidney $\bar{x}$ (ppb) ± s %
-	-	1638 ± 17
100	-	1040 ± 17
200	-	803 ± 19
200	1.0	497 ± 16

Since higher Cd-contaminated wheat is not available we only compared $CdCl_2$ and Cd_3phytate at a concentration of 3 ppm and 9 ppm. Again the binding form did not significantly influence Cd-retention in the kidney. Similar results were obtained, when the Cd-content of the liver was measured. While the increase of Cd in the feed from 1 ppm to 3 ppm (factor 3) lead to an increase in the kidney by a factor 4, the increase from 3 ppm to 9 ppm in the feed increased Cd in the kidney only by the factor of approximately 3.

The influence of iron on renal Cd-concentration can be seen from Table 1. Addition of 200 ppm Fe^{2+} lowers the Cd-burden of the kidney by 50 %. Fe^{3+} (not shown in the Table) has little or no effect. A further dramatic decrease of the heavy metal content in the kidney occurs when 1 % vitamin C is added to the diet. This is probably due to the positive influence of vitamin C on iron absorption via a soluble chelate complex. The effect of Fe^{2+} on Cd may be explained by competition of the two metals during absorption and during binding to the common transport system. Since calcium influences Cd-retention, we also investigated the effect of pectine, which - on its part - is known to influence calcium bioavailability. However, neither at 1 % nor at 2.5 % pectine in the chicken diet, Cd values of the kidney were different from the control. Selenium, given as selenite, was also investigated for its influence on Cd-bioavailability (Table 2).

Table 2. Influence of Se in the Diet on Feed Conversion and on Cd-Retention in the Kidney. (n = 12)

Se concentration in the diet ppm	feed conversion	Cd-concentration in the kidney $\bar{x}$ (ppb) $\pm$ s %
0.5	0.62	605 $\pm$ 15
5.0	0.62	414 $\pm$ 15
10.0	0.39	360 $\pm$ 19

We were especially interested, which selenium concentration in the chick diet reduces Cd-retention without being toxic. Feed conversion was used as a parameter of toxicity. 5 ppm of selenium were tolerated by the animals without a decrease of feed conversion. This concentration of selenium is already high enough to reduce the Cd-burden by one third. Obviously, it is of no use to increase selenium even more, because feed conversion drops dramatically at 10 ppm while the Cd-content of the kidney decreases only slightly. Summarizing, these studies demonstrate that the Cd bioavailability is influenced not only by feed components such as fiber, fat and protein, but also by a large number of trace elements and some vitamins.

References

1. Winkler, C., W.A. Rambeck, W.E. Kollmer, H. Zucker. 1984. Z. Tierphysiol. Tierernährg. und Futtermittelkde. 51, 250.

2. Bundscherer, B., W.A. Rambeck, W.E. Kollmer, H. Zucker. 1985. Z. Ernährungswiss. 24, 73.

3. Jackl, G.A., W.A. Rambeck, W.E. Kollmer.1985. Biol. Trace. Res. 7, 69.

4. Rambeck, W.A., G. Meiringer, W.E. Kollmer, H. Zucker. 1987. VDLUFA Schriftenreihe 20, 635.

5. Morris, E.R., R. Ellis. 1980. J. Nutr. 110, 1037.

6. Hettich, R., W.A. Rambeck, W.E. Kollmer, H. Zucker. 1986, GSF-Ber. 29/86.

TRACE ELEMENT ANALYSIS IN DIAGNOSIS AND IN PATHOLOGICAL STATES

Chairmen:

J.D. Kruse-Jarres
A.M. Reichelmayr-Lais
G. Tölg

Poster Reports:

F. Alt, P. Brätter, G. Knapp
E. Steinnes, G. Tölg

TRACE ELEMENT REGULATION OF IMMUNITY AND INFECTION; BASIC CONCEPTS AND PRACTICAL APPLICATIONS.

R. K. Chandra
Janeway Child Health Centre, St. John's, NF A1A 1R8, Canada

Abstract

Epidemiologic and clinical studies have documented that nutritional deficiencies result in increased risk of infection. Among other factors, impaired immunity is responsible for the enhanced severity and incidence of infection. Protein-energy malnutrition as well as deficiencies of single nutrients such as trace elements reduce cell-mediated immunity, phagocyte function, mucosal immunity, complement system, and antibody affinity. There are changes in levels of thymic hormone as well as immunoregulatory network. The production of interleukins and interferons is decreased. These basic changes in immunity have several practical applications.

Introduction

Nutrition and infection are a conjugate pair. Nutritional deficiencies result in reduced host resistance. It is the commonest cause of secondary immunodeficiency worldwide. The problem of malnutrition is not confined to the developing countries. Recent surveys in North America have revealed a surprisingly high prevalence of mild to moderate deficiencies, particularly those of trace elements such as iron and zinc. Among children attending hospitals, as many as 18 percent may show evidence of nutritional problems (1).

Trace Element Analytical Chemistry in Medicine and Biology

In this brief review, the effects of trace element deficiencies and excesses on immune responses and risk of infection would be discussed. In order to understand these relationships, it is useful to examine the interactions of various mechanisms of host protection and to summarize the effects of protein-energy malnutrition on the immune system. The subject has been extensively reviewed in various monographs and articles and reference should be made to these for specific citations (2-10).

Immunocompetence

There are many protective mechanisms that ward off microorganisms and other noxious agents. The immune system is critical for human survival. If the main tiers of immunity do not function, the individual will suffer from a variety of serious and prolonged infections that can prove fatal. The two main components of immunity are antigen-specific and antigen-nonspecific. The former consists of antibodies, which are immunoglobulins of five different isotypes, and of cell-mediated immunity that depends upon T lymphocytes. The nonspecific component consists of physical barriers such as the skin and mucous membranes and phagocytes, complement system, lysozyme, interferon, and others. These immune responses do not act in isolation, Instead, there is a complex but systematic interaction between various aspects of immunocompetence. For example, antibody response to several antigens depends upon help given by T lymphocytes and its products. In other situations, a variety of T cells, the suppressor cells, may inhibit antibody production by B lymphocytes. The functional significance of various protective factors and their interdependence is best shown by various human immunodeficiency syndromes. For example, in the condition known as common varied immunodeficiency, the hypoimmunoglobulinemia may be the result of one or more of the following processes: decreased number of B cells, reduced synthesis of immunoglobulin, reduction in the number and/or function of T helper cells, increase in the number and/or activity of T suppressor cells, or inactivation by autoantibody. In some instances, reduced function of one protective mechanism may be compensated by other normal responses. This is illustrated by the asymptomatic nature of selective IgA deficiency in the majority of affected individuals.

Immune Responses in Protein-Energy Malnutrition

Almost all aspects of immunity are impaired in severe nutritional deficiency syndromes of kwashiorkor and marasmus. The most consistent changes are observed in cell-mediated immunity, microbicidal activity of phagocytes, complement system, mucosal secretory IgA antibody response, and antibody affinity.

The number of mature rosette-forming T cells is decreased to less than 50 percent of control values. This is partly the result of decreased thymic hormone production, thereby impeding the normal maturation and differentiation of thymocytes. The ability of lymphocytes to respond to alloantigens and mitogens is decreased, particularly when culture is carried out in the presence of the patient's own plasma. The latter may contain a variety of inhibitory substances; alternatively, it may lack various nutrients that promote lymphocyte proliferation. Recent studies have shown a reduction in the production of interleukin-2 and of gamma interferon by leukocytes from malnourished children.

The serum level of immunoglobulins is generally high or normal, the former the probable consequence of repeated gastrointestinal and respiratory infections. In addition, reduced suppressor T cell activity may increase immunoglobulin levels, specially IgA and IgE. On the other hand, secretory IgA in various mucosal locations is decreased in malnutrition. This is true for total levels as well as specific antibody response after viral vaccine administration. These changes in mucosal immunity are of considerable clinical significance. For example, the incidence of septicemia is increased in malnutrition.

Neutrophils migrate slowly and their ability to kill ingested bacteria and fungi is decreased in protein-energy malnutrition. Rarely, neutropenia may be observed.

The impact of moderate and marginal malnutrition on immunity is less well defined. Several studies point to the possibility that even mild nutritional deficits may decrease immunologic responses. For instance, the number of T cells may be decreased and delayed cutaneous hypersensitivity responses are impaired in frequency and size.Tuberculin conversion following BCG vaccination is decreased. A variety of other responses is reduced. Understandably, the proportion of individuals with mild and moderate malnutrition who show immunologic deficits is relatively small. Moreover, the impairment in immunocompetence is less marked and may not be demonstrable if insensitive techniques are employed.

Trace elements and immunity

Many of the trace elements are deemed essential because a reduced intake produces impaired function, the abnormality being corrected or prevented by physiological amounts of the specific element. Such findings are generally reproducible in more than one mammalian species. The following trace elements are now considered essential for the growth, development and function of animals, and in many instances, of man(11). Many trace elements act as structural or catalytic components of larger molecules such as hormones and enzymes. It is not surprising, then, to find that deficiencies of several trace elements influence immune responses since the latter depend upon an efficient turnover of cells and synthesis of proteins.

Zinc

Zinc deficiency is best illustrated by the well recognized but rare inherited syndrome of acrodermatitis enteropathica. These patients fail to absorb adequate amounts of zinc and show the characteristic skin changes, abnormalities in hair, growth failure, recurrent infection and diarrhoea. Moderate amounts of zinc given orally reverses all the abnormalities. Patients with this condition show decreased lymphocyte proliferation responses to mitogens, reduced chemotaxis, and impaired delayed cutaneous hypersensitivity reactions to common recall antigens such as Candida, trichophyton, streptokinase-dornase, mumps, and purified protein derivative of tuberculin (5). Zinc supplements correct all the immunologic changes. The presence of zinc is important for thymic hormone activity (Fig. 1).

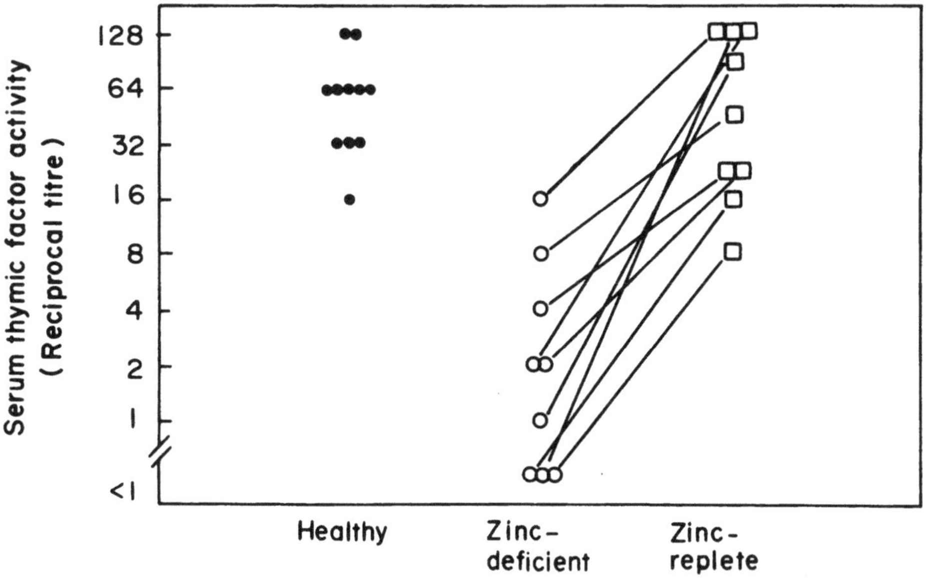

Serum thymic factor loses much of its biological activity after passage through a chelating column(6); the activity is restored after addition of zinc and to a lesser extent, by other trace elements.

Zinc deficiency associated with protein-energy malnutrition may enhance the immunologic abnormalities seen in undernourished subjects; the prevalence of zinc deficiency in such patients varies from more than 70 percent in the West Indies to less than 15 percent in India. In industrialized countries, zinc deficiency has been observed in patients on prolonged total parenteral nutrition, those with chronic malabsorption, Crohn's disease, and short gut syndrome following surgery.

In laboratory animals, zinc deficiency results in profound and long-lasting effects on the immune system. There is rapid and pronounced atrophy of the thymus and other lymphoid organs. Lymphopenia may be seen. Both primary and secondary antibody responses are decreased. Impaired T cell function in zinc deficient animals can be restored by zinc supplements. Delayed cutaneous hypersensitivity reactions are decreased and the generation of cytotoxic cells is impaired. Natural killer cell function is significantly reduced.

There is recent evidence to suggest that large doses of zinc given to healthy volunteers resulted in impaired lymphocyte and granulocyte functions(7). It was postulated that this may be the consequence of altered cell membrane function, contributed to by an increase in serum and membrane-bound low density lipoprotein.

Iron

Iron deficiency is common all over the world. Infants and adolescents are particular risk. In experiments conducted on laboratory animals, both deficiency and excess of iron result in increased susceptibility to infection. In man, mucocutaneous candidiasis and herpes simplex (Fig. 2) occur more frequently in the iron-deficient than among iron-replete subjects.

Iron deficiency impairs cell-mediated immunity as well as neutrophil function. T cell number is reduced slightly and lymphocyte response to mitogens is decreased. Intracellular killing of bacteria is decreased. In contrast, there is a minimum effect of iron deficiency on the complement system and antibody production.

The cellular and molecular basis of immunologic changes in iron deficiency has been the subject of much investighation. Iron is essential for the activity of ribonucleotidyl transferase enzyme that is critical for DNA synthesis and cell proliferation. Iron-binding lactoferrin has

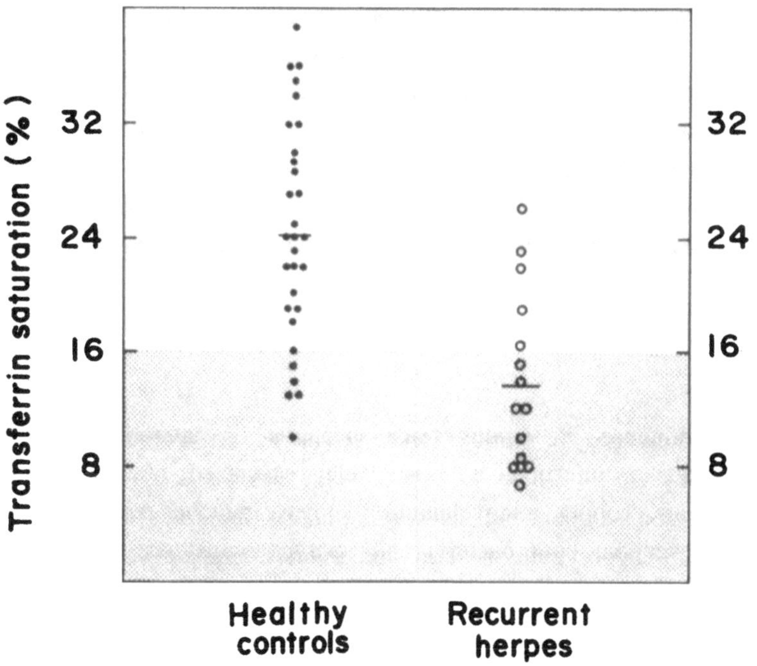

bactericidal properties and interacts with neutrophils to kill bacteria. Myeloperoxidase requires iron for optimum bactericidal function.

Copper

Copper deficiency is extremely rare in clinical practice. Children with the rare condition of Menkes disease show frequent infections such as pneumonia. Cell-mediated immunity is impaired, as judged by lymphocyte proliferation response to mitogens and antigens. Neutropenia may be seen. In copper-deficient animals, reticuloendothelial system function is decreased and granulocyte function is reduced. Antibody response to heterologous red cells is impaired, the abnormality correlates with the extent of reduction in ceruloplasmin levels. Thymic hormone activity is slightly reduced.

Selenium

There is paucity of data on the effects of selenium on immunity. To some extent, the influence of selenium deficiency on immune responses can

can be minimized by adequate or moderate excess of vitamin E; both have a recognized antioxidant function, preventing the peroxidation of lymphocyte membrane. Animals challenged with viral infection are more likely to develop cardiomyopathy if they are selenium-deficient. These observations may be pertinent to the pathogenesis of the unique cardiomyopathy of man and animals encountered in the Keshan region of China where the soil and water content of selenium is extremely low.

Conclusions

The critical influence of various trace elements on immunocompetence and susceptibility to infection is now well recognized. Among others, deficiencies of zinc, copper and selenium, depress immune responses. Iron is double-edged weapon; both bacteria and immunocompetence cells require iron for optimal growth and function. It is important to bear in mind that excessive intake of trace elements may also produce detrimental effects on the immune system.

Acknowledgements

I thank Dr Shakuntala Puri and Dr Devhuti Vyas for assistance. Parts of this paper were presented at the International Symposium on Pediatric Nutrition held in Modena, Italy, October 1987.

References

1. Puri, S., R.K. Chandra: 1985. Pediatr. Clinics N. Amer. 32, 499.
2. Chandra, R.K., D.H. Dayton. 1982. Nutr. Res. 2, 721.
3. Beisel, W.R. 1982. Amer. J. Clin. Nutr. 35: 417-68.
4. Gross, R.L., P.M. Newberne. 1980. Physiol. Rev. 66, 188.
5. Chandra, R.K. 1980. Pediatrics 66, 786.

6. Dardenne, M., J.M. Pleau, P. LeFrancier, J-F Bach.1981. CR Acad. Sci. Paris, 292, 793.
7. Chandra, R.K. 1984. J.Amer. Med. Ass. 252, 1443.
8. Joynson, D.H.M., A. Jacob, D. Murray Walker, A.E. Dolby. 1972. Lancet 2, 1058.
9. Bagchi, K., M. Mohanram, V. Reddy. 1980. Brit med J. 280, 1249.
10. Chandra, R.K. 1975. J. Pediat. 86, 899.
11. Mertz, W. 1981. Science 213, 1332.

TUBE FEEDING AND LOW SELENIUM STATE

D. Frosch *, H.J. Lübke **, H. Menzel ***, M. Laryea *, I. Lombeck *
* Children's and ** Medical Hospital and *** Institute of Toxicology, University of Düsseldorf

Abstract

Long-term tube feeding may result in low selenium state. In a 43 year old man selenium was reduced to 13 ng/ml in plasma, 21 ng/ml in whole blood, 188 ng/g in hair after 15 months of tube feeding; the daily urinary selenium excretion amounted to 5.3 µg. The patient's daily selenium intake was between 6.3 and 8.4 µg. Selenium contents of 29 formulas for tube feeding amounted to 3 - 17 ng/ml, i.e. 0.7 - 4.1 µg/MJ. Three short-term selenium supplementation periods resulted in a rapid increase of all selenium values estimated. Muscular symptoms like those reported in other patients with low selenium state responded to selenium as well as to placebo.

Introduction

Selenium deficiency in humans and its clinical symptoms are difficult to define. Besides endemic cardiomyopathy in China (Keshan Disease) (1) cardiomyopathy and selenium responsive myopathies have been reported (2, 3, 4). Recently in addition macrocytosis and pseudoalbinism have been observed in two patients (5).

The clinical symptoms of skeletal myopathies in patients receiving long-term parenteral nutrition have been described as pain, weakness and rigidity mainly in the legs but also in arms and shoulders. Selenium contents of plasma were between 3 and less than 20 ng/ml. Selenium was given orally or intravenously as selenite, selenious acid or seleno-

Trace Element Analytical Chemistry in Medicine and Biology

methionine in daily doses between 40 and 250 µg selenium. Symptoms disappeared within days or months as selenium contents rose to normal levels.

Case Report

We observed muscular symptoms in a 43 year old man who suffered for twenty years from various gastrointestinal diseases: gastric and duodenal ulcers led to vagotomy and subtotal gastrectomy, he developed a steatosis hepatis, chronic relapsing pancreatitis resulted in diabetes and steatorrhea. Due to malnutrition and malabsorption he developed severe underweight, 38 kg at 180 cm height, therefore intermittent oral hyperalimentation was started three years ago. Finally the patient exclusively lived on tube feeding.

After six months of total tube feeding the patient noticed progressive muscular weakness and cramps in his legs, there was no loss of sensibility. He also showed intermittent erythemas, the remainder of the physical examination was unremarkable. There were no signs of cardiomyopathy. Detailed neurological examination led to the diagnosis of encephalo-myelopathy probably due to malnutrition. Therefore vitamins and micronutrients were estimated and deficiencies of iron and zinc were diagnosed and treated.

After fifteen months of tube feeding the patient was unable to walk, therefore tube feeding was finished. At this time all micronutrients estimated were normal except the selenium levels which were determined for the first time.

Methods

The detection limit of our method was 0.5 ng, the accuracy of the method was checked by measuring the selenium content of NBS Bovine Liver SRM 1577 (certificate value 1.1 ± 0.1 µg/g, our value 1.15 ± 0.04 µg/g). The standard error of analysis was less than 7 % (15 ng/ml). Glutathione peroxidase activity was assayed by the method of Flohé (6), using t-butyl-hydroperoxide as acceptor substrate.

Results

The selenium content of plasma amounted to 13 ng/ml, of whole blood to 21 ng/ml, of hair 188 ng/g; daily urinary selenium excretion was only 5.3 µg (control values in healthy adults (n = 20): plasma 80 ± 15 ng/ml, whole blood 94 ± 16 ng/ml, urine 16 ± 6 µg/d; range in controls: hair 195 - 633 ng/g (n = 36)).

Glutathione peroxidase activity in plasma was reduced to 8 mU/ml (range in healthy adults: 110 - 150 mU/ml (n = 21)).

In order to calculate the patients selenium intake we estimated the selenium content of the formula used for his treatment, it was only 3 to 4 ng/ml. According to his mean caloric intake (8.8 MJ) his daily selenium intake was between 6.3 and 8.4 µg.

In comparison to normal food, the content of polyunsaturated fatty acids was two times higher.

Formula		Selenium Content	
		ng / ml	ng / MJ
N.S.	neutral	13	3,1
	schoko	15	3,3
	vanille	12	2,9
N.I.		17	3,8
Su.	OPD	9	2,3
Su.	neutral	4	1,0
	banane	3	0,7
Sa.	liquid	16	3,8
	toffee	15	3,6
B. 1500	neutral	13	2,2
B.S.		12	2,9
B.D.	neutral	9	2,3
	kakao	10	2,4
	vanille	12	2,9
F.	nuss	15	3,6
	schoko	16	3,8
	plus müsli	15	3,6
	vanille	15	3,6
	pfirsich	17	4,1

Table 1: Selenium contents of formulas available for long-term tube feeding

The selenium content of 29 formulas available for total tube feeding in our hospital ranged between 3 and 17 ng/ml (estimation in duplicates), that is 0.7 to 4.1 µg/MJ (Table 1). There was no difference in selenium content between different flavours and charges of the same brand. The results show no differences between meat-, milk- and soy-based formulas.

Selenium Supplementation and Discussion

During tube feeding the selenium intake in the reported case was in the range of adults from Keshan Disease affected areas. Since we found muscular symptoms, otherwise unexplained, coincident with low selenium state and reduced glutathione peroxidase activity we decided to start an oral selenium supplementation.

300 µg selenium / m^2 x d given as sodium selenite for ten days resulted in a rapid increase of all selenium values estimated: plasma selenium rose to 66 ng/ml at the tenth and last day of this first supplementation period, then decreased first rapidly, then more slowly. The selenium content of whole blood rose less fast, maximum at day fifteen with 49 ng/ml, maximum glutathione peroxidase activity in plasma was also noticed at the fifteenth day with 73 mU/ml. Hair selenium rose slightly to 288 ng/g.

The patient reported an improvement of his muscular symptoms and asked for further supplementation. One month later a ten day placebo course was performed and again he reported an improvement, although there was no change in selenium state.

Two and six months later we repeated the selenium supplementation. If we compare the three supplementations in the same patient with the same dose given as selenite (figure 1), there was always a steep increase of the plasma selenium values which decreased also rapidly to a somewhat higher plateau. The second supplementation led to a higher peak but with a similar slope and the third peak is nearly identical to the second one. The whole blood selenium contents showed a slower increase and a flat plateau. Selenium excretion in urine reflected the increase of plasma selenium levels: during the first supplementation daily urinary excretion was below 10 µg, during the second and third one it rose to 20 µg.

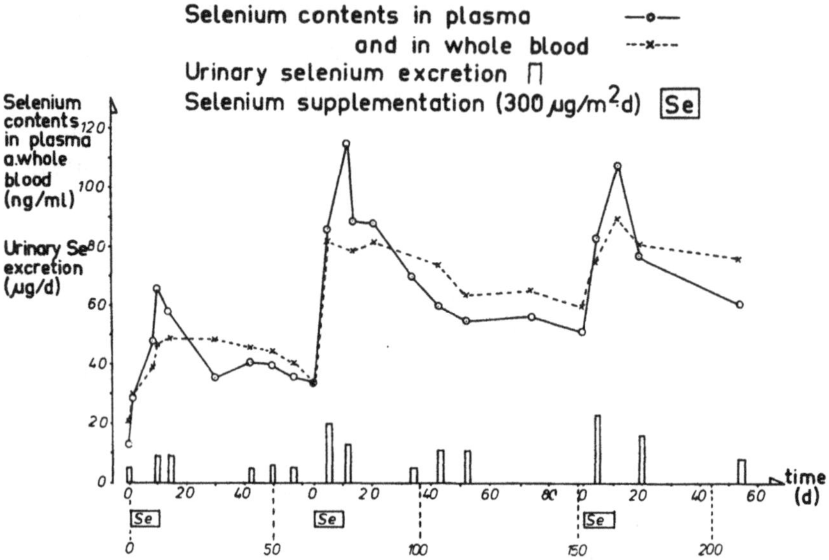

figure 1: Selenium contents in plasma and whole blood, urinary selenium excretion per day

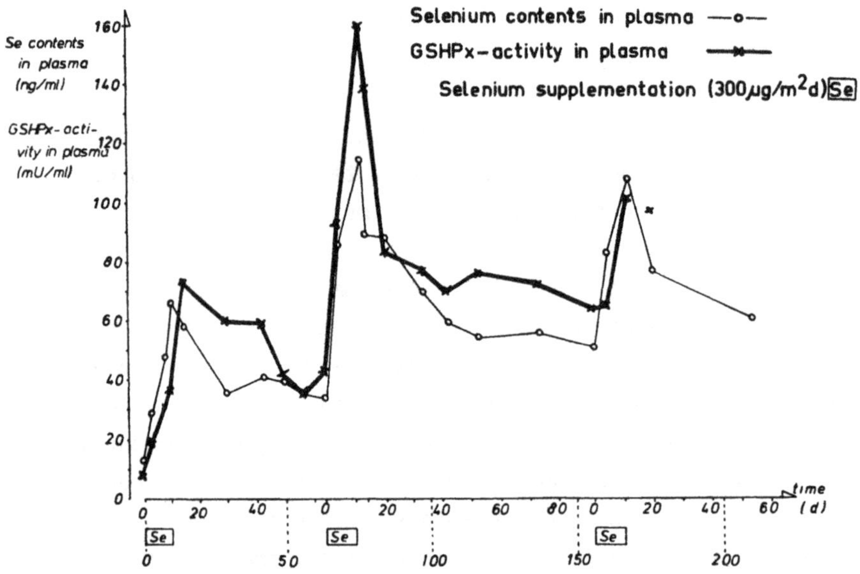

figure 2: Plasma selenium contents and glutathione peroxidase activity in plasma

Glutathione peroxidase activity in plasma showed a very steep increase, too. The values double within three days, the shape of the curve is similar to that of plasma selenium, but the peak values are from one to five days later (figure 2).
The patients muscular symptoms stopped worsening but the improvement he felt during the first supplementation and the placebo period did not result in a normal gait.
By recall methods his caloric, protein and carbohydrate intake was normal and he started slowly to gain weight. It remains questionable, why his selenium levels decreased again during normal food intake. Is that caused by specific malabsorption or still a sign of reduced body selenium stores?

Conclusion

Not only parenteral nutrition but also long-term tube feeding may result in low selenium state due to low selenium intake. Selenium supplementations should be performed in controlled clinical studies excluding placebo effects. Further investigations will be necessary to decide whether selenium supplementation of formulas, used for tube feeding, should be recommended.

References

1. Chen, X., Yang, G., Chen, J., Chen, X., Wen, Z., Ge, K. 1981, Biol. Trace Elem. Res. 2, 91-107

2. Kien, C.L., Ganther, H.E. 1983, Am. J. Clin. Nutr. 37, 319-328

3. Watson, D.R., Cannon, R.A., Kurland, G.S., Cox, K.L., Frates, R.C. 1985, J.P.E.N. 9, 58-60

4. Brown, M.R., Cohen, H.J., Lyons, J.M., Cutis, T.W., Thunberg, B., Cochran, W.J., Klish, W.J. 1986, Am. J. Clin. Nutr. 43, 549-554

5. Vinton, E.N., Dahlstrom, K.A., Strohl, C.T., Ament, M.E. 1987, J. Pediatr. 111, 711-717

6. Flohé, L., Günzler, W.A. 1984, Methods Enzymol. 105, 114-121

AN APPROACH TO MOLYBDENUM IN HEMODIALYSIS PATIENTS

A. Himmel, Y. Schmitt and J.D. Kruse-Jarres
Institute of Clinical Chemistry and Laboratory Medicine
Katharinenhospital Stuttgart, FRG

Abstract

Molybdenum, an essential trace element in man, is required in low quantities only. The daily Mo consumption is covered by a balanced nutrition. Negative balances result from an absorption of less than 0,1 ug/day. Mo is resorbed easily by the gastrointestinal tract and is excreted mainly via urine. Though having rather poor information about a Mo deficiency or intoxication we had an interest in knowing how the Mo concentrations change in hemodialysis patients with kidney diseases as been observed with respect to other trace elements.

In this study Mo levels in plasma of 20 hemodialysis patients were determined before and after dialysis using flameless atomic absorption spectrometry. The results were compared with Mo levels of healthy blood donors. Contamination of syringes and puncture needles were low and could be neglected.

69 % of the healthy blood donors had Mo levels lower than the detection limit of 1 µg/l. All patients before as well as after dialysis had values higher than the detection limit with a mean value of 9.9 ug/l +/- 6.9 before and 4.73 ug/l +/- 3.3 after dialysis. A decrease of Mo in plasma depends on an excretion into the dialysis fluids and, in addition, on a transfer to other intracorporeal compartments. No correlation between Mo levels and the duration of dialysis or age of the patients has been observed. Further investigation on the Mo mass balance during hemodialysis will be done.

Trace Element Analytical Chemistry in Medicine and Biology

Introduction

Molybdenum, an essential trace element in man, is required in low quantities only. With an intracorporal Mo content of about 5 mg the daily need is approximately 0.1 mg. With common nutrition the consumption rate is 0.1-0.3 mg/day. The richest sources of Mo are meats, cereal grain and legumes, the poorest are leafy and root vegetables, fruits, sugars, oils, and fats (1). The resorption scatters considerably depending on the different diet. It is excreted mainly via urine and in part via the bile. Proteins used especially for Mo transportation are not known yet (2). In man the trace element Mo is an essential component of the three enzymes sulfite oxidase, xanthin oxidase, and aldehyde oxidase. It is present in all three enzymes in a non covalently bound cofactor which can be released from the carrier proteins. Mo deficiency will lead to a deficiency of all three enzymes (3). A xanthin oxidase deficiency restrains the oxidation of purine. The shortage of aldehyde oxidase leads to a defect in the conversion of fatty acids from carbon hydrates (4). The mitochondrial sulfite oxidase serves as a terminal enzyme in the reduction of sulphureous amino acids.

The extreme frequency of endemic gout in populations who nourish mainly from food cultivated by themselves was correlated to the abnormally high Mo concentration in soils and plants. The Mo consumption in the refered regions was estimated at 10-15 mg/day compared to 1-2 mg/day in nearby regions with reduced gout incidence (2). The increased Mo consumption of 0.54 mg/day in balance experiments resulted in a distinctly increased excretion of copper. It might be possible that there is a prevention of gout if the Copper concentration is as well high as the Mo concentration (5).

Though having rather poor information about Mo deficiency and acute or chronic Mo intoxication (anemia, coma, liver and kidney disease) we had an interest in knowing how the Mo concentrations change in hemodialysis patients in comparison to healthy blood donors as been observed with respect to other trace elements like aluminum and chromium.

Material and Methods

In this study Mo levels in plasma and dialysis water of 20 hemodialysis patients were determined before and after dialysis using flameless atomic absorption spectrometry. The results were compared to over hundred Mo levels of healthy blood donors. The duration of dialysis of the patients was 240 min.The dialysis membranes were not standardized. The calibration of the measuring equipment was carried out using aqueous solutions. A reference curve was determined by the correlation of the extinctions of a dummy H20 specimen and three MO solutions with different volumes (10,15,20 µl). The aqueous solutions of the blood and dialysis water specimens were injected into the graphite tube using microliter pipettes and the specimen apparatus Perkin-Elmer AS60. At the inner surface the graphite tube was coated by pyro-carbon. By stepwise increasing the temperature (Tab.1) the stages drying, thermal dissociation of the matrix and thermal dissociation into free atoms (atomization) can be separated. An advantage of this method is the separation of accompanying substances from the specimen and , therefore, the uninfluenced measurement of the specimen in the atomized stage.

The results from the different measurements were evaluated with the aid of the reference curve. The compensation of the background through the Zeeman effect was good. The used materials for specimen withdrawal and preparation (syringes, puncture needles and solution) were checked with respect to their Mo contamination. Since they all were below the detection limit an influence of these materials on the results could be neglected.

Table 1: Graphite Furnace Operating Parameters

	Oven Temperature (degrees centigrade)	ramp (seconds)	hold time (seconds)	Gas Flow (ml/min)
1	80	5	5	300
2	120	60	20	300
3	1800	90	20	300
4	1800	1	3	0
5	2650	0	3	0
6	2700	1	5	300
7	20	1	10	300

Results

All hemodialysis patients had serum Mo values higher than the detection limit in comparison to the blood donors. 95 % of the blood donors had Mo levels lower than 2 µg/l and 69 % lying below the detection limit of 1 µg/l.

As shown in figure 1 the hemodialysis patients Mo values before dialysis had an average of 9.9 µg/l (SEM 6.87) in contrast to the values after dialysis which had an average of 4.73 µg/l (SEM 3.27). The values in the dialysis fluids during the first minutes of dialysis had a median of 1.6 µg/l while 5 being lower than the detection limit. During the last minutes of dialysis 18 patients showed levels lower than the detection limit (figure 2).

No correlation between the Mo levels in serum before and after the dialysis and the duration of hemodialysis, the age of the patients or the copper values has been observed. Only one patient showed extremely high Mo plasma levels (28.8 µg/l before and 13.69 µg/l after dialysis).

Figure 1: Molybdenum Levels in Serum before and after Dialysis

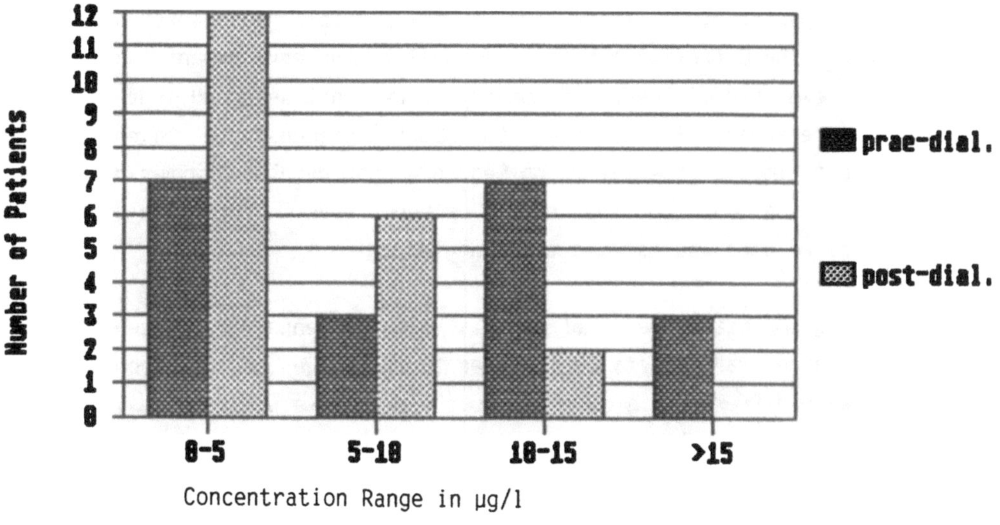

Figure 2: Molybdenum Levels in the Dialysis Fluids during the first and the last Minutes of Dialysis

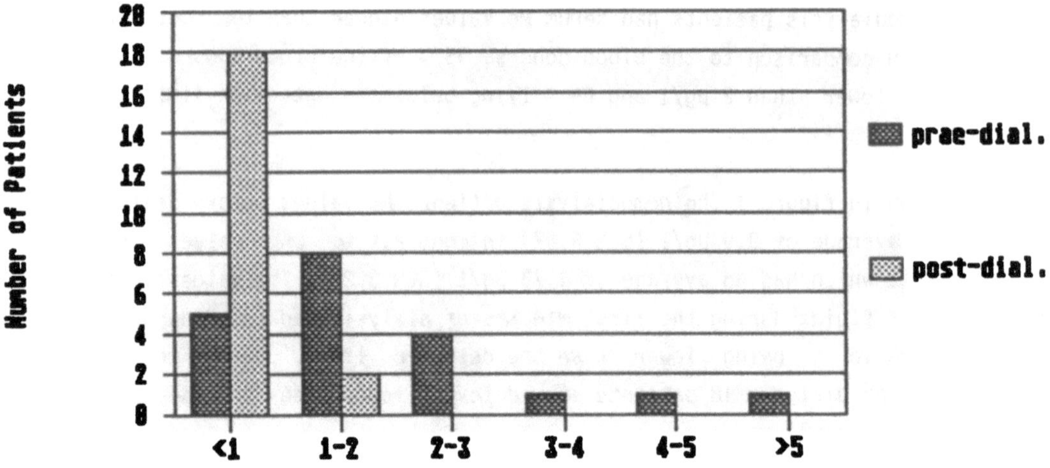

Discussion

The Mo concentration below 2 µg/l for 95 % of the blood donors is contrasted to a mean value of 0.6 µg/l in human serum reported by (6). This value was determined from selected healthy, non-exposed probands using atomic spectrometry. There are a few possible reasons for this difference. The differing nutrition as well as the environmental and vocational exposure of the blood donors who were not selected in this study must be regarded. Furthermore, the blood specimens were taken during different times of the day which led to a considerable scatter of the Mo concentration due to the reason given above. Contamination of the material we used was assumed to be neglectable.

Comparing the measurements of the dialysis patients Mo concentrations show enlarged variability as compared to the values of the blood donors. There were differences between various times Mo values were measured. By this fact, it can be presumed that the Mo serum content of hemodialysis patients depends more on different food consumption and environmental exposition than in healthy people because normal way of excretion via urine being reduced. In this case it must be discussed that increased Mo excretion through the bile occurs with kidney insufficiency. An enterohepatic circle might prevent Mo from being excreted. Furthermore a decrease of Mo in serum might depend on an excretion into dialysis fluid and, in addition, on a transfer into other intracorporal compartments.

The physiological antagonism between Mo and copper could not be observed in our blood donors because there was no correlation between these trace elements in serum. Whether there was an increased urinary copper loss with increasing Mo levels, as described by (7), we cannot explain.

We have to consider the patient with the high Mo level. It seems to be possible that this results from other diseases or lots of drugs from which he had more than the other patients. Further investigations of the influence of drugs on the Mo content will be done.

Literature

1. Hazell, T. 1985. In: Minerals in Food and Nutritional Topics (H. Borne, ed.). Karger, Basel, New York. p.82-87

2. Versiek, J. 1983. In: Spurenelemente (H. Zumkley, ed.). Thieme Verlag, Stuttgart, New York. pp. 152-165

3. Wadmann, S.K., M.Duran, P.K. de Bree, F.A. Beemer. 1985. In: Spurenelemente: Stoffwechsel, Ernährung, Imbalancen, Ultra-Trace-Elemente (E. Gladke, ed.). Thieme Stuttgart, New York. p. 92-95

4. Davis, G.K. 1984. In: Metalle in der Umwelt (E. Merian, ed.). Verlag Chemie, Weinheim. pp. 480-485

5. Merz, W. 1979. In: Spurenelemente (E. Gladke, ed.). Thieme Verlag, Stuttgart. pp. 153-160

6. Welz, B. 1983. Atomabsorptionspektrometrie. Verlag Chemie, Weinheim. p. 370

7. Perlman, M. 1984. In: Metabolism of Trace Metals in Man (O.M. Rennert, W.Y. Chan, eds.). CRC Press, Boca Raton. p. 51-62

METAL CONTENT IN HUMAN ALVEOLAR CELLS OF IRON-STEEL WORKERS AND CONTROL SUBJECTS STUDIED BY PIXE.

J-L. Corhay, J-P. Delavignette, M.F. Radermecker
CHU Pneumology Department

I. Roelandts, G. Robaye, G. Weber*
Experimental Nuclear Physics Department

University of Liège, Sart-Tilman B15, B-4000 Liège, Belgium.

Abstract

Most of the particles which have made their way into the alveoli will be processed by the alveolar macrophages. We studied with the PIXE method, the elementary composition of macrophages found in the bronchoalveolar lavage fluid of: twenty-one healthy subjects with no evidence of occupational exposure and thirty-two blue collarmen who work or have worked in an iron-steel factory. Alveolar macrophages of iron-steel workers from the Liège area contain more Fe, Cu and Ti than alveolar cells of non-exposed controls. Furthermore, we found a trend towards a Br, Zn, Ni, Cr and Sr increase in alveolar cells of steel workers.

Introduction

The etiology of many bronchopulmonary diseases remains obscure or totally unknown. Some of them have been associated with the elevated levels of certain elements in the lung tissue (1).With the cyclotron as a source of accelerated charged particles, we can easily measure trace-elements in the bronchoalveolar lavage fluid (BALF) using the PIXE method (particle induced X-ray emission)(2). It is known that most of the particles which have made their way into the alveoli will be processed by the alveolar macrophages (3,4,5,6). The exact measurement of the humoral elements of the BALF is difficult because it is altered by the amount of physiological fluid injected during fibroscopy (7). Iron, for instance, is present at high level in lung tissue, but its concentration can be considerably influenced by the amount of residual blood (5).

---

*Research Associate of the National Found for Scientific Research.

Trace Element Analytical Chemistry in Medicine and Biology

Trace element determinations on a given number of alveolar cells ($1x10^6$) would give us a more precise expression of the results and thus, renders us less vulnerable to fluctuations provoked by the injection of physiological fluid into the alveoli, and by the amount of residual blood at autopsy or biopsy. Therefore, we studied with the PIXE method, the elementary composition of macrophages found in the BALF of control subjects (white collarmen) and iron-steel workers (blue collarmen).The many foundries and steel plants in the Liège area and a high rate of lung fibrosis, bronchitis,... among the steel-workers, were additional reasons for initiating this study by these groups.

Material and methods

Patients

Thirty-two blue collar-men (mean age 54, range 35-80) who work or have worked in three different sectors of an iron-steel factory, blast-furnace (n=19), coke-oven (n=11), foundry (n=2) ,were studied. They worked for an average of 26 years. This group contained 18 smokers (18cig/day) and 14 non-smokers. Among the blue collarmen, twenty-two were asymptomatic, had normal chest-X ray and lung function, and, ten were suffering from chronic obstructive pulmonary disease (COPD). Twenty-one age-matched healthy subjects (10 smokers and 11 non-smokers) with no evidence of occupational exposure (white collarmen) were used as controls.

Bronchoalveolar lavage

Bronchoalveolar lavage fluid (BALF) was obtained by flexible fiberoptic bronchoscopy according to the conventional technique (8).It consists of instilling into a bronchopulmonary segment (e.g.middle lobe), in a careful and progressive way, 250 ml of sterile saline at 37°C, in fractions of 50 ml; and to recover immediately the solution by soft aspiration. Recovery is 40 to 80% of the instilled fluid, but sometimes less when the patient is bronchospastic or emphysematous. This procedure is well tolerated. A slowspeed centrifugation of the BALF is done as soon as possible in order to separate cells and supernatant. After red cells destruction (NH_4Cl), cells were washed three times with saline and suspended to $1x10^6$ cells/ml RPMI.

PIXE.method

The tube containing the sample was first deep frozen and lyophilised for 24 hours. The residue was suspended in 250 µl of Yttrium solution (internal standard; 400 µlY/ml) by using a Paramix II mixer. Thin targets were prepared by pipetting 25 µl aliquots of the resulting suspension onto a 3 µm-thick polypropylene film, stretched on a commercial 24X36 mm slide frame. This type of backing was found to be free from any serious contamination. The droplet was then air-dried at room temperature.
Then, the sample was bombarded with 2.5 MeV protons and the induced characteristic X-rays were recorded with a Si(Li) detector. After analysing and counting the events by means of a multichannel analyser, the concentrations were expressed as $\mu g/10^6$ cells from the peak areas and the calibration curve obtained from standard solutions.

Statistical analysis

The Student's t-test was used for statistical analysis. All data are expressed as the mean +/- SD and a probability level of less than 0.05 was accepted for statistical significance.

Results

Cell yields were not different in exposed versus non-exposed subjects but were higher in smokers ($X \approx 23\ 10^4$/ml, 90% macrophages) than in non-smokers ($X \approx 7\ 10^4$/ml, 84% macrophages) in both groups of subjects. The elements have been divided into two sets according to their frequency of occurrence.

1) In all BALF: K,Ca,Ti,Fe,Cu,Zn and Br.
2) In some BALF: Ni,Mn,Cr,Sr,Pb,As and Rb

No significant difference, of metal content in macrophages ($\mu g/10^6$ alveolar cells) was observed between smokers and no-smokers in healthy subjects of both groups (Table 1).

	Ti	Fe	Cu	Zn	Br
		WHITE COLLARMEN			
Smokers	**0.10**	**0.37**	**0.04**	**0.10**	**0.07**
(n=10) +/-	0.03	0.52	0.2	0.1	0.07
Non smokers	**0.09**	**0.42**	**0.04**	**0.10**	**0.10**
(n=11) +/-	0.08	0.38	0.2	0.03	0.07
		BLUE COLLARMEN			
Smokers	**0.12**	**0.93**	**0.06**	**0.11**	**0.11**
(n=14) +/-	0.04	0.47	0.05	0.11	0.09
Non smokers	**0.11**	**0.79**	**0.07**	**0.08**	**0.08**
(n=8) +/-	0.02	0.73	0.06	0.02	0.05
		STUDENT'S TEST			
ρ	**NS**	**NS**	**NS**	**NS**	**NS**

TABLE 1:Comparison of the metal content in the alveolar cells of the healthy smokers and non-smokers (microg/10^6 alveolar cells).

When comparing the content of these elements in the alveolar cells in healthy workers versus non-exposed controls, we found:

- for the first set of elements(Table 2) a significant Fe, Cu and Ti increase and a trend to Zn and Br increase in macrophages of blue collarmen;
- for the second set of elements, Ni, Cr and Sr were most frequently found in iron-steel workers;the detection limit (DL) of these elements and their frequency are described in Table 3;no values were upper than the detection limit data in the alveolar cells of white collarmen.

	Ti	Fe	Cu	Zn	Br
	WHITE COLLARMEN (n=21)				
	0.09	**0.40**	**0.04**	**0.10**	**0.09**
+/-	0.03	0.44	0.02	0.07	0.07
	BLUE COLLARMEN (n=22)				
	0.12	**0.88**	**0.06**	**0.10**	**0.10**
+/-	0.04	0.57	0.05	0.09	0.08
	STUDENT'S TEST				
ρ	**<0.05**	**<0.01**	**<0.05**	**NS**	**NS**

TABLE 2.Comparison of the metal content (microg/10^6 alveolar cells) of macrophages in healthy white collarmen and iron-steel workers for the first set of elements.

	Ni	Cr	Sr
DETECTION LIMIT (DL) (μg/10^6 cells)	0.005	0.010	0.030
% BLUE COLLARMEN >DL	**13%**	**1%**	**10%**
% WHITE COLLARMEN >DL	**0%**	**0%**	**0%**

TABLE 3.Comparison of the metal occurrence in macrophages of blue and white collarmen for the second set of elements.

Table 4 shows the comparison of macrophage content (microg/10^6 alveolar cells) between healthy and COPD iron-steel workers.

	Ti	Fe	Cu	Zn	Br
		HEALTHY BLUE COLLARMEN (N=22)			
	0.12	**0.88**	**0.06**	**0.10**	0.10
+/-	0.04	0.57	0.05	0.09	0.08
		COPD BLUE COLLARMEN (N=10)			
	0.14	**1.31**	**0.10**	**0.12**	0.21
+/-	0.02	1.41	0.11	0.12	0.20
		STUDENT'S TEST			
p	**NS**	**NS**	**NS**	**NS**	<0.05

TABLE 4.Comparison of the metal content (microg/10^6 alveolar cells) in macrophages of healthy and COPD iron-steel workers.

Conclusion

Alveolar macrophages of iron-steel workers from the Liège(Belgium) area contain more Fe, Cu and Ti than alveolar cells of non-exposed controls. Furthermore, we found a trend towards a Br, Zn, Ni, Cr and Sr increase in alveolar cells of steel workers. Iron increase can be easily explained by chronic exposure of the workers to iron-ores (9). In a preliminary study (2), we have already found a very high content of iron in the BALF of a steel-worker. An increase of the other elements (Zn,Ti,...) correlates with their presence in iron-ores, procedures of steel production and work environment:nearness of welders and maintenance technicians working in the same area. Increasing Fe in macrophages of COPD workers was probably due to the impaired macrophage clearance, or repetitive intrabronchial bleeding or both.The measurement of trace elements in human alveolar macrophages collected by BALF represents a valuable index of occupational exposure to potentially toxic metals

Acknowledgement

Study supported by grant n° 7248-11-035 of the ECSC.
We are indebted to the Institut Interuniversitaire des Sciences Nucléaires for financial support.

References

1. Sabbioni, E., R. Pietra, F. Mousty, F. Colombo, G. Rizzato and G. Scansetti.J.Radioanal.Nucl.Chem.110,595,601.

2. Weber, G., G. Robaye, J.M. Delbrouck, I. Roelandts, O. Dideberg, P. Bartsch and M.C. Depauw.1980.Nucl.Instr.and Methods.168,551-556.

3. Brain, J.D.1970.Arch.Intern.Med.126,477-487.

4. Fraser, R.G. and J.A.P. Paré.1977. Diagnosis of diseases of the chest. W.B. Saunders Compagny.

5. Vanoeteren, C., R. Cornelis and R. Sabbioni.1986. Critical evaluation of normal levels of major and trace elements in human lung tissue. Eur-10440-EN, CEC-Luxembourg.pp.57-58.

6. Voisin, C. and L. Lenoir.1985.Notions générales sur les pneumoconioses.1985. In: Encycl.Med.Chir.7,pp.1-6.

7. Gaudichet, A., J.C. Pairon, O. Malandain, B. Couste, P. Brochard and J. Bignon 1987.Rev.Mal.Resp.4,237-243.

8. Radermecker, M., M. Gustin, P. Saint-Remy, Y. Sambon and R. Broux.1983.Acta.Clinica.Belgica 38,1,35-42.

9. Desoille, H., J.Scherrer and R.Truhaut.1980.In: Precis de medecine du travail, (Masson), pp.595-618.

MULTIELEMENT ANALYSES IN HUMAN LUNG TISSUE CORRELATED WITH SMOKING, EMPHYSEMA AND LUNG CANCER

P.-L. Kalliomäki[1], P. Pääkkö[2], K. Malmqvist[3], J. Pallon[3], S. Anttila[1], H. Vainio[1], K. Kalliomäki[4]

Institute of Occupational Health, Helsinki Finland[1]

Department of Pathology[2] & Electrical Measuring Technique[4], University of Oulu, Oulu Finland

Department of Nuclear Physics, Institute of Technology, Lund Sweden[3]

Introduction

Fragmental knowledge has been published on the tissue levels of major and minor trace elements in the human lungs in relation e.g. to the exposure to cigarette smoke and pulmonary tissue reactions (1). Modern analytical instrumentation reveals new possibilities for detecting a number of elements in the same sample.

The aim of this study is to determine the concentrations of light elements derived mainly from minerals, heavy metals and essential trace elements in human lung tissue. The elemental concentrations found in lung tissue are related to smoking, emphysema and lung cancer.

Material

The lung specimens were collected from subjects without any malignant disease and from the lung cancer patients. The group without malignancies consisted of 41 lungs (alternatively the right or the left lung) from consecutive autopsies of male subjects with or without emphysema (Table 1). Because emphysema is an indicator of smoking, the subjects without lung cancer were selected to include cases with different grades of emphysema (Table 2).

Trace Element Analytical Chemistry in Medicine and Biology

Table 1. Age, smoking habits and grade of emphysema of the subjects without cancer (autopsy lungs) and of the lung cancer patients.

	autopsy lungs (N=41)	lung cancer patients (N=19)
Age, years	68 ± 9	64 ± 9
Smoking time, years	36 ± 15	40 ± 9
Stopped, years ago	7 ± 10	5 ± 7
Grade of emphysema	2,5 ± 0,9	2,6 ± 0,8

Table 2. Age and smoking habits of the subjects without cancer grouped as a function of emphysema.

	grade I (normal) (N=4)	grade II (mild) (N=20)	grade III (moderate) (N=9)	grade IV (severe) (N=8)
Age, years	75 ± 2	64 ± 9	65 ± 8	75 ± 5
Smoking time, years	26 ± 9	32 ± 12	37 ± 13	49 ± 13
Stopped, years ago	6 ± 12	6 ± 11	6 ± 10	10 ± 5
Stopped	(N=4)	(N=16)	(N=9)	(N=6)

No significant difference in right/left lung/lung lobe in each group.

Smoking habits and occupational history were ascertained from the next of kin. Two nonsmokers, 19 active smokers, and 16 ex-smokers were among the subjects without cancer (Tables 3 and 4).

The lungs or lung lobes have been obtained from 19 male patients operated for lung cancer (Table 1). Smoking habits and occupational history have been acquired by personal interview.

Table 3. Age, smoking habits and grade of emphysema of the subjects without cancer (autopsy lungs) grouped as a function of smoking time (including smokers and ex-smokers).

	group I (N=10)	group II (N=8)	group III (N=9)	group IV (N=8)
Age, years	66 ± 10	64 ± 11	70 ± 8	70 ± 9
Smoking time, years	18 ± 10	33 ± 2	42 ± 2	53 ± 8
Stopped, years ago	7 ± 10	12 ± 16	5 ± 7	3 ± 4
Grade of emphysema	2,1 ± 0,7	2,3 ± 0,9	2,6 ± 1,0	3,1 ± 0,8

Table 4. Age, smoking habits and grade of emphysema in the non- smokers, smokers and ex-smokers without cancer (autopsy lungs).

	non-smokers	smokers	ex-smokers
Age, years	75 + 3	64 + 9	72*± 8
Smoking time, years	0	33 ± 15	38 ± 14
Stopped, years ago		0	15 ± 10
Grade of emphysema	1,5 ± 0,7	2,1 ± 0,7	2,9*± 1,0

Methods

The lungs and lobes of lungs were radiographed during continuous air-inflation, and the severity of emphysema was graded from the radiographs into normal lung and mild, moderate and severe emphysema (4). In addition, the grade of emphysema (3) and the histological types of cancer were determined from histological sections (1).

A piece of fresh lung tissue was taken after radiographing from the (apico) posterior and anterior segment of the upper lobe and from the superior (apical) part and basal segment of the lower lobe (1,3). The sample, measuring 0.5 - 2 g and containing no pleural surface or cancer tissue. Equal portions of samples representing left and right lungs were unanalyzed in the nonmalignant and cancer material. The fresh sample was weighed, dried for three hours in vacuum, and ashed in glass cups with a blank and NBS Bovine Liver (SMR 1577) for 48 - 72 hours in a low temperature oxygen plasma asher (100 - 120°C).

The ashed samples (1 - 10 mg) were digested in 10 ul 7 M HNO_3 (suprapur) and diluted in 100 ul ultrapure (MilliQ) water and 10 ug yttrium (as YNO_3) was added as an internal standard. The solution was spotted (20 ul) on thin polycarbonate foils [Kimfol (R), 2 um] which was freeze-dried and the samples are irradiated with 2.5 MeV protons (10 nA) with a totally accumulated charge of 3 uC. The X-rays produced were analysed in a Si(Li) detector at 135 degrees relative to the beam direction. A 55 um thick aluminium absorber with a small hole was used to optimize the sensitivity in various regions of the PIXE spectra. The samples thus prepared were thin (< 1 mg/cm^2) and consequently do not require any correction for X-ray spectra are stored on magnetic tape and evaluated off-line using the HEX code.

Normalization in the PIXE analysis was made by comparing with the added internal standard which is not present in the samples above the detection level. The results are corrected for the results from analyses of blank foils. When calculated the original tissue concentrations the weighed masses were used.

Results

These results are preliminary, the study is still under review. The concentrations of 15 elements in lung tissue were analyzed (Tables 5 and 6). Table 5 shows that the concentrations of P, Cl, Ca, Ti, Cr and Rb are typically twice higher in the surgical specimens of lung cancer patients than in autopsy lung tissue of nonmalignant subjects. It can be also observed that the scatter in the concentrations of P, S, Cl, K, Ca, Ti and Mn are significantly larger in the surgical specimens than those in the autopsy lung tissue of nonmalignant subjects (Table 5).

Table 5. The element concentrations in lung tissue of the subjects without malignant disease and in surgical specimens from lung cancer patients.

element	ug/g dry weight				p
	autopsy lungs		surgical specimens		
	mean	95 % intervall	mean	95 % intervall	
P	3.300	2.300 - 4.700	11.000	4.000 - 30.000	0,000
S	4.300	1.400 - 12.000	5.500	400 - 80.000	
Cl	1.200	260 - 5.300	500	40 - 7.000	0,16
K	6.600	4.400 - 10.000	7.400	2.200 - 2.500	
Ca	740	440 - 1.300	1.100	350 - 3.400	0,07
Ti	10	5 - 21	21	6 - 80	0,026
Cr	4	2 - 10	9	4 - 22	0,002
Mn	2	1,4 - 3,4	2,2	0,9 - 6	
Fe	1.400	900 - 2.200	1.100	500 - 2.300	
Cu	8	5 - 12	7	5 - 10	
Zn	66	40 - 110	65	50 - 90	
Sn	30	20 - 45	30	20 - 50	
Se	1,0	0,7 - 1,5	1,0	0,7 - 1,5	
Rb	12	8 - 18	16	12 - 21	0,022
Sr	1,2	0,9 - 1,6	0,8	0,6 - 1,1	0,012

Table 6. Effect of smoking habits upon the element concentrations in non-malignant autopsy lungs.

element	mean, ug/g dry weight		
	non-smokers (N=2)	smokers (N=19)	ex-smokers (N=16)
P		3.100	
S		4.200	
Cl		1.200	
K		6.300	
Ca		780	
Ti	<detec.lim.	13	11
Cr	<detec.lim.	2,9	5,4
Mn		2,1	
Fe		1.400	
Cu		8	
Zn		63	
Sn	<detec.lim.	30	26
Se		1,0	
Rb		12	
Sr		1,2	

Figure 1. Element concentrations in lung tissue of nonmalignant autopsy lungs in relation to the severity of emphysema.

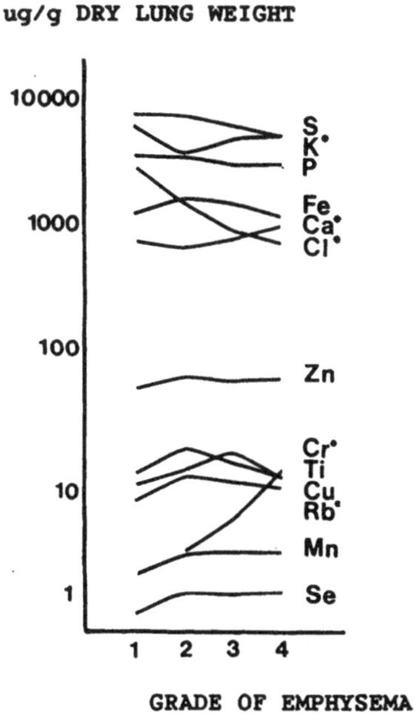

Figure 2. Element concentrations in lung tissue in relation to the length of smoking time among non-malignant autopsy lungs.

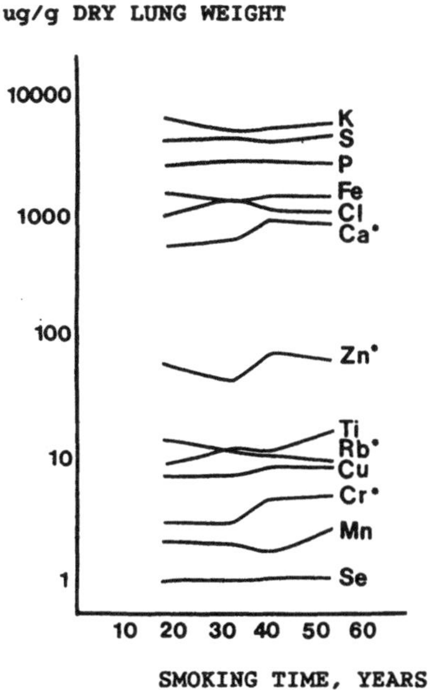

Figure 1 reveals that severity of emphysema is related to the increase of the concentrations of Cr and Ca and to the decrease of the concentrations of K and Cl in lung tissue of nonmalignant autopsy lungs. No essential difference regarding the age or the smoking time existed in the subjects categorized according to severity of emphysema. The effect of severity of emphysema upon the any elements was not significant in surgical specimens from the lung cancer patients.

The effect of cigarette smoke upon the elements in the lungs was estimated (Table 6, Figure 2). The concentrations of other elements except those of Cr, Ti and Sn were at the same level in lung tissue of nonsmokers, smokers and ex-smokers without cancer (Table 6). The concentrations of Cr, Ti and Sn were below the detection limit in lung tissue of nonsmokers' nonmalignant subjects, the Cr concentration had an increased trend in smokers' lungs and it was also found to be increased among ex-smokers' lungs (Table 6) indicating that the Cr, accumulating in the lungs from the cigarette smoke, does not clear from the lungs. The concentrations of Cr, Ca and Zn increased and that of Rb decreased as a function of smoking time in lung tissue of nonmalignant autopsy lungs (Figure 2) (3).

References

1. Anttila, S., P. Kokkonen, P. Pääkkö, T. Kerola, P. Rainio, P.-L. Kalliomäki, S. Sutinen. 1987. In: Extended Abstracts: Trace Elements in Human Health and Disease, 2nd Nordic Symposium, 17-21 August 1987, Odense Denmark. p. 173.

2. International Agency for Research on Cancer. 1987. IARC Monograph on the Evaluation of the Carcinogenic Risk of Chemicals to Humans: Tobacco Smoking, Vol. 38. IARC, Lyon France. 116 p.

3. Pääkkö, P., P. Kokkonen, S. Anttila, T. Kerola, P.-L. Kalliomäki, S. Sutinen. 1987. In: Extended Abstracts: Trace Elements in Human Health and Disease, 2nd Nordic Symposium, 17-21 August 1987, Odense Denmark. p. 148.

4. Sutinen, S., P. Pääkkö, P. Lohela, R. Lahti. 1981. Eur. J. Resp. Dis. 62, 297.

5. Vanoeteren, C., R. Cornelis, E. Sabbioni. 1986. Critical Evaluation of Normal Levels of Major and Trace Elements in Human Lung Tissue. Commission of the European Communities, Luxemburg.

EFFECTS OF SMOKING ON IRON, COPPER, ZINC, CADMIUM, COERULOPLASMIN AND TOTAL PROTEIN CONTENTS OF AMNIOTIC FLUID AND SERUM OF PREGNANT WOMEN AND NEONATES

L. Salgó

Central Laboratory of Clinical Chemistry, Albert Szent-Györgyi Medical School, Szeged, Hungary

Abstract

Investigations were performed on the amniotic fluid of 237 pregnant women, and on 1542 maternal blood samples of 1170 pregnancies and 1170 blood samples from the umbilical vein at birth. The smoking mothers (n=147) were classified in two groups: those smoking not more than 9; and those smoking 10 or more cigarettes daily (strong smokers). The serum Cd level in strong smokers, 56±10.4 nmol/l (mean±s range), was significantly higher than in the non-smokers, 37.2±8.2 nmol/l, and the coeruloplasmin activity was higher too: 442±54 vs. 356±47 U/l. Zn and Fe levels in the amniotic fluid in the case of strong smokers, 3.4±1.0 µmol/l and 12.0±4.7 µmol/l, were significantly lower than the corresponding levels for non-smokers: Zn 5.0±1.3 µmol/l and Fe 15.3±5.9 µmol/l.

Introduction

One of the factors most seriously endangering the health of the population is the increase in smoking. Unfortunately, many people tend to the view that smoking is a harmless habit, in contrast with the consumption of drugs. The seriousness of the problem is intensified by the fact that most smokers begin smoking at a very early age: before they are 16 years old.

The harmful effects of smoking during pregnancy are very complex (1,4). It is well known and very important that maternal smoking during preg-

Trace Element Analytical Chemistry in Medicine and Biology

nancy impairs fetal growth (decreased placental blood flow, reduced oxygen transport capacity because of carboxyhemoglobin formation in mother and fetus, toxic effects of metabolites absorbed from the smoke, etc.); the resulting reduction in birth weight is closely related to the maternal smoking level (2,3,6,7).

The intrauterine fetus has a fundamental need for most trace elements (Fe, Zn, Cu, etc.), but at the same time the fetus may come into contact with certain trace metals (Cd, Ni, Pb, etc.) which, above a certain level, are toxic. There is a correlation between the levels of these elements and the extent of smoking or the degree of pollution of the environment (5,9). We have carried out investigations of the following questions:

1. *How do smoking and the extent of smoking influence the concentrations of Fe, Cu, Zn, Cd, coeruloplasmin and total protein in the serum of pregnant women and in the amniotic fluid?*
2. *To what extent does maternal smoking influence these concentrations in the serum of neonates, separate consideration being paid to premature neonates and full-term neonates?*

Material and Methods

In 1983-86 we analyzed 237 amniotic fluid samples (237 pregnancies) and 1542 maternal blood samples at different stages of gestation (14 to 42 weeks) (1170 pregnancies) and 1170 blood samples from the umbilical vein at birth. The mothers ranged in age between 15 and 40 years. All subjects were healthy. The smoking mothers (n=147) were classified in two groups: those smoking not more than 9 cigarettes a day (moderate smokers), and those smoking 10 or more cigarettes daily (strong smokers).

The amniotic fluid samples were obtained by transabdominal amniocentesis or cesarean section or by puncture of the amniotic sac before term delivery. All samples of amniotic fluid contaminated with meconium and/or blood were excluded. The amniotic fluid and the serum samples were centrifuged for 30 minutes at 10000 x g in order to separate corpuscular and cellular components.

Cadmium (Cd), zinc (Zn), copper (Cu) and iron (Fe) were determined by atomic absorption spectrophotometry (Perkin-Elmer); total protein (pr) by the biuret method; and coeruloplasmin (CPL) by a standard photometric method. Statistical analysis was performed by the two-tailed "t"-test.

Results and Discussion

As concerns the age distribution of the pregnant women, the proportion of those aged below 25 years was 52.4% for the non-smokers, 57.5% for the moderate smokers, and 58,9% for the strong smokers. Premature neonates (-2500 g) were born to 10.5% of the non-smokers, 17.5% of the moderate smokers, and 40% of the strong smokers. The variations in the concentrations of Cu, Zn, Cd, pr and CPL in the maternal serum during pregnancy are to be seen in Table 1.

Parameters of serum		time of gestation							
		I. trimester n=97 x̄ ± SD		II. trimester n=148 x̄ ± SD		III. trimester n = 127 x̄ ± SD		at birth n = 1170 x̄ ± SD	
Cu (µmol/l)	non-smoking	26.1	1.0	30.9	1.4	37.3	3.0	38.1	3.3
	moderate smokers	27.4	1.3	32.1	1.6	39.0	3.3	39.6	4.0
	strong smokers	28.0	1.4	33.5	1.6	41.7	4.4	42.4	4.2
Zn (µmol/l)	non-smoking	14.7	1.3	12.9	0.9	11.3	1.4	10.9	2.1
	moderate smokers	14.2	1.2	12.1	1.0	10.8	2.0	10.2	2.0
	strong smokers	14.2	1.4	12.0	1.0	10.3	2.5	10.0	2.0
Fe (µmol/l)	non-smoking	18.9	3.9	17.3	3.0	15.9	2.8	15.2	3.0
	moderate smokers	18.6	4.3	16.9	3.0	15.6	3.5	15.0	3.2
	strong smokers	18.8	4.0	16.5	4.2	15.7	4.0	14.8	3.5
Cd (nmol/l)	non-smoking	50.2	13.1	42.8	8.6	38.0	8.8	37.2	8.2
	moderate smokers	54.9	17.3	49.7	11.3	44.7	9.0	42.9	8.7
	strong smokers	67.0	16.4	58.3	12.3	56.3	13.1	56.0	10.4
pr (g/l)	non-smoking	66.2	3.7	64.8	4.0	60.5	3.5	59.7	3.0
	moderate smokers	66.5	3.8	65.1	4.0	61.1	3.2	61.9	3.1
	strong smokers	66.6	3.5	65.0	4.0	61.4	3.3	62.1	3.0
CPL (U/l)	non-smoking	179.0	22.0	224.0	29.1	312.4	40.2	356.0	47.0
	moderate smokers	198.8	31.0	251.0	32.9	350.1	44.8	381.2	50.0
	srong smokers	221.5	30.0	290.0	34.0	407.4	43.1	442.0	54.1

Table 1. Variations in Concentrations of Cu, Zn, Fe, Cd, Pr, and CPL in the Maternal Serum During Pregnancy, Categorized on the Basis of the Extent of Smoking by the Mother

The concentrations of Cd, Fe, Cu and CPL in the serum of the strong smokers were significantly higher than the corresponding levels for the non-smokers.
The equivalent variations in the amniotic fluid during pregnancy are shown in Table 2.

Parameters of amniotic fluid		I. trimester n=12 $\bar{x} \pm SD$		II. trimester n = 8 $\bar{x} \pm SD$		III. trimester n = 74 $\bar{x} \pm SD$		at birth n = 143 $\bar{x} \pm SD$	
Cu (µmol/l)	non-smoking	2.6	0.5	2.6	0.7	2.8	0.8	2.8	0.6
	moderate smokers	2.6	0.6	2.9	0.7	2.8	1.0	3.0	0.8
	strong smokers	2.9	0.6	3.2	0.0	3.2	1.0	3.2	1.1
Zn (µmol/l)	non-smoking	1.7	0.3	1.2	0.2	5.9	1.6	5.0	1.3
	moderate smokers	1.4	0.5	1.2	0.2	4.7	2.0	4.5	1.5
	strong smokers	1.4	0.5	0.8	0.0	4.3	1.3	3.4	1.0
Fe (µmol/l)	non-smoking	18.7	3.9	18.3	2.9	16.2	2.3	15.3	5.9
	moderate smokers	18.3	4.4	18.0	3.1	15.3	1.9	13.8	6.2
	strong smokers	18.2	4.2	16.9	0.0	14.2	2.0	12.0	4.7
Cd (nmol/l)	non-smoking	175.0	73.0	103.3	54.1	68.6	32.2	65.0	29.1
	moderate smokers	237.4	105.1	149.5	71.0	84.0	29.1	83.8	33.0
	strong smokers	285.7	103.0	190.1	0.0	117.3	41.0	108.0	32.0
pr (g/l)	non-smoking	4.2	1.0	5.3	1.9	2.9	0.6	1.9	0.2
	moderate smokers	4.3	1.0	5.9	2.2	3.1	1.0	1.9	0.3
	strong smokers	4.3	1.0	6.0	0.0	3.0	0.8	2.1	0.5

Table 2. Variations in Concentrations of Cu, Zn, Fe, Cd and pr in the Amniotic Fluid During Pregnancy, Categorized on the Basis of the Extent of Smoking by the Mother

In the group of strong smokers, the Cd concentration was significantly elevated, while the Zn and Fe levels were significantly depressed, relative to those in the amniotic fluid in the case of the non-smoking mothers. The low Zn concentration in the amniotic fluid in the group of smoking mothers is of very great importance, for the antibacterial effect of the amniotic fluid is Zn-dependent. Accordingly, in the event of a depressed Zn level, these fetuses are particularly exposed to the risk of intrauterine infection (8).
The variations of the above parameters in the serum of premature and full-term neonates are given in Table 3, categorized on the basis of the

smoking habits of the mothers.

Parameters of serum	neonates							
	premature (n=160)				full-term (n=1010)			
	smoking mothers		non-smoking mothers		smoking mothers		non-smoking mothers	
	$\bar{x}$	± SD	$\bar{x}$	± SD	$\bar{x}$	± SD	$\bar{x}$	± SD
Cu (μmol/l)	4.6	2.2	5.0	1.1	5.8	2.4	6.5	1.5
Zn (μmol/l)	11.2	5.4	12.2	6.1	11.9	4.2	12.6	3.7
Fe (μmol/l)	14.6	5.1	16.3	3.9	15.8	4.8	17.4	4.0
Cd (nmol/l)	52.5	20.0	43.0	13.2	40.3	12.5	31.0	8.2
pr (g/l)	57.3	2.4	60.1	2.3	62.7	1.8	65.6	0.7
CPL (U/l)	44.8	40.3	39.3	27.1	48.0	23.5	42.7	18.9

Table 3. Levels of Cu, Zn, Fe, Cd, Pr and CPL in the Serum of Premature Neonates and of Full-Term Neonates, Categorized on the Basis of the Extent of Smoking by the Mother

Independently of the smoking habit, the serum Cd and pr concentrations were significantly higher in the serum of premature neonates than in that of full-term neonates.

Conclusions

1. *Neonates born to mothers who smoke during their pregnancy do not attain the birth weight to which they are genetically determined.*
2. *The Cd, Fe, Cu and CPL concentrations in the serum of smoking mothers (and particularly strong smokers) are significantly elevated.*
3. *Smoking significantly enhances the Cd level and depresses the Zn concentration of the amniotic fluid, thereby considerably increasing the possibility of intrauterine infection. The Fe level in the amniotic fluid decreases more strongly during pregnancy in the heavy-smoker*

mothers than in the non-smokers.

4. *The Cd and total protein concentrations are significantly higher in the serum of premature neonates than in that of full-term neonates.*
5. *The total protein concentration in the serum of full-term neonates born to smoking-mothers is significantly lower than the corresponding level in the case of non-smoking mothers.*

The results lend support to the view that complete abstention from active smoking (and, if possible, passive smoking too) is fully justified in the course of pregnancy.

All investigations were performed with the advised consent of the mothers, and conformed in all respects to the requirements of the University Ethics Committee.

References

1. Andrews, J., J.M. Mc Garry. 1972. J.Obstet.Gynaecol.Br.Commonw. 79, 1057.

2. Bottoms, S.F., B.R. Kuhnert, P.M. Kuhnert, A.L. Reese. 1982. Am.J.Obstet.Gynecol. 144, 787.

3. Cole, P.V., L.H. Hawkins, D. Roberts. 1972. Br.J.Obstet.Gynaecol. 79, 782.

4. Himmelberger, D.V., B.W. Brown, E.N. Cohen. 1978. Am.J.Epidemiol. 1978, 470.

5. Kuhnert, P.M., B.R. Kuhnert, S.F. Bottoms, P. Erhard. 1982. Am.J.Obstet.Gynecol. 142, 1021.

6. Lethovirta, P., M. Forss. 1978. Br.J.Obstet.Gynaecol. 85, 729.

7. Meberg, A., S. Marstein. 1986. Acta Paediatr.Scand. 75, 762.

8. Salgó,L., A. Pál, K. Moholi, K. Gyurkovits, L. Kovács. 1984. In: Trace Element-Anal Chem. in Med. and Biol. Vol.3. (P. Brätter and P. Schramel, ads.) Walter de Gruyter, p. 677.

9. Siegers, C.-P., J.R. Jungblut, F. Klink, F. Oberheuser. 1983. Toxicol.Lett. 19, 327.

DETERMINATION OF TRACE ELEMENTS IN PATIENTS AND CARRIERS OF DUCHENNE MUSCULAR DYSTROPHY. A NEW PHYSIOPATHOLOGICAL APPROACH

J. Reyes

Universidad de Valparaíso

L.J. Holmgren

Instituto de Rehabilitación Infantil-Santiago, Sociedad Pro Ayuda del Niño Lisiado

N. Gras, V. Cassorla

Comisión Chilena de Energía Nuclear

T. Barros, M. Colombo

Instituto de Nutrición y Tecnología de los Alimentos, Universidad de Chile

Abstract

The present preliminary study measures the levels of Zn and Se by the method of neutronic activation, in serum and hair of patients and carriers of DMD, comparing them with control groups in order to contribute to define the phenotype of this pathology and discuss the role that these trace elements would have in the physiopathology of the disease.

Sixteen children with DMD and 16 carriers belonging to 18 families, whose first consultation was done at the "Instituto de Rehabilitación Infantil" of Santiago, Chile, during the period 1979-1987 were analyzed.

In children with DMD the levels of Zn and Se measured were higher for serum and lower for hair when compared to their respective well paired controls. Levels of serum CPK, LDH and ceruloplasmine were also measured. Children with DMD showed higher levels of these enzymes when compared to their controls.

Trace Element Analytical Chemistry in Medicine and Biology

In carriers, only CPK showed a significant increase.

Introduction

Duchenne muscular dystrophy(DMD) is one of the most frequent of hereditary myopathies. The gene determinating it is located in the short arm of the X chromosome. It has been proposed that approximately one third of the cases are the product of a new mutation (1,2,3). The incidence is 1:3,500 male births (4). At present, the most accepted theory to explain the pathogenesis of this disease is an abnormality of the sarcoplasmatic membrane, inducing a disarrangement in the Calcium transport(5,6). This alteration in the muscular fiber would produce variations in the levels of some circulating enzymes such as: creatinphosphokinase(CPK)(2,7); lactate dehydrogenase (LDH)(1); ceruloplasmine (CP)(8) and at the same time changes in the levels of some trace elements (Cu,Zn,Se)(9).
High levels of Zn and Se in urine of DMD patients have been found, compared to well paired control individuals (10).

Material and Method

Four groups of subjects were analyzed: patients (N=16, $\bar{X}$ age=8.2); control patients (N=16, $\bar{X}$ age=9.2); carriers (N=16, $\bar{X}$ age=27.77); control carriers (N=16, $\bar{X}$ age=21.88).
Samples of blood and hair were taken to the individuals under study and their serum enzyme levels (CPK, LDH and CP) and traçe elements (Zn, Se, and Cu).
CPK determination was performed by Sigma 661 technique (11).
LDH was measured with the UV kinetic technique with Biochemical Corporation reagents (L8501). CP was determined by the technique proposed by Sheimber in 1952. Hair levels of Zn and Cu were determined by neutronic activation, as well as Zn and Se in serum. Samples were irradiated with a neutronic flow of

$1 \times 10^{13} n/cm^2 \times S$ during 24 hours. Cu in serum was determined by atomic absorption spectrophotometry.

Results and discussion

Zn and Se levels are higher for serum and lower for hair compared to their controls (table 1). These differences are not statistically significative. Cu levels are higher for serum and hair in patients.

TABLE 1. ZINC, SELENIUM AND COPPER IN SERUM AND HAIR OF DMD PATIENTS (N=16) AND CONTROLS (N=16)

		Patients		Controls	
		$\bar{\bar{X}}$		$\bar{\bar{X}}$	
SERUM (ug/dl)	Zn	90.78	84.34	76.93	56.68
	Se	6.78	6.15	6.20	4.57
	Cu	96.81	37.07	94.26	39.56
HAIR (ppm)	Zn	130.41	56.02	142.7	38.24
	Se	0.401	0.09	0.442	0.09
	Cu	12.98	38.24	11.55	3.70

In carriers, the results show higher levels of Zn and Se in serum than in their respective controls (table 2). In hair, levels are higher for Zn and lower for Se. Cu levels are lower both in serum and hair of carriers (table 2). These differences are not statistically significative.

TABLE 2. ZINC, SELENIUM AND COPPER IN SERUM (N=16) AND HAIR (N=16) IN DMD CARRIERS AND CONTROLS (N=16)

		Carriers		Controls	
		$\bar{X}$	S	$\bar{X}$	S
SERUM (ug/dl)	Zn	65.43	35.15	64.84	37.42
	Se	8.24	5.18	7.43	6.63
	Cu	52.31	20.41	53.53	42.42
HAIR (ppm)	Zn	169.82	38.65	164.53	49.34
	Se	0.485	0.04	0.521	0.10
	Cu	16.98	12.60	14.76	6.10

The preliminary results with respect to trace elements do not show clear differences in the study groups, but we did observe a tendency to high hair levels than those seen in control groups. With regards to CPK levels, the differences found both in patients (table 3) and in carriers (table 4), com - pared to their control groups, are significative and confirm their diagnostic value.

TABLE 3. CPK, LDH AND CP IN DMD PATIENTS AND CONTROLS

	Patients (N=16)		Controls (N=16)	
	$\bar{X}$	S	$\bar{X}$	S
CPK	157.87	92.73	5.42	5.13
LDH	999.75	456.82	247.6	60.39
CP	26.09	8.25	24.92	7.09

LDH is very high for patients (table 3) and not so for carriers (table 4), with respect to their respective controls.

TABLE 4. CPK, LDH AND CP IN CARRIERS AND CONTROLS

	Carriers (N=16)		Controls (N=16)	
	$\bar{X}$	S	$\bar{X}$	S
CPK	17.5	18.05	3.03	4.58
LDH	236.3	38.25	209.6	32.74
CP	16.43	5.12	22.64	6.09

CP is high for patients, compared to their controls, and not so for carriers. These differences are not statistically significative.

In our opinion, it is important to stand out the increase of LDH and Zn, CP and Cu in patients, which is not so in carriers, this allows us to state that this increase is due to membrane permeability alteration of the muscular fiber, as a consequence of the disease and would not have a physiopathological role in the disease.

The CP has an antioxidant function (12) and could be significantly high in the first stages of the disease, in which it would play a protective role over the membrane.

The preliminary results obtained make necessary to us to continue with this study in order to confirm the observed tendencies.

References

1. Rodríguez, J., Ferriere, G. Las distrofias musculares progresivas en la infancia. Rev. Chil. Ped. 53(4):379, 1982.
2. Devilat, M. et al. Detección de portadoras sanas en distrofia muscular progresiva: 24 familias. Rev. Chil. Ped. 46:313, 1975.
3. Ferrer, S. Aspectos genéticos de las enfermedades musculares. Rev. Med. Chile 105:253, 1977.
4. Hoffman, E. et al. Subcellular fractionation of dystrophin to the triads of skeletal muscle. Nature 330:754-757, 1987.

5. Schotland, D., Bonilla, E., Wakayema, I. Freeze fracture studies of muscle plasma membrane in human muscular dystrophy. Acta. Neuropathol. 54:189, 1981.

6. Carpenter, S., Karpeti, G. Duchenne muscular dystrophy: plasma membrane loss initiates muscle cell necrosis unless it is repaired. Brain 102:147, 1979.

7. Orfanos Adam, P. and Naylos Edwin, E. A rapid screening test for Duchenne muscular dystrophy using dried blood specimens. Clin. Chim. Acta 138:267, 1984.

8. Cousins, R.J. Absorption, transport and hepatic metabolism of copper and zinc: special reference to metallothionein and ceruloplasmine. Physiological Review 65: 238-309, 1986.

9. Revis, N.W., Hortonm, C.I., Curtis, S. Metabolism of selenium in skeletal and liver of mice with Genetic Muscular Dystrophy. Proc. Soc. Exp. Biol. Med. 160:139, 1977.

10. Westerlund Brit Alhrot and Carlmark, Bjorn. Urinary excretion of selenium and other minerals in patients with Duchenne muscular dystrophy and Werdnig Hoffman Spinal atrophy. Nutr. Res. Supply. 1; 406, 1985.

11. Sigma Chemical Co. Ltd. Reagent for the colorimeter determination of creatinekinase in serum, plasma or cerebro spinal fluid. Procedure 661. Catalogue 1986, page 1396.

12. Hunter, M., Ian, S., Mohamed, J.B. Plasma antioxidants and lipid peroxidation products in Duchenne muscular dystrophy. Clin. Chim. Acta 155: 123-131, 1986.

ZINC, COPPER AND SELENIUM IN CHRONIC AND ACUTE INFLAMMATORY RHEUMATIC DISEASES.

A. Peretz [*], J. Nève [**], J.P. Famaey [*]

Free University of Brussels, [*]Department of Rheumatology and Physical Medicine, Saint-Pierre Hospital, rue Haute, 322, 1000 Brussels - [**]Toxicology and Bioanalytical Chemistry Unit, Campus Plaine,205-1, 1000 Brussels,Belgium.

Abstract

Zinc (Zn), copper (Cu) and selenium (Se) concentrations were determined in different biological compartments of patients suffering from chronic (CI) and acute (AI) rheumatic diseases, and the results compared with those of non inflammatory patients (NI). Low plasma Zn and high plasma Cu were found in CI and AI, and were inversely correlated. Plasma Zn and Cu were correlated with biochemical indices of inflammation (ESR and CRP) but not with their main blood carriers, albumin and ceruloplasmin. Plasma Se was not modified. Urine Zn, Cu and Se levels were significantly higher in AI, but erythrocyte levels were not modified. Synovial fluid trace element levels were elevated in inflammatory patients : correlations were found between Zn levels and white blood cell count. These results suggest a redistribution of trace elements in inflammatory patients rather than a deficiency state. Drug treatment, especially corticosteroids, precipitates the decrease in plasma Zn and Se.

Introduction

Zinc (Zn), copper (Cu) and selenium (Se) are trace elements involved in the inflammatory and immune responses mainly as protectors against the damaging effect to cell membranes and constituents of free radicals and peroxides (1). These species are generated in great amount by activated phagocytes during the inflammatory process ; they contribute to some extent to the pathogenesis of rheumatic diseases (2). Zn is the constituent of ubiquitous enzymes, involved in DNA and protein synthesis and repair. It modulates the functional capacity of phagocytes and immunocompetent cells (3). Cu, as an essential constituent of the enzyme superoxide dismu-

Trace Element Analytical Chemistry in Medicine and Biology

tase, takes part in the dismutation of the agressive superoxide anion ($O_2^{\bar{\cdot}}$) and is a component of ceruloplasmin which possess antiinflammatory properties (4). Se, is not only the active site of the peroxide destroying enzyme glutathione peroxidase, but it also regulates the production of prostaglandins and leukotrienes, which are mediators of inflammation (5).

In these conditions, it is important to determine whether trace elements deficiency occurs in inflammatory states and to study the possible consequences on host defense mechanisms. On the other hand, the inflammatory process by itself can modify the trace elements metabolism, what has sometimes been interpreted as a protective mechanism (6).

This study wants to assess trace elements status of patients suffering from chronic and acute rheumatic diseases and to correlate the modifications to clinical and biochemical indices of inflammation. The effect of drug treatment is also considered.

Patients and methods

Patients were divided into 3 groups according to their inflammatory status. Firstly, chronic inflammation (CI) : 29 rheumatoid arthritis patients (RA), mean age 57 y. (range: 21-82), 1 man, 28 women, with active disease. Twenty onewere treated with NSAID, 8 with gold salts. The effet of corticosteroid treatment was investigated in another group of 21 patients. Secondly acute inflammation (AI) : 24 patients with chrystal-induced arthritis (n=13) or reactive arthritis (n=11), mean age 57 y. (range : 16-76), 13 men, 11 women. Thirdly, non inflammatory (NI) rheumatic diseases : 27 patients with mild osteoarthritis, mean age 54 (range: 47-84), 3 men, 24 women.

Erythrocyte sedimentation rate (ESR) haemoglobin (Hb), C-reactive protein (CRP), albumin (Alb) were measured by classical procedures. Somatomedin- C (SMC) was measured by radioimmunoassay (7) ; Retinol binding protein (RBP), Prealbumin (Prealb) and Ceruloplasmin (CP) by turbidimetry (RA MILL-Technicon). Plasma (P), erythrocyte (E), urine (U) and synovial fluid (SF) Zn were determined by flame atomic spectroscopy (AAS) after dilution in 1 % HCl.P, E, SF-Cu by flame AAS after deproteinization with 20 % trichloracetic acid. U-Cu was determined by graphite furnace AAS after extraction with APDC in methylisobutylketone. P and SF-Se were determined by direct GFAAS with Zeeman background correction. Finally E and U-Se were determined by GFAAS after wet-ashing and extraction with an aromatic o-diamine (8).

Student's t test or Jaffe procedure were applied to test differences between means after one way variance analysis. Mann-Whitney U test was used to test differences between non normally distributed data.

Results

Table 1 shows clinical and biochemical parameters of inflammation in the three studied groups. Nutritional status was normal as attested by the BMI and SMC levels. Patients with CI and AI presented increased ESR, CRP and CP (positive acute phase reactants); RBP and Prealb (negative acute phase reactants) were decreased only in patients with AI, and Hb and Alb were low both in CI and AI.

Table 1: clinical and biochemical parameters ($\bar{X} \pm$ SD or range).

Parameter	CI	AI	NI
BMI, kg/m^2	25 ± 5	24 ± 3	25 ± 4
SMC, U/ml	0.6 (0.2-1.0)	0.6 (0.4-1.6)	0.5 (0.2-1.4)
ESR, mm/h	40 ± 28*	44 ± 31*	8 ± 5
CRP, mg/dl	3 (1-10)	6 (2.5-35)*	< 1
CP, mg/dl	34 ± 11*	41 ± 15*	18 ± 4
Hb, mg/dl	11.8 ± 1.4*	12.1 ± 2.0*	13.5 ± 1.3
Alb, g/dl	35.6 ± 3.8*	35.0 ± 5.0*	41.2 ± 5.4
RBP, mg/dl	5.0 ± 2.5	2.5 ± 1.8*	4.6 ± 1.5
Prealb, mg/dl	21 ± 9	15 ± 4*	19 ± 3

*$p < 0.001$, CI or AI versus NI.

Such as indicated in table 2, low P-Zn and high P-Cu were found in CI, and more especially in AI. An inverse correlation could be calculated between the two elements in inflammation ($r = -0.42$, $p < 0.01$). Significant correlations were also observed between Zn or Cu and ESR ($r = -0.53$ and $r = 0.48$, $p < 0.001$) or CRP ($r = -0.61$ and $r = 0.70$, $p < 0.001$). However, no correlation between these elements and their blood carriers, Alb or CP, could be disclosed. Taking into account a 0.32 % w/w Cu content of CP, it appeared that an important part of P-Cu (30-40 %) was not bound to CP in inflammatory groups, contrary to what is usually observed in non inflammatory patients (1-5 %).

E-Zn, Cu and Se remained similar in the three groups. U-Zn, Cu and Se were increased in AI. Finally, SF-Zn, Cu and Se were elevated in CI and AI, but the modifications were not in the same range (3.6 fold for Zn, 2.1 fold for Cu, and 1.3 fold for Se) . SF proteins were correlated with the three elements only in the non inflammatory group. In the inflammatory group, Zn and white blood cell count were correlated ($r = 0.55$, $p < 0.01$).

Table 2: Zinc, copper and selenium levels ($\bar{X} \pm SD$ or range).

Parameter	CI	AI	NI
Plasma			
Zn, µg/dl	80 ± 14**	70 ± 16***	93 ± 14
Cu, µg/dl	151 ± 34**	181 ± 54***	117 ± 18
Se, µg/l	86 ± 14	89 ± 19	84 ± 14
Erythrocyte			
Zn, µg/g Hb	34.6 ± 5.2	37.2 ± 7.2	35.2 ± 5.9
Cu, µg/g Hb	2.5 ± 0.4	2.7 ± 0.5	2.5 ± 0.6
Se, µg/g Hb	0.42 ± 0.08	0.38 ± 0.10	0.40 ± 0.10
Urine			
Zn, µg/g creat.	84 (8-280)	135 (13-412)*	61 (18-169)
Cu, µg/g creat.	3 (1-10)	5 (1-12)*	2 (1-5)
Se, µg/g creat.	36 (8-95)	51 (14-130)*	18 (4-54)
Synovial Fluid ✦			
Zn, µg/dl	69 ± 41 ***		19 ± 13
Cu, µg/dl	97 ± 46 ***		46 ± 19
Se, µg/l	50 ± 14**		39 ± 10

***$p < 0.001$, ** $p < 0.01$, *$p < 0.05$ (CI or AI versus NI).
✦ CI and AI pooled for results in synovial fluid.

Table 3 reports the modifications induced by drug treatment in RA patients with similar inflammatory state. Corticosteroid administration is associated with low P-Zn and Se, when compared to treatment with NSAID.

Table 3 : Effect of chronic drug treatment ($\bar{X} \pm SD$).

Parameter	NSAID	Corticosteroids	Gold salts
Zn, µg/dl	80 ± 14	68 ± 10 ***	79 ± 5
Cu, µg/dl	151 ± 34	159 ± 29	147 ± 20
Se, µg/l	86 ± 15	66 ± 21 ***	84 ± 13
ESR, mm/h	41 ± 29	43 ± 22	38 ± 17

*** $p < 0.001$, corticosteroids versus NSAID.

Discussion

This study confirms the low P-Zn and high P-Cu previously reported in RA and other inflammatory diseases (9). Active inflammation was evidenced both in CI and AI, but was more intense in the last group.The correlation between Zn, Cu and inflammatory parameters suggests that these changes are secondary to the inflammatory process rather than to modifications in carriers proteins (10). Additionaly, the absence of correlation between P-Zn and Alb is in favour of this last hypothesis. A similar conclusion can be drawn for Cu and CP : although both levels were increased, they were not correlated. We established that an appreciable amount of plasma Cu was not bound to CP but more probably to low molecular wheight compounds or to albumin. The role played by this fraction remains controversial (11). Mechanisms underlying the changes in Zn and Cu levels can be attributed to the cytokine, Interleukin I (IL_1), synthetized by activated phagocytes. In vitro and in vivo studies indicated that IL_1 enhances the synthesis of liver metallothionein, that causes an accumulation of Zn in the liver and a decrease in P-Zn. An enhancement of acute phase reactant proteins synthesis including CP simultaneously occurs (12). P-Se was not modified in our RA patients, contrary to the reports from Nordic countries where Se deficiency is frequent (13). An altered nutritional status of patients with inflammatory diseases could partly explain these discrepancies. Moreover, in AI no modification in P-Se is observed suggesting the absence of influence of the inflammatory process on P-Se.

Erythrocyte levels were scarcely studied (14); in fact, they did not appear as useful indicators of trace element modifications related to inflammation.

Urine excretion of Zn, Cu and Se was modified in AI. The increase of non CP-bound Cu corresponding to the ultrafiltrable fraction could account for the increased cupruria (15). Enhanced zincuria has been reported in association with protein and muscle catabolism in severe diseases (16). A similar mechanism could be evoked for the increase in U-Se. In chronic diseases one might expect that a regulatory process prevent excessive excretion of the elements.

In synovial fluid the increase in trace elements is partly due to the increased permeability of the synovial membrane and parly to the release from damaged white blood cells , synoviocytes and cartilage (17). The importance of these changes in the pathogenesis of synovial inflammation remains poorly understood.

In addition to inflammation, the administration of corticosteroids alters Zn and Se levels. The mechanism underlying the decrease in P-Zn corresponds to the corticosteroid-induced synthesis of metallothionein in the liver and probably also in the intestine (12). The renal handling of the element is also modified by the therapy (18). Concerning Se, no explanation is available for the effects of corticosteroid therapy.

In conclusion, this study suggests that modifications in trace elements metabolism during the inflammatory process in probably the consequence of a redistribution of the elements through body fluids and tissues rather than a deficiency state. The treatment with corticosteroids provides additional modifications, mainly in Zn and Se levels.

Acnowledgements: This work was supported by the Fonds National de la Recherche Scientifique et Médicale de Belgique and by Labcatal, SA, France.

References

1. Aggett, P.J., 1985. Clin. End. Metabol. 14, 513-543.
2. Halliwel, B., Gutteridge, J.M., 1985. Mol. Aspect. Med. 8, 89-193.
3. Vallee, B.L., 1959. Physiol. Rev. 39, 443-490.
4. Evans, G.W., 1973. Physiol. Rev. 53, 535-570.
5. Lands, W.E.M., 1984. Prostagland. Leukotr. Med. 13, 35-46.
6. Klasing, K.C., 1984. Am. J. Physiol. 247, R 901-914.
7. Van Vliet, G., Bosson, D., Rummens, E., Robyn, C., Wolter, R., 1986. Acta Endocrinol. 112 (suppl. 279), 403-410.
8. Nève, J., Fontaine, J., Peretz, A., Famaey, J.P., 1988. Agents and Actions, in press.
9. Aaseth, J., Munthe, E., 1978. Scand. J. Rheumatol. 7, 237-240.
10. Sorenson, J.R.J., Di Tommaso, D., 1976. Ann. Rheum. Dis. 35, 317-331.
11. Lewis, A.J., 1984. Agents and Actions. 15, 504-513.
12. Cousins, R.J., 1985. Physiol. Rev. 65, 238-309.
13. Tarp, U., Overvad, K., Hanssens, J.C., Thorling, E.B., 1985. Scand. J. Rheuatol. 14, 97-101.
14. Peretz, A., Nève, J., Famaey, J.P., Molle, L., 1987. J. Pharm. Belg. 42, 395-403.
15. Milanino, R., Cassini, A., Franco, L., Conforti, A., Marella, M., Velo, G.P., 1985. Int. J. Tiss. Reac. 7, 469-474.
16. Fell, G.S., Fleck, A., Cuthbertson, D.P., 1973. Lancet 1, 280-282.
17. Peretz, A., Nève, J., Vertongen, F., Famaey, J.P., 1987. In "Biology of Copper complexes". J.R.J. Sorenson Ed., Humana Press, pp. 583-589.
18. Flynn, A., Strain, W.H., Hill, O.A., Frantianne, R.B., 1971. Lancet, 2, 1169-1171.

CHANGES OF COPPER, ZINC AND IRON VALUES IN PATIENTS WITH ACUTE AND CHRONIC ISCHEMIC HEART DISEASE

E.Acartürk, G.T.Yüregir, Z.N.Alparslan

Çukurova University Medical School, Cardiology, Biochemistry and Public Health Departments, Adana, Turkey

Abstract

Serum copper, iron and plasma zinc concentrations were measured in patients with acute myocardial infarction (Group I, n=45), acute myocardial ischemia (Group II, n=8), old myocardial infarction (Group III, n=14) and in healthy subjects (Group IV, n=30). The mean copper values were significantly higher in patients groups than the mean control values ($p<0.001$). The mean zinc levels were higher in Group I and lower in Group III than the mean control levels. The mean iron concentrations were significantly lower in Group I than the values in Groups II, III and IV ($p<0,001$). Thus, serum iron might be a helpful additional parameter for diagnosis of acute myocardial infarction (AMI) and for differential diagnosis between AMI and acute mycardial ischemia and high copper levels could be searched for as a risk factor for ischemic heart diseases.

Trace Element Analytical Chemistry in Medicine and Biology

Introduction

Although changes in the concentrations of some trace elements in patients with acute myocardial infarction (AMI) have been reported, only recently, considerable attention has been focused on trace elements in patients with acute and chronic ischemic heart diseases (1,2,3,4).
It is suggested that, in the diagnosis of AMI, estimation of serum copper is simple and superior to enzyme estimations (5) and serum iron might prove useful when used in association with enzyme activity assays (6). In this study, changes of serum copper, iron and plasma zinc were investigated in patients with acute and chronic ischemic heart disease and diagnostic significance of these trace elements in AMI was evaluated.

Materials and Methods

Forty five patients with AMI, 41 males,4 females, aged 32-83, were admitted to the coronary - care unit of Çukurova University Medical School (GROUP I). AMI was diagnosed when two of the three criteria were met : 1-Severe chest pain with a duration of more than 30 minutes, 2-Typical ST segment and T wave changes and pathological Q waves on electrocardiogram (ECG) and 3-Typical increases of serum glutamic oxalacetic transaminase (SGOT) concentrations. Eight patients, 7 males and 1 female, aged 49-70, were also admitted to the coronary-care unit with severe chest pain and a diagnosis of suspected AMI (GROUP II). All of these patients had ST segment and T wave changes of their ECGs but no characteristic alterations on serial ECGs. None of them showed typical increases of SGOT concentrations. Fourteen asymptomatic patients, 12 males and 2 females, aged 37-78, who had been hospitalized at least 3 months ago with AMI, were included in GROUP III and 30 healthy subjects, 20 males and 10 females, aged 35-70, served as GROUP IV.

Blood samples for copper, zinc, iron and total iron binding capacity (TIBC) determinations were taken once on the first day of ischemic chest pain in Group I and II. Blood samples, in Group III and in Group IV, were also collected only once. Serum copper, plasma zinc and serum iron and TIBC were determined by Landers and Zak (7), Johnson (8) and Ramsay (9, 10) methods respectively. Statistical analysis of data was performed by using the Student's t and paired t tests. The correlation coefficients calculated were the Pearson coefficients.

Results

Table 1 shows the characteristics of the patients and the controls. The mean SGOT values in patients with AMI were significantly higher than the mean values in patients with acute myocardial ischemia.

Table 2 shows the mean $\pm$ standard deviations (SD) of copper, zinc and iron values of the patients groups and the healthy subjects. The mean copper values were higher in patients with AMI, acute myocardial ischemia and old myocardial infarction compared to the mean control values ($p<0.001$). The mean iron values in Group I were significantly lower than the mean values in Group II, III and IV ($p<0.001$). TIBC and hematocrit values did not differ between groups. There were no correlations between copper, zinc and iron values and infarct localisations in Group I and III. In Group I, 4 patients died with left ventricular failure due to AMI. None of them had the lowest iron or the highest zinc and copper values.

Table 1 : Charateristic of the Patients and the Control Subjects

Groups	n	Mean age (Range)	Sex M:F	Mean Hct. (%)	Infarct Site
I	45	54.8±10.8 (32-83)	41:4	42.5±4.8	Inferior : 24 Anteroseptal: 15 Anterior : 6
II	8	56.5±6.9 (49-70)	7:1	41.1±4.7	-
III	14	55.1±11.5 (37-78)	12:2	41.5±3.9	Inferior : 7 Anteroseptal: 6 Anterior : 1
IV	30	56.2±9.8 (35-70)	20:10	44.4±4.4	-

Table 2 : Copper, Zinc and Iron Values (mean±SD) of the Patients and the Control Subjects.

Groups	Copper(µg/dl)	Zinc(µg/dl)	Iron(µg/dl)
I	123.2±45.2	121.9±36.7	78.0±36.6
II	127.0±82.2	15.3±22.0	101.4±51.8
III	125.4±32.7	84.2±31.4	103.8±36.8
IV	112.0±24.0	103.0±22.0	102.0±32.0

Discussion

Serum concentrations of certain trace elements have been investigated by various authors in patients with acute and chronic ischemic heart disease (11, 12). Khan (11) found higher copper levels in patients with AMI, acute myocardial ischemia and old myocardial infarction than the control values, the highest copper values were in patients with AMI. Woods (12) observed a trend to increasing copper values for 72 hours after AMI. We also found high copper levels in patients with all type of ischemic heart disease. In our study copper concentrations did not differ significantly between patients groups as Khan et al (11) found. Some authors observed low zinc levels after AMI (11, 12). Khan (11) showed higher zinc values in patients with acute myocardial ischemia and old myocardial infarction than the control values,but differences were not significant. In our patients with AMI zinc levels were higher than the control values. Decreased serum iron concentrations have been reported after AMI (1, 6). It has been suggested than serum iron determination might prove useful in diagnosing myocardial infarction (6). Khan (11) has found low iron levels in patients with AMI and no significant differences in iron levels between acute myocardial ischemia, old myocardial infarction and the control values. We also found significantly lower iron levels in AMI than the other groups. This study shows a significant decrease in serum iron concentrations following AMI and a high copper concentrations in patients with all type of ischemic heart disease. In conclusion,serum iron might be a helpful additional parameter for diagnosis of AMI and for differential diagnosis between AMI and acute myocardial ischemia. Further investigations are needed to find out the role and the origin of high copper concentrations in patients with ischemic heart diseases.

References

1. Griffiths, J.D., L.J. Campbell, I.W. Woodruff. 1985. Am J. Clin.Pathol.84, 649.

2. Gupta, R. 1981. JAPI. 29, 987.

3. Medeiros, D.M. 1985. Biochem. Arch. 1, 67.

4. Rector, W.G., M.A. DeWood, R.V. Williams. 1981. Clin. Chem. 21, 476.

5. Singh, M.M., R.Singh, M.C. Gupta. 1985. J.Vascular Diseases. 36, 504.

6. Fitzsimons, E.J., K.Kaplan. 1980. Am. J. Clin. Pathol. 73, 552.

7. Landers, J.W.,B. Zak. 1985. Am.J. Clin. Pathol. 29, 509.

8. Johnson, D.J., Y.Y. Djuh. 1977. Clin. Chem. 23, 1321.

9. Ramsay, W.N.M. 1957. Clin. Chem. Acta. 2, 214.

10. Ramsay, W.N.M. 1957. Clin. Chem. Acta. 2, 211.

11. Khan, S.N., M.A. Rahman, A. Samad. 1984. Clin. Chem. 30, 644.

12. Woods, J.B., J.H. Holt, W.T. Woods. 1983. J. Electrocardiology 16, 379.

TRACE ELEMENTS IN HUMAN BRAIN : REGIONAL DISTRIBUTION AND ALTERATIONS IN STRUCTURES AFFECTED BY CEREBRAL INFARCTION

H. Duflou, W. Maenhaut

Instituut voor Nucleaire Wetenschappen, Proeftuinstraat 86,
B-9000 Gent, Belgium.

J. De Reuck

Kliniek voor Neurologie, De Pintelaan 185, B-9000 Gent, Belgium.

Abstract

Particle-induced X-ray emission was used to measure K, Ca, Mn, Fe, Cu, Zn, Se, and Rb in up to 50 different regions of 12 normal human brains and in several structures of 7 brains affected by a recent cerebral infarction. Two of these brains also showed damage due to old infarctions. In the recent infarcts Ca, Fe, and Se were raised to 400, 160, and 115% of the control values, whereas K, Cu, and Rb were lowered to 40, 70, and 30%. In the old infarcts, on the other hand, Mn, Zn, and Se were enhanced above control levels, whereas K and Rb showed a tendency to return to normal.

Introduction

Several diseases of the central nervous system are attended with alterations of trace element concentrations in the brain. For instance, Wilson's disease, a disorder with progressive extrapyramidal movements, involves an elevated Cu concentration in the brain (1). Brain tumours exhibit higher than normal Ca, Fe, Zn and Se levels (2), and elevated Al concentrations have been measured in brains of Alzheimer patients (3).

In order to evaluate the elemental alterations associated with neurological diseases, reliable concentration data for normal and pathological brain are needed. Bulk analysis of the whole brain is often not adequate.

Trace Element Analytical Chemistry in Medicine and Biology

Since the trace elements are heterogeneously distributed in the brain, and certain diseases affect only small brain areas, extensive elemental mappings of the different brain structures are needed.
In a first part of our research project we measured 8 elements (i.e. K, Ca, Mn, Fe, Cu, Zn, Se, and Rb) in about 600 samples, which originated from up to 50 different regions in 12 normal human brains. We subsequently started with the investigation of brains which were affected by a cerebral infarction. Until now we have analyzed a number of regions from 7 brains with recent infarcts. Two of these brains also showed damage due to old infarctions. All analyses were carried out by particle-induced X-ray emission analysis (PIXE).

Methods

The normal brains originated from 12 persons aged between 7 and 69 years who died from acute respiratory distress, acute cardiac failure, bronchus carcinoma or anemic anoxia. Macroscopic examination revealed no abnormalities. The brains affected by an infarction were taken from 7 persons with ages ranging from 54 to 93 years. All brains were removed at autopsy within 72 h after death by means of standard procedures. Tissue samples were dissected using molybdenum knives, as described elsewhere (4). For the normal brains, samples were taken from corresponding regions in both the left and the right cerebral hemisphere, the brainstem and the cerebellum (Table 1). However, only the tissues from the left cerebral hemisphere, the brainstem and cerebellum of each brain were normally analyzed by PIXE. For the cerebral infarctions, several samples were collected of the affected regions and the surrounding tissue and of the corresponding healthy regions in the other cerebral hemisphere. The samples were weighed to determine the wet weight, freeze-dried for 72 h in a Leybold-Heraeus vacuum apparatus, oven dried for 12 h at 75 °C, and digested in 9 M nitric acid in a closed Teflon vessel. Typically, 4 PIXE targets were prepared and analyzed for each brain sample. The PIXE analysis method, experimental set-up, and target preparation procedure are described in detail elsewhere (4-6).

Table 1 : Brain regions studied for normal human brains.

Cerebral hemispheres :

- Cerebral cortex

(1)[a] (G)[b] gyrus frontalis superior	(17) (M) uncus hippocampi
(2) (G) gyrus cinguli	(18) (G) gyrus parahippocampalis
(4) (G) gyrus rectus	(23) (G) gyrus precentralis
(9) (G) gyrus temporalis inferior	(24) (G) gyrus postcentralis
(10) (G) insula	(38) (G) gyrus lingualis

- Cerebral white matter

(3) (W) genu corporis callosi	(21) (W) fasciculus mamillothalamicus
(5) (W) septum pellucidum	(25) (W) capsula interna
(11) (W) chiasma opticum	(37) (W) radiatio optica
(14) (W) corpus fornicis	

- Basal ganglia

(6) (M) nucleus caudatus	(28) (M) nucleus centromedianus thalami
(7) (M) putamen	(29) (M) nucleus medialis dorsalis thalami
(8) (M) globus pallidus	(30) (M) nucleus lateralis posterior thalami
(12) (M) corpus amygdaloideum	(31) (M) nucleus ventralis lateralis thalami
(13) (M) infundibulum	(32) (M) nucleus pulvinaris thalami
(15) (M) claustrum	
(19) (M) nucleus ventralis anterior thalami	
(20) (M) corpora mamillaria	
(27) (M) nucleus ruber	

Brainstem :

(22) (M) substantia nigra	(39) (M) tractus corticospinalis
(26) (W) pedunculus cerebri	(40) (M) dorsum pontis
(33) (M) anulus aquaeductus cerebri	(41) (M) nucleus olivaris inferior
(34) (M) colliculi superiores	(42) (W) pyramid
(35) (M) colliculi inferiores	(43) (M) dorsum medullae

Cerebellum :

(44) (G) paraflocculus ventralis cerebelli	(46) (G) vermis cerebelli
(45) (G) lobulus flocculonodularis	(47) (M) nucleus dentatus
	(48) (W) corpus medullare cerebelli

Other brain areas :

(16) plexus choroideus	(50) blood vessel
(36) corpus pineale	(51) meninx

[a]The numbers in parentheses indicate the dissection numbers.

[b]G, M or W indicate that the region has been classified as gray matter, as a structure of mixed composition or as white matter, respectively.

Results and Discussion

Trace element distribution in normal human brain

The mean concentration and associated standard deviation (both expressed in $\mu g/g$ dry weight) were calculated for each element in each of the 50 different brain regions by averaging the individual results of the 12 normal brains. Additionally, the average concentrations for gray, mixed, and white matter were derived for each brain, taking into account the weight fraction of the different brain regions relative to the whole brain. These individual brain data were subsequently averaged over all 12 brains to calculate overall mean gray, mixed and white matter elemental concentrations and associated standard deviations. It was found that K, Fe, Cu, Zn, Se, and Rb exhibit consistently higher levels for gray (with overall mean concentrations of 17200 ± 800 $\mu g/g$, 240 ± 50 $\mu g/g$, 26 ± 5 $\mu g/g$, 71 ± 4 $\mu g/g$, 0.92 ± 0.11 $\mu g/g$, and 14.6 ± 2.3 $\mu g/g$, respectively) than for white matter areas (10700 ± 1000 $\mu g/g$, 134 ± 26 $\mu g/g$, 10.1 ± 2.1 $\mu g/g$, 28 ± 4 $\mu g/g$, 0.45 ± 0.08 $\mu g/g$, and 10.5 ± 1.8 $\mu g/g$, respectively). The K, Zn, and Se concentrations for the regions of mixed composition (with overall mean concentrations of 14200 ± 1400 $\mu g/g$, 56 ± 5 $\mu g/g$, and 0.81 ± 0.08 $\mu g/g$, respectively) are intermediate between the gray and white matter values, and they tend to decrease with decreasing neuron density. This is, to a lesser extent, also the case for Rb with an overall mean concentration in the mixed matter regions of 13.5 ± 2.1 $\mu g/g$. The mean K, Ca, Zn, Se, and Rb concentrations of the 46 regions of gray, mixed or white matter are highly correlated with the wet-to-dry weight ratio of the brain structures indicating that the dry weight concentration variability of these elements may to a large extent be explained by variations in fluid content.

Manganese, Fe, and Cu, on the other hand, show a clearly different picture, and have the highest concentration in the regions with a mixed composition. Mn is particularly elevated in the nucleus caudatus (2.48 ± 0.26 $\mu g/g$), putamen (2.67 ± 0.33 $\mu g/g$) and globus pallidus (2.18 ± 0.55 $\mu g/g$), regions that are preferentially damaged in Mn intoxication (7). Even higher Mn concentrations are found in the corpus pineale (3.5 ± 2.8 $\mu g/g$), indicating that the element could play an important role in the endocrine functions of the brain. It was further noticed that the gray matter regions of

the cerebellum contain more Mn (1.82±0.16 μg/g) than the cerebral gray matter structures (1.25±0.07 μg/g).
Iron is concentrated in the nucleus caudatus (520±180 μg/g), putamen (610±240 μg/g), globus pallidus (820±200 μg/g), and substantia nigra (710±320 μg/g), structures involved in the inhibition or facilitation of movement (8). This suggests that Fe may be involved in the motoric function of the brain. The reduced motor activity in Fe deficient rats (9) seems to support this idea. Fe also shows elevated concentrations in the plexus choroideus (830±340 μg/g) and in blood vessels (560±570 μg/g).
Although the Cu level is fairly constant in the gray matter areas, Cu is like Mn somewhat more concentrated in the cortex of the cerebellum (33±6 μg/g) than in the gray matter structures of the cerebrum (24±4 μg/g). A similar phenomenon is observed for the white matter areas, where the Cu concentration in the cerebellum (19.2± 5.8 μg/g) is 2.1 times higher than the cerebral Cu level of 9.0±1.9 μg/g. Of the 50 brain structures analyzed, the substantia nigra has by far the highest Cu level (68±32 μg/g). Szerdahelyi and Kása (10) demonstrated, by using a staining procedure, that in the substantia nigra of rat brains the element is exclusively present in the glial cells. They suggested that Cu may play an important role in normal physiological functioning of the glial cells, and, via glia-neuron interactions, in neuronal processes.

Concentration alterations associated with cerebral infarctions

The elemental concentrations (in dry weight) and wet-to-dry weight ratios (reduction factors) of the various regions from the affected brains were compared with the normal values for the same region, as observed in the study on the 12 normal brains. The median percentage deviations, with respect to the control values, for the recent infarcts, surrounding tissue, healthy tissue from the other hemisphere, and old infarcts are listed in Table 2. Additionally, Fig. 1 shows the individual percentage Cu deviations for the various recent cerebral infarcts and the surrounding tissue. Table 2 indicates that in the recent infarcts, K is lowered to 40% of the control values, while Ca is raised to 400% of the normal level. Similar concentration changes have been reported for ischemic gerbil (11) and cat brain (12). The alterations seem to be induced by membrane failure, resulting in a movement of K^+ from the cytoplasm into the extracellular fluid and a massive influx of extracellular Ca^{2+} into the cell (13). The loss of Ca homeostasis may even be the cause of cell necrosis after cerebral ischemia (13,14). Rb shows a similar behavior as K and is lowered to 30% of the control levels, suggesting that Rb is also released from the cytoplasm. Also Cu exhibits lower than normal values (-30%). Fe, on the other hand, is raised to 160% of the control data. This may be due to a dysfunction of the blood-brain barrier allowing blood iron to enter the infarcted tissue. Se and the reduction factor are only slightly enhanced (+15 and +30%, respectively). The percentage deviation from the normal values is, however, higher for the white matter areas (+70 and + 90%, respectively). The increase of the reduction factor can be related to the accumulation of water in ischemic brain tissue. O'Brien et al. (15), who measured the water content in ischemic and infarcted cat brain, also found a greater increase of water content in white matter areas. Mn and Zn show no systematic deviation from the control values. However, in the regions with highest Mn levels (nos. 6, 7, and 8) the element is systematically lower than normal (-40%).

The alterations are much smaller in the surrounding tissue than in the recent infarcts, and the deviation from the normal level is generally (except for Ca) of the same order as for the healthy tissue samples of the unaffected hemisphere. The difference between the healthy tissue levels and the control data may be due to the generally higher age of the infarcted brains than the control cases. It has indeed been reported that elements like Fe increase with aging, while K and Rb decrease (16).

In the few samples studied so far for structures affected by an old cerebral infarction, K and Rb show a tendency to return to normal values. Mn, Zn, and Se, on the other hand, are now highly raised above the control levels. An increased Zn content has also been observed in necrotic liver tissue (17). Since Zn is closely related to DNA synthesis and growth, the increase in Zn may be linked to cell regeneration and gliosis. Kogure et al. (18) suggested that peroxidative decomposition of membrane lipids is one of the mechanisms through which cellular damage occurs in ischemia. Therefore, the raised Se concentration could be related to the protective function of this element against lipid peroxidation, as Se is a constituent of glutathione peroxidase.

Table 2 : Median Value of Percentage Deviation, with Respect to Control Value, for Eight Elements and the Wet-to-dry Weight Ratio in Tissues of Brains Affected by a Cerebral Infarction.

Element	Recent infarct (N=7, n=36)[a]	Surrounding tissue (N=5, n=6)	Healthy tissue (N=5, n=15)	Old infarct (N=2, n=4)
K	- 60%	- 30%	- 20%	- 3%
Ca	300%	110%	40%	80%
Mn	- 7%	- 1%	10%	70%
Fe	60%	20%	30%	200%
Cu	- 30%	- 7%	14%	5%
Zn	1%	20%	20%	50%
Se	15%	13%	8%	90%
Rb	- 70%	- 40%	- 30%	- 30%
Red. fact.	30%	10%	13%	30%

[a]Number of brains (N) and total number of samples (n) analyzed.

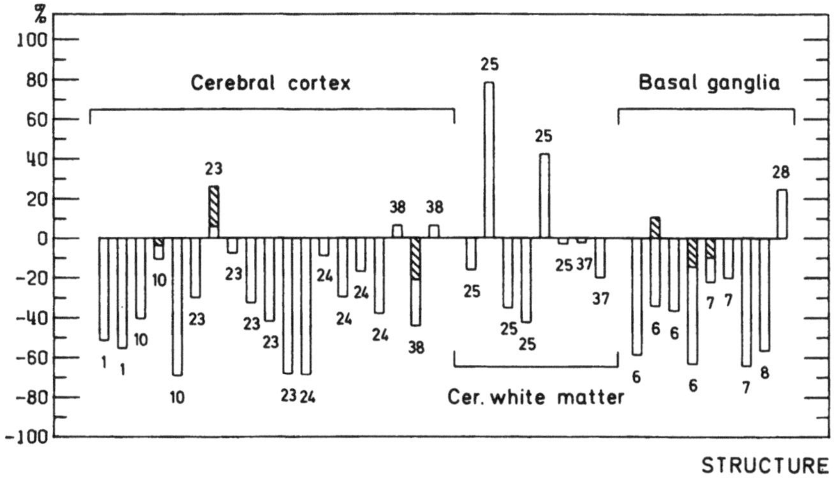

Fig. 1 : Percentage deviation, with respect to control value, for the Cu concentration of brain tissue affected by a recent cerebral infarction (□) and surrounding tissue (▨). The numbers indicate the brain regions analyzed (see Table 1).

References

1. Underwood, E.J.. 1977. Trace Elements in Human and Animal Nutrition. Academic Press, New York.

2. Tapper, U.A.S., K.G. Malmqvist , A. Brun and L.G. Salford. 1987. Nucl. Instr. and Meth. B22, 176-178.

3. Krishnan, S.S., J.E. Harrison and D.R. Crapper McLachlan. 1987. Biol. Trace Elem. Res. 13, 35-42.

4. Duflou, H., W. Maenhaut and J. De Reuck. 1987. Biol. Trace Elem. Res. 13, 1-17.

5. Maenhaut, W., L. De Reu, H.A. Van Rinsvelt, J. Cafmeyer J. and P. Van Espen. 1980. Nucl. Instr. and Meth. 168, 557-562.

6. Maenhaut, W. and H. Raemdonck. 1984. Nucl. Instr. and Meth. B1, 123-136.

7. Bonilla, E., S. Levine and E. De Salazar. 1978. Acta Cient. Venezolana 29, 332-337.

8. Höck, A., U. Demmel, H. Schicha, K. Kasperek and L.E. Feinendegen. 1975. Brain 98, 49-64.

9. Youdim, M.B.H., S. Yehuda, D. Ben-Shachar and R. Ashkenazi. 1982. In: Iron Deficiency: Brain Biochemistry and Behavior (E. Pollitt and R.L. Leibel, eds.). Raven Press, New York. pp. 39-56.

10. Szerdahelyi, P. and P. Kása. 1986. Histochem. 85, 341-347.

11. Yanagihara, T. and J.T. McCall. 1982. Life Sci. 30, 1921-1925.

12. Hossmann, K.-A., W. Paschen and L. Csiba. 1983. J. Cereb. Blood Flow Metabol. 3, 346-353.

13. Siesjö, B.K.. 1986. Eur. Neurol. 25, 45-56.

14. Farber, J.L.. 1981. Life Sci. 29, 1289-1295.

15. O'Brien, M.D., A.G. Waltz and M.M. Jordan. 1974. Arch. Neurol. 30, 456-460.

16. Markesbery, W.R., W.D. Ehmann, M. Alauddin and T.I.M. Hossain. 1984. Neurobiol. Aging 5, 19-28.

17. Vandenhaute, J.. 1987. PhD Thesis, University of Gent, Belgium.

18. Kogure, K., B.D. Watson, R. Busto and K. Abe. 1982. Neurochem. Res. 7, 437-454.

USE OF ENRICHED STABLE ISOTOPES TO DETERMINE ZINC AND COPPER ABSORPTION IN PREMATURE INFANTS AND PATIENTS WITH M. WILSON

Klaus Dörner, M. Ali Soltani-Neshan, Erika Sievers and Jürgen Schaub

University Children's Hospital Kiel
Schwanenweg 20, D 2300 Kiel 1

Abstract

The zinc absorption was studied by fecal monitoring in 19 preterm infants by oral load with 0.7 mg ^{70}Zn/kg body weight. The gestational age was 33.9±2.5 weeks, the postnatal age at the time of zinc loading 3.6±1.3 weeks. The analytical technique comprised chelate gaschromatography/mass spectrometry for determination of isotope ratios and measurement of the total zinc fecal excretion per hour. The mean absorption was 43.8±17.1%, being lower in the 1.5-3.5 weeks old infants (36.3±14.9%) than in the older group of 4-6.5 weeks (56.8±12.5%). Gestational age and weight did not correlate with the relative zinc absorption.- Copper loads with 0.2 mg ^{65}Cu/kg were performed with two patients with M. Wilson: one patient under conventional D-penicillamine therapy showed a 79.7% absorption rate, the other patient with an oral copper displacement therapy (150 mg zinc/d) a lower rate of 49.9%.

Trace Element Analytical Chemistry in Medicine and Biology

Introduction

The catch-up growth of preterm infants demands a balanced supply of all nutrients, especially of trace elements because of the limited quantity stored. Two thirds of foetal copper and zinc are transferred to the foetus during the last 10-12 weeks of gestation (1). Intake recommendations are generally based upon metabolic balance studies performed under steady state conditions. Positive balances are regarded as an indicator for adequate uptake. They enable the calculation of minimal requirements in absolute terms (2). This concept was recently criticized, on the basis that balance studies only describe the intake required to maintain the existing pool (3). Tracer studies with stable isotopes overcome the limitations of the classical balance studies by measuring the "true" absorption (which is not really true because of potential fast entero-hepatic circulations).

Material and Methods

The data presented here are based on 19 preterm infants (age 3.6±1.3 weeks [mean±SD], gestational age 33.9±2.5 weeks, post-conceptional age 37.6±1.8 weeks, actual weight 2228±366 g). Their average daily zinc intake was 190±20 µg/kg body weight (adapted formula). The adaption time (period of exclusively enteral nutrition before zinc load) was at least 3 days (mostly >5 days). The infants were hospitalized for neonatal nursing and received medications for preterms as usual(i.e. calcium

Table 1: Sampling and Analytical Procedure of ^{70}Zn-Absorption Studies.

ISOTOPE MARKING Extrinsic tag 0.7mg ^{70}Zn/kg body weight added to milk together with appr. 15 mg carmine as a stool passage marker
SAMPLING Collection of stool samples from at least 15 hours before until 72 hours after zinc load; collection of urines 12 hours before until 24 hours afterwards
Weighing of stool and urine samples and freeze drying
ANALYTICAL PRELIMINARIES Dry ashing for 3 days at 500°C (copper: wet ashing!) of 200 mg stool sample and 10 ml urine respectively, resolubilisation with 400 µl HNO_3 and dilution with water
Complexing of metals with 15 µmol sodium-bis(trifluorethyl)dithiocarbamate (4) at pH 6 (borate buffer 0.2 mol/l and extraction with $CHCl_3$; final volume 50-100 µl
DETERMINATION OF ISOTOPE CONCENTRATIONS Metal chelate gaschromatography/mass spectrometry(MAT 44S; EI-mode): Silica capillary DB5 6m, temperature program 140-240°C, 8°C/min, electron energy 74 eV, ionizing current 0.6mA. Isotope ratios are read from a calibration curve prepared from mixtures of natural zinc and the enriched ^{70}Zn
Determination of total zinc in the samples by atomic absorption spectroscopy
Calculation of the zinc excretions (total and ^{70}Zn) per hour

and partly iron supplementation, vit. D and multi-vitamine preparations; 3 received 2x2-3 mg theophylline). The parents gave their informed consent. - The 2 patients with M. Wilson were 16 and 32 years old. The younger one had a conventional D-penicillamine therapy (3x300 mg/d), which was omitted during copper load, the other received an oral zinc therapy (150 mg/d) because of penicillamine intolerance.

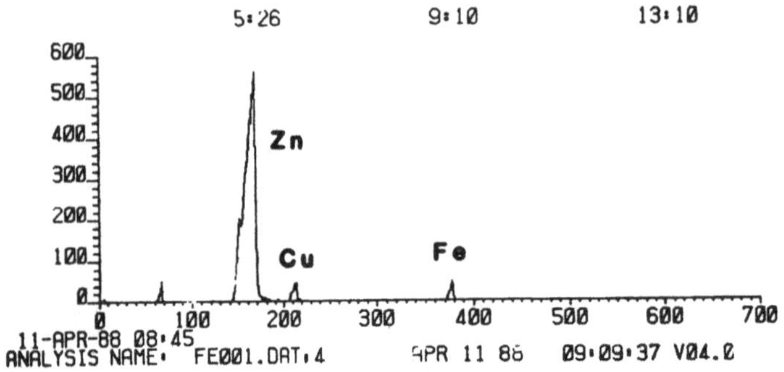

Fig. 1: Gaschromatogram (Total Ion Current) of the Metal Chelates extracted from a Preterm Infants's Stool

Zinc dose (0.7 mg ^{70}Zn/kg, supplier Hempel GmbH & Co, Düsseldorf) was given as a mixture with milk during regular feeding. ^{65}Cu was given in a dosage of 0.2 mg/kg with juice before breakfast. The sampling and the analytical procedures are summarized in table 1.

Results and Discussion

Analytical imprecisions of total metal determinations by atomic absorption spectroscopy and the isotope ratio measurements are typically less than 5%. The detection limit of ^{70}Zn is 0.5 µg. Our GC/MS technique affords a reasonable chromatographic separation of the chelates of zinc, copper and iron (fig.1) and the precise determination of their isotope ratios on the basis of one single preanalytical procedure. This approach enables us to study metal interactions more easily. Up to now only single metal absorptions (zinc or copper) have been studied.

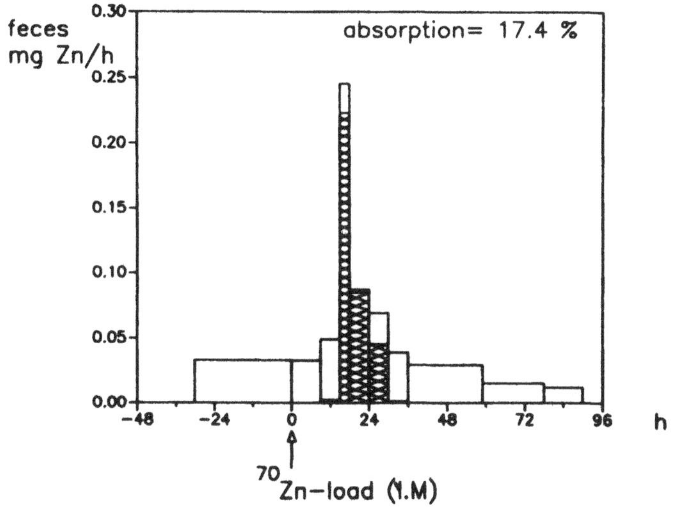

Fig. 2: Fecal Zinc Excretion Pattern after an oral Load with ^{70}Zn; the hatched Area presents the relative Amount of ^{70}Zn

The excretion pattern of an infant with a low absorption rate (17.4%) is shown in fig. 2. This example was selected because of the relatively high defecation frequency. The heights of bars show the fecal excretion per hour, the hatched area the portion of ^{70}Zn. The highest excretion occurs at 15 hours after the load, falling continuously until hour 35. In later stool samples no more ^{70}Zn is detectable. The relative absorption of 17.4% is calculated from the hatched area in relation to the oral dose.

Our study revealed a significant dependency of relative absorption and age of the infants (fig. 3). Absorption at age 1.5-3.5 weeks is 36.3±14.9% and at 4-6.5 weeks 56.8±12.5%. The difference is statistically significant (t-test $p \leq 0.01$). No other parameter rendered a statistically significant influence on absorption rates.

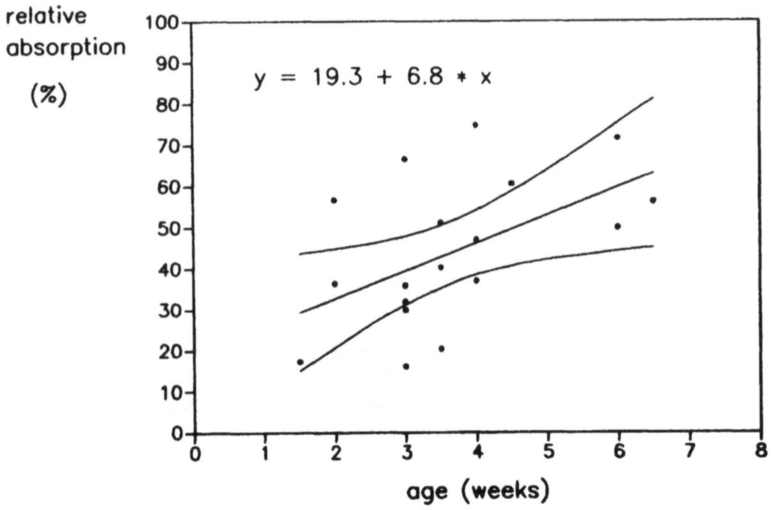

Fig. 3: Correlation of Infants' postnatal Age and the relative Absorption of ^{70}Zn

This result clarifies earlier findings from our group(5), in which classical metabolic zinc balances in preterm triplets led to negative zinc balances at the age of 3 weeks which became positive later in life. From our recent results with stable zinc isotope load, these negative balances seem to be more affected by a low absorption than by a high loss of zinc by pancreatic and intestinal secretions and discarded epithelial cells. Ehrenkranz et al.(6) have published the only comparable study with 13 preterms, 5 of whom were fed with formula milk. Their zinc load, however, was 14 % of the daily intake and we gave about 350%. The absorption rate in the group cited was somewhat higher, namely 66(48-79)%, which might be explained by the lower loading dose. The low absorption rate in very young preterms is a new finding. A recommendation for a strongly increased zinc supplementation in preterm formula milk can not yet be derived from these early results.

The copper absorption in the patient with M. Wilson and oral zinc therapy was 49.9%, in the patient without zinc therapy 79.9%. These results support the otherwise clinically proven usefulness of zinc therapy in M. Wilson (7).

Acknowledgements

This study was generously supported by the Deutsche Forschungsgemeinschaft (Do 167/4) and partly by Milupa AG, Friedrichsdorf.- For perfect technical assistance we are grateful to the technical assistants Mrs. Beate Kühn and Mrs. Maria Rendtorff and to the nurse Annette Vöge.

References

1. Shaw, J.C.L..1979. Am.J.Dis.Child. 133, 1260-1268; Shaw, J.C.L..1980. Am.J.Dis.Child. 134, 74-81

2. Beisel, W.R..1979. Am.J.Clin.Nutr. 32, 271-274

3. Mertz, W..1987. J.Nutr. 117, 1811-1813

4. Neeb, R..1982. Pure & Appl.Chem. 54, 847-852

5. Sievers, E., H.-D. Oldigs, J. Schaub. 1986. Pediatr.Res. 20, 1048

6. Ehrenkranz, R.A., B.A. Ackerman, C.M. Nelli, M. Janghorbani. 1984. Am.J.Clin.Nutr. 40, 72-81

7. Cossack, Z.T.. 1987. In: Trace Element Analytical Chemistry in Medicine and Biology, Vol. 4 (P. Brätter, P. Schramel, eds.). Walter de Gruyter, Berlin pp. 237-246

COPPER UPTAKE IN SHORT TIME CULTIVATED RAT HEPATOCYTES AND COPPER ACCUMULATION WITHIN THE RAT LIVER DURING CHOLESTASIS STUDIED BY PROTON INDUCED X-RAY EMISSION (PIXE)

A. Ochs, H. P. Buscher, W. Gerok

Medizinische Universitätsklinik II, 7800 Freiburg, FRG

D.Heck

Institut für Kernphysik III, Kernforschungszentrum, D 7500 Karlsruhe, FRG

H. Thom

Institut für organische Chemie und Biochemie der Universität, D 7800 Freiburg, FRG

Abstract

With proton induced X-ray emission (PIXE) an analytical tool for the direct trace element analysis is available, which is used for the study of copper-uptake in short-time cultivated hepatocytes. The hepatocytes were exposed to a 30 µM $CuSO_4$-solution. PIXE revealed uptake-rates comparable to those obtained utilizing ^{64}Cu. Though the marked mean variation during the two first minutes of uptake made it more difficult to prove the linearity, these results are consistent with the present view of a facilitated transport by a special carrier protein combined with diffusional uptake.

In a second experiment the bile ducts of rats were ligated for various time periods. Within livers treated this way, a continuous increase of copper-concentration was observed, prefering the acinar zone I (around the portal tracts). PIXE revealed copper-concentrations up to 7000 µg/g in localized maxima (normal <50 µg/g DW) after 60 days of ligation. These data throw some doubt on the assumption, that copper-accumulation during extrahepatic cholestasis is non-toxic.

Introduction

The copper-uptake by hepatocytes is thought to be a combination of a diffusion process and transport by a special carrier protein. (1, 11), which also transports zinc and can be inhibited by sulfhydryl agents (11, 12). L-histidine increases while albumin decreases the uptake process, showing saturation kinetics and marked temperature dependency. In recent studies the radioactive tracer ^{64}Cu was primarily used (11). Extrahepatic cholestasis leads to copper-accumulation in the liver, because 80 % of the hepatic copper is excreted into the bile (1). This accumulation is believed to be non-toxic, though some authors emphasized, that in the case of advanced cholestasis copper-concentrations

Trace Element Analytical Chemistry in Medicine and Biology

reach levels comparable to those in Wilson's disease (13,15, 16). This was demonstrated by bulk analysis when relatively large amounts of material were available. Bulk analytical methods fail however to visualize the distribution of copper, histological stains for copper are known to lack specificity and sensitivity (2, 15, 16). With PIXE we studied copper-uptake by short-time cultivated rat hepatocytes and the copper-accumulation during extrahepatic cholestasis in rats.

Material and Methods

Hepatocytes were taken from 300 g male wistar rats anaesthetized with pentobarbital by intraperitoneal injection (60 mg/kg bodyweight). The liver was perfused with collagenase according to the method of Berry and Friend (17). Aluminum holders with a Formvar® foil stretched over a central hole were coated with collagen, to enable the hepatocytes to attach. The isolated hepatocytes were suspended in a modified Leibowitz L15 medium, containing 12.5 mM HEPES, 12.5 mM $NaHCO_3$, 8.3 mM D(+)glucose, 5 mM L(+)lactate, 1% (v/v) antibiotic/antimycotic solution and 10% (v/v) fetal calf serum. Hepatocytes were plated on Formvar® foils with a density of 900 cells/mm^2 and cultivated in an humidified incubator at 37° C with 5 % CO_2 and 95 % air. For the first 6 hours of cultivation 1 µU/ml dexamethason and 25 mU/ml insulin were added. After 6 hours the medium was changed to contain additionally (1 µmol each) testosterone, 3,3',5-triiodo-L-thyronine, (-)-epinephrine, epidermal growth factor, β-estradiol and glucagon, but not insulin. Uptake-studies were conducted with 30 µM $CuSO_4$ and 30 µM L-histidine in a modified L-15 medium at 37° C, pH 7.4. The viability of the cells was tested using trypan blue (0.34 mM, 2 min in L-15) prior to uptake. Under continuous stirring the hepatocytes on Formvar® foils were exposed for periods varying from 0 to 30 minutes to the Cu^{++}-solution. Immediately after removing the cells from the solution, the remaining medium was sucked off and the hepatocytes were washed twice with Cu^{++}-free medium at 4° C to prevent further Cu^{++}-uptake. After drying, 60 x 70 μm^2 areas of interest were selected containing groups of 4 - 6 unstained hepatocytes, using Normarski interference contrast. These selected areas underwent trace element analysis with PIXE as described below. (Fig. 1)

In the second experiment 300 g male wistar rats were anaesthetized with pentobarbital (60 mg / kg body weight) by an intraperitoneal injection. The bile ducts were ligated twice and cut. After 0, 4, 10, 20, 40 and 60 d of bile duct ligation the rats were anaesthetized again and blood samples were collected from the vena cava inferior. After removing the blood, the liver was carefully excised. The highest copper concentration is found in the median right lobe (1), where the main specimen was excised for further PIXE and light microscopy examination. The preparation procedure has been described elsewhere in detail (3, 7, 8, 9, 14). Briefly the snap frozen specimens were cut by a cryomicrotom. Cuts 1 and 3 from three consecutive cuts were placed on glass slides and stained either with hematoxylin eosin or rubeanic acid to make the copper visible (13, 16, 17). For the PIXE analysis the middle cut was put on Formvar® foil without staining to avoid any trace element contamination from the dyes. An area of 0.6 x 0.6 mm^2 was selected, which contained a periportal tract and a terminal hepatic venule. This area was scanned using a proton beam focussed to a diameter of 3 µm, provided by the 3.75 MV van de Graaff accelerator at Karlsruhe (9, 10). The raster pattern yields 1024 impacts within the scanned area. X-ray emission is induced in each impact characterizing the trace element contents. Bar maps for various trace elements are plotted and compared with the neighbouring stained specimens in order to differentiate them with respect to their histological architecture.

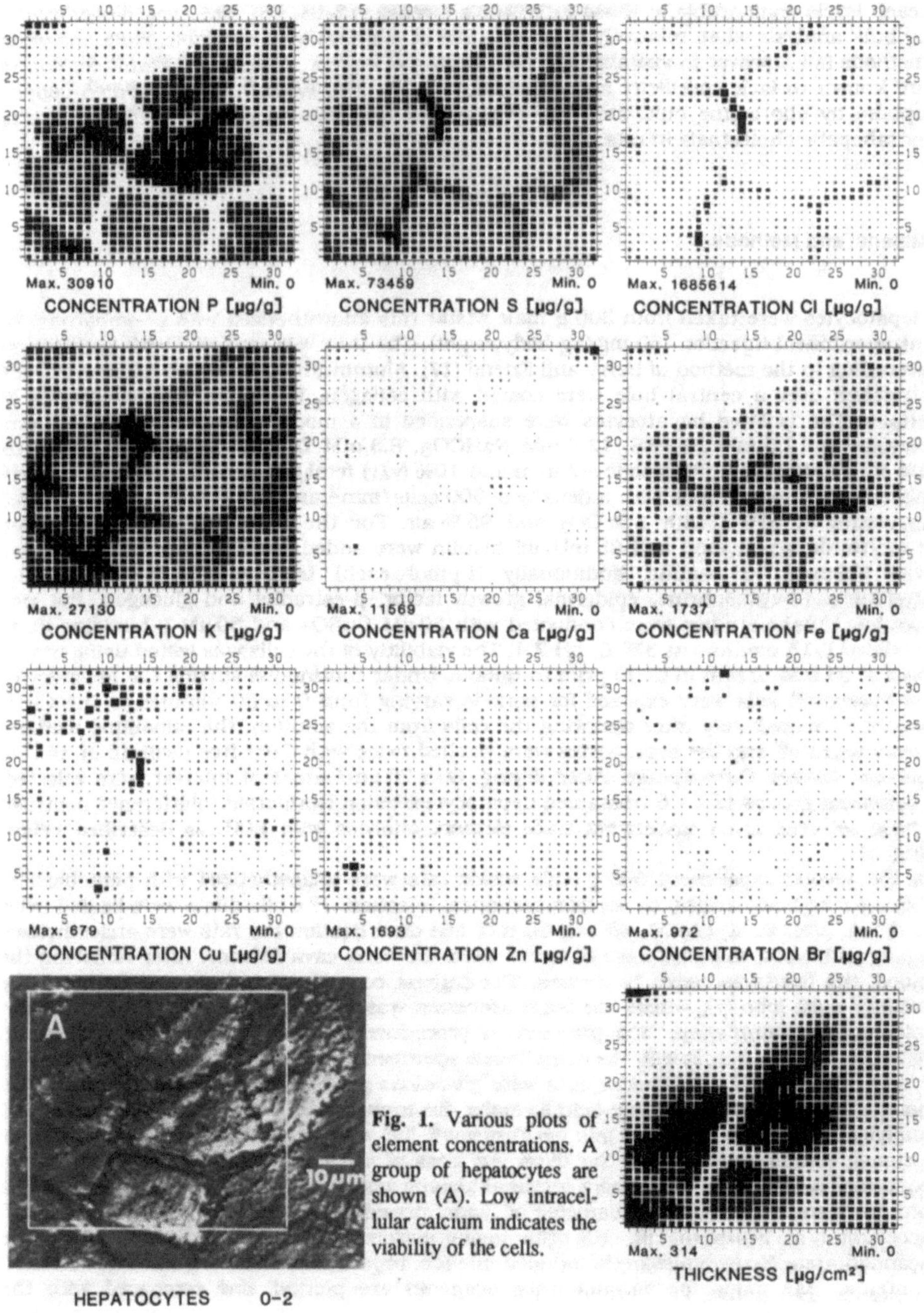

Fig. 1. Various plots of element concentrations. A group of hepatocytes are shown (A). Low intracellular calcium indicates the viability of the cells.

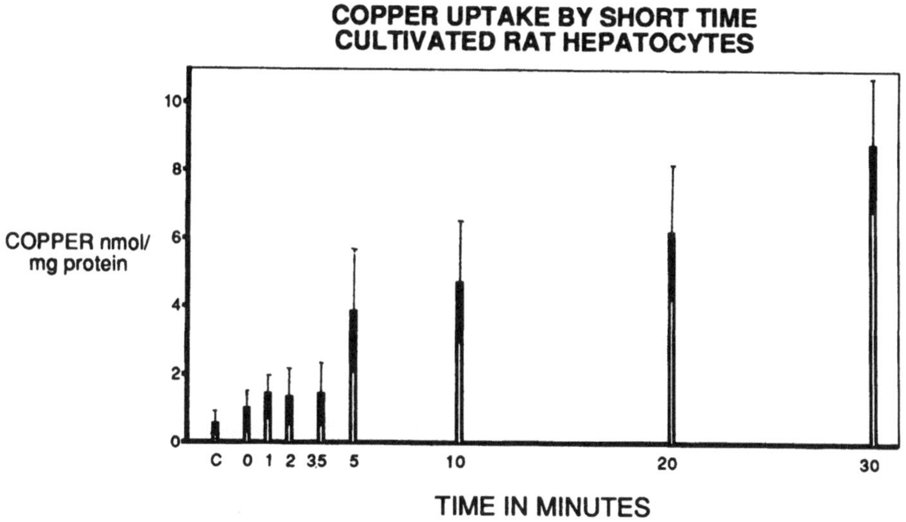

Fig. 2. Copper uptake by short time cultivated rat hepatocytes, exposed to a 30 µM $CuSO_4$ and equimolar L-histidin solution. Small differences and marked mean variation of the copper concentration during the first minutes made it difficult to prove the linearity.

Results

Copper uptake

The PIXE analysis (10) reveals concentrations in µg/g DW. Conversion to nmol copper/mg protein was made using a factor, obtained by determining the wet/dry weight ratio and protein content (Bio-Rad). In the light microscopy examination nonviable cells are clearly identified by staining their nuclei, caused by the uninhibited influx of trypanblue. In addition these cells displayed a striking increased calcium content (up to 9000 µg/g DW) compared with vital cells (400 - 800 µg/g DW). The results of Cu^{++}-uptake are shown in Fig. 2. These results are in agreement with published data (11, 12). The Cu^{++}-uptake is not finished after 30 minutes. Iron and zinc concentrations remain unchanged during the whole experiment.

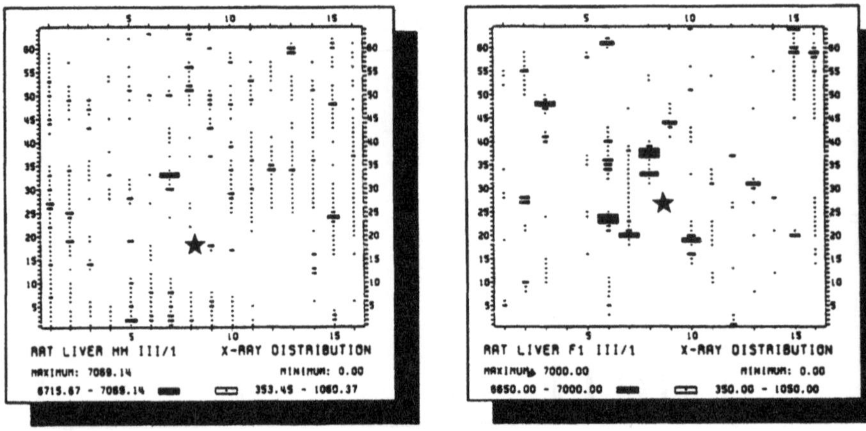

Fig. 3. Two examples of copper concentration after 60 d bile duct ligation are shown. The plots correspond to a scanned area of 0.6 x 0.6 mm^2. The scale of copper concentration is given below the plots. Large bars demonstrate about 7000 µg/g copper, the small ones about 350 µg/g. The portal tracts are indicated by an asterisk.

Extrahepatic cholestasis

In the second experiment, there was a clear correlation of copper accumulation proportionally to the time of bile duct ligation. The mean copper concentration was 35 - 50 µg/g DW within the livers without ligation. After 60 d of ligation copper levels increased up to 1200 µg/g DW. Especially around the acinar portal tracts (zone I) we observed localized peaks of up to 7000 µg/g DW! (Fig. 3)

Discussion:

The Cu^{++}-uptake experiment demonstrated a continuous increase of Cu^{++} concentration depending on the time of exposure. The small differences between the first four measurements made a definite linearity determination of uptake difficult. To study the early phases of copper uptake using concentrations occuring in vivo, a more sensitive method is required.
Surprisingly high copper concentrations observed after 60 d of bile duct ligation, cause doubts whether copper is really non-toxic in the case of extrahepatic cholestasis. The results exceed those obtained in Wilson's disease (14).

Acknowledgment

Many thanks go to the van de Graaff operational staff for their continuous effort to prepare a high quality proton beam. The generous help of Mrs MacNelly has been greatly appreciated.

References

1. Winge, D. R., 1984. Semin. Liv. Dis. 4. 239-251.

2. Fuentealba, I. C., S. Haywood and J. Trafford. 1987. Liver. 7.277-282.

3. Heck, D., 1985. Kerntechnik. 46. 187-192.

4. Folkmann, F., 1975. J. Phys. E.: Sci. Instrum. 8. 429-444.

5. Wiesner, R. H., S.S. Barham and E.R. Dickson. 1979. Gastroenterolgogy 77, A47.

6. Dyson, N. A., A. E. Simpson and J. T. Dabek. 1978. J. Radioanal. Chem. 46. 309-319.

7. Ochs, A., D. Heck, C. Kratt, G. Schneider and K. P. Maier. 1984. Verhandl. Deutsch. Gesellschaft Innere Medizin. 90. 1531-1534.

8. Heck, D., A. Ochs, A. Klempnow, K. P. Maier and C. Kratt. 1987. In: Trace Element - Analytical Chemistry in Medicine and Biology,(P. Brätter P. Schramel, eds.) Walter de Gruyter & Co., Berlin, New York. pp. 367-374.

9. Heck, D., A. Ochs, A. Klempnow, K. P. Maier and C. Kratt. 1987. Nucl. Instr. and Meth. B22. 196-200.

10. Heck, D., A. Ochs, H. Thom and H. P. Buscher. 1988. Nucl. Instr. Meth. B30. 420-423.

11. Schmitt, R., H. Darwish, J. Cheney and M. Ettinger. 1983. Am. J. Physiol. 244. G183-G191.

12. Darwish, H., R. Schmitt, J. Cheney and M. Ettinger. 1984. Am. J. Physiol. 246. G48-G55.

13. Fleming, C. R., E. R. Dickson, A. H. Baggenstoss and J. T. McCall. 1974. Gastroenterology. 67. 1182-1187.

14. Ochs, A., D. Heck, B. Rogaczewski, K. P. Maier, H. E. Schäfer and W. Gerok. 1987. Gastroenterology. 92. A1761.

15. Lesna, M., A. N. Hamlyn and A. J. Watson. 1981. Digestion. 22. 113-118.

16. Goldfischer, S., H. Popper and I. Sternlieb. 1980. Am. J. Pathol. 99. 715-730.

17. Berry, M. N., D. S. Friend. 1969. J. Cell. Biol. 43. 506-520.

INAA OF RABBIT TISSUES AND ORGANS AFTER APPLICATION OF TANTALUM IMPLANTS

R. Michel

Zentraleinrichtung fuer Strahlenschutz, Universitaet Hannover, Am Kleinen Felde 30, 3000 Hannover 1, F.R.G.

M. Reich, M. Nolte

Abt. Nuklearchemie der Universitaet zu Koeln, Zuelpicher Str. 47, 5000 Koeln 1, F.R.G.

L. Rabenseifner

Orthopaedische Klinik, König-Ludwig-Haus,Brettreichstr. 11, 8700 Wuerzburg, F.R.G.

E.-M. Horn, R. Holm

Bayer AG, 5090 Leverkusen, F.R.G.

Abstract

Local and systemic effects of tantalum intramedullary nails were investigated by instrumental neutron activation analysis of rabbit tissues and organs after implantation times up to 13 months. In addition, corrosion rates of Ta in Ringer's solution were determined using radiotracer techniques. The analyses of tissues from the vicinity of the implants show massive Ta burdening, correlated with an increase of Se for Ta concentrations higher than 10^{-6} g/g dry weight. The results for distant tissues and organs demonstrate that Ta is not inert in the body and that its corrosion products undergo transport processes.

Trace Element Analytical Chemistry in Medicine and Biology

Introduction

Up to now, no metal or alloy has been found which fullfills all requirements for medical implants. All known materials undergo corrosion and destruction processes in the body, thereby giving rise to an internal burdening by elements which normally are present in the organism at trace or ultra-trace levels, only. The question of biocompatibility of these corrosion products is not yet answered [1]. The search for new extremely corrosion resistant biomaterials implies the analytical description of trace elements in biological systems also for those elements for which neither normal concentrations in man nor the physiological action are known. During the last years the elements tantalum and niobium were discussed in orthopaedic surgery as implant materials [2,3]. This work describes an investigation of the behaviour of tantalum intramedullary nails in rabbits using INAA as analytical tool. The investigation is complemented by radiotracer experiments on the corrosion of tantalum in Ringer's solution.

Experimental

Tantalum intramedulary nails [4] were implanted into the legs of 16 rabbits of the same breed. Three different groups were distinguished. For 5 rabbits the nails were embedded into predrilled ducts, in order to avoid any physical stress during implantation (group I). For 11 rabbits the nails were driven in as in common implantation techniques (group II: N = 5 and group III: N = 6). The animals of groups I and II were killed after 4 weeks of implantation, those of group III after an implantation time of 13 months. Samples of various tissues (connective tissue in contact with the implants, proximal muscle, liver, spleen, kidneys, heart and lymphatic tissues from the lower pelvis) were dissected by common stainless steel surgical instruments. A group of 5 rabbits living under identical conditions served as controls. The dissected tissues were freeze-dried and sealed in Suprasil AN quartz ampoules. The samples were irradiated together with a large number of single- and multielement standards for 36 h at neutron flux of $6 * 10^{13}$ cm^{-2} s^{-1}. After a decay time of 3 d the samples were repeatedly measured by gamma-spectrometry over a period of 6 weeks. 14 elements (Na, K, Sc, Cr, Fe, Co, Zn, Se, Br, Rb, Sb, Cs, and Ta) could be determined. The results of the blanc analyses and of SRM's can be found elsewhere [5,6].

For the tracer experiments quadratic Ta foils with a size of 1 cm^2 and a thickness of 0.1 mm were used. Three different samples were investigated, one as delivered by Goodfellow Metals Ltd, U.K., and two with artificial passive layers of 500 A and 1000 A, respectively. These samples were packed in high-purity Al-foil and irradiated together with suitable Ta-standards for 36 h at a flux of $7 * 10^{13}$ cm^{-2} s^{-1}. After a cooling time of 8 d, the samples were immersed in Ringer's solution at 37° C in polyethylene-containers, which were changed after appropriate time intervalls. The ^{182}Ta in the Ringer's solution was measured by gamma-spectrometry. Absolute Ta-amounts were determined by comparison with the Ta-standards.

Results and Discussion

The corrosion rates of Ta "as delivered" (fig. 1) are slowly oscillating with time over a period of 80 d between extreme values of $3 * 10^{-7}$ g/d m^2 and $8 * 10^{-6}$ g/d m^{-2}. This demonstrates that no marked passivation process is going on, but

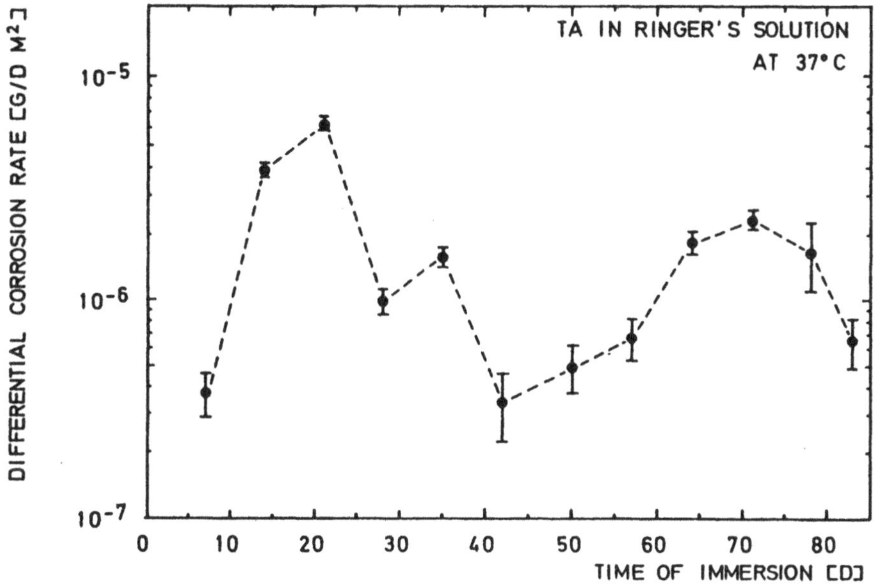

Fig. 1: Differential corrosion rates of tantalum in Ringer's solution at 37° C.

rather that there is a reaction of the Ta_2O_5 passive layer built during storage in air to the immersion in Ringer's solution. For the Ta-samples with well-defined passive layers of 500 A and 1000 A, respectively, there is a slight trend to decrease from $1.5 * 10^{-6}$ g/d m^{-2} to $5 * 10^{-7}$ g/d m^{-2} over a period of 130 d, indicating a very long time scale for the rearrangement of Ta passive layers (fig. 2). The final corrosion rates are equal to the lowest ones observed for Ta in HNO_3 at 22° C, where a marked passivation process occurs with the actual corrosion rates decreasing from $3 * 10^{-5}$ g/d m^{-2} to $5 * 10^{-7}$ g/d m^{-2} [6]. The maximum in the corrosion rates observed for the 500 A sample was caused by an accident at day 42. At this day the foil fell down from the Pt wire by which it was fixed through a small hole. By rearranging the fixation slight mechanical stress may have occured, thereby possibly damaging the passive layer. After this event a marked passivation process is seen lasting for about 30 d until the corrosion rates

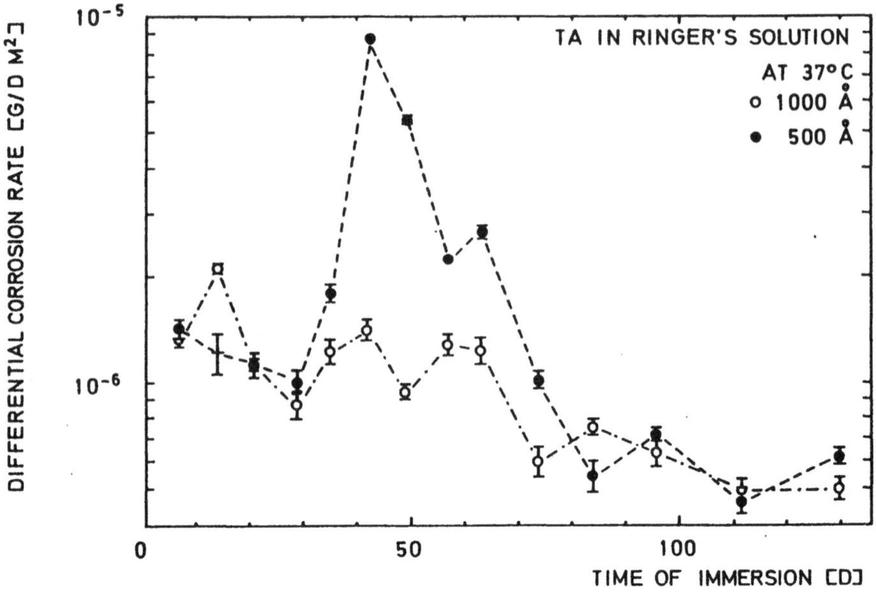

Fig. 2: Differential corrosion rates of prepassivated tantalum samples with passive layers of 1000 and 500 A in Ringer's solution at 37° C.

reached the same values as for the 1000 A sample. This sensitivity against physical stress emphasizes that the corrosion rates derived from immersion experiments only give lower limits for the material flux from loaded implants to the organism.

According to Thull [7] the Ta_2O_5 passive layer of Ta is ion conducting. The slow diffusion velocity of ions through this material is giving rise to the low velocity of passivation observed here. Moreover, under in body conditons, with possibly reduced oxigen availablity, an incomplete passivation under contribution of lower valence oxides and hydroxides has also to be accounted for [8].

The average corrosion rates of Ta in Ringer's solution are about two orders of magnitude lower than those of Co-Cr-alloys and of stainless steel under the same conditions [9,10]. An evaluation of these findings with respect to the biocompatibility of these materials is, however, not possible, since the pysiological action of Ta is completely unknown. Moreover, the actual flux of material from materials can be strongly rised by wear and destruction for loaded implants. Therefore, the actual consequences of implantation only can be determined by the analysis of tissues and body fluids after application of functionating implants. This was done in our animal experiments, where the burdening by corrosion and wear products from Ta intramedullary nails in rabbits was derived by comparing the trace element concentrations of animals with and without implants.

The results for the "normal" rabbits are indistinguishable from those reported earlier from our group [11]. Ta could not be detected by INAA in normal rabbit tissues, the detection limits (L_D) according to Currie [12] being between 4 and 0.3 ng/g dry weight. Since not all the elements investigated can be dealt with here, for a more detailed discussion we refer to [6].

The Ta concentrations in the connective tissue surrounding the implants exhibit massive local effects (fig. 3). The actual concentrations go up to 10^{-4} g/g dry weight. All three implant groups showed such extreme values. Groups I and II with 4 weeks implantation time, however, exhibited also low Ta concentrations in the order of 10 ng/g only, while the 13 months group exclusively showed Ta concentrations above 10^{-6} g/g in the connective tissue from the vicinity of the implants. This indicates an increase of local burdening with time. The proximal muscles showed marginal effects. Ta was only seen in one sample of the 13 month group with 4.2 ng/g, while all others were below 1 ng/g.

The local effects of the Ta implants are accompanied by a response of selenium (fig. 3). For samples with Ta concentrations

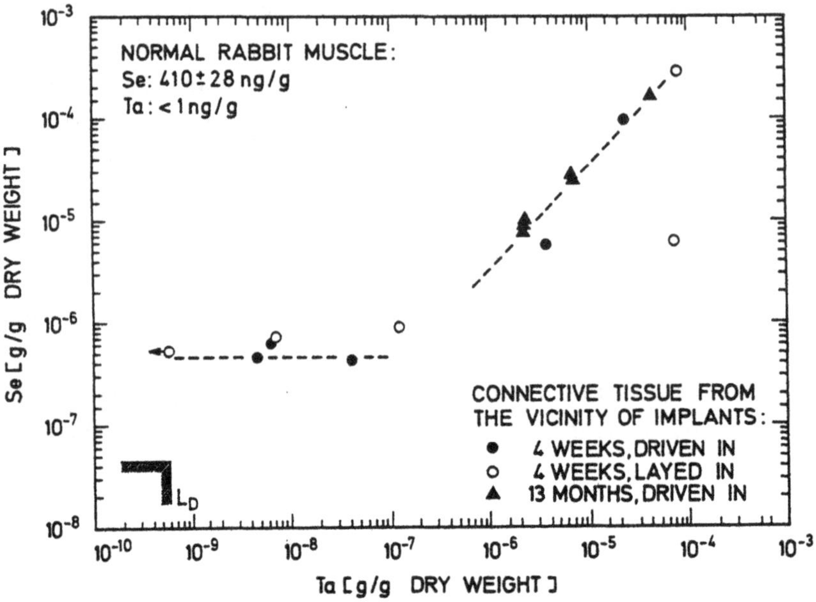

Fig. 3: Tantalum and selenium in connective tissue encapsulating tantalum intramedullary nails implanted into the legs of rabbits.

above 10^{-6} g/g the selenium concentrations are significantely rised. For the largest Ta values they are higher by two orders of magnitude than the comparable normal value for muscle of 410 $\pm$ 28 ng/g. For Ta below 10^{-6} g/g Se is fairly constant in the connective tissue with 625 $\pm$ 180 ng/g. This effect is not yet understood and has to be further investigated.

The analysis of tissues and organs distant from the implants also showed a response to tantalum, which, however, is by increased Ta concentrations only. All other elements were indistinguishable from normal [11]. The systemic burdening of Ta corrosion and/or wear products is seen in samples of the lymphatic tissue and in liver and spleen. For lymphatic tissue Ta up to 40 ng/g was found in 5 samples [6], 4 of them belonging to the 13 months group. All other samples were below 4 ng/g.

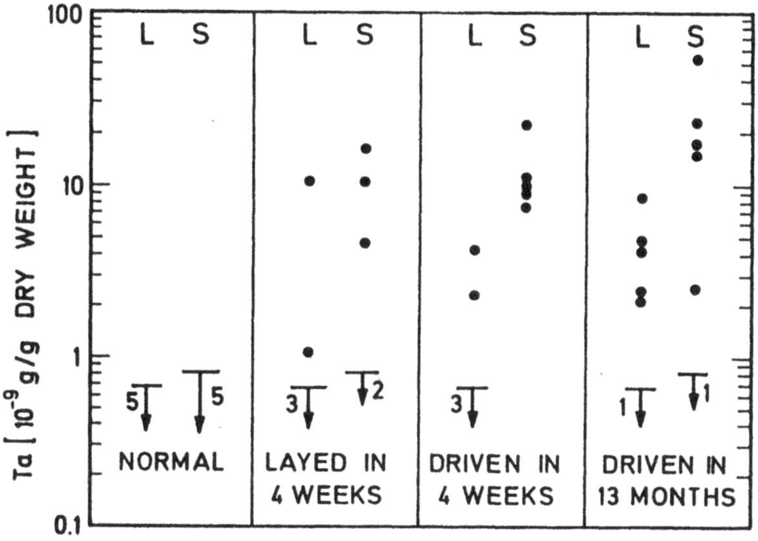

Fig. 4: Tantalum in liver and spleen of rabbits with and without tantalum intramedullary nails.

More clear cut results were obtained for liver and spleen (fig. 4). They exhibit Ta contents up to two orders above the detection limits. The fact that systemic Ta burdening is found mainly in sample of tissues from the reticulo-endothelian system points to a particulate transport of Ta debris. Here, histological investigations combined with elemental analysis are required in the future.

Though Ta shows extremely low corrosion rates in in vitro experiments with unloaded implants, the analysis of tissues and organs from animal experiments demonstrates that there are massive local and systemic effects of tantalum. These effects have to be accounted for when applying tantalum as medical implant.

Acknowledgement

This work was partially supported by the Deutsche Forschungsgemeinschaft. Our thanks are due to the staff of the research reactors of the KFA Juelich, where the irradiations were performed.

References

1. Michel, R.. 1987. CRC Critical Reviews in Biocompatibility 3, 235.

2. Rabenseifner, L.. 1986. Tantal und Niob als Implantatwerkstoffe. Enke Copythek, Ferdinand Enke Verlag, Stuttgart.

3. Schider, S. and H. Bilstein. 1982. In: Proc. First World Biomaterials Congress (G.D. Winter, D.F. Gibbons and H. Plenk, Jr., eds.). John Wiley and Sons, Chichester.

4. Rabenseifner, L., W.D. Otte and S. Schider. 1981. Langenbecks Arch. Chir. 355, 618.

5. Michel, R., F. Löer, M. Nolte, M. Reich and J. Zilkens. 1987. J. Radioanal. Nucl. Chem. 113, 83.

6. Reich, M. 1987. Thesis, Universität zu Köln.

7. Thull,R.. 1983. Biomed. Technik 28 (Suppl.), 38.

9. Holm, R. and S. Storp. 1976. Vakuum-Technik 25, 172.

10. Hofmann, J., R. Michel, R. Holm and J. Zilkens. 1981. Surf. Interf. Anal. 3, 110.

11. Hofmann, J., N. Wiehl, R. Michel and J. Zilkens. 1982. J. Radioanal. Chem. 70, 85.

12. Michel, R., J. Hofmann and J. Zilkens. 1980. In: Trace Element Analytical Chemistry in Medicine and Biology (P. Brätter and P. Schramel, eds.) Walter de Gruyter & Co., Berlin, p. 137.

13. Currie, L.A.. 1968. Anal. Chem. 40, 586.

TRACE ELEMENT STATUS OF CEREBROSPINAL FLUID OF INDIVIDUALS WITH NEUROLOGICAL DISEASES BY ICP-MS

Neil I. Ward, Nicola Walker, Ann E. Ward

Department of Chemistry, University of Surrey, Guildford, Surrey, GU2 5XH, England

Margaret A. Hall

Clinical Biochemistry, St. Luke's Hospital, Warren Road, Guildford, Surrey, England

Abstract

The elemental assessment of cerebrospinal fluid collected from individuals with no known neurological disorders (controls, n=10), Parkinson's disease (n=28), motor neurone disease (n=27), and multiple sclerosis (n=40), is presented. Elemental analysis was performed by inductively coupled plasma-source mass spectrometry (ICP-MS). Elements are typically present in CSF at levels of: > 10 µg ml^{-1} (Na, Mg, P, Cl, K, Ca, Br); <1000->10 ng ml^{-1} (Al, Si, Fe, Cu, Zn, Rb, Sr, Ba); <10->1 ng ml^{-1} (B, Ni, Co, As, Se, Ag, Cd, Sn, Sb, I, Pb); and <1 ng ml^{-1} (Li, Sc, V, Cr, Mn, Co, Mo, Cs, Hg). Wide variations are found in CSF copper concentrations with low values for multiple sclerosis (35.63 ± 32.68 ng ml^{-1}) and motor neurone disease (35.13 ± 16.90 ng ml^{-1}) compared with control (57.14 ± 43.16 ng ml^{-1}) individuals. Reduced CSF Fe values are found in Parkinson's disease (77.4 ± 33.6 ng ml^{-1}) compared with controls (98.9 ± 33.8 ng ml^{-1}).

Introduction

The importance of trace elements in the physiology and

Trace Element Analytical Chemistry in Medicine and Biology

pathology of the central nervous system (CNS) is well established. Cerebrospinal fluid (CSF) may change significantly in physical characteristics, cytological constituents or chemical content as a result of certain neurological diseases. Basic clinical measurements on CSF samples involve examination of the 'physical appearance', that is, for example, the protein, glucose, and globulin content. Specific analysis of the chemical constituents of CSF may be useful for differential diagnosis (1).

Many trace elements are present in CSF (2). Neurochemical and neurophysiological evidence indicates that trace elements markedly affect the metabolism of neurotransmitters in the brain. Investigation of pathological changes in CSF elemental content have yielded conflicting results with little agreement for establishing an acceptable normal range. Analytical techniques employed for the measurement of CSF trace elements include neutron activation analysis (3,4), electrothermal atomisation atomic absorption spectrophotometry (5), flow injection atomic absorption spectrophotometry (6), flame atomic absorption spectrophotometry (7), spark-source mass spectrometry (8), and more recently the use of an inductively coupled plasma as an emission source for atomic absorption spectrophotometry (9). This variety of techniques, together with contamination problems associated with sample collection and pre-analysis treatment processes has led to a wide variation in reported concentration ranges for CSF trace elements. In addition, most CSF studies have reported only one or two elements; the number of multi-element analyses are few (8, 10). This paper presents a study on a multi-element basis for a number of CSF samples from patients with neurological diseases. Analysis was performed by inductively coupled plasma-source mass spectrometry (ICP-MS).

Materials and Methods

Sample collection

CSF samples were taken as part of a routine lumbar puncture procedure. The control group consisted of individuals having no known neurological disorder (admitted due to symptoms of headache, dizziness, prolapsed disc, etc) and whose CSF was 'normal' with respect to appearance (macroscopically clear), protein content of less than 0.5 g l^{-1}, and globulin/ glucose content. Any blood contaminated samples were excluded from the study. The diseased group patients were classified according to their neurological diagnosis on admission.

The CSF collected for trace element analysis was the last sample drawn from the spinal tap to reduce the contamination risk. Samples were drawn directly into 'trace element free' tubes (Teklab Medical Laboratories Ltd) and stored at -20°C prior to analysis. Samples were neither centrifuged nor transferred to any other containers.

Experimental

Inductively coupled plasma-source mass spectrometry (ICP-MS) comprises a plasma which produces charged particles consisting of the parent ion and ionic fragments of the original molecule and a quadrupole mass analyser which separates the ions according to their mass/charge ratio (11). All measurements were undertaken on the Surrey VG Isotopes Ltd PLASMAQUAD instrument. A typical run sequence involved scanning over 5 to 240 m/z mass range, 2048 channels with a dwell time of 250 µs on each channel, and 120 sweeps. CSF samples were introduced directly with no prior dilution steps. A series of three multi-element standards (Ventron AAS solutions) were prepared at 10 and 100 ng ml^{-1} levels.

Results and discussion

The elemental mean and (standard deviation) values for human cerebrospinal fluid from individuals with no known neurological disorders; controls (n=10), Parkinson's disease (n=28), motor neurone disease (n=27), and multiple sclerosis (n=40) are shown in Tables 1 and 2 respectively. All elemental values are expressed in nanograms per millilitre (ng ml^{-1}) or (***) micrograms per millilitre (μg ml^{-1}).

In general, the range of elemental content present in CSF are at levels of:

> 10 μg ml^{-1}	Na, Mg, P, Cl, K, Ca, Br
<1000->10 ng ml^{-1}	Al, Si, Fe, Cu, Zn, Rb, Sr, Ba
<10 - >1 ng ml^{-1}	B, Ni, As, Se, Ag, Cd, Sn, Sb, I, Pb
<1 ng ml^{-1}	Li, Sc, V, Cr, Mn, Co, Mo, Cs, Hg

The enhanced ultra-trace sensitivity of ICP-MS is clearly demonstrated by the number of elements measured at or below the nanogram level in CSF.

Investigation of pathological changes in CSF copper have yielded conflicting results in the literature. Its role in the myelination of neural tissue suggest a possible role in the etiology of multiple sclerosis. Studies on the copper level of CSF from multiple sclerosis patients have resulted in not significantly increased from normal (12), or conversely, very high (13) values. In this study there is a wide variation (min.-max.) in copper CSF values: controls 20.3-139.8, Parkinson's disease 14.1-162.8, motor neurone disease 15.1-100.6, and multiple sclerosis 9.6-151.4, ng ml^{-1}, respectively. Mean (S.D.) values between controls 57.14(43.16) and multiple sclerosis 35.63(32.68) ng ml^{-1} individuals implies a possibly significant difference (C>MS) at the $p < 0.10$ probability level. Low copper scalp hair

Table 1. Elemental Mean+ and (Standard Deviation) of CSF: Control* and Parkinson's Disease** Individuals

Element	Elemental Concentration: mean ± (SD) Control*		Parkinson's disease**	
Li	0.28	(0.07)	0.29	(0.06)
B	1.5	(0.8)	1.2	(0.5)
Na***	3232	(27)	3236	(27)
Mg***	30.8	(8.9)	35.9	(7.0)
Al	22.4	(8.6)	24.3	(10.0)
Si	230.4	(48.8)	202.8	(37.2)
P***	15.4	(3.2)	14.8	(1.3)
Cl***	4372	(64)	4382	(47)
K***	173	(54)	180	(58)
Ca***	44	(7)	41	(7)
Sc	0.42	(0.11)	0.45	(0.13)
V	0.66	(0.13)	0.67	(0.16)
Cr	0.67	(0.09)	0.61	(0.16)
Mn	0.60	(0.08)	0.61	(0.09)
Fe	98.9	(33.8)	77.4	(33.6)
Co	0.64	(0.22)	0.72	(0.27)
Ni	1.4	(0.6)	1.6	(1.1)
Cu	57.1	(43.2)	56.1	(41.2)
Zn	75.7	(10.6)	57.0	(14.0)
As	1.6	(0.3)	1.9	(0.7)
Se	6.4	(2.7)	4.4	(1.6)
Br***	2.36	(0.57)	2.83	(0.87)
Rb	64.9	(18.2)	59.4	(15.6)
Sr	7.5	(4.0)	8.8	(4.1)
Mo	0.67	(0.15)	0.76	(0.18)
Ag	1.2	(0.6)	1.1	(0.4)
Cd	3.27	(1.72)	2.21	(1.08)
Sn	3.6	(2.0)	3.6	(1.4)
Sb	3.2	(1.2)	3.4	(1.3)
I	3.1	(0.6)	3.3	(1.1)
Cs	0.64	(0.15)	0.62	(0.21)
Ba	6.0	(2.0)	5.9	(2.3)
Hg	0.7	(0.1)	0.8	(0.1)
Pb	4.8	(2.6)	4.3	(2.4)

+ Elemental content; ng ml^{-1} or *** µg ml^{-1}

* Control (n = 10)

** Parkinson's disease (n = 28)

Table 2. Elemental mean+ and (Standard Deviation) of CSF: Motor Neurone Disease* and Multiple Sclerosis** Individuals

Element	Motor neurone disease* mean	(SD)	Multiple sclerosis** mean	(SD)
Li	0.29	(0.05)	0.28	(0.05)
B	1.0	(0.2)	1.0	(0.1)
Na***	3252	(34)	3250	(38)
Mg***	29.5	(8.8)	29.1	(3.6)
Al	24.6	(8.7)	19.7	(3.0)
Si	168.4	(86.1)	228.1	(48.5)
P***	13.9	(1.4)	14.6	(1.7)
Cl***	4045	(64)	4385	(49)
K***	147	(73)	175	(39)
Ca***	43	(4)	44	(6)
Sc	0.68	(0.15)	0.54	(0.26)
V	0.63	(0.14)	0.69	(0.14)
Cr	0.67	(0.13)	0.69	(0.20)
Mn	0.73	(0.24)	0.60	(0.11)
Fe	96.6	(48.7)	82.7	(34.6)
Co	0.85	(0.15)	0.81	(0.18)
Ni	1.3	(0.7)	1.8	(0.8)
Cu	35.1	(16.9)	35.6	(32.7)
Zn	58.8	(21.6)	77.5	(23.7)
As	1.6	(0.5)	2.1	(0.7)
Se	5.5	(2.6)	4.7	(1.7)
Br***	2.85	(0.83)	2.39	(0.62)
Rb	56.8	(18.1)	64.1	(19.7)
Sr	8.8	(2.6)	9.2	(4.7)
Mo	0.77	(0.24)	0.76	(0.18)
Ag	1.1	(0.5)	1.4	(0.5)
Cd	2.13	(0.73)	2.48	(1.14)
Sn	2.6	(0.9)	4.19	(1.72)
Sb	4.7	(1.9)	2.9	(1.3)
I	2.9	(0.5)	3.2	(0.9)
Cs	0.57	(0.14)	0.72	(0.18)
Ba	6.0	(2.2)	6.4	(2.5)
Hg	0.6	(0.3)	0.7	(0.2)
Pb	5.6	(3.8)	3.1	(1.4)

Column heading: Elemental Concentration: mean ± (SD)

+ Elemental content; ng ml^{-1} or *** µg ml^{-1}

* Motor neurone disease (n = 27)

** Multiple sclerosis (n = 40)

values have been reported for multiple sclerosis individuals (14). Other disorders in which raised CSF copper values have been found are cerebral infarction (1), cerebrovascular accident and anxiety states (15), and Huntingtons chorea (13). This study also shows low CSF copper values for motor neurone disease individuals (35.13 ± 16.90 ng ml^{-1}).

Reduced CSF iron values have also been reported in Parkinsonism (16). The respective values (min.-max: mean ± SD) for this study are controls, 46.3-146.4: 98.9 ± 33.8 ng ml^{-1}; and Parkinson's disease, 24.8-167.8: 77.4 ± 33.6 ng ml^{-1}.

Few measurements of CSF manganese have been made, and those that have show a wide range in values (17,18). Evidence suggests that chromic manganese poisoning can injure the dopaminergic components of the extrapyramidal system in a manner paralleling Parkinson's disease (19). There is no variation in CSF iron content between control, Parkinson's or multiple sclerosis individuals.

In conclusion, this paper has shown that a modern multi-element analytical technique such as ICP-MS is capable of establishing accurate values for thirty-four elements, of which a large number at or below the nanogram level have not been reported before in cerebrospinal fluid.

References

1. Bogden, J.D., R.A. Troiano, M.M. Joselow. 1977. Clin. Chem. 23, 485-89.

2. Gooddy, W., E.I. Hamilton, T.R. Williams. 1975. Brain 98, 65-70.

3. Kanabrocki, E.L., L.F. Case, E.B. Miller, E. Kaplan, Y.T. Oester. 1964. J. Nuc. Med. 5, 643-8.

4. Kjellin, K. 1981. Clin. Toxicol. 18, 1237-45.

5. Agarwal, R.P., R.I. Henkin. 1982. Biol. Trace Elem. Res. 4, 117-24.

6. Burguera, J.L., M. Burguera, O.M. Alarcon. 1986. JAAS 1, 79-83.

7. Palm, R., R. Sjöström, G. Hallmans. 1983. Clin. Chem. 29 486-491.

8. Gooddy, W., T.R. Williams, D. Nicholas. 1974. Brain 97, 327-36.

9. Ebdon, L. 1984. Lab. Practice, May, 9-12.

10. El-Yazigi, A., Iman-Al-Saleh, O. Al-Meefty. 1984. Clin. Chem. 30, 1358-60.

11. Gray, A.L. 1975. Analyst 100, 289.

12. Plum, C.M. 1960. Acta. Psychiat. Neurol. Scand. 35, 79-83.

13. McCall, J.T., N.P. Goldstein, L.H. Smith. 1971. Fed. Proc. 30, 1011-1015.

14. Ryan, D.E., J. Holzbecher, D.C. Stuart. 1978. Clin. Chem. 24, 1996.

15. Madera-Orsini, F., A.O. Jibril, F.I. Volini. 1976. Clin. Chem. 22, 1210-11.

16. Lehman, H.E. V.A. Kral. 1951. Arch. Neurol. Psychiatr. 65, 326-36.

17. Cotzias G.C., P.S. Papavasiliou. 1962. Nature 195, 823-24.

18. D'Amico, O.J., H.L. Klawans. 1976. Anal. Chem. 48, 1469-72.

19. Cotzias, G.C., K. Horiuchi, S. Fuenzalida, I. Mena. 1968. Neurology 18, 376-82.

TRACE ELEMENT METABOLISM

Plenary Lecture given by:

P.J. Agett

Chairman:

N.I. Ward

Poster Reports:

P.J . Aggett, J.D. Kruse-Jarres

THE METABOLISM OF TRACE ELEMENTS IN PREGNANCY AND LACTATION

Peter J. Aggett
Department of Child Health, Polwarth Buildings, University of Aberdeen, Foresterhill, Aberdeen, AB9 2ZD, Scotland, UK.

ABSTRACT

Current knowledge of the metabolism of zinc, copper, selenium and chromium in pregnancy and lactation is considered. Most literature has been devoted to zinc nevertheless it is apparent that the metabolic consequences of pregnancy outweigh many nutritional factors in altering the metabolism of all these metals.

INTRODUCTION

The metabolic changes which occur during pregnancy include increased intestinal absorption of nutrients, an expansion of the plasma volume, altered metabolism of protein, carbohydrate, calcium and bone, increased renal blood flow and glomerular filtration rate with possible alterations in tubular reabsorption, and increased blood cell numbers with altered proportions of circulating subsets (1). These changes which are effected by endocrine changes are seen as being a means of ensuring an adequate supply of nutrients to the placenta for transfer to the fetus and to protect the mother from the trauma associated with delivery. Most, if not all, of these changes are anticipatory in as much as they occur in early pregnancy before any appreciable nutrient demand is imposed by the products of conception.

These metabolic changes and altered partitioning of nutrients distort the customary indices of nutrient status,

Trace Element Analytical Chemistry in Medicine and Biology

particularly those of minerals including the trace elements, and, unfortunately, these distortions are mistaken frequently as evidence of altered nutritional status. Another misconception is that a pregnant woman needs "to eat for two". In fact the extra nutrient requirements of pregnancy are relatively small; in women the weight of the products of conception expressed as a percentage of the mother's preconceptional weight is about 6%. This compares with 40% in the mouse, 25% in the rat, and 9% in the rhesus monkey (2). The relative metabolic demands represented by these proportions should be considered when extrapolating to humans data derived from animal models. Even so the relative requirements for minerals during pregnancy may differ from that represented by maternal body composition because the trace element content of some fetal tissues (eg. relative high contents of copper and iron in the liver and lower zinc in muscle) as well as the mass relative to body size of the organs themselves differs from that in adults (3,4).

The metabolic changes of pregnancy could be construed as representing a maternal sacrifice for the benefit of offspring. However, progressive failure of reproductive efficiency accompanies increasingly severe deficiencies of the major nutrients. This is manifest as abnormal nursing and puerperal care of the newborn, inefficient labour, intrauterine growth retardation, fetal resorption, congenital malformations, failure of implantation, and failure to conceive. This spectrum of defects has been encountered in animal models of pre-natal trace element deficiency and is summarised in Table 1. Both sustained and transitory deficiencies of trace elements can be terratogenic.

In humans 2-5% and 4% of newborns respectively have a major or a minor congenital malformation. The causes of major anomalies are unknown in about 30% of cases and this percentage is probably higher for minor defects. Most of these anomalies arise in early pregnancy during organogenesis. Although this precedes the period of increased fetal nutrient

TABLE 1

Defects arising in animal models from pre-natal trace metal deficiencies (5,5a,6)

ZINC

Impaired implantation,
Embryonic and fetal death and resorption,
Congenital malformations
  Cleft lip and/or palate;
  Anencephaly, hydrocephaly, neural tube defects;
  Spina bifida, Syndactyly,
  Urogenital defects, Cardiac malformations,
  Pulmonary malformations, altered surfactant,
  Endocrine and exocrine pancreatic insufficiency.
  Immune defects.
Low birthweight
Inefficient labour, prolonged bleeding
Abnormal postnatal behavioural in mother and offspring

COPPER

Fetal and early neonatal death,
Neurological abnormalities,
  fits, defective myelin-synthesis,
  cerebral and/or cerebellar hypotrophy
Cardiovascular; aneurysms, variocosities, vascular fragility
Skeletal matrix defective,
Altered metabolism of energy and phospholipid.
Impaired growth.

requirements associated with growth it does not exclude the possibility that in man as well as in animal models terratogens may operate by altering the metabolism of trace elements and that marginal tracel element "status" may act as a co-terratogen. Thus in animal models zinc deprivation exacerbates the terratogenicity of EDTA, vitamin A deficiency, 6-mercaptopurne, thalidomide (5), and alcohol (7). Although zinc deprivation is co-terratogenic with these agents, there is no evidence to indicate that supra-physiological intakes of zinc would prevent their effects. However a possible exception to this generalisation may appertain when terratogenicity arises from the chelation effects of some agents. A possible example of this is that antenatal penicillamine therapy and low copper intakes in rats enhance the risk of fetal copper deficiency but the sustained use of this drug is usually safe in pregnant women with Wilson's disease in whom the body burden of copper is above normal (5,8). In order to identify any possible causal association between abnormal metabolism of trace elements and congenital defects in man a different strategy to that of performing trace element analyses during later pregnancy, the puerperium or even later (eg. 9) is needed. Some of these have been interpreted as providing evidence of possible deficiencies occurring shortly after conception. Such studies can only generate hypotheses which should be challenged by appropriate prospective epidemiological studies.

Another comment on animal studies is that the nutritional deprivations employed often exceed those encountered in normal clinical practice. For example, recent studies in sub-human primates (rhesus monkeys) employed a diet providing 4 µg of zinc per gram of diet dry matter (10). This would correspond to a daily dietary intake of about 2 mg of zinc in human adults and is much less than the putative marginal intakes reported in studies to date.

It is against this general background that I would like to comment on some aspects of trace element metabolism in human pregnancy and lactation.

ZINC

The consequences of zinc deprivation during reproduction are summarised in Table 1. A negative correlation between birthweight and food intake during the third trimester of pregnancy was noted in the zinc deficient rhesus monkey model (11). It was surmised that, as has been encountered in other instances of zinc deficiency, catabolised tissue releases zinc and other nutrients into the plasma; these would then be available for transfer to the fetus. This protective, albeit adventitious, release of zinc has been studied in pregnant rats (12,13). It was found that the accumulation of zinc in the products of conception exceeded the quantity ingested during pregnancy by the dams on a restricted and zinc deficient intake; the extra zinc was shown to have come from the mothers' bones and muscle. This shows the inter-relationship between the intakes of other nutrients such as energy, protein and calcium relative to that of zinc in influencing whether or not marginal intakes of zinc alter reproductive efficiency.

Nutrient interactions such as those outlined above may well have influenced the outcome of pregnancies in women with Acrodermatitis enteropathica, in whom zinc deficiency occurs probably secondary to a defect in their intestinal absorption of the element. Fourteen pregnancies have been described in women with this condition who have not been treated with zinc. One woman had an achondroplastic dwarf and when treated subsequently with the chelating agent, 5,7-diiodo-8-hydroxyquinoline (DIH), had three normal infants (14); another mother who was only 1.51m tall and weighed 40 kg needed a caesarean section to deliver a 2.52 kg infant because cephalopelvic disproportion (15). One other women was reported to have had a spontaneous abortion and to have delivered an anencephalic infant in a subsequent pregnancy, she was being treated with DIH (16). A woman who was totally untreated, experienced severe exacerbations of her symptoms throughout her four pregnancies, the first two of which had a normal outcome but the others were terminated because of her severe symptoms (17); these resolved promptly following the end of each of these pregnancies. The final woman had similar experience; in her first pregnancy she delivered a normal infant but suffered a severe relapse, this recurred during the next pregnancy and she had a therapeutic termination and an immediate remission. However she endured her symptoms during a third pregnancy at the end of which she was diagnosed and she was treated with zinc, despite the earlier severe zinc deficiency she again had a normal child (18). With adequate zinc therapy, women with AE can have uneventful pregnancies (19).

It seems possible that metabolic phenomena similar to those described in models contribute to these varied responses to severe zinc deficiency. This anecdotal experience offers at least three tentative messages which are (i) that zinc deficiency may be terratogenic in man, (ii) altered metabolism in pregnancy is such that it can induce symptomatic zinc deficiency in women with Acrodermatitis enteropathica in whom marginal zinc deficiency may already be present, (iii) severely symptomatic zinc deficient women can have normal offspring.

In normal, presumed healthy, women the amount of zinc which is required to be absorbed is not accurately known irrespective of whether they are pregnant or not. The accumulation of extra zinc during pregnancy can be calculated from a factorial approach allowing for the zinc content of amniotic fluid (20), increased blood components and plasma volume (21), and the accretion of lean tissue by the mother (22) and of zinc by the fetus (3). The computed figures and dietary intakes based on an effective 'worst case' efficiency of intestinal absorption of 20% are given in Table 2. The efficiency of the intestinal absorption of zinc is, of course, probably under estimated.

TABLE 2
Calculated zinc and copper requirements during human pregnancy

(a) Cumulative theoretical increments

Period of gestation (weeks)	10	20	30	40
Protein (g)	36	165	498	925
Zn (mg)	4.6	21.1	63.7	118.4
Cu (mg)	1.2	5.5	16.6	30.8

(b) Calculated need (mg/d)

Between weeks	1-10	11-20	21-30	31-40
Zn retention	0.066	0.236	0.609	0.780
Cu retention	0.017	0.061	0.158	0.283
Zn requirement*	0.32	1.18	3.04	3.90
Cu requirement*	0.09	0.30	0.80	1.4

* Assuming 20% availability of Zn and Cu from the diet.

Various recommended intakes of zinc during pregnancy have been promulgated (Table 3)(23). These recommendations are not strictly comparable because their terms of reference differ. Even so, as can be seen in Table 4, it is noteworthy that few

TABLE 3
Some national recommended intakes of Zinc (mg/d) for women (23).

Country	Non-pregnant	Pregnant	Lactating
Australia	14	18.5	20
Canada	10	13	17
Czechoslovakia	8	16	18
DDR	12	13-16	25
USA*	15	20	25

* Recommendations adopted by Italy, Spain, New Zealand, and Uruguay.

assessed intakes by pregnant women match even the lowest recommendations. However, the degree of risk of zinc deficiency cannot yet be related to any failure to achieve any of these recommended intakes. Before this can be done much more accurate information is required about zinc metabolism and requirements during pregnancy and the influence on these of changes in the systemic metabolism of other nutrients.

The acquisition of such information is difficult. Balance studies have been used to assess the metabolism of zinc; approaches based on compositional and functional assays have been used to assess zinc status during pregnancy; and both of these approaches have been combined with investigations of the effect of zinc supplementation.

Unfortunately many balance studies have used zinc intakes in excess of those customarily observed. Thus the reported negative zinc balances on intakes of 26-32 mg of zinc (30) may represent homeostatic adjustments to such high dietary levels. One study of women at mid-pregnancy on their natural intakes of 8.9 ($\pm$ 3.2 SD) mg of zinc per day found a net retention of 1.9 ($\pm$ 3.4) mg per day (31). On supplementation to 23 mg of zinc per day the mean retention (2.3 $\pm$ 7.6 mg) was not altered substantially. Other balance studies using an intake of 20 mg of zinc per day found a mean difference of daily zinc retention between pregnant and non-pregnant women of 1.3 mg which may represent the effect of pregnancy (32), but this difference

TABLE 4

Some reported daily intakes during pregnancy

Population (stage)	Zn (mg)	Cu (mg)	Energy (MJ)	Protein (g)	Method
Navajo (Mid)(24)	12 (6.9)	1.57 (0.7)	10.2 (37)	95 (41)	1 dR
Middle Income (L)(25)	9.8 (0.6)	-	8.5 (0.3)	80 (3)	3d dup
(37 w) (NL)	7.9 (0.5)	-	6.8 (0.4)	63 (4)	
Denver (throughout)(26)	11.3 (4.1)	-	8.5 (2.1)	85 (25)	1 3 dR
Hispanics (mid)(27)	9.7 (5.0)	1.39 (1.7)	6.9 (2.5)	68 (30)	1 dR
Aberdeen (30w)(21)	9.1 (2.0)	1.5 (0.3)	8.4 (1.9)	72 (14)	7d WI
USA (28)					
Vegetarian (37w)	12.6 (0.9)	2.8 (0.3)	10.2 (0.6)	93 (7)	3dRec
Non-vegetarian (37w)	14.4 (0.6)	2.1 (0.4)	8.4 (0.6)	97 (5)	
Non-pregnant Vegetarian	6.4 (0.8)	1.3 (0.2)	5.6 (0.6)	51 (6)	
North London (Mid)(29)					
Vegetarian	7.5 (2.1)	1.53 (0.7)	8.2 (2.2)	57.3 (16)	7 dR
Non-vegetarian	10.2 (3.0)	1.8 (0.8)	9.1 (1.9)	75 (19)	
European	11.6 (2.8)	1.5 (0.3)	8.5 (1.1)	80 (20)	

1 (3 or 7) dR: 24h (3 or 7 day) recall; 3d dup: analysis fo 3 day duplicate diet;

7 dWI: 7 day weighed intake.

L: Subsequent lactators; NL: Non-lactators

was not detected when the same group studied mid-term pregnant women on meat or vegetable based diets containing 16 mg zinc per day (33).

An effective and unequivocal, compositional or functional indicator of zinc status is not available. In pregnancy plasma zinc is even more unreliable than in the non-pregnant state. Plasma zinc concentrations fall early in pregnancy (26,34), and continue to fall throughout pregnancy (21,26,35). Some of the later fall in plasma zinc concentrations may be attributable to the expansion in plasma volume, and allowing for this, the total amount of zinc in the plasma compartment may increase during pregnancy (21). An equation has been derived to enable normalisation and interpretation of plasma zinc determinations during pregnancy (26), but the women studied were on iron supplements which may have interfered with the metabolism of zinc (see below).

Since increments in plasma volume are smaller in abnormal pregnancies and infants of low birth weight, it is not surprising that an inverse correlation between plasma zinc concentrations and birth weight has been reported (36,37,38). However this relationship has not been consistent and some authors have described associations between decreased plasma zinc concentrations and abnormal pregnancies (35) whereas others have not noticed any particular relationship. Since the cause of such abnormalities are multifactorial these reports need not be contradictory; they may represent different pathogeneses which have differing effects on the metabolism of zinc, and few of these studies allowed for independent factors which may alter plasma zinc or fetal birthweight (21,38).

A reduced zinc content of maternal mixed leucocytes has been associated with intrauterine growth retardation (39,40). Leucocytes derived from the cord blood of growth retarded infants also have a reduced zinc content (41). However changes in the relative proportions of white cell subsets, each of which have a different zinc content, and the other possibility that their zinc content like that of other nutrients, may vary secondary to other factors (38) means that the low cellular content of zinc may not represent zinc deprivation, but studies of the zinc content of leucocyte subsets have provided more evidence of altered zinc metabolism in association with growth retarded fetuses (42). The zinc content of erythrocytes has been reported as increased during late pregnancy (25).

There is no clear evidence that zinc supplementation is necessarily beneficial during human pregnancy. Such studies are difficult to undertake and supervise and most of those reported have been small, difficult to control for causes of low birthweight and abnormal deliveries. The timing of the initiation of zinc supplementation is of obvious importance if the effect on malformations is to be examined. Thus reports that supplements of 90 and 45 mg of elemental zinc reduce the duration and incidence of abnormal labour as well as reducing maternal blood loss at delivery need to be further substantiated (43). In a double blind study of Hispanic women zinc supplementation (20 mg/d) reduced the incidence of women with serum zinc concentrations below 8.4 µmol/L (44). In another study on a similar population the incidence of toxaemia in adults, but not in teenagers, was less in those women receiving zinc supplements, however in this study and others the overall outcome of pregnancy was not altered by zinc supplementation (45,46). Another study showed that zinc supplements (15 mg/d) increased the zinc content of hair and saliva and the serum alkaline phosphatase activity, but found no effect on serum zinc concentration (26). A recent German intervention study (47) found a marginally lower mean birth weight (3236 g) in unsupplemented mothers, and a possibly higher incidence of newborn abnormalities in women not receiving a daily 20 mg supplement of zinc. However this, as with most of the other studies, demonstrates a need for accurate control for the extraneous factors which may affect the outcome of pregnancy.

Little is known about the transfer of zinc across the placenta. Investigations using guinea pig placenta (48) and the isolated dually perfused human placental lobule (49) suggest that the placental uptake of zinc is "regulated" at the maternal surface, zinc accumulates in the placenta, presumably in some bound form, and enters the fetal circulation by diffusion.

Although a dietary deficiency of zinc has not been identified in human pregnancy there is increasing evidence that the metabolism of zinc may be perturbed by other nutritional supplements (eg. iron and folate), alcohol, and smoking. Short courses of folate (350 μg) with or without iron (50 mg) impair the absorption of zinc following the administration of 25 mg of the element orally one day after the end of supplementation (50). The clinical implication of this phenomenon is uncertain. Some studies have shown that in clinical practice iron supplements may alter the metabolism of zinc (26,34,51) but this has been disputed (52).

Maternal smoking reduces the birthweight of infants, and has been associated with reduced zinc content of some white cell subsets (42). Recent studies have shown that the zinc: cadmium ratio in placental tissue is inversely related to maternal age, and more closely with maternal smoking (53). Both infant birth weight and cord blood erythrocyte zinc content were lower in those born to mothers who smoked (54,55). Thus it seems that a cadmium:zinc interaction occurs in the feto-placental unit which may compromise the zinc supply to the fetus and contribute to intrauterine growth retardation.

Animal data have shown that, as with other nutrients, even a brief exposure to alcohol impairs the placental uptake and transfer of zinc. This defect was not overcome by zinc supplementation of the alcohol treated dams (56). This raises interest in the possible role of altered zinc metabolism contributing to the pathogenesis of fetal alcohol syndrome (7). An apparent correlation has been reported between the plasma zinc concentrations of alcoholic mothers at delivery and the number of malformations in their offspring (56a), but the plasma zinc values would be particularly difficult to interpret in such circumstances. Nonetheless, residual defects of zinc metabolism may occur in the offspring of alcoholic mothers (7).

COPPER

It has been calculated that a pregnant woman requires 75 μg (1.2 μmol) of elemental copper daily to meet the requirements of pregnancy (57). The calculated daily copper accumulation by a fetus if 51 μg (0.8 μmol)/kg body weight/d (4). At term the total amount of copper in the fetus and placenta may be as much as 17% of that in the non-pregnant woman. This represents an accretion of about 45 mg of copper throughout pregnancy and constitutes an increased daily retention of 4%. Assessments of the daily dietary intakes of copper are shown in Table 2. Recorded intakes (Table 4) are less than the USA estimated safe and adequate range of 2-3 mg/d (58). Although, as with zinc, this comparison is of little value one recent report suggests that to achieve a reliable copper retention the diet of mid-term pregnant women should contain 3 mg of copper daily (31). Balance studies conducted in women consuming animal and plant based diets respectively supplying 1.4 and 2.5 mg of copper showed that the intestinal absorption of copper was marginally higher in pregnant compared with non-pregnant women; this difference however only just achieved statistical significance with the diet based on vegetable protein and its biological significance is less clear (57).

The plasma or serum copper concentrations and that of caeruloplasmin increase steadily throughout pregnancy (21,34). After delivery these concentrations return rapidly to those observed in the non-pregnant state, sensitivity of these to endocrine influences is reflected in their reduction in association with placental insufficiency (59), intrauterine death (60), and threatened abortion (61). In contrast high concentrations are associated with toxaemia. The reduced hepatic copper contents observed in anecdotal studies on pregnant women who have died in road traffic accidents (62) or with late toxaemia (63) suggests that the increased plasma concentrations may arise from mobilisation of copper from the liver. Mid pregnancy plasma or serum copper concentrations do not correlate with neonatal birthweight or the outcome of pregnancy. The contribution of copper concentrations interacting with those of some amino acids in a prediction equation for newborn birthweight may well reflect endocrine and metabolic rather than nutritional influences (38).

Although there is no clear description of the effect of maternal copper deficiency on human development, there are two case reports of abnormalities in neonates delivered to women who were being treated with the chelating agent Penicillamine (64,65). Both these children had abnormalities of collagen and elastin as manifest by the presence of hernias, cutis laxa; one infant had hypotonia, hyperflexibility of the joints, vascular anomalies, and died following a Candida septicaemia. Before attributing these phenomena entirely to copper deficiency one should remember that Penicillamine can chelate other trace elements.

The concentration of copper in the maternal plasma exceeds that in the fetal circulation (66), but the influence of this gradient on the translocation of copper across the placenta to the fetus has not been established.

SELENIUM

Pregnant women have been estimated to accumulate 3.5-5 μg of selenium daily (67). Since selenium is an anionic element homeostasis is predominantly regulated by renal rather than by hepato-alimentary mechanisms. Thus the finding in balance studies of women consuming 150 μg of selenium per day of a progressive decrease of urinary excretion of selenium during pregnancy is suggestive of some adaptive mechanism which is, as yet, uncharacterised. Pregnant women retained up to 34 $\pm$ 2 μg of selenium daily compared with 11 $\pm$ 2 μg in non-pregnant women (67). Plasma concentrations of selenium may be lower in pregnant women whereas those in erythrocytes are unaltered (68, 69); maternal levels exceed those in cord blood (68). Intriguingly plasma and red blood cell selenium levels were lower in late pregnancy in women who planned to breast feed than those in women who had decided not to breast feed (70). This difference was maintained during lactation, Although the dietary intakes of selenium were higher in the lactators both before and after parturition, these intakes were similar after allowance was made for the higher energy intakes of the lactators. Data on the activities in plasma and erythrocytes of the seleno-enzyme glutathione peroxidase are conflicting (68,69,71).

CHROMIUM

Interest in the metabolism of chromium during pregnancy centres on the role which this element have in carbohydrate metabolism (72). Fasting plasma chromium in pregnant women (57 ± 2 nmol/L) is reported to be less than that in non-pregnant women (90 ± 3 nmol/L). In non-pregnant women intravenous and oral glucose loads depress the plasma chromium concentration, this was not observed in some pregnant women (73), but was detected in late pregnancy in 33% of another cohort of women (74), it remains to be substantiated if this represents incipient chromium deficiency. However hair content of chromium is reduced in multiparous women (75) but again this finding is difficult to interpret, and the analytical determination of chromium was notoriously difficult at the time of these studies.

LACTATION

Little is known about metabolic adaptation in lactating women in respect to trace elements. After parturition plasma zinc levels increase and those of copper fall. Ideal dietary requirements and intakes have been suggested for such women but as with pregnancy the recorded intakes bear little relation to those recommended. As yet there is little evidence that low trace element intakes alter the breast milk content of zinc and copper. However a supplementation (15 mg of zinc) study has shown that the progressive decline in zinc content of human breast milk can be attenuated (76). Currently manganese is the only cation for which a correlation between dietary intake (2.3 - 9.4 mg/d) and its content in breast milk has been noted (77). The metabolic significance of this relationship is not clear but this may be related to the role of manganese in glycosyl transferase activities and therefore in the synthesis of lactose.

Indices of selenium nutritional adequacy are lower in lactating than in non-lactating mothers (70,78) and maternal plasma selenium content may correlate with that in breast milk (70,78).

Estimates of the daily loss of elements during lactation varies. In early lactation up to 5 mg of zinc per day could be secreted in breast milk but because the elemental content of milk declines with lactation it is possible that at three months this daily loss would be reduced to the order of 1 mg of zinc and 0.13 mg of copper (77). It should be noted however that these estimations remain to be verified.

CONCLUSION

It should be apparent from this review that considerable research effort is still required to establish the nature of the metabolism of trace elements during pregnancy and lactation. Only when this has been achieved could it be possible to devise means of detecting deficiencies of these minerals and the effects of such deficiencies on reproductive efficiency. It would then be possible to develop strategies to assess more accurately the requirements for trace elements during pregnancy and lactation.

REFERENCES

1. Hytten, F., and G. Chamberlain (eds). 1980. Clinical Physiology in Obstetrics. Blackwell Scientific Publications, Oxford.

2. Widdowson, E.M. In+ Maternal Nutrition in Pregnancy; Eating for Two? (J. Dobbing, ed). Academic Press, London, pp. 1-17.

3. Shaw, J.C.L. 1979. Am. J. Dis. Child. 133, 1260-1268.

4. Shaw, J.C.L. 1980. Am. J. Dis. Child. 134, 74-81.

5. Hurley, L.S. 1981. Physiol. Rev. 61, 249-295.

5a. Beach, R.S., M.E. Gershwin, and L.S. Hurley. 1982. Scinece 218, 469-471.

6. Apgar, J. 1985. An. Rev. Nutr. 5, 43-68.

7. Nutrition Reviews. 1986. 44, 359-360.

8. Walshe, J.M. 1977. Quart. J. Med. 46, 73-83.

9. Soltan, M.H., and D.M. Jenkins. 1982. Br. J. Obstet. Gynaecol. 89, 56-58.

10. Golub, M.S., M.E. Gershwin, L.S. Hurley, D. Baly, and A. G. Hendrickx. 1984. Am. J. Clin. Nutr. 39, 265-280.

11. Golub, M.S., M.E. Gershwin, L.S. Hurley, D.L. Baly, and A.G. Hendrickx. 1984. Am. J. Clin. Nutr. 39, 879-887.

12. Masters, D.G., C.L. Keen, B. Lonnerdal, and L.S. Hurley. 1985. J. Nutr. 113, 905-912.

13. Masters, D.G., C.L. Keen, B. Lonnerdal, and L.S. Hurley. 1986. J. Nutr. 116, 2148-2154.

14. Epstein, S., and J.S. Vedder. 1960. Arch. Dermatol. Syph. 82, 189-190.

15. Verburg, D.J., L.I. Burd, E.O. Hoxtell, and L.K. Merrill. 1974. Obstet. Gynecol. 44, 233-237.

16. Neldner, K.H., and K.M. Hambidge. 1975. N. Engl. J. Med. 292, 879-882.

17. Olham-Larson, P. 1978. Dermatologica 156, 155-166.

18. Bronson, D.M., R. Barksky, and S. Barksky. 1983. J. Am. Acad. Dermatol. 9, 140-144.

19. Brenton, D.P., M.J. Jackson, and A. Young. 1981. Lancet 2, 500-502.

20. Gardiner, P.E., E. Rosick, U. ROsick, P. Bratter, and G. Kynast. 1982. Clin. Chim. Acta. 120, 103-117.

21. Tuttle, S., P.J. Aggett, D.M. Campbell, and I. MacGillivray. 1985. Am. J. Clin. Nutr. 41, 1032-1041.

22. Hytten, F., and I. Leitch. 1971. The Physiology of human pregnancy. 2nd edition. Blackwell Scientific Publications, Oxford.

23. International Union of Nutritional Sciences. 1983. Nutr. Abstr. Rev. Clin. Nutr. Series A. 53, 1109-1110.

24. Butte, N., D.H. Calloway, and J.L. Van Dryen. 1981. Am. J. Clin. Nutr. 34, 2216-2228.

25. Moser, P.B., and R.D. Reynolds. 1983. Am. J. Clin. Nutr. 38, 101-108.

26. Hambidge, K.M., N.T. Krebs, M.K. Jacobs, A. Favier, L. Guyette, and D.N. Ikle. 1983. Am. J. Clin. Nutr. 37, 429-442.

27. Hunt, I.F., N.J. Murphy, J. Gonez, J.C. Smith. 1979. Am. J. Clin. Nutr. 32, 1511-1518.

28. King, J.C., T. Stein, and M. Doyle. 1981. Am. J. Clin. Nutr. 34, 1649-1655.

29. Campbell-Brown, M., R.J. Ward, A.P. Haines, W.R.S. North, R. Abraham, I.R. McFadyen, J.R. Turnland, and J.C. King. 1985. Br. J. Obstet. Gynaecol. 92, 875-885.

30. Schraer, K.K., and D.H. Calloway. 1974. Nutrition and Metabolism 17, 205-212.

31. Taper, L.J., J.T. Oliva, and S.J. Ritchy. 1985. Am. J. Clin. Nutr. 41, 1184-1192.

32. Swanson, C.A., J.R. Turnland, and J.C. King. 1982. J. Nutr. 112, 697-707.

33. Swanson, C.A., J.R. Turnland, and J.C. King. 1983. J. Nutr. 113, 2557-2567.

34. Breskin, M.W., B.S. Worthington-Roberts, and R.H. Knopp. 1983. Am. J. Clin. Nutr. 38, 943-953.

35. Jameson, S. 1976. Acta Medica Scandinavica 593, Suppl., 1-89.

36. McMichael, A.J., I.E. Dreosh, G.T. Gibson, J.M. Hartshorne, R.A. Buckley, and D.P. Colley. 1982. Early Human Development 7, 59-69.

37. Cherry, F.A., E.A. Bennett, G.S. Bazzano, L.K. Johnson, G.J. Fosmire, and H.K. Batson. 1981. Am. J. Clin. Nutr. 34, 2367-2375.

38. Metcoff, J., J.P. Costiloe, W. Crosby, L. Bentle, D. Seshachalam, H.H. Sandstead, L.E. Bodwell, F. Weaver, and P. McClain. 1981. Am. J. Clin. Nutr. 34, 708-721.

39. Meadows, N.J., W. Ruse, M.F. Smith, J. Day, P.W.N. Keeling, J.W. Scopes, R.P.H. Thompson and D.L. Bloxam. 1981. Lancet 2, 1135-1137.

40. Wells, J.L., D.K. James, R. Luxton, and C.A. Pennock. 1987. Br. Med. J. 294, 1054-1056.

41. Meadows, N.J., W. Ruse, P.W.N. Keeling, J.W. Scopes, and R.P.H. Thompson. 1983. Arch. Dis. Child, 58, 807-809.

42. Simmer, K., and R.P.H. Thompson. 1985. Clinical Science 68, 395-399.

43. Jameson, S. 1979. In: Zinc and the Environment II. Health Effects. Wiley, New York. pp. 183-197.

44. Hunt, I.F., N.J. Murphy, A.S. Cleaver, B. Faraji, M.E. Swendseid, A.H. Coulson, V.A. Clark, N. Laine, C.A. Davis, J.C. Smith. 1983. Am. J. Clin. Nutr. 37, 572-582.

45. Hunt, I.F., N.J. Murphy, A.E. Cleaver, B. Faraji, M.E. Swendseid, A.H. Coulson, V.A. Clark, B.L. Bravdy, T. Cabalman, and J.C. Smith. 1984. Am. J. Clin. Nutr. 40, 508-521.

46. Hunt, I.F., N.J. Murphy, A.E. Cleaver, B. Faraji, M.E. Swendseid, B.L. Browdy, A.H. Coulson, V.A. Clark, R.H. Settlage, and J.C. Smith. 1985. Am. J. Clin. Nutr. 42, 815-828.

47. Kynast, G., and E. Saling. 1986. Gynecol. Obstet. Invest. 21, 117-123.

48. Simmer, K., J.StJ. Dwight, I.M.H. Brown, and R.P.H. Thompson. 1985. Biol. Neonate. 48, 114-121.

49. Page, K.R., D.R. Abramovich, P.J. Aggett, A. Todd, and C.G. Dacke. 1988. Quart. J. Expl. Physiol. In press.

50. Simmer, K., C.A. Iles, C. James, and R.P.H. Thompson. 1987. Am. J. Clin. Nutr. 45, 122-125.

51. Hambidge, K.M., N.F. Krebbs, L. Sibbley, J. & J. English. 1987. Obst. Gynecol. 70, 593-596.

52. Sheldon, W.L., M.O. Aspillaga, P.A. Smith, and T. Lind. 1985. Br. J. Obstet. Gynaec. 92, 892-898.

53. Kuhnert, B.R., P.M. Kuhnert, and T.J. Zarlingo. 1988. Obstet. Gynecol. 71, 67-60.

54. Kuhnert, P.M., B.R. Kuhnert, P. Erhard, W.T. Barshear, S.L. Groh-wargo, and S. Webster. 1987. Am. J. Obstet. Gyneco. 157, 1241-1246.

55. Kuhnert, B.R., P.M. Kuhnert, S. Debanne, and T.G. williams. 1987. Am. J. Obstet. Gynecol. 157, 1247-1251.

56. Gishan, F.K., and H.L. Greene. 1983. Pediatr. Res. 15, 529-531.

56a. Flynn, A., S.I. Miller, S.S. Martier, N.L. Goldern, R.J. Sokol, and B.C. Del Villano. 1981. Lancet 1, 572-575.

57. Turnlund, J.R., C.A. Swanson, and J.C. King. 1983. J. Nutr. 113, 2346-2352.

58. Committee on Dietary Allowances, Food and Nutrition Board. 1980. Recommended Dietary Allowances, 9th Edition. National Academy of Sciences, Washington DC, USA.

59. Freidman, S., C. Baharg, B. Eckerling, and B. Gans. 1969. Obstet. Gynecol. 33, 189-194.

60. Schenker, J.G., E. Jiengies, and W.Z. Polishiek. 1969. Am. J. Obstet. Gynecol. 105, 933-937.

61. O'Leary, J.A., G.S. Novalis, and G.J. Vosburgh. 1966. Obstet. Gynecol. 28, 112-117.

62. Mirzakarimov, M.G. 1953. Akush. Ginekol. 1, 55-85.

63. Rasuli, Z.M. 1963. Akush. Ginekol. 39, 63-66.

64. Linares, A., J.J. Larranz, J. Rodriguez-Aluscon, and J.L. Diaz-Perez. 1979. Lancet 2, 43-44.

65. Mjolnerod, O.K., K. Rasmussen, S.A. Dommerad, and S.T. Gjeraldsen. 1971. Lancet 1, 673-675.

66. Henkin, R.I., J.R. Marshall, and S. Meret. 1971. Am. J. Obstet. Gynecol. 103, 320-344.

67. Swanson, C.A., D.C. Reames, C. Veillon, J.C. King, and O.A. Levander. 1983. Am. J. Clin. Nutr. 38, 169-180.

68. Rudolph, N., and Wong, S.L. 1978. Ped. Res. 12, 789-792.

69. Behne, D., and W. Walters. 1979. J. Clin. Chem. Clin. Biochem. 17, 133-135.

70. Levander, O.A., P.B. Moser, and V.C. Morris. 1987. Am. J. Clin. Nutr. 46, 694-698.

71. Butler, J.A., P.D. Whanger, M.J. Tripp. 1982. Am. J. Clin. Nutr. 36, 15-23.

72. Anderson, R.A. 1986. Clin. Physiol. Biochem. 4, 31-41.

73. Davidson, I.W.F., and R.C. Burt. 1973. Am. J. Obstet. Gynecol. 116, 601-608.

74. Hambidge, K.M., and W. Dioegemueller. 1974. Obstet. Gynecol. 44, 666-672.

75. Mahalko, J.R., and M. Bennion. 1976. Am. J. Clin. Nutr. 29, 1069-1072.

76. Krebs, N.F., K.M. Hambidge, M.A. Jacobs, and J.O. Rasback. 1985. Am. J. Clin. Nutr. 41, 560-570.

77. Vuori, E., S.M. Makinen, R.Kara, and P. Kuitunen. 1980. Am. J. Clin. Nutr. 33, 227-231.

78. Mannan, S., and M.F. Picciano. 1987. Am. J. Clin. Nutr. 46, 95-100.

THE REGULATION OF LIVER IRON METABOLISM

Y. Beguin, H.A. Huebers, C.A. Finch

Division of Hematology, University of Washington, Seattle, USA

G. Weber, I. Roelandts

Department of Experimental Nuclear Physics, University of Liege, Belgium

Abstract

^{55}Fe-transferrin and ^{59}Fe-ferritin were injected simultaneously in rats and storage iron release by the hepatocyte was calculated from the cumulative incorporation of the two isotopes in the red cell mass over 2 weeks. About 6 % of hepatocyte storage iron was released daily in normal rats, but a non-mobilizable pool was identified in iron overload. Iron release was regulated by the rate of erythropoiesis and iron status of the animal, while acute inflammation blocked iron release only transiently.

Introduction

Hepatocyte iron release to the plasma is less well characterized than iron uptake by the liver (1). Two phases of iron release can be distinguished, i.e. (a) an immediate release during the initial processing of transferrin, hemoglobin, ferritin, or "free" iron (early release, ER), and (b) iron release from stores (late release, LR). We present the results of studies of storage iron release in normal rats and animals with various disorders of iron status and erythropoiesis, using ^{59}Fe-labelled ferritin as a selective tag of hepatocytes (2,3,4).

Trace Element Analytical Chemistry in Medicine and Biology

Methods

Rat transferrin was purified by saturating plasma transferrin with ferrous sulfate, followed by ammonium sulfate precipitation, ion-exchange chromatography using a DEAE Sephacel column and gel chromatography using a S-200 Sephacryl column. Apotransferrin was prepared by incubating diferric transferrin with desferrioxamine, which was removed afterwards by ultrafiltration. Labeling of apotransferrin with ^{55}Fe ferrous sulfate was achieved in the presence of 0.05 M Na bicarbonate and 0.1 M Tris/HCl. The transferrin solution was brought to the point of saturation by adding cold ferrous sulfate. Ferritin was purified from rat liver by heating at 75°C for 10 minutes, ammonium sulfate precipitation, gel chromatography using a Sepharose 6B column, and ultracentrifugation (3). In vitro tagging of ferritin was achieved by adding $^{59}FeSO_4$ to a normal rat ferritin preparation in 0.1 M Hepes buffer (pH 7.5). The labelled ferritin was purified by ion-exchange chromatography using a DEAE Sephacel column (3). Male Sprague-Dawley rats 8 to 10 weeks of age were used in most experiments. Animals to be made iron deficient were placed at 4 weeks on a diet containing 4-8 mg of iron/Kg, or alternatively were given an iron deficient diet at 6 weeks of age and bled 2-3 ml 4 to 5 times over two weeks to accelerate iron depletion. Female rats to be iron-loaded were placed on a 1% carbonyl iron diet for > 1 year with occasional reintroduction of a normal diet to maintain animals in good health. In order to suppress erythropoiesis, rats were transfused with 3 ml packed red cells approximately twice a week to maintain an hematocrit of 60-65 %, and ^{59}Fe ferritin was injected on the 5th day after the first transfusion. Hemolytic anemia was produced by the intraperitoneal injection of 50 mg acetylphenylhydrazine per Kg body weight, and ^{59}Fe ferritin was injected 5 days later. Inflammation was produced by injecting 0.25 ml turpentine oil in each thigh, and ^{59}Fe ferritin was injected 16 hours later.
Plasma was drawn from a rat prepared in the same way as the experimental animals and labelled by adding 0.5 µCi ^{55}Fe as $FeSO_4$ in 0.9% saline (pH 2) to 0.5 ml plasma. The rate of radioiron disappearance (t-1/2) was determined and the plasma iron turnover (PIT) was calculated in µg/Kg/day:

$$PIT = \frac{SeFe\ (\mu g/dl) \times (100 - Hct \times 0.92) \times 7}{t\text{-}1/2\ (min)}$$

Release of storage iron by the hepatocyte (i.e. late release) was studied over a period of 14 days. The cumulative incorporation of ^{59}Fe (injected as ferritin) and ^{55}Fe (injected as diferric transferrin) in the red cell mass was determined over a 2 wk period. The ratio of ^{59}Fe to ^{55}Fe activity in the red cell mass at any time represents the total fraction of ^{59}Fe released by the hepatocyte (total release at time t = early release + late release at time t). When the complement of this ratio was plotted against time on a semilogarithmic paper, late release appeared to be a single exponential curve. The proportion of the injected dose released in the early and late phases was obtained by extrapolating the curve to zero time. The t-1/2 of the curve, determined by least squares, represents the speed of late release and was expressed as % of the dose/day. Formal proof of this method has been published (5). Labelled compounds were injected into a tail vein and blood samples were obtained from the tail vein on the opposite side. At the end of each experiment, rats were exsanguinated from the abdominal aorta, and perfused with 20-30 ml saline. Whole blood, plasma, and liver activities were counted. Total plasma and total red cell activities were calculated from the hematocrit and assuming a blood volume of 0.07 ml/g body weight. Non-heme iron was extracted from rat livers and iron was determined by Proton-Induced X-ray Emission (PIXE). Assuming that ferritin ^{59}Fe uniformly labelled non-heme iron stores in the hepatocyte, the amount of iron released daily by the hepatic parenchyma was calculated as the product of radioiron release (%/day) and hepatocyte iron stores (µg).

Results

Measurements of the PIT and late release are presented in table 1.

Table 1. Plasma iron turnover and storage iron release.

Group	PIT (μg/Kg/d)	LR (%)	LR (%/day)	LR (μg/Kg/d)
Normal males	1815±381	78±8	6.3±0.9	207±48
Normal females	1612±202	91±2	3.7±0.3	347±32
Iron deficiency	738±3	38±3	17.0±4.6	79±26
Iron overload	1395±316	96±1	1.6±0.5	1584±650
Phenylhydrazine	6802±1961	56±14	14.9±6.2	331±75
Hypertransfusion	654±222	96±1	0.9±0.1	81±3
Inflammation	1322±60	93±2	6.4±2.7	193±96

Normal males released 17.5% of the ferritin ^{59}Fe in the early phase. Thereafter, 6.3% of the residual hepatocyte radioiron was released daily, representing a total of 207 μg Fe/Kg/day, i.e. about 11% of the PIT. Normal females had a slower rate of late release. However, as their iron stores were consistently higher, the contribution of hepatocyte iron to the PIT was found to be higher than in males (21%). Iron turnover in the hepatocyte of iron deficient rats is extremely rapid (56% ER and thereafter 17%/day of LR) but involves relatively small amounts of iron, a little more than 10% of the PIT. By contrast, iron overloaded rats had a turnover about 10 times slower (6% ER and 1.6%/day of LR), amounting to a calculated value of 1584 μg/kg/day, which is an overestimation. Rats treated with phenylhydrazine had an increased amount and rate of iron turnover in the hepatocyte. However, the participation of the parenchymal cell in the PIT was rather small (5%). On the other hand, hypertransfused rats had a slow turnover involving small amounts of iron which represented 12% of the PIT. Inflammation reduced considerably the early release of radioiron. However, release of hepatocyte iron in the following days was normal. There was a very significant correlation between the percent ER and the rate of LR ($R=0.90$, $P<0.001$). Iron release was inversely related to the hematocrit and positively to the PIT. Plasma iron and radioiron t-1/2, as well as hepatic iron, correlated with the rate but not the absolute amount of iron release.

Discussion

After uptake by the hepatocyte, ferritin is catabolized and its iron enters an intermediate pool before being incorporated into stores (3,4). Because it is based on the assumptions that ferritin ^{59}Fe labels all hepatocyte iron stores uniformly and that all non-heme iron is mobilizable in the same way, the calculation of the amount of hepatocyte iron released daily to the plasma is less precise than the measurement of radioiron release. The rate of radioiron release was considerably increased in iron deficient rats but 6.2% of the ferritin ^{59}Fe remained in the liver after 2 wks, indicating that not all hepatic iron is mobilizable by increased marrow requirements. Another pool of non-exchangeable iron, presumably hemosiderin, was identified in iron overloaded rats in which the calculation of iron released by the liver was grossly overestimated. In iron overload, hemosiderin iron represents a large proportion of hepatic iron, which is not exchangeable at least over the time and in the conditions of the present study. Nevertheless, the values obtained in non-deficient and non-overloaded rats are reasonable. Iron release in normal animals amounted to 207 µg Fe/Kg/day, which represents 11.4% of the PIT. These findings are similar to values reported by others (4). This is also in accordance with hepatocyte iron uptake, as approximately 10% of transferrin radioiron is localized in the liver after the early release phase. With iron deficient and iron overloaded rats not included in the analysis, there was a good correlation between hepatic storage iron and the rate of LR, but not with the absolute amount of iron released daily. This observation confirms that ^{59}Fe and cold iron are released in a similar fashion. Inflammation produced a significant decrease in the rate of early radioiron release. However, late release was not affected. These observations confirm a previous study in which the effect of turpentine-induced inflammation on storage iron mobilization was also transient and was no longer apparent beyond 48 hrs (6). However, human studies have shown that chronic inflammation is associated with a blockade of iron release from RE stores (7). The effect of erythropoiesis on hepatocyte iron mobilization was examined in phenylhydrazine-injected and hypertransfused rats. Iron release was considerably modified in parallel to changes in erythropoiesis. The amount of iron released by the hepatic parenchyma seemed appropriate (12% of the PIT) in hypertransfused rats. However, it accounted for only 4.9% of the PIT in phenylhydrazine-injected animals, indicating that recycling of red cell iron through the reticuloendothelial system contributed more than hepatocyte iron mobilization to increased marrow iron requirements.

Acknowledgement

Supported in part by NIH (USA) Grant No. HL06242 and by IISN (Belgium) Grant No. 4450865. Y. Beguin was a Research Assistant of the FNRS (Belgium) and supported in part by a Fogarty International Research Fellowship from the NIH. G. Weber is a Research Associate of the FNRS.

References

1. Morgan, E.H., E. Baker. 1986. Fed.Proc. **45**, 2810.

2. Kim, B.-K., H.A. Huebers, C.A. Finch. 1987. Am.J.Hematol. **24**, 277.

3. Pippard, M.J., D.K. Johnson, C.A. Finch. 1982. Brit.J.Haematol. **52**, 211.

4. Unger, A., C. Hershko. 1974. Brit.J.Haematol. **28**, 169.

5. Fillet, G., J.D. Cook, C.A. Finch. 1974. J.Clin.Invest. **53**, 1527.

6. Hershko, C., J.D. Cook, C.A. Finch. 1974. Brit.J.Haematol. **28**, 67.

7. Dresch, C., Y. Najean. 1972. Eur.J.Clin.Biol.Res. **17**, 930.

NEW DATA ON THE HYPOTHESIS OF THE BRAIN PARTICIPATION IN IRON HOMEOSTASIS

B. Ribas, J.F. Pelayo, N. L. Rodrigues.
Institute of Biochemistry, CSIC-Complutense University, Faculty of Pharmacy, 28040-Madrid

Abstract

Ferropenic posthemorrhagic rats were used when hematocrite reached 30% and hemoglobin was lower than 8g/100ml. The brain is homogenized (1:4,w:v), sonicated, centrifuged and the supernatant, is incubated with 2µCi ^{59}Fe for a period of 20h. The subsequent elution of two peaks, supposedly ferritin and metallothionein marked with ^{59}Fe justifies the hypothesis before exposed of the brain participation in iron homeostasis.

Introduction

Metallothionein, a well known low relative molecular weight protein, with approximately 6000 D, 30% cystein and without aromatic aminoacid residues, associates between 5 to 11 atoms of metals (1,2). Its signification in metabolism is not well understood, it is suggested that it could participate as regulatory protein (3). Some data are published before, comparing the increase of transferrin and metallothionein in the intestinal mucosa of ferropenic rats in the brain (5,6).

The object of this work is to furnish new data which justify the hypothesis of the brain participation in iron homeostasis (7,8), exposed after the experimental results on the partici-

Trace Element Analytical Chemistry in Medicine and Biology

pation of brain metallothionein in the anemic ferropenic disease (9,10). Patients with toxic, psychiatric, nutritional, genetic syndromes and others with unknown etiology some of which improve with empiric iron therapy, suffer from neurological symptomatology. In children, some of anemic syndromes involve psychological and behavioural changes (11-13), which could justify also the connexion of the molecular pathology of iron with the brain.

Material and Methods

Male Wistar rats of 200±10 g were used distributed in control and anemic ferropenic groups, the last submitted to iron deficient diet and to two blood extractions a week. Anemic condition established, animals are killed, brain and systemic organs dissected, homogenized, sonicated and centrifuged. Protein concentration in the supernatant is determined (14) and it is incubated with 2 µCi ^{59}Fe during 20h in refrigerator 0ºC. An aliquot of 6mg protein is deposited on a Sephadex G-150 column (39x2.7 cm), buffer: (N,N-bis-2-hydroxyethyl-glycine)/(2-amine,2-methyl-1,3-propanediol), 0.05M, pH 7.8, fraction volume of 7 ml, flow rate of 8 ml/h. The elution of which gives two radioactive peaks of high and low molecular weight, compared to authentic compounds used as reference molecules: ferritin, transferrin, ribonuclease and tryptophan.

Anemic condition is established after the physiological parameters of animals reached to: hemoglobin 6g/100ml and hematocrite under 30%. Rat control values were respectively of 15g/100ml and 45%. Out of routine are also determined proteinemia, total iron binding capacity and iron.

Results

The classical isolation and purification of proteins of the brain extracts incubated here "in vitro" with ^{59}Fe and eluted from the Sephadex G-150 give similar elution profiles as those of before "in vivo" treated animals injected through the intracerebroventricular way (7,8). Figure 1 shows two peaks of high and low molecular weight proteins, compared with the calibration profile of authentic compounds showed in figure 2: ferritin, transferrin, ribonuclease and tryptophan. Two peaks are eluted from brain cytosol in both, anemic and control animals, quantitatively higher in the first group of animals with anemic condition (figure 1).

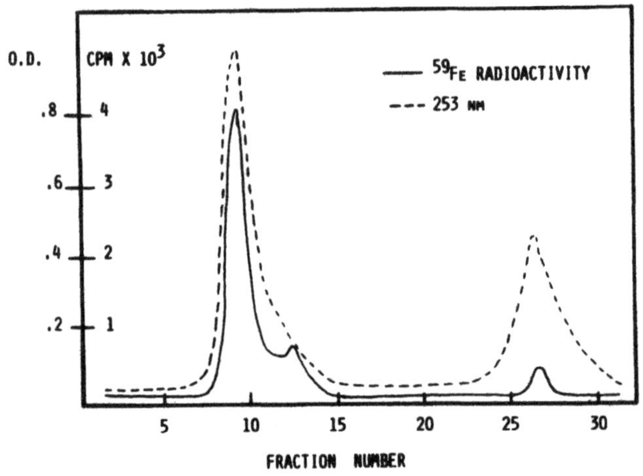

Figure 1. Sephadex G-150 elution profile of brain cytosol of anemic rats after "in vitro" incubation with ^{59}Fe during 20h at 0ºC with the same conditions as expressed in Material and Methods.

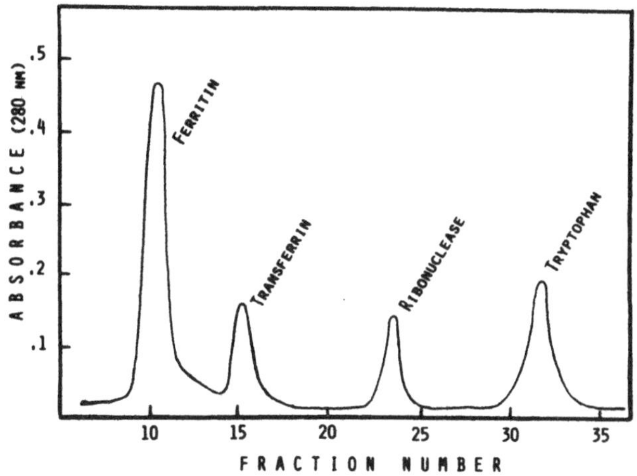

Figure 2. Elution profile of authentic molecules with known Mr used for calibration of the column 39x2.7 cm, buffer as in Material and Methods.

Table 1. Neurological Syndromes Connected with Anemia, Iron Deficiency or Thesaurusmosis of Iron.

---

a). Syndromes with anemic symptomatology, some of which of unknown ethiology improve with empiric iron therapy
   1. Neurotoxic Syndrome
   2. Psychoanemic Syndrome
   3. Restless legs (Ekbom) Syndrome
   4. Plummer-Vinson (Kelly-Patterson) Syndrome

b). Syndromes with iron overload
   1. Haemochromatosis
   2. Zellweger Syndrome
   3. Ceelen disease (Idiopatic Pulmonar Hemosiderosis)
   4. Goodpasture Syndrome (Pulmonar Haemosiderosis with Segmentary Necrotic Glomerulonephritis)

Discussion

Previous results showing the heterogeneity and specificity of ^{59}Fe accumulation in cerebral structures of anemic and control rats (10,15), lead us to study the possible isolation of iron proteins in the brain. Others works deal with the detection of proteins in other systemic tissues, principally from the intestinal mucosa of rat are detected several proteins among this are signalized ferritin, transferrin (16,17), a mucoprotein of high molecular weight (18) a microvillar iron binding glycoprotein (19) and other non identified iron binding substances (17,18,20,21). The high molecular weight protein detected in our study is eluted with the similar physico-chemical characteristics as ferritin, and this protein could be of greater significance as actually assumed, for its new projection in the brain on the mechanism of regulation of iron homeostasis, like transferrin, receptors of which have a neuropeptide like pattern and relationship to iron distribution (22). ferritin was signalized as limiting factor in iron absorption, and also in steady state with transferrin in the intestinal mucosa (23). This protein is a indicator of total body iron storage, and its presence in brain should not be strange, participating in iron homeostasis through the neurochemical changes in the brain, and being more active in some pathologic conditions. Metallothionein was suggested to be involved in iron metabolism as a regulatory protein (3), the concentration of which increases concomitantly in the intestinal mucosa and in the brain of anemic ferropenic rats (10,15). The peak of low molecular weight protein, marked with ^{59}Fe detected in our experiment, shows analogous characteristics with metallothionein. The role of which could be significant as afore mentioned for the brain participation in the mechanism of regulation of iron metabolism. The label of metallothionein with ^{59}Fe shows the signification of this protein in iron homeostasis as before proposed (7,8,10).

One of the roles of metallothionein in its participation of metabolism could be the reverse chelation and releasing of metals. Iron as other essential metals in the human body, is related to the extent of their dietary absorption by the small intestine.
Cadmium induces anemia, blocking the metallothionein in the intestinal mucosa cells, which could not participate in iron absorption. In this sense, it could be hypothetized that metallothionein could participate in the intestinal absorption of essential bivalent cations.
The neuropathology of iron is little understood, it is known that iron is necessary for a normal nervous function (24), its concentration as for other trace elements varies specifica-lly in diferent regions of the human brain in relation to the associated functions. In human brain iron increases during the first decade, and remains constant until it decreases at the age of 74 (24). Several Syndromes connected with iron pathology, with deficiency and with thesaurusmosis, could be connected with a disturbance of iron metabolism as before suggested (10) table 1. Patients with neurotoxic and neuroane-mic Syndromes suffer from anemia and low iron absorption, probably the first as the result of high liposolubility of certain Xenobiotics as organic solvents, DDT and others, affec-ting the normal neurochemistry of the CNS, sulphur metabolism, oxidoreductive processes and mieloproliferation. We propose other Syndromes with unknown ethiology which improve with empiric iron therapy like the Restless legs (Ekbom) Syndrome (25), and Plummer-Vinson (Kelly-Patterson) Syndrome. In both cases the neurochemistry of the molecular pathology of iron homeostasis could be suggested. Psychiatric, nutritional and genetic Syndromes of iron, some of which adscribed to iron overload, show neuropathy and or convulsions. Such distur-bances could be in connexion with the metallothionein polymor-phism, each isoform originating from a gene from which theore-tically could depend a function or a pathological defect.

References

1. Kägi, J.H.R., M. Nordberg, eds. Metallothionein, 1st Intern. Meeting on Metallothionein, Zurich 1978, Birkhause Verlag, Basel-Boston-Stuttgart.

2. Kägi, J.H.R., Y. Kojima: 2nd Intern. Meeting on Metallothionein, Zurich 1985, University Zurich, Switzerland.

3. Ribas, B., I. Basagoiti, M.A. Brenes, M.P. Iniesta, M.I. Sánchez Reus. In: Trace Elements, 5, 413-420, 1986, (M. Anke, W. Bauman, H. Braunlich, Chr. Brückner, B. Groppel, eds) A.W.P. Friedrich Schiller Universität Jena, GDR.

4. Iniesta, M.P., C. Rubio, B. Ribas. 2nd Intern. Meeting on Metallothionein, Zurich, 1985, Universität Zurich Switzeland.

5. Ribas, B. 1981. An Med. 67, 941-953.

6. Ribas, B. In: Trace Elements Anal. Chem. Med. Biol., 1985, vol. 2, 585-597. (P. Brätter and P. Schramel, eds) Walter de Gruyter Co., Berlín-New York.

7. Ribas, B., J.F. Pelayo. 1988. An. Real Acad. Farm. 54 (in press).

8. Ribas, B., J.F. Pelayo. 1988. An. Med. Intern. 5 (in press).

9. Ribas, B., M.A. Brenes, I. Basagoiti, M.I. Sánchez Reus. 1986. Trace Elements 5, 405-412. (M. Anke, W. Baumann, H. Bräunlich, Chr. Brückner und M. Groppel eds.), A.W.P. Friedrich Schiller Universität Jena, GDR.

10. Ribas, B., M.A. Brenes, F.J. de Pascual, J. del Rio, M.I. Sánchez Reus. 1987. Trace Elements Anal. Chem. Med. Biol. vol. 4, 317-324 (P. Brätter and P. Schramel, eds.) Walter de Gruyter Co., Berlín-New York.

11. Tucker, D.M., R.A. Swenson, H.H. Sandstead. 1983. In: Neurobiology of the Trace Elements, Chapter 8, vol. 1, 269-291.

12. Tucker, D.M., H.H. Sandstead. 1982. In: Iron deficiency: Brain Biochemistry and Behaviour (E. Pollit and R.L. Leibel, eds.). Rasen Press, New York, 161-181.

13. Lofoff, B., G. Brittenham, F.E. Viteri, J.J. Urrutia. 1982. In: Iron deficiency, Brain Biochemistry and Behavior. Raven Press, New York, (E. Pollit, R. Leibel, eds.).

14. Lowry, O.H., N.J. Rosenbrough, A.L. Farr, R.J. Randall. 1951. J. Biol. Chem. 193, 265-275.

15. Ribas, B., M.A. Brenes, F.J. de Pascual, J. del Rio, N. Lobato. 1986. An. Real Acad. Farm.52, 341-348.

16. Pearson, W.N., M.B. Reich. 1969. J. Nutr. 99, 137-140.

17. Huebers, H., E. Huebers, W. Forth, W. Rummel. 1971. Life Sci. 10, Part. I, 1141-1148.

18. Worwood, M., A. Jacobs. 1971. Life Sci. 10, Part I, 1363-1373.

19. Odonell, M.W., T.M. Cox. 1982. Biochem. J. 202, 107-115.

20. Boulard, M., M. Selin, Y.C. Najear. 1972. Proc. Soc. Exptl. Biol. Med. 139, 1379-1384.

21. Johnson, G., P. Jacobs, L.R. Purves. 1983. J. Clin. Invest. 71, 1467-1476.

22. Hill, J.M., R. Michael, R.J. Weber, C.B. Pert. 1985. Proc. Natl. Acad. Sci. USA, 82, 4553-4557.

23. Savin, M.A., J.D. Cook. 1980. Blood. 56, 1029-1035.

24. Höck, A., U. Demmel, H. Schicha, K. Kasperek, Feinendegen. 1975. Brain 98, 49-64.

25. Ekbom, K.A. 1960. Neurol., 10, 868-873.

STRAIN-DEPENDENT DIFFERENCES IN COPPER ACCUMULATION AND EXCRETION UPON DIETARY COPPER LOADING IN THE RAT

Janine H.H. Verheesen, H. Nederbragt.

Department of Pathology, Faculty of Veterinary Medicine, University of Utrecht, Yalelaan 1, 3584 CL Utrecht, the Netherlands.

Abstract

After dietary administration of 1500 mg/kg Cu, considerable strain differences can be found in Brown Norway and Fischer rats in the handling of the high dose of Cu.

Introduction

Copper (Cu) is a trace element, necessary for the maintenance of many cellular functions. Homeostasis is maintained by excretion of Cu in the bile. Incorporation of Cu in hepatic metallothionein (MT) is thought to protect the organism against Cu toxicity. MT is a low molecular weight, cysteine rich protein that binds metals and is assumed to play a role in metabolism, storage and detoxification of metals (1).
Former investigations at our institute have shown that following dietary administration of Cu, Fischer (F) rats show a more severe liver damage than Brown Norway (BN) rats, notwithstanding a lower liver Cu concentration (2).

Trace Element Analytical Chemistry in Medicine and Biology

This study was performed to investigate whether there are strain differences in Cu distribution throughout the body after dietary Cu loading.

Materials and Methods

Inbred male F and BN rats were given commercial food to which Cu was added (1500 mg/kg).

Once a week the rats were placed in metabolism cages and urine was collected during 24 hours.

At week 4 bile cannulation and bile sampling was performed as described before (3) except that the body temperature was maintained by placing the rats on a heating pad. Following the cannulation the rats were killed by exsanguination by aorta puncture with heparinized syringes. The blood was centrifuged to obtain plasma.

Several organs were excised and used for Cu determination. Each time the same part of liver, kidney and jejunum was used for centrifugal fractionation. A portion of the supernatant of the homogenate (60 min., 105,000 g) was applied to a Superose-12 column (30 x 1.0 cm) and eluted at room temperature with a Tris-acetate buffer (0.025 M, pH 7.4) with 0.02% sodiumazide and 0.01 M mercaptoethanol by fast protein liquid chromatography (FPLC). The elution rate was 30 ml/h. Fractions of 0.5 ml

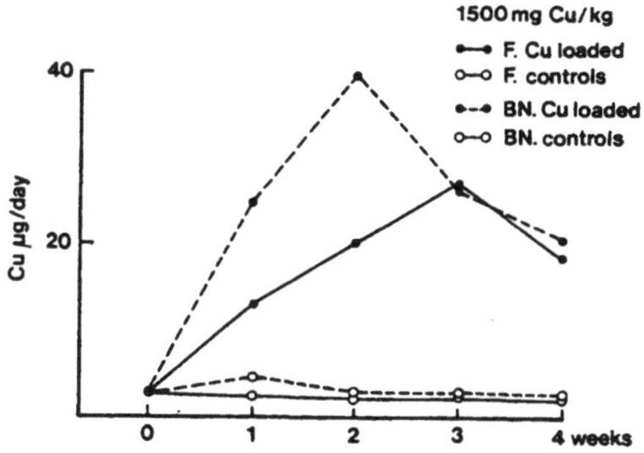

Fig.1. Urinary Cu excretion after dietary administration of 1500 mg Cu/kg in Brown Norway (BN) and Fischer (F) rats.

were collected and monitored directly for Cu.
For Cu determination samples of bile and pplasma were diluted 10x with 6% butanol; samples of the organs were digested with nitric acid under high pressure; urine and gel filtration fractions were used directly. Cu was determined by atomic absorption spectrophotometry, using an air-acetylene flame.

Results

The urinary Cu excretion of BN and F rats is shown in fig. 1. From week 1 to 3 BN rats excreted more Cu in the urine than F rats, but after 3 weeks Cu excretion in the urine was similar in BN and F rats. Results of Cu determinations of other body fluids and some tissues are

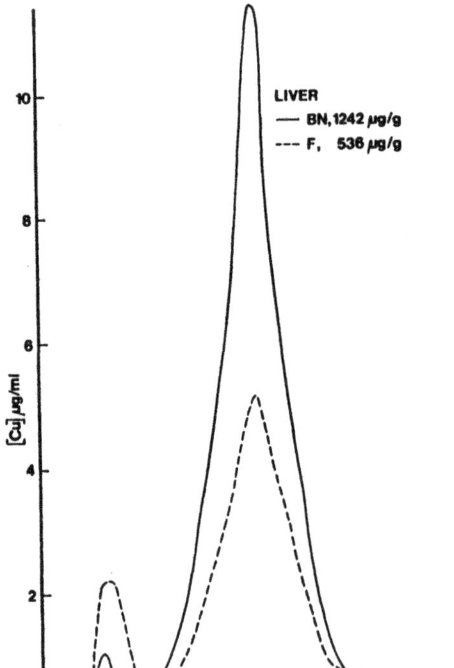

Fig.2. Gel filtration profile of Cu-containing proteins in the cytosol of the liver of Brown Norway (BN) and Fischer (F) rats after dietary administration of 1500 mg Cu/kg.

Table 1. Cu concentrations of several tissues in Brown Norway (BN) and Fischer (F) rats, after dietary administration of 1500 mg Cu/kg.

	CONTROLS		CU-LOADED	
	BN	F	BN	F
PLASMA µg/ml	0.91 ± 0.04	0.89 ± 0.04	1.60 ± 0.14	1.46 ± 0.15
BILE µg/g liver/90 min.	0.14 ± 0.02	0.09 ± 0.02	7.15 ± 0.61	6.52 ± 0.17
LIVER µg/g	4.6 ± 0.6	5.4 ± 0.6	1242 ± 70	536 ± 45
KIDNEY µg/g	5.9 ± 0.6	8.3 ± 0.5	250 ± 65	344 ± 17
SPLEEN µg/g	1.2 ± 0.1	1.4 ± 0.2	218 ± 60	14 ± 2
BONE µg/g	2.8 ± 0.3	2.4 ± 0.1	9.6 ± 2.5	3.4 ± 0.1
LUNGS µg/g	1.7 ± 0.4	1.6 ± 0.0	14.1 ± 1.2	4.8 ± 1.5
DUODENUM µg/g	2.1 ± 0.2	1.2 ± 0.1	14.9 ± 2.7	5.3 ± 0.8
JEJUNUM µg/g	1.3 ± 0.2	0.9 ± 0.3	34.3 ± 2.6	14.6 ± 4.2

values are given in mean ± SEM

given in table 1. The Cu concentration of plasma and bile was equal in both strains. BN rats accumulated more Cu than F rats in several tissues. F rats accumulated more Cu than BN rats in the kidneys only.
Gel filtration of the cytosol of the liver revealed two Cu-binding protein peaks; peak 1 (Ve/Vo=1.0) and peak 2 (Ve/Vo=1.8) (fig.2). The second peak is assumed to be MT, because of its lack of absorbance at 280 nm and its elution volume. The amount of Cu associated with peak 2, relative to total liver Cu, was the same in both strains of rats. The amount of Cu associated with peak 1 was higher in F rats than in BN rats. The same gel filtration profiles were obtained for the cytosol of kidney and jejunum (results not shown). In these tissues the Cu level of both peak 1 and peak 2 was higher in F rats.

Discussion

Investigations at our institute have shown that comparing the effects of the dietary administration of Cu to BN and F rats offers a suitable model for the investigation of Cu toxicity.
The study described here has been performed to obtain a more complete picture of the existing strain differences in Cu distribution after high dietary Cu loading.
Our results show that BN rats accumulate more Cu in the body than F rats. A metabolism study of Cu in rats was already performed as early as 1929 by Lindow et al. (4). They found after dietary administration of Cu, a Cu accumulation in the same tissues as we did, although they did not investigate Cu accumulation in the intestinal tract. We also found a remarkable high accumulation of Cu in the spleen, which has not been described elsewhere.
Cu excretion in the urine was similar in both strains of rats. At the beginning of the study BN rats excreted more Cu in the urine than F rats. However, the urinary Cu excretion is of minor importance, for the biliary Cu excretion is 20 times higher. The major pathway for Cu excretion is via

the bile through fecal excretion. Cu excretion in the bile was similar in both strains of rats.

The higher Cu accumulation in the body of BN rats may be the result of a higher Cu absorption (duodenum and jejunum). It is thought that in most species Cu absorption takes place in the duodenum and jejunum, but in rats Cu is mainly absorbed in the duodenum (5).

Our results do not confirm the hypothesis that association of Cu with hepatic MT protects the animal against liver damage, for the Cu level of peak 2 (MT) relative to total liver Cu is not lower in F rats than in BN rats. The most striking difference between the two strains is that F rats incorporate more Cu in the HMW protein fraction of the cytosol of liver, kidney and jejunum. It has been proposed that Cu bound to this protein in the jejunum is available for serosal transfer, whereas Cu associated with MT is more tightly bound and is not available for transfer in the body (6). It is also possible that the Cu bound to the HMW protein fraction plays a role in Cu toxicity in the liver.

Acknowledgements-The investigations were supported by the Foundation for Fundamental Biological Research (BION), which is subsidized by the Netherlands Organization for the Advancement of Pure Research (NWO).

References

1. Cousins, R.J. 1985. Physiol.Rev. 65, 238-309.

2. Nederbragt, H., M. Gijbels and A.J. Lagerwerf. 1987. In: Trace Element - Analytical Chemistry in Medicine and Biology, Vol. 4 (P. Braetter and P. Schramel, eds.). Walter de Gruyter, Berlin . New York. pp. 589-596.

3. Nederbragt, H. 1985. Comp.Biochem.Physiol. 81C, 425-431.

4. Lindow, C.W., W.H. Peterson and H. Steenbock. 1929. J.Biol.Chem. 84, 419-436.

5. Owen, C.A. 1964. Am.J.Physiol. 207, 446-448.

6. Fischer, P.W.F., A. Giroux and M.R. L'Abbe. 1983. J.Nutr. 113, 462-469.

METALLOTHIONEIN INDUCTION BY METHOTREXATE IN LIVER AND INTESTINAL MUCOSA

M.I. Sánchez Reus, M.P. Iniesta, B. Ribas
Institute of Biochemistry, CSIC-Complutense University, Faculty of Pharmacy, 28040-Madrid, Spain.

Abstract

In human clinical treatment, chimiotherapy with Methotrexate (MTX) is one of the most used in Oncology. It is interesting to know its influence with the diet and the mineral distribution in humans.

Male Wistar rats of 200±10 g were used. Thymusses of rats treated with 6 doses of MTX, through intraperitoneal route 1mg/Kg/day, have a weight of 0.170±0.03 g/100 g rat, and 0.277±0.046 the controls ($p < 0.001$).

The liver of MTX treated rats show 37.75±5.20 µg Cu/g fresh tissue (controls 41.88±2.63 µg). 16.75±2.75 µg Zn/g fresh tissue (control 20.81±1.99 $p < 0.02$) and 67.12±3.40 µg Fe/g fresh tissue (control 61.28±11.67). Metallothionein concentrations in the intestinal mucosa are established as 2.72±0.25 µg MT/g fresh tissue for controls and 3.32±0.41 µg MT/g fresh tissue for MTX treated animals. Liver MT concentration is 14.74±2.75 µg MT/g fresh tissue for controls and of 36.57±4.12 µg MT/g fresh tissue for MTX treated animals ($p < 0.02$). Hematocrite and hemoglobin are both decreased in MTX treated animals compared to control ones.

Trace Element Analytical Chemistry in Medicine and Biology

Introduction

The chimiotherapy of methotrexate alone but generally in combination with other pharmaceutical agents used to an increasingly great extent by humans, is the clearest indication that it is regarded as an effective and established treatment in Oncology.
The Chemistry of MTX is studied with MT (1), because both molecules can interact with each other as also MTX with other exogenous molecules used in polypharmacy, as planified in oncological therapy. Different parts of the molecules can be associated interacting with different physiological and biochemical consequences, so that in single drug one has the elements of polypharmacy. Methotrexate (4-amino,4-deoxi-N^{10}-methylpteroil-L-glutamic acid), is used moreover as anticarcinogenic agent, in Immunology and surgical situations such as transplantation (2). As structural analogue of folic acid, condenses chemically together with the enzyme molecule dihydrofolate reductase (3,4), inhibiting it and the DNA synthesis (5,6), provoking anemia. The object of this work is to study the interaction of MTX with MT (1) and the concentrations of Zn, Cu and Fe in liver and intestinal mucosa, in relation to the induced anemia.

Material and Methods

To male Wistar rats of 200±10 g were administered 1mg MTX/Kg during 6 days through the intraperitoneal route. The induction by MTX of the anemic condition is controled, bleeding the animals by cardiac puncture under ether anesthesia, with the evolution of hemoglobin, hematocrite and proteinemia with the classical analytical methods.

Protein is determined in liver and intestinal mucosa homogenates (8). Metallothionein is determined in both organs with the cadmium saturation and hemoglobin method (7), using atomic absorption spectrophotometer Beckman mod. 1248. Zinc, copper and iron are determined in above mentioned tissues with the same aparate. All the animals are fed with standard food and water "ad libitum" as in other experiments (1).

Results

Hemoglobin in MTX treated animals was of 9g/100 ml blood and of 13.5g/100 ml in control ones. Hematocrite value were for MTX treated rats of 35% and of 46% in control ones (Table 1). Proteinemia in MTX treated rats was of 5g/100 ml blood and of 5.5g/100 ml in control ones, as signalized in figure 1A. This physiological parameter is maintained constant but with a little diminution (table 1). Protein concentration in homogenates of rats treated with MTX diminishes in both, liver and intestinal mucosa compared to the controls as it is shown in figure 1B. The weight of thymus is strongly diminished from control 0.277 ± 0.04g/100 g rat to the MTX treated animals 0.170 ± 0.03 g/100 g rat, which difference is statistically significant ($p < 0.001$). The weight of lungs shows a little diminution from 0.570 ± 0.04 to 0.549 ± 0.01 g/100 g total animal weight. Table 1 shows different physiological parameters of control and MTX treated rats.

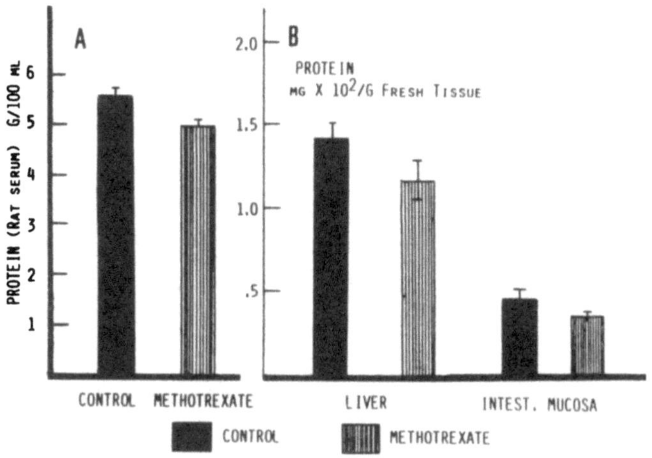

Figure 1.- 1A: Proteinemia of rats treated with 1mg MTX/Kg during 6 days compared to control animals.
1B: Protein concentration in liver and intestinal mucosa homogenates of rats treated with MTX compared to control animals.

Table 1.- Physiological parameters of rats treated during 6 days with 1mg MTX/kg rat/day by intraperitoneal way.

Parameter	Control	MTX treatment
Hematocrite	46+4%	35+2%
Hemoglobin	13+0,6g/100ml	9+0.8g/100ml
Proteinemia	5.5g/100ml	5g/100ml
Zinc in liver	20.8µg/g.f.t.	16.7µg/g.f.t.
Copper in liver	41.8µg/g.f.t.	37.7µg/g.f.t.
Iron in liver	61.8µg/g.f.t.	67.1µg/g.f.t.
Liver MT	14.74µg/g.f.t.	36.57µg/g.f.t.
Intes. mucosa MT	2.72µg/g.f.t.	3.32µg/g.f.t.

g.f.t.: gramme fresh tissue
MTX: Methotrexate
MT: Metallothionein

Discussion

The induction of the synthesis of MT after the administration of MTX to rats by intraperitoneal route suggested the formation of the complex MT-MTX-Me (1). In this case, the condensation of n molecules of MTX with one molecule of MT goes through the release of several bivalent cations from MT. This fact could induce to minor concentration of metals in the tissues which MTX accumulates (9): liver (10), muscle (11) and serum (12), and MT principally in liver and kidney are induced or its concentrations are higher than in other tissues (13).
Thymus is another tissue the weight and volume of which diminishes in a statistically significant way in MTX treated rats as already demonstrated by cadmium (15), another toxic agent for the thymic gland. The toxic action of MTX is already known, inducing human anergic condition, through the inhibition of the defense mechanisms, in which thymus plays an important role (16). Nevertheless zinc, an essential trace element, localized in the thymic cellular reticulum (17), is inhibited competitively and sustituted by cadmium in its binding sites, inducing a great regression of the thymus cortical area (15) a mechanism in which takes part corticol from suprarrhenal gland. MT in the thymus is already characterized (18) and induces two proteins (19), but in human systemic organs as liver, its polymorphysm is established in six isoforms (20).
The mechanism of action of MTX shows a great spectrum of scientific options (2), agonists and antagonists are also known but there are metals cited neither affecting the action or efficiency of MTX as anticarcinogenic agent, nor the participating with the MT in carcinogenic mechanism, exception in before published results (1). MT synthesis is induced by both MTX and cadmium. Here the increase

of MT synthesis in liver by MTX is statistically significant, and in the intestinal mucosa is also positiv but it is not significant, compared to the control values (21,22). In this work Zn, Cu and Fe are determined, in order to establish the possibility of forming the complex before proposed MT-MTX-Me (1). Excepted for Fe, which values are greater in liver of animals treated with MTX, the concentrations of Zn and Cu in liver of MTX treated animals are higher than in control ones. The lower levels of Zn and Cu could explain the formation of the aforementioned complex MT-MTX-Me, which release both metals from the MT molecule. The difference between Fe which is lower, with Zn and Cu could be consequent with the anemic disease, which ii is induced the iron absortion, being Fe levels in blood and other reticuloendotelial tissues greater than in control animals. Cystein and transferrin in anemic rats, are higher than in control animals (1). Transferrin in rat intestinal mucosa with anemic disease is increased as it is well known, compared to the control animals. Metallothionein is determined before as cystein (1) by the method of Zahler and Clelland (23) and here by the saturation with cadmium (7) in both experimental results, MT is increased after MTX treatment inducing anemia (21,22). MT is considered for Fe as also for Zn and Cu, an iron binding (24,25) and homeostatic regulation protein (26). Methotrexate could condenses with native MT and induce toi the partial depletion of Zn and Cu, which could be one of the possible ways to inhibit the biological activity in living cells, as in the case for tumor growth.

It could be interesting to supplement the human diet, during the treatment with MTX, with Zinc and Copper, and to establish the absorption and distribution of other essential trace metals of the diet in several systemic organs. The evaluation of the effect of several metals on the development of tumors is also unknown as also unknown the effect of anticarcinogenic agents on the levels of essential cations in systemic organs.

References

1. González Barón, M., M.P. Iniesta, M.I. Sánchez Reus, B. Ribas, 1987, Toxicol. Env. Chem., 13, 161-170.

2. Schornagel, J.H., J.G. McVie, 1983, Academic Press London, 53-75.

3. Werkheiser, W.C. 1961. J. Biol. Chem. 236, 888.

3. Bleyer, W.A. 1978. Cancer 36.

5. Godman, I.D. 1975. Cancer Chemother. Rep. 6, 51.

6. Waltham, M.C., J.V. Holland, P.F. Nixon, D.I. Winzor. 1988. Biochem. Pharmacol. 37, 88.

7. Heilmaier, H.E., P. Schramel, K.H. Summer. 1984. vol.3, (P. Brätter and P. Schramel, eds) Walter de Gruyter Co, Berlín-New York, 669-676.

8. Lowry, O.H., N.J. Rosenbrough, A.L. Farr, R.J. Randall. 1951. J. Biol. Chem. 193, 265.

9. Pimm, M.V., J.E. Caten, J.A. Clegg, E. Jacobs, R.W. Baldwin. 1987. J. Pharm. Pharmacol. 39, 764.

10. Rademarker, M., J.A.N. Webb, D.G. Lowe, R.H. Meyrickthomas, J.D.T. Kirby, D.D. Munro. 1987. Brit. Med. Dermatol. 117, 311.

11. Klinov, S.V., N.A. Cheborateva, B.M. Sheiman, E.M. Birinberg, B.I. Kurganov, I.P. Rudakova. 1987. Bioorganicheskaya Khimiya 13, 908.

12. Williams, L.H., K.B. Udupa, D.A. Lipschitz. 1988. Exptl. Hematol. 16.

13. Kägi, J.H.R., M. Nordberg, eds. Metallothionein, 1st. Intern. Meeting on Metallothionein, Birkhauser Verlag, Basel-Boston-Stuttgart, 1979.

14. Mcbride, R.K., C. Harper, I.A. Siegel, 1987. J. Dental Res. 66, 1445.

15. Ribas, B., F.J. de Pascual, J. del Rio, M.I. Sãnchez Reus. 1987. Trace Element Anal. Chem. Med. Biol. vol. 4 (P. Brätter and P. Schramel, eds.) Walter de Gruyter Co., Berlĩn-New York, 325-333.

16. Blakley, B.R. 1985. Canad. J. Compar. Med. 49, 104-108.

17. Nabarra, B., S. Halpern, D. Kaiserlian, M. Dardenne. 1984. Cell. Tissue Res. 238, 209-212.

18. Olafson, R.W. 1985. Canad. J. Biochem. Cell. Biol. 63, 91-106.

19. Maytin, E.V., D.A. Young. 1983. J. Biol. Chem. 258, 12718-12722.

20. Hunziker, P.E., J.H.R. Kägi. 1985. Biochem. J. 231, 375-382.

21. Iniesta, M.P., M.C. Rubio, B. Ribas. 1985. Anal. Real Acad. Farm. 51, 357.

22. Iniesta, M.P., M.C. Rubio, B. Ribas. 1985. 2nd Intern. Meeting on Metallothionein, University Zürich, Switzerland.

23. Zahler, W.L., W.W. Clelland. 1968. J. Biol. Chem. 243, 716.

24. Ribas, B., M.A. Brenes, I. Basagoiti, M.I. Sãnchez Reus. Trace Elements 5, 405-412. 1986. (M. Anke, W. Baumann, H. Bräunlich, Chr. Brückner, M. Groppel, eds.) A.W.P. Friedrich Schiller Universität Jena, GDR.

25. Ribas, B., M.A. Brenes, F.J. de Pascual, J. del Rio, M.I. Sãnchez Reus. Trace Elements Anal. Chem. Med. Biol. vol. 4, 317-324, 1987 (P. Brätter, P. Schramel, eds.) Walter de Gruyter Co., Berlĩn-New York.

26. Ribas, B., I. Basagoiti, M.A. Brenes, M.P. Iniesta, M.I. Sãnchez Reus. 1986. (M. Anke, W. Baumann, H. Bräunlich, Chr. Brückner, B. Groppel, eds) A.W.P. Friedrich Schiller Universität Jena, GDR.

EFFECT OF MERCURY ON THE SUBCELLULAR DISTRIBUTION OF ENDOGENOUS COPPER AND ZINC AND THE PRESENCE OF Hg, Cu, Zn -METALLOTHIONEIN IN THE KIDNEY OF RATS EXPOSED TO MERCURY VAPOUR

M. Škreblin

Centre for Marine Research, Institute "R.Bošković", Rovinj, Yugoslavia

P. Stegnar, I. Kregar

"J. Stefan" Institute, "E. Kardelj" University, Ljubljana, Yugoslavia

Abstract

The subcellular distribution of Hg, Cu, and Zn between the soluble and particulate phase of rat kidney and their binding to metallothionein (MT) during three weeks of exposure to elementary mercury vapour at 0.5-0.8 mg Hg^0/m^3 in the Idrija mercury mine, Slovenia, were investigated in adult male Wistar rats. The composite kidney samples of two rats were used for analysis at each stage of the experiment. During the exposure period the increase of Hg in both the soluble and particulate fraction of kidney was followed by elevation in endogenous Cu until the 10th day of exposure; afterwards Hg and Cu in the soluble phase as well as Cu in the particulate phase remained relatively unchanged, while Hg in the particulate fraction continued to increase until the end of the experiment when kidney contained 132 µg Hg/g, 97 µg Cu/g and 69 µg Zn/g (wet weight). In comparison to control values, the levels of Hg, Cu and Zn increased 4400, 8 and 1.5 times, respectively. The distribution of total Hg and Cu content varied during the exposure period: 56-75 % of mercury and 48-75 % of copper were found in the soluble fraction. Zn concentrations did not change until the end of the experiment when enrichment of the soluble fraction was observed. Gel chromatographic analysis (Sephadex G-75) revealed that Zn was mainly bound to high molecular weight (HMW) proteins, while Hg and Cu were preferentially bound to MT fraction, where

Trace Element Analytical Chemistry in Medicine and Biology

their content increased with time of exposure. After 3 weeks exposure, a redistribution of Hg and Cu was observed from HMW protein to MT and the low molecular weight Hg, Cu-binding protein related to MT which occurred at the end of the experiment.

Introduction

High renal accumulation of mercury in the rat has been found after repeated administration of $HgCl_2$(1) or after inhalation of elemental mercury vapour (2). Tissue fractionation studies have shown that mercury is enriched in the soluble fraction where it is preferentially bound to the low molecular weight, cysteine-rich protein, metallothionein (MT), for which a protective role towards heavy metal toxicity was proposed (1,2,3). Rat renal MT normally contains Cu and Zn (4), and some suggestions were made that mercury could alter their metabolism (5). In this study we investigated the distribution of mercury between the soluble and particulate phase of rat kidney and its binding to MT, as well as the effect of mercury on essential Cu and Zn during 3 weeks of exposure to elemental mercury vapour.

Materials and methods

Male Wistar rats (two months old, body weight about 200 g) were continuously exposed to elementary mercury vapour for 3 weeks of the winter period in the Idrija mercury mine, Slovenia, at an air concentration of 0.5-0.8 mg Hg^0/m^3 and at temperatures ranging from $+4^oC$ to -3^oC. The animals were fed with a standard laboratory food (Hg, Cu and Zn were 0.12, 20 and 50 µg/g dry weight) twice a day, outside the mine (for 1/2 hr) to minimise oral ingestion of Hg. Before the experiment started the animals were adapted to low temperatures and the feeding regime. At determined intervals, two rats were sacrificed, the kidneys were removed and deep-frozen. The homogenation was performed using deaerated, ice-cold buffer (0.02 M Tris-HCl, 0.25 M saccharose, pH 7.6) in a glass homogenizer with motorized teflon pestle. The 10 % homogenates prepared were passed through a 200 µm nylon net and ultracentrifuged at 35000 rpm for 60 min. The 5 ml supernate was applied to a calibrated Sephadex G-75 column (70 x 2.4 cm) and eluted with 0,02 M Tris-HCl, pH 8.6, at 4°C in an atmosphere of nitrogen. The chromatographed fractions of 5-6 ml were analysed for UV absorbance (at 230, 254 and 280 nm) and for metal content (Hg, Cu, Zn). The concentrations of Zn and Cu were measured by flame AAS directly by aspirating the eluate, or in the case of homogenate and supernate, after wet digestion with mixed acid. Determination of mercury in the tissue samples was done using the method of neutron activation analysis (NAA) (6); for other samples the method of cold vapour atomic absorption spectrophotometry (CVAAS) (7) was applied. Absorption spectra of the central fractions of peaks were obtained by automatic scanning on a Perkin Elmer λ-3 spectrophotometer.

Results and discussion

Mercury readily accumulated in the kidney during the exposure time and after 3 weeks it was near the steady state, reaching a concentration of 132 µg/g, which is 4400 times the control values. At the same time endogenous Cu and Zn were increased approximately 8 and 1.5 fold, respectively, taking into account the mean concentration of induvidual values from Table 1.

Table 1: Concentrations of Metals in Rat Kidney
(µg/g wet weight)

TREATMENT	Hg	Cu	Zn
Control	0.03	10.4	56.8
(n = 2)		13.8	36.8
After 3 weeks		135.5	95.6
of exposure	131.8	94.2	50.6
(n=3)		62.1	61.2

Similar results were found after repeated subcutaneous injections of $HgCl_2$: Szymanska and Zelazowski (5) found more than a sixfold increase of Cu in rat kidney, while Nicholson et. al. (8) did not find any change in zinc level in the kidney of mercury-dosed mice. The metal concentrations in homogenate, particulate and soluble phase of kidney throughout the course of the exposure are shown in Fig.1.

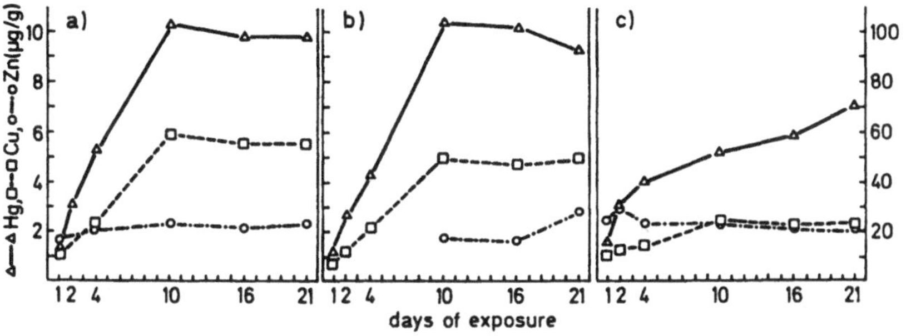

Fig. 1.: Changes in Hg,Cu and Zn concentrations in rat kidney homogenate (a); soluble (b) and particulate phase (c) during 3 weeks of in vivo exposure to mercury vapour. Values are expressed as µg Me/g wet weight of 10 % homogenate, supernate or wet sediment.

In the soluble fraction there was a linear increase of Hg and Cu from the 1st to 10th day: afterwards they remained relatively unchanged until the end of the experiment. In the particulate phase Cu increased until the 10th day and then remained at the same level, while Hg increased continuously over the entire exposure time. Zn concentrations did not change markedly; some enrichment in the soluble fraction was observed at the end of the experiment. Subcellular distribution of each total metal content (obtained by summing up the particular metal content of both cellular compartments) varied during the exposure period: 56-75 % Hg, 48-75 % Cu and 50-66 % Zn were found in the soluble kidney fraction. The described pattern of accumulation for mercury is in agreement with its accumulation in kidney tissue (unpublished results) where slow but continous increase of mercury was observed between the 10th day and the end of exposure. Discrepant results for the homogenate probably arose from methodological errors.
Fractionation of the kidney soluble phase on the Sephadex G-75 column (Fig.2) revealed that during the exposure period both Hg and Cu preferentially bound to MT (peak IV), where their content increased markedly with time of exposure, while Zn remained aproximately at the same level and mainly accumulated in HMW proteins (peak I, II). From 1st to 10th day of exposure, besides the MT fraction, Hg also incorporated in HMW protein, as well as in the non-MT

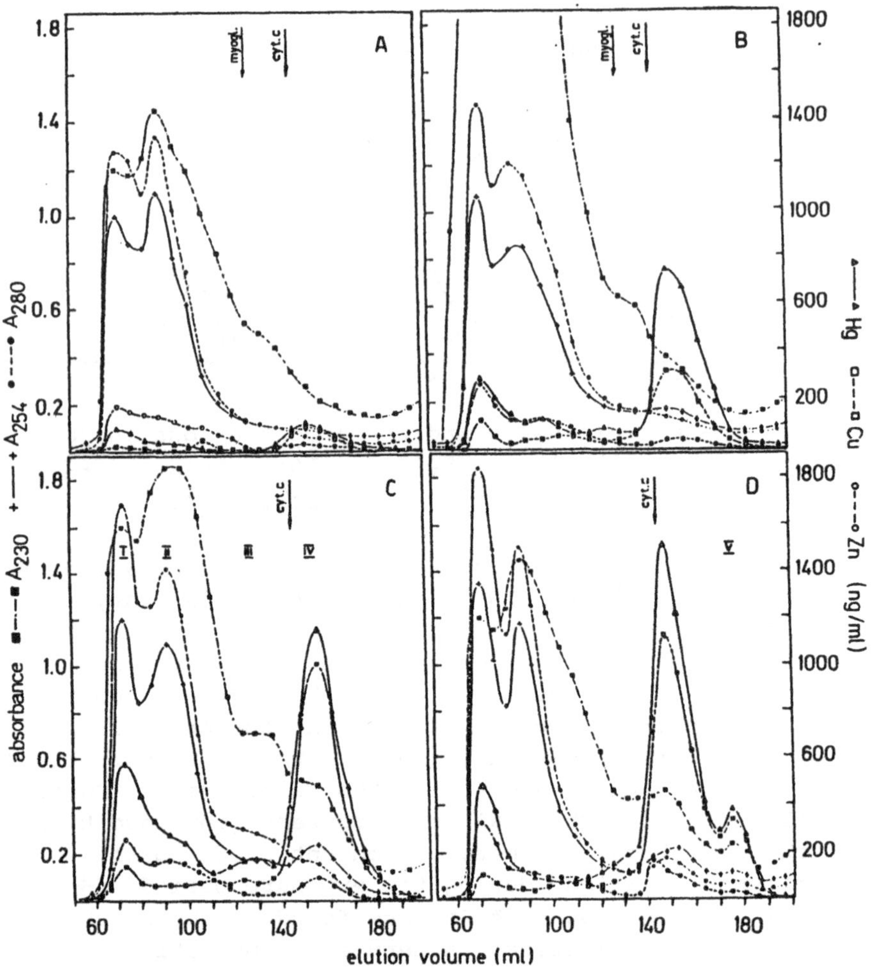

Fig. 2: Sephadex G-75 elution profiles of supernates obtained by ultracentrifugation of 10 % kidney homogenates of rats exposed to mercury vapour for 1, 4, 10 and 21 days (A, B, C, D), respectively.

protein (peak III) at an elution position close to myoglobin(note its elevated absorption at 230 and 280 nm at the 10th day which disappeared at the 21st day). Between the 10th and 21st day of the experiment, when the soluble phase was found to be saturated, a redistribution of Hg and Cu was observed from HMW protein to MT and low molecular weight Hg, Cu-binding protein (peak V) related to MT which occurred at the end of the experiment. Fig.3 shows the absorption spectra of the central fractions of MT peaks.

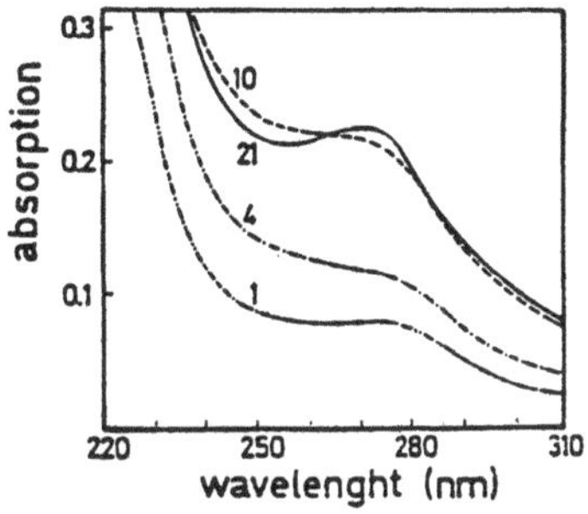

Fig. 3: Absorption spectra of the central fractions of MT peaks isolated from rat kidney exposed in vivo to mercury vapour for 1, 4, 10 and 21 days.

Their A_{254}/A_{280} ratio changed with time and were 1.1 (1st day); 1.3 (4th day); 1.5 (10th day) and 1.1 (21st day). By careful observation of the MT peak from the 21st day it can be seen it was not homogeneous; there was a shift in 254 nm absorbance maximum from the central fraction (see Fig. 2d) and its A_{254}/A_{280} ratio allowing for this shift was still 1.5. It could be supposed that a saturation of MT and a degradation of HMW proteins were occuring in the last stage of the exposure period. The role of MT in the interaction of mercury with essential metals requires futher investigations.

Acknowledgement:

The assistance of Miss A. Prosenc and Mrs. M. Špenko in performing some of the analyses is gratefully acknowledged.

References

1. Piotrowski, J.K., B.Trojanowska, J.M. Wisniewska-Knypl, W.Bolanowska. 1974. Toxicol. Appl. Pharmacol. 27, 11.

2. Cherian, M.G., T.W.Clarkson. 1976. Chem-Biol. Interactions, 12, 109.

3. Piscator, M. 1964. Nord. Hyg. Tidskr. 45, 76

4. Zelazowski, A.J., J.K. Piotrowski, 1980. Biochim. Biophys. Acta, 625, 89.

5. Szymanska J.A., A.J. Zelazowski. 1979. Environ. Res. 19, 121.

6. Byrne, A.R., L. Kosta. 1974. Talanta, 21, 1083.

7. Horvat, M., T. Zvonarić, P. Stegnar. 1986, Vestn. Slov. Kem. Druš. 33, 475.

8. Nicholson, J.K., D. Osborn, M.D. Kendal. 1984. Comp. Biochem. Physiol. 77C, 249.

EFFECT OF ZINC DEFICIENCY ON THE GUINEA PIG DURING PREGNANCY

J. Apgar and G.A. Everett

U.S. Plant, Soil, and Nutrition Laboratory, USDA-ARS, Ithaca, NY 14853

Severe effects of inadequate zinc (Zn) intake on pregnant animals suggest that adequate Zn status is important in human pregnancy. Plasma/serum Zn concentration, the most common measure of Zn status, is not a sensitive indicator and is particularly difficult to interpret during pregnancy. We have therefore investigated plasma enzymes as indicators. The guinea pig was used because its size and 68 day gestation permit sampling the same individual during a developing Zn deficiency. Half the animals (-Zn) were given a low Zn intake beginning day 30 (d30). Zn-supplemented animals (+ZnRI) were individually paired with -Zn and feed intake restricted to maintain the same weight gain. Reduced weight gain, skin lesions, sensitivity to handling, peculiar stance, and inability to right were observed in -Zn. +ZnRI looked healthy, but abortion/delivery occurred in both groups about d55. +Zn ad libitum-fed delivered living young at term. Failure of -Zn to maintain pregnancy limits the usefulness of the guinea pig as a model; but mannosidase and angiotensin-converting enzyme, enzymes that reflect Zn status in the ewe, were also affected by low Zn intake in the guinea pig.

Introduction

Because the deleterious effects of zinc deficiency during pregnancy in animals suggest that inadequate zinc status may have adverse effects on human pregnancy (1), a good method of measuring Zn status is needed. Plasma/serum Zn, the most commonly used criterion, is not sufficiently sensitive and is affected by factors other than Zn status. It is particularly difficult to interpret during pregnancy because it normally decreases at this time (2). In order to develop better

Trace Element Analytical Chemistry in Medicine and Biology

indicators, a good animal model is needed. The short gestation and large relative conceptus size of the rat limit its usefulness. The ewe with its longer gestation (150 days) and smaller relative conceptus size is a better model but the special diet and caging required are expensive. Although the guinea pig also has a large relative conceptus size, it has a 68 day gestation and sufficiently large size to permit serial sampling of an individual during a developing Zn deficiency. We therefore induced Zn deficiency in the pregnant guinea pig and measured plasma enzymes that appeared to be indicative of Zn status in other species. Zinc concentration in plasma, red blood cells, placenta, and fetal liver was also determined.

Methods

Eighteen adult guinea pigs were fed a low-Zn ($\leq$3 ppm) casein-agar diet supplemented with 15 ppm Zn as Zn acetate in drinking water. Zinc supplement was reduced to 1 ppm for half the females (-Zn) starting d30 of gestation. The remaining animals (+ZnRI) were individually paired with a -Zn, and food intake was restricted to that of the -Zn or, if necessary, to 80% of that of the -Zn to maintain equivalent weight gains. Five ml heparinized blood was collected weekly from the ear vein. Assays for plasma Zn concentration, mannosidase (MANN), angiotensin-converting enzyme (ACE), alkaline phosphatase, and 5'nucleotidase have been described (3). Zn concentration of fetal liver and placenta was determined by flame atomic absorption spectrometry after digestion in hot nitric-sulfuric acid. Data were analyzed either by the General Linear Models using SAS software (4) with treatment, pairs, days, and treatment by day interaction as variables or by two-tailed unpaired Student's t-test. Data are expressed as mean±SEM.

Results and Discussion

Weight gain by -Zn was small; weights of +ZnRI were approximately the same (Table 1). Weights are expressed relative to weight d30 to reduce variation due to premature delivery. Weight gain of four +Zn ad libitum-fed was 1.45±.04 at delivery (d68-69). In addition to reduced weight gain,

Table 1. Outcome of Pregnancy in Zn-Deficient Guinea Pigs

	-Zn	+ZnRI
Body Weight, wt/wt d30		
day 32-38	1.04±.01	1.02±.01
day 39-48	1.10±.02	1.07±.01
day 49-58	1.05±.03	1.09±.02
Day of Delivery	57±1	55±2
Litter Weight, g	95±7	112±13
Litter size	2.9±.3	3.4±.3
Zn, ug/g dry wt.		
Fetal liver	38±4*	54±7
Placenta	87±5**	100±3

* Different from +ZnRI, P≤.06

** Different from +ZnRI, P≤.05

severe skin lesions, sensitivity to handling, inability to right when placed on its side, and peculiar stance were observed in -Zn; +ZnRI appeared healthy. Although low Zn intake was not started until d30, none of the -Zn maintained pregnancy to term. Abortion/delivery occurred d57±1. Since all +ZnRI females also delivered prematurely, d55±2, premature delivery was presumably due to the effect of low Zn intake on feed intake and/or weight gain. None of the young were viable; four -Zn litters apparently died in utero. Total weight of pups delivered by -Zn and +ZnRI did not differ nor did litter size (Table 1). Ad libitum-fed +Zn delivered living young at term. Zinc concentration of -Zn placentas was less than that of +ZnRI ($P \leq .05$). Fetal liver Zn concentration was also lower in -Zn than in +ZnRI, but the difference was significant only at $P \leq .06$. Fetal guinea pig liver contained less Zn than other species we have studied (480, 300, and 230 ug/g dry wt for lamb (5), rat (6), and rabbit (7), respectively, compared to 40±5 for +Zn ad libitum-fed guinea pigs).

Zinc treatment significantly affected plasma Zn concentration and MANN and ACE activity with and without added Zn in vitro (Table 2). Treatment x day interactions were also significant for plasma Zn and for MANN activity with added Zn in vitro.

Table 2. Indicators of Zinc Status in Plasma of Pregnant Guinea Pigs

		d30-38	d39-48	d49-58	Mean
Zn, ug/ml	-Zn	1.0±.08	.58±.06	.45±.03	.71±.06
	+ZnRI	1.4±.08	1.4±.10	1.1±.07	1.3±.06[1]
MANN[2], Basal	-Zn	.40±.03	.34±.02	.26±.01	.34±.02
	+ZnRI	.44±.03	.43±.02	.40±.02	.42±.01[1]
Increase[3]	-Zn	.19±.03	.24±.03	.18±.02	.20±.01
	+ZnRI	.17±.02	.14±.02	.10±.02	.14±.01[1]
ACE[4], Basal	-Zn	.85±.04	.70±.03	.31±.04	.59±.05
	+ZnRI	.95±.06	.77±.10	.54±.06	.82±.05[1]
Increase[3]	-Zn	.44±.04	.52±.03	.26±.03	.40±.03
	+ZnRI	.35±.03	.41±.02	.24±.03	.35±.02[1]

[1] Different from -Zn, $P \leq .01$

[2] Mannosidase activity, Absorbance at 400 nM

[3] Increase in activity measured with Zn added in vitro

[4] Angiotensin-converting enzyme, Absorbance at 228 nM

In vitro Zn stimulated activity of both MANN and ACE in +ZnRI as well as in -Zn. The stimulation of MANN activity by in vitro Zn was smaller in the -Zn guinea pig than in the -Zn ewe (5) compared to their respective controls. The smaller difference was probably due to the decreased basal activity in -Zn guinea pigs; basal MANN was not affected in -Zn ewes. If the lower basal activity in -Zn guinea pigs is taken into account, stimulation in -Zn plasma is more easily seen than by comparison of the absolute increases. MANN activity in -Zn plasma increased 60%; in +ZnRI, 35%. Results with ACE were similar. Alkaline phosphatase activity was somewhat lower in -Zn than in +ZnRI, 2.2±.1 Sigma units/ml and 2.5±.1, respectively; but the difference was significant only at $P \leq .06$. There was no difference in 5'-nucleotidase activity or in Zn concentration of red blood cells.

Conclusion: Failure of -Zn to maintain pregnancy limits the usefulness of the guinea pig as a model; but MANN and ACE, enzymes that reflect Zn status in the ewe and rat (8), were also affected by low Zn intake in the guinea pig.

References

1. Apgar, J. 1985. Ann.Rev.Nutr. 5, 43.

2. Hambidge, K.M., N.F. Krebs, M.A. Jacobs, A. Favier, L. Guyette, et al. 1983. Am.J.Clin.Nutr. 37, 429.

3. Everett, G., J. Apgar. 1986. Acta Pharm.Toxicol. 59(suppl. VII), 163.

4. SAS Institute, Inc. SAS User's Guide: Statistics, 1982 Edition. Cary, NC: SAS Institute, Inc., 1982. 584 pp.

5. Apgar, J., J.A. Fitzgerald. 1987. Nutr.Res. 7, 1281.

6. Apgar, J. 1971. J.Anim.Sci. 33, 1255.

7. Apgar, J. 1972. J.Nutr. 102, 343.

8. White, C.L., J. Pschoor, I.C.M. Jacob, N.V. Lutterotti, H. Dahlheim. 1986. Biochem.Pharm. 35, 2489.

EFFECT OF ARSENIC ON MITOCHONDRIAL METABOLISM AFTER SINGLE OR REPEATED INJECTION IN GUINEA PIGS

F.X. Reichl, L. Szinicz, H. Kreppel, B. Fichtl and W. Forth

Walther-Straub-Institut für Pharmakologie und Toxikologie der LMU München, Nußbaumstr. 26, D-8000 München 2, FRG

ABSTRACT

As was reported recently the effects of arsenic on glycolysis intermediates after repeated administration may differ from those in acute poisoning (2)(3). The present experiments were performed to further elucidate the effect of arsenic on cellular metabolism.
Male guinea pigs received As_2O_3 s.c. either as a single dose or two times daily on 5 consecutive days. The animals were sacrificed in anaesthesia and the livers removed by a freeze stop procedure for the determination of substrates. The arsenic content of liver ($\mu g\ As/g_{(w.w.)}$) as measured by hydride atomic absorption spectrometry was 18,5 ± 0,4 (single dose) and 3,3 ± 0,6 (repeated treatment).
In the single dose group an increase in the content of pyruvate, citrate and malate; and a decrease in hydroxybutyrate was found.
In the repeated treatment group only the content of pyruvate was decreased.
Whether this indicates a different mechanism of action remains to be established.
No correlation was found between metabolic changes and arsenic burden of liver after single and repeated treatment with As_2O_3.

Trace Element Analytical Chemistry in Medicine and Biology

INTRODUCTION

The toxicity of trivalent arsenicals is thought to be due to the inhibition of pyruvate dehydrogenase activity resulting in pyruvate accumulation (Peters et al. 1955). Some findings not compatible with this hypothesis were observed in recent experiments, where a different glycolysis intermediate pattern was observed in livers of guinea pigs after single or repeated injection of As_2O_3 (Reichl et al. 1988).
The present experiments were performed to further elucidate the effect of arsenic on cellular metabolism.

METHODS

Male guinea pigs, 350 - 450 g, received As_2O_3 s.c. either as a single dose of 10 mg/kg at 9 a.m. or in repeated doses of 2,5 mg/kg b.i.d. on 5 consecutive days (at 9 a.m. and 4 p.m.).
One hour after the single dose or 16 hours after the last injection in the repeated treatment group the animals were sacrificed in anaesthesia (urethane + ether) and the livers removed by a freeze stop procedure for the determination of pyruvate (Pyr), citrate (Cit), α-ketoglutarate (α-Ket.Glut.), malate (Mal), hydroxybutyrate (OH-But) and acetoacetate (AcAc).
The arsenic content of livers was measured by hydride atomic absorption spectrometry (MHS1, Perkin-Elmer, Überlingen).

RESULTS

a. Substrate contents in livers of guinea pigs after a single dose of As_2O_3.

With respect to control animals there was an increase in the content (μmol substrate/$g_{liver(w.w.)}$, $\bar{x}$ ± SEM, n = 5-8) of pyruvate (0,31 ± 0,02 vs 0,43 ± 0,02), citrate (0,72 ± 0,05 vs 1,22 ± 0,09) and malate (0,32 ± 0,04 vs 0,58 ± 0,11) and a decrease of hydroxybutyrate (0,12 ± 0,02 vs 0,02 ± 0,01) . No change in α-ketoglutarate (0,33 ± 0,03 vs 0,32 ± 0,02) and acetoacetate (0,38 ± 0,02 vs 0,41 ± 0,02) was found.
The ratio of OH-But/AcAc was significantly diminished in treated animals (0,33 ± 0,03 vs 0,05 ± 0,01).
The arsenic content of the liver (μgAs/$g_{liver(w.w.)}$) was 18,5 ± 0,4 (Fig. 1).

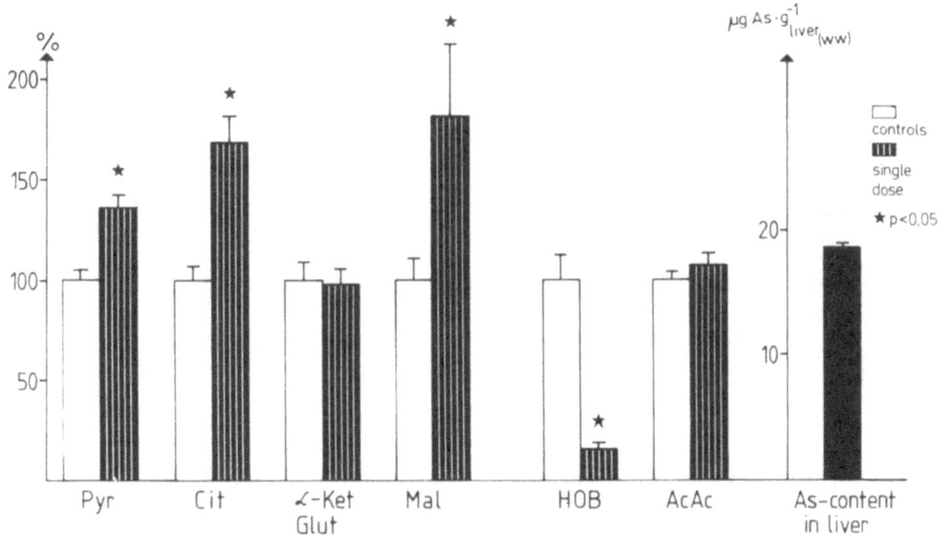

Fig 1 : Substrate content in livers of guinea pigs after a single dose of As_2O_3 (percent of controls).

b. Substrate contents in livers of guinea pigs after repeated treatment with As_2O_3.

Contrary to the single dose experiments a decrease ($\bar{x}$ ± SEM, n = 5) in pyruvate content (0,31 ± 0,02 vs 0,20 ± 0,01) was found with respect to controls. No change in citrate (0,72 ± 0,05 vs 0,67 ± 0,05), α-ketoglutarate (0,33 ± 0,03 vs 0,25 ± 0,03), malate (0,32 ± 0,04 vs 0,45 ± 0,11), hydroxybutyrate (0,12 ± 0,02 vs 0,08 ± 0,02) and acetoacetate (0,38 ± 0,02 vs 0,37 ± 0,01) content was observed.
The ratio of OH-But/AcAc was 0,20 ± 0,04 .
The arsenic content of the liver ($\mu gAs/g_{liver(w.w.)}$) was 3,3 ± 0,6 (Fig. 2).

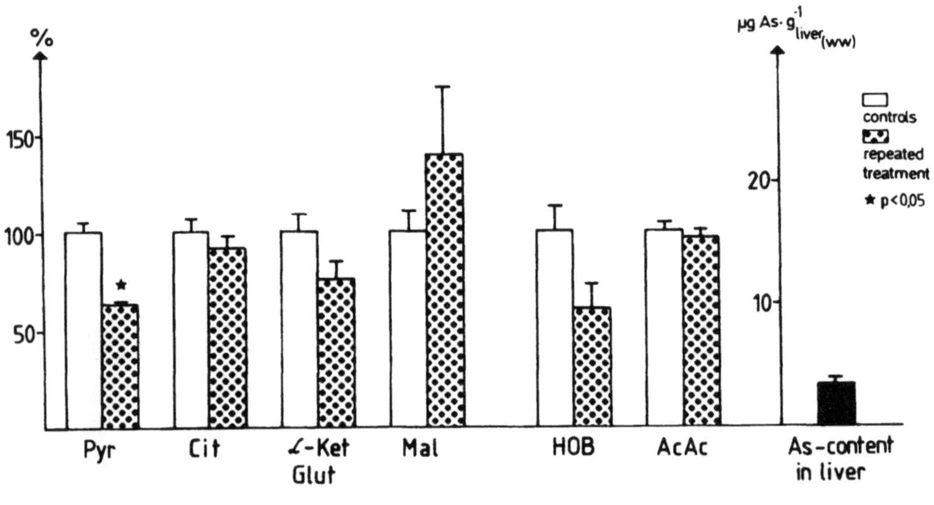

Fig 2 : Substrate contents in livers of guinea pigs after repeated treatment with As_2O_3 (percent of controls).

c. Relation between pyruvate content and arsenic burden of the guinea pig liver.

Results of additional experiments, in which the livers were removed one hour after the last As_2O_3 injection show that no correlation exists between pyruvate content and arsenic burden (Fig. 3).
The same was found with citrate (not shown).
This indicates, that the metabolic differences observed between single dose and repeated treatment were not due to the differences in the arsenic content in livers.

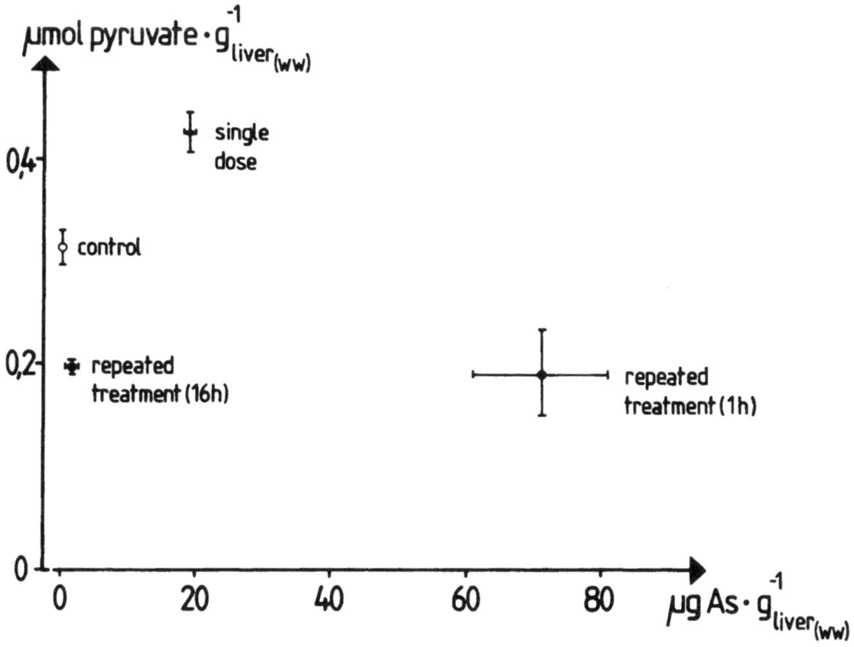

Fig 3 : Relation between pyruvate content and arsenic burden of the guinea pig liver.

CONCLUSIONS

The present results support previous findings where differences in metabolite contents were found after single or repeated treatment with As_2O_3.
Whether this indicates differences in the mechanism of action remains to be determined.
Analysis of dose-effect relationships based on overall organ burden of the toxicant may lead to misinterpretations if the dose regimen is not taken into account.

REFERENCES

1. Peters, R.A.
   Bull. John Hopkins Hosp. 97, 1 - 20 (1955)

2. Reichl, F.X.; Szinicz, L.; Kreppel, H. and Forth, W.;
   Naunyn-Schmiedeberg's Arch. Pharmacol. 335 :
   suppl., R 17 (1987)

3. Reichl, F.X.; Szinicz, L.; Kreppel, H.; Fichtl, B.
   and Forth, W.;
   The Toxicologist, 8, No. 1; 20 (1988)

INFLUENCE OF SOME ORGANIC ACIDS ON CADMIUM RETENTION IN PIGS

H. Schenkel, B.C. Matthes
Institute of Animal Nutrition, University of Hohenheim,
PB 700562, D-7000 Stuttgart 70, FRG

Abstract

Addition of citric acid to the diet of young pigs caused an increase of cadmium concentration in liver and kidneys. No effect on cadmium concentration in liver, kidney, muscle and bile could be found after addition of fumaric acid, sorbic acid or calcium propionate. Some of these supplements also caused a decrease of zinc concentration in muscle and bile, as well as copper concentration in kidneys.

Introduction

Organic acids and their salts are widely used as food additives in human and animal nutrition.

In animal nutrition several organic acids were used for feed preservation. Furthermore in several investigations an improvement of animal performance could be stated (1, 4, 7, 8, 14, 15).

An influence on mineral and trace-element metabolism has been demonstrated for several organic acids. For example, additions of citric acid increased iron (5) and aluminium (12) absorption and/or retention humans and rats, respectively. According to experiments in vitro (2) as well as in vivo (3, 13) citric acid may also increase lead absorption. In in vitro and in situ experiments done by Kiyozumi et al. (6) the addition of

Trace Element Analytical Chemistry in Medicine and Biology

citric acid decreased cadmium uptake in intestinal tissue.

Conflicting results were reported about the effect of the supplementation of fumaric acid on retention of different minerals and trace elements in pigs and rats (9, 11).

For investigations on effects of organic acid supplementation on heavy metal metabolism are scarse, in the experiment presented the influence of some of these additives on cadmium retention was investigated in growing pigs.

Experiment

The feeding experiments were carried out with young male castrated pigs with a mean body mass of 16 kg at the beginning. The animals were kept individually in stainless steel metabolism cages under controlled environmental conditions. They received diets based on corn, wheat, cassava meal and soybean meal. Diets were supplemented with a premixture for minerals, trace elements and vitamins. By addition of a cadmium chloride-glucose-mixture the cadmium concentration was adjusted at 6 mg per kg dry matter in all diets. As shown in table 1 acids or their salts were added in 4 diets within the range of 0,2 (sorbic acid) and 5 percent (citric acid). In experiment I the addition of citric acid and fumaric acid was investigated, whereas in experiment II calcium propionate and sorbic acid were added. Each treatment group consisted of 8 animals. The feeding level was at 70 g feed per kg metabolic body weight ($W^{0.75}$) during the first 9 days and at 80 g per $W^{0.75}$ till the end of the experiment. After 28 days the animals were slaughtered and samples of different tissues were prepared for trace element determination. Bile fluid was sampled with plastic sygringes from gall-bladder after slaughtering.

Tissue samples were freeze-dried, grinded and digested with conc. nitric acid in pressure bombs. Cadmium was extracted with dithizone and reextracted with 0,5 n HCl (10). Cadmium was determined by graphite furnace atomic spectroscopy with deuterium background correction. NBS bovine liver (1577) was used for quality control. With 5 subsamples a mean cadmium concentration of 0.268 ± 0.008 mg/kg (cert. 0.27 mg/kg) could be found.

Table 1: Treatment Groups and Feed Additives

Experiment	Group	n	Additive	Concentration
I	1	8	none	
	2	8	citric acid	5 %
	3	8	fumaric acid	2 %
II	4	8	none	
	5	8	calcium propionate	0.5 %
	6	8	sorbic acid	0.2 %

Results

Addition of the different acids and salts had no influence on animal performance in this study. No significant differences could be stated with respect to daily weight gain and feed efficiency. Also no treatment effect was found on luminal pH-values in duodenum and ileum at slaughtering.

As shown in table 2 the addition of citric acid caused a significant increase of the cadmium concentration in kidney and liver. The mean cadmium concentration in livers of animals from group were 3.6 times higher than in the control group. The increase in the kidneys was less pronounced. No significant effect on cadmium concentration in the different tissues could be stated for the other additives.

Table 2: Cadmium Concentrations in Kidney, Liver, Muscle, and Bile (n: 8; x ± s.d.)

Experiment	Group	Kidney	Liver	Bile	Muscle
		mg/kg d.m.			
I	1 control	22.0 ± 6.0	2.31 ±0.60	0.24 ±0.14	0.13 ±0.03
	2 citric acid	50.8[a] ±19.7	8.46[a] ±3.68	0.25 ±0.09	0.15 ±0.07
	3 fumaric acid	26.8 ± 5.1	3.56 ±1.05	0.17 ±0.11	0.11 ±0.04
	4 control	25.1 ± 3.4	3.39 ±0.69	0.34 ±0.16	0.17 ±0.05
	5 Ca-propionate	25.9 ± 8.7	3.01 ±1.03	0.32 ±0.14	0.13 ±0.02
	6 sorbic acid	28.6 ± 5.2	3.82 ±0.64	0.25 ±0.18	0.14 0.04

a) significantly ($p \leq 0.05$) different from group 1 and 3

With respect to the organ mass of liver and kidneys the total cadmium content in these two organs was at 0.81 mg and 0.86 mg for the control groups 1 and 4, respectively. Mean cadmium intake throughout the whole experiment was at 138 and 131 mg in these groups. Therefore cadmium content in liver and kidneys related to cadmium intake during the experiment was at 0.6 and 0.7 percent, respectively. Addition of citric acid caused an increase of total cadmium in liver and kidneys up to 2.38 mg. Hence cadmium retention in these organs related to cadmium intake during the experiment was nearly three times higher than in the control group. In experiment II an increase of this relation could also be found for the sorbic acid group. Total cadmium content in liver and kidneys in this animals (group 6) was at 1.06 mg corresponding with 1.1 % of the cadmium intake during the experiment.

As shown in table 2 supplementation of the different additives affected also the metabolism of zinc and copper.

In comparison with the control group lower zinc concentrations in the muscles could be found after addition of citric acid and fumaric acid. In experiment II a lower zinc concentration in muscle could also be found in the calcium propionate group (group 5) in comparison with the corresponding control group (group 4). In group 5 a significant decrease of copper concentration in kidneys could also be stated.

Table 3: Zinc and Copper Concentrations in Kidney, Liver, Muscle, and Bile (n: 8; x ± s.d.)

Experiment	Group	Kidney		Liver		Bile		Muscle
		Zn	Cu	Zn	Cu	Zn	Cu	Zn
		mg/kg d.m.						
I	1 control	158 ±17	38 ±7	194 ±42	41 ±24	10 ±4	31 ±12	124 ±13
	2 citric acid	173 ±20	44 ±7	173 ±26	28 ±8	11 ±1	26 ±14	104[c] ±12
	3 fumaric acid	158 ±11	39 ±7	198 ±48	34 ±18	11 ±3	32 ±9	110[c] ±13
II	4 control	167 ±23	49 ±13	209 ±27	35 ±15	19 ±8	29 ±11	140 ±31
	5 Ca-propionate	154 ±12	37[a] ±6	204 ±27	28 ±8	12 ±5	24 ±7	109[d] ±23
	6 sorbic acid	165 ±17	37 ±13	216 ±30	31 ±17	12[b] ±3	21 ±7	117 ±22

a), b), d) significantly different ($p \leq 0.05$) from group 4
c) significantly different ($p \leq 0.05$) from group 1

Results of these experiments show that common food additives in human and animal nutrition can influence heavy metal metabolism. Therefore further evaluation experiments with different doses and exposure times will be carried out.

Acknowledgement

A part of this investigation was supported by the Ministerium für Ernährung, Landwirtschaft, Umwelt und Forsten, Baden-Württemberg. The authors wish to thank S. Plank und A. Marquardt for their technical assistance.

References

1. Cole, D.J.A., R.M. Beal, J.R. Luscombe. 1970. Vet.Rec. 86, 400.

2. Coleman, I.P.L., J.A. Blair, M.E. Hilburn. 1982. Int.J.Environ.Studies 18, 187.

3. Garber, B.T., E. Wei. 1974. Toxicol.Appl.Pharmacol. 27, 685.

4. Giesting, D.W., R.A. Easter. 1985. J.Anim.Sci. 60, 1288.

5. Gillooly, M., T.H. Bothwell, J.T. Torrance. 1983. Brit.J. Nutr. 49, 331.

6. Kiyozumi, M., M. Yamane, T. Honda, S. Kojima. 1986. Eisei Kagaku 32, 312.

7. Kirchgessner, M., F.X. Roth. 1978a. Züchtungskde. 50, 329.

8. Kirchgessner, M., F.X. Roth. 1978b. Z.Tierphysiol.Tierernährg.Futtermittelkde. 34, 3.

9. Kirchgessner, M., F.X. Roth. 1980. Z.Tierphysiol.Tierernährg.Futtermittelkde. 44, 239.

10. Oelschläger, W., W. Lautenschläger. 1977. Fres.Zschr.Anal. Chem. 287, 28.

11. Roth, F.X., M. Kirchgessner, E. Grassmann, A. Tschierschwitz. 1982. Z.Tierphysiol.Tierernährg.Futtermittelkde. 48, 16.

12. Slanina, P., Y. Falkeborn, W. Frech, A. Cedergren. 1984. Food Chem. Toxicol. 22, 391.

13. Spickett, J.T., R.R. Bell, J. Stawell, S. Polan. 1984. Agents and Actions 15, 459.

14. Vogt, H., S. Matthes, S. Harnisch. 1981. Arch.Geflügelkde. 45, 221.

15. Vogt, H., S. Matthes, S. Harnisch. 1982. Arch.Geflügelkde. 46, 223.

PRETREATMENT WITH LOW DOSES OF CADMIUM OR ZINK DECREASES LETHALITY IN MICE ACUTELY POISONED WITH ARSENIC

H.Kreppel, K.Kolb, F.X.Reichl, B.Fichtl, and W.Forth
Walther-Straub-Institut für Pharmakologie und Toxikologie der LMU München, Nußbaumstr. 26, 8 München 2, FRG

Abstract

The effects of Cd or Zn on the acute toxicity of arsenic were studied in NMRI mice. 7 d pretreatment of mice with $CdCl_2$ (4...18 µmol Cd/kg and d) resulted in a marked reduction of arsenic toxicity as assessed by the 30 d survival interval. A similar pretreatment with $ZnCl_2$ decreased the lethality due to arsenic only when the Zn dosage was increased to the maximal tolerated levels (~100 µmol Zn/kg and d).
The respective metal contents in liver, kidneys and intestinal mucosa showed a positive dose-dependent behaviour during pretreatment as assessed by atomic emission spectrometry (ICP-AES). Interestingly, in Cd treated mice there was a concomitant rise of Zn in the liver. The elevated metal loads were accompanied by a simultaneous rise in liver MT measured by DISC gel electrophoresis and immunoblotting.
The possible role of MT in arsenic poisoning deserves further investigation.

Introduction

Metallothionein (MT) is a small protein (MW 6000 D), first isolated from the equine renal cortex (1), and is found in various tissues of several species. It has a high content of cystein (approximately 30 % - (2)) and hence a strong affinity for me-

Trace Element Analytical Chemistry in Medicine and Biology

tal ions. One molecule MT binds seven metal ions, e.g., Cd, Hg, and Zn (3). Exposure to heavy metals, e.g., Cd and Zn, induces a rapid production of MT (4). The physiological role of MT may be storage of biological metals and/or protection against metal toxicity. We have studied the effect of MT induction by pre-treatment with low-dose Cd or Zn on the acute experimental toxicity of arsenic in mice.

Material and methods

Male NMRI mice (Süddeutsche Versuchstierfarm, Tuttlingen, FRG) were supplied with a standard diet (As: 0.0; Cd: 0.0; Cu: 7.1; Fe: 163; Zn: 49 µg/g; "C 1000"; Altromin, Lage, FRG) and tap water ad lib. After seven days groups of 7 mice received 0,1, 2,3,4,8,12 or 18 µmol $CdCl_2$/kg and d (in saline; 1 µl/g b.w.) i.p. on day 2, 4, 6, and 8; on day 1 only 1/4 of the dose was administered. On day 9 all the mice were injected 12.9 mg/kg As_2O_3 s.c. (approximately LD_{80} - (5)).
In a second experiment mice (n = 7/group) received a Zn-depleted diet (As: 0.0; Cd: 0.0; Cu: 17.5; Fe: 169; Zn: 3.1 µg/g diet; "C 1040", Altromin, Lage, FRG) and distilled water ad lib. After 7 days the animals were given Zn (0,25,50,100,or 200 µmol $ZnCl_2$/kg and d in saline) in the above mentioned dosing regimen, followed by s.c. injection of 12.9 mg/kg arsenic on day 9. Body weight and survival of mice were recorded over a period of 30 d after arsenic poisoning.
In another experiment groups of 5 mice were treated with Cd or Zn as mentioned above but killed in ether on day 9 instead of being poisoned with arsenic; liver, kidneys and mucosa from the proximal half of the small intestine were removed, homogenized with dist. water (1 : 5), and stored frozen (-18°C).
After addition of nitric acid homogenates were digested (176°C, 2.4 h) in a pressure bomb (Berghof, Tübingen, FRG) and the content of As, Cd, Cu, Fe, and Zn was measured by atomic emission spectroscopy (ICP-AES; Jarrell-Ash, Waltham, MA, USA). The ex-

perimental groups were tested for statistical significant ($p < 0.05$) differences with respect to the controls by U-test (Mann and Whitney).
After heat denaturation (85°C, 5 min) and centrifugation (20000 x g; 15 min) the metallothionein content in the liver homogenate supernatants was measured by cylindric discontinuous gel electrophoresis (DISC gel electrophoresis; 150 V, 4 mA/sample) (6) and immune blotting technique (7) using anti-rat liver MT antibodies prepared in the authors' laboratory (8). Standard MT I+II (2.5 µg/µl; Sigma, St.Louis, MO, USA) was used as a control with known content.

Results

NMRI mice pretreated with 4,8,12 or 18 µmol Cd/kg and d showed a marked protection against the lethal effects of arsenic (fig. 1a). Only one out of seven animals died within 30 d after poisoning with arsenic (LD_{80}). Animals pretreated with 12 or 18 µmol Cd/kg and d had a stagnation or decrease, respectively, in body weight, whereas lower doses of this metal led not to altered body weights.

Fig. 1a: Influence of pretreatment with Cd on the survival rate (30 d) of mice in acute arsenic poisoning The mice were pretreated with 0,1,2,3,4,8, 12 or 18 µmol Cd/kg and d s.c. followed by experimental arsenic poisoning i.p. (12.9 mg/kg As_2O_3) (see methods); n = 7/group.

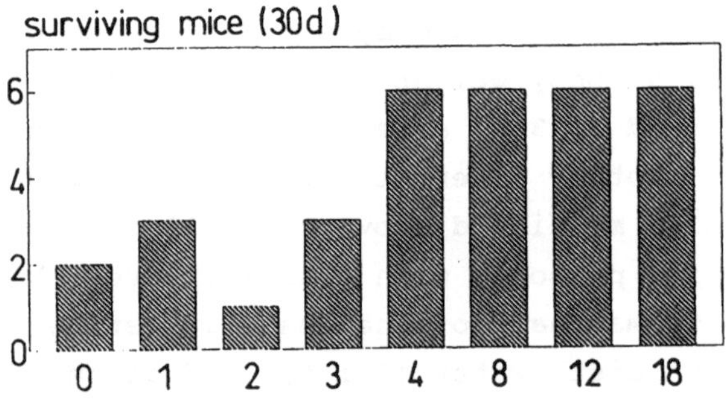

Fig. 1b: Influence of pretreatment with Zn on the survival rate (30 d) of mice in acute arsenic poisoning
The mice were pretreated with 0,25,50,100 or 200 μmol Zn/kg and d s.c. followed by experimental arsenic poisoning i.p. (12.9 mg/kg As_2O_3) (see methods); n = 7/group.

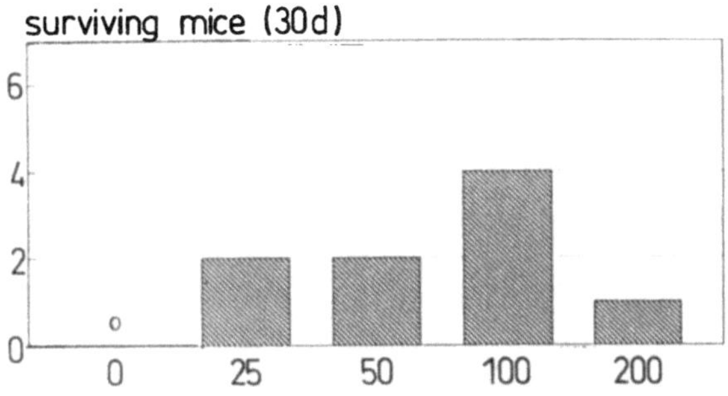

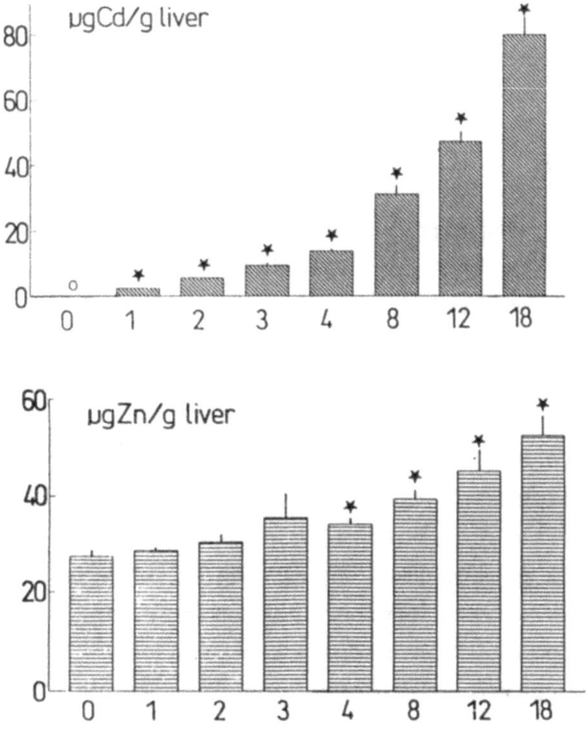

Fig. 2: Content of Cd or Zn in the liver
The mice were pretreated with 0,1,2,3,4,8, 12 or 18 μmol Cd/kg and d s.c. (see methods); n = 5/group; $\bar{x} \pm$ SE
* : sig. difference ($p < 0.05$) to the controls.

Fig. 3: **Cylindric discontinuous gel electrophoresis of standard MT I+II or liver homogenates**

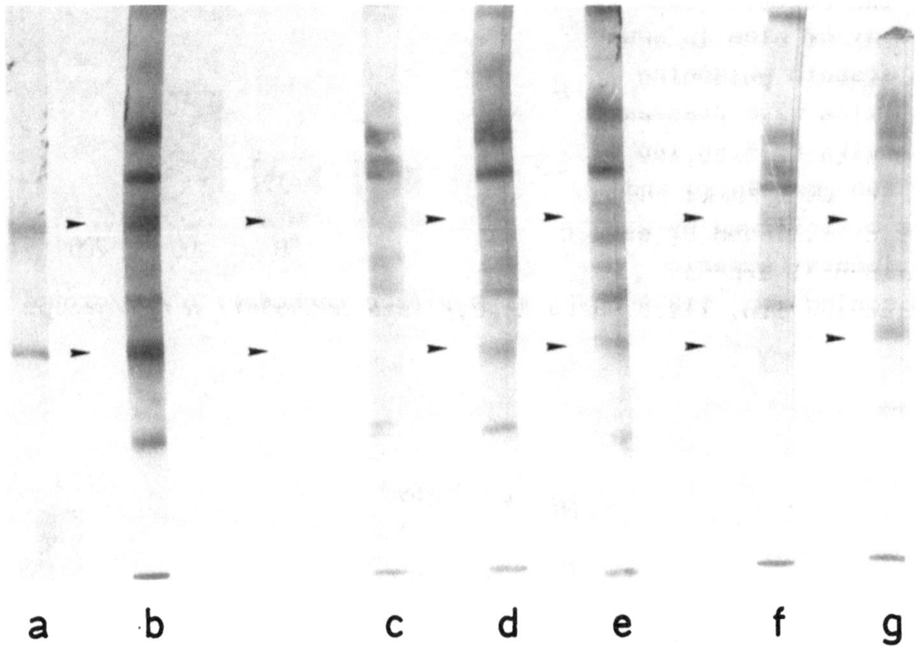

a) standard MT I+II
b) control and standard MT I+II
c) control
d) 12 μM Cd/kg and d
e) 18 μM Cd/kg and d
f) 100 μM Zn/kg and d
g) 200 μM Zn/kg and d

In NMRI mice pretreated with Zn after seven days on a low-Zn diet a distinct protective effect against arsenic was observed with 100 μmol Zn/kg and d (fig. 1b). Higher concentrations (e.g. 200 μmol Zn/kg and d) were toxic - four out of seven animals died during pretreatment and all animals had severe weight losses. The respective metal content of the organs studied increased in a dose dependent manner during pretreatment (fig. 2). With Cd there was also a concomitant rise in liver Zn proportional to the administered Cd dose. Cu and Fe contents did not vary under these conditions; no sample had detectable amounts of As.

The elevated metal contents of liver homogenates corresponded to a marked increase of MT concentrations, ascertained by semi-quantitative DISC gel electrophoresis, when Cd (12 or 18 μmol/kg and d) or Zn (200 μmol/kg and d) were given (fig. 3). These findings were confirmed by immunoblotting (fig. 4).

Fig. 4: **Immune blot of standard MT I+II or liver homogenates**

standard MT I+II	1c
standard MT I+II *	2c
control	1a,b
control *	2a,b
12 μM Cd/kg and	2d,e
12 μM Cd/kg and d *	1d,e
18 μM Cd/kg and d	2f
18 μM Cd/kg and d *	1f
100 μM Zn/kg and d	3a-d
100 μM Zn/kg and d *	4a-d
200 μM Zn/kg and d	3e,f
200 μM Zn/kg and d *	4e,f

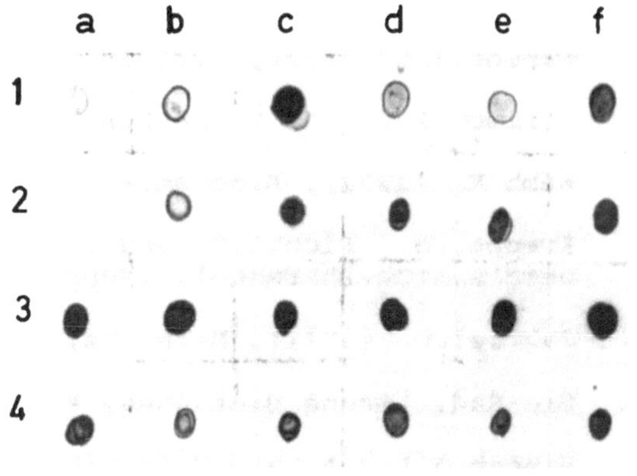

*: diluted with dist. water (1 : 10)

Discussion

Pretreatment of NMRI mice with low doses of Cd afforded a marked protection against the acute toxicity of arsenic. On the other hand, Zn, a chemical homologue of Cd and important biological trace element, was far less effective than Cd, and amounts for efficient pretreatment were in the range of the maximum tolerated dose.

At present, no sound explanation can be offered. The observed increase of MT content in the liver is a concomitant finding and may or may not be related to the reduction of As toxicity. Further studies are necessary to substantiate a possible role of MT in arsenic poisoning.

Acknowledgement

The skillful assistence of Mrs. E.Doklea is greatfully acknowledged.

References

1. Margoshes,M. and Vallee,B.L. (1957), J.Americ.Chem.Soc. 79, 4831

2. Kägi,J.H.R., Himmeljoch,S.R. Whanger,P.D. Bethune,J.L. and Vallee,B.L. (1974),J.Biol.Chem. 249, 3537

3. Nielson,K.B. and Winge,D.R. (1983), J.Biol.Chem. 258, 13063

4. Webb,M. (1972), Biochem.Pharmacol. 21, 2751

5. Kreppel,H., Fichtl,B. and Forth,W. (1987), Naun-Schmiedeberg's Arch.Pharmacol. (Suppl.) 334, 17

6. Gabriel,O. (1971), Meth.Enzym. 22, 565

7. Bio-Rad, Immune Blot Assay Kit, No 1706500

8. Habeeb,A.F.S.A. and Hiramoto,R. (1968), Arch.Biochem.Biophys. 126, 16

DETERMINATION OF THE SUPEROXIDE SCAVENGING ACTIVITY OF VARIOUS MANGANESE COMPOUNDS BY USE OF OPTICAL AND ESR TECHNIQUES

F. Nepveu

Laboratoire de Chimie Bioinorganique - INSERM U-305,
Université Paul Sabatier,38 rue des 36-Ponts, F-31400 Toulouse

A. Ducrocq, P. Derache
Université Paul Sabatier - INSERM U-87, 2, rue F. Magendie,
31400 Toulouse.

Abstract

The reactivity of some manganese complexes with the superoxide anion ($O_2^{\bar{\cdot}}$) was tested with different methods. No significant differences in the reactivity of the various compounds were observed using acetaldehyde/xanthine oxidase/cytochrome-c spectrophotometric assays or activated macrophages/luminol system with chemiluminescence measurements. Electron spin resonance combined with the technique of spin-trapping shows a ligand-dependent activity of the Mn complexes. Mn lactate, known as being a very effective scavenger of $O_2^{\bar{\cdot}}$ and taken as reference, exhibits the highest catalytic scavenging effect, then followed by Mn anthranilate. Mn complexes with malonic, tartaric, citric and picolinic acids exhibit a smaller reactivity upon disproportionation of the $O_2^{\bar{\cdot}}$ radical.

Trace Element Analytical Chemistry in Medicine and Biology

Introduction

The deleterious consequences of oxygen-derived radical production in aerobic organisms include lipid peroxidation, aging, arthritis and cancer. This production can be inhibited by specific metallo-enzymes (Fe, Mn, Cu/Zn superoxide dismutases (SOD)) but also by free hydrated metal ions and low molecular weight metal complexes. The interest in such metal compounds for various therapies explains the intensive search to define the essential properties that a metal complex must have to be an effective mimic of SOD. Parallel with these basic characteristics, there is still the question of reliable methods for detecting superoxide anion, an unstable free radical species, without ambiguity as to the nature of this radical. The multiplicity of available assays shows the difficulty of this analytical aspect (1,2,3).
In this work, the catalytic effect of some manganese complexes on the disproportionation of $O_2^{\cdot-}$ radical has been tested with three analytical methods: (i) spectrophotometric assay with the acetaldehyde (Ach)/xanthine oxidase (XO)/cytochrome c (Cyt-c) method, (ii) chemiluminescence (CL) measurements with activated macrophages/luminol system and (iii) free radical assays by electron spin resonance (ESR) and spin trapping technique. The performances of these methods and their aptitude for rapid availability are compared.

Materials and methods

Enzymes, zymosan, luminol, 5,5-dimethyl-1-pyrroline-N-oxide (DMPO), diethylenetriaminepentacetic acid (DETAPAC) were purchased from Sigma. Ach (Fluka), metallic salts (Merck), ligands (Aldrich), buffer reagents (Merck, Prolabo) were standard analytical grade chemicals used without further purification. The manganese complexes were prepared by mixing aqueous solutions of manganese(II) chloride and ligands. Precipitated complexes were filtered, washed with alcohol, ether and dried. The manganese compounds studied have been prepared from various carboxylic acid : lactic ($Cl(C_3H_5O_3)Mn.2H_2O$), malonic ($(C_3H_2O_4)Mn.2H_2O$), tartaric ($(C_4H_4O_6)Mn.2H_2O$), citric ($(C_6H_6O_7)Mn.1H_2O$), picolinic ($(C_{12}H_8N_2O_4)Mn.4H_2O$) and anthranilic ($(C_{14}H_{12}N_2O_4)Mn$). Analytical data are consistent with these formulas. These complexes will be termed as "Mn carboxylate" in the text.

Spectrophotometric measurements were performed with a Perkin-Elmer 554 spectrometer. SOD-like activity was determined from the difference in absorption at 550 nm using the Ach/XO/Cyt-c assay. According to previous work (4), assay conditions were: Ach 80 mM, XO 12 mU/ml, Cyt-c 100 µM ; final volume 1.5 ml with phosphate buffer 0.1 M, pH 7.4. The reaction was started by the addition of XO.

CL measurements were performed with a luminometer LKB. Peritoneal mice macrophages were prepared as previously described (5). Incubation conditions were: macrophages 4.10^6 cells per ml in Hank's balanced salts solution (HBSS) at 37°C, luminol 100mM, zymosan 100 µl (3mg/ml) ; final volume 600 µl obtained with HBSS. The reaction was started with zymosan.

ESR spectra were recorded at room temperature using a Bruker ER200D X-band spectrometer. DMPO was employed as a spin trap for superoxide anion. The reaction mixture contained: Ach 10 mM, XO 12 mU/ml, DMPO 50 mM, DETAPAC 15 µM; final volume 2 ml with phosphate buffer 0.1 M pH 7.4. The reaction was started by addition of Ach. Spin adduct concentration (DMPO-OOH) was calculated by measuring the signal intensity of the last but one downfield hyperfine component (six lines) at g=2.0006. The instrument settings were: power 19.7 mW, microwave frequency 9.72 GHz, central field set 3500 G, field scan 200 G, gain 4 to 6.10^6 (depending on the daily control), modulation amplitude 1 G, modulation frequency 100 kHz, time constant 0.5 s. The tested concentrations of the manganese compounds are given in the text and in Table 1.

Results

With the Ach/XO/Cyt-c system, the reduction of ferricytochrome is decreased by adding a manganese derivative. In the tested concentration range 1 µM to 100 µM, no significant difference of reactivity between the compounds has been observed with our experimental conditions. Archibald and Fridovich (4) showed that low levels of Mn(II) inhibited the reduction of ferricytochrome by O_2^- and higher levels caused a reoxidation of the ferrocytochrome when the superoxide anion was generated by the xanthine/XO reaction. As suggested by these authors, Ach has been used as substrate and others experimental details have been taken into account. Nevertheless, competitions between parallel oxido-reduction reactions cannot be completely avoided. These interferences obviously show the limits of these spectrophometric assays.

Table 1 compares the results obtained from CL measurements. About 50 % inhibition of luminescence is obtained with a 120 µM concentration, whatever the tested compound is. This could be due to a toxic effect and/or an interaction between luminol and the metal ion. Work is in progress to test this last hypothesis. At lower concentration studied the well-known high catalytic activity of Mn lactate is not observed. The complexity of these cellular systems makes difficult the interpretation of these results.

As presented in Table 1, the ESR assay with spin trapping technique gives the expected result for Mn(II) lactate, reported as being a very effective scavenger of $O_2^{\overline{\cdot}}$ among other manganese complexes derived from β-hydroxy acids (4,7). Anthranilic acid has no β-hydroxy function but its activity is nearest to Mn lactate. Others compounds exhibit weaker but comparable $O_2^{\overline{\cdot}}$ scavenging activities. This observation is encouraging for preparing new metal complexes with SOD-like properties. Some tests mixing solutions of hydroxy-acids and solutions of metallic salts have been carried out. The results were different and the $O_2^{\overline{\cdot}}$ scavenging effect was smaller.

Thus, free radical assay by ESR-spin trapping technique appears to be a rapid and sensitive method for testing $O_2^{\overline{\cdot}}$ scavenging properties of isolated metallic compounds introduced in various biological model systems.

	Chemiluminescence		ESR - Spin trapping	
Compounds	concentration µM	% inhibition	concentration µM	% inhibition
Mn Lactate	30 µM 60 µM 120 µM	7 23 52	9.1 µM 18.5 µM -	50 100 -
Mn anthranilate	30 µM 120 µM	12 50	27 µM 54 µM	50 100
Mn malonate	30 µM 120 µM	21 52	46 µM 90 µM	50 100
Mn picolinate	30 µM 120 µM	10 63	50 µM 100 µM	50 100
Mn tartrate	30 µM 120 µM	29 45	60 µM 120 µM	50 100
Mn citrate	30 µM 120 µM	12 52	63 µM 126 µM	50 100
Mn chloride	30 µM 60 µM 120 µM	23 33 55	90 µM 180 µM -	50 100 -

Acknowledgement

We are grateful to F. Anglade and J. Vandaele for technical assistance and to INSERM for financial support.

References

1. Fridovich I. 1982.In : Superoxide Dismutase, Vol.1 (L.W.Oberley, ed.) CRC Press, Inc, pp. 69-77

2. Bors W., M. Saran, C. Michel In: Superoxide Dismutase. Vol.2 (L.W Oberley, ed.) CRC Press, Inc.,pp 31-62 ; G.R Buettner, ibid, pp. 63-81.

3. Halliwell B., J.M.C. Gutteridge (eds). 1985. In: Free Radicals in Biology and Medecine, Oxford University Press, New York.

4. Archibald F.S., I. Fridovich. 1982. Arch. Biochem. Biophys.214,452-463.

5. De Maroussem D., B. Pipy, M. Beraud, P. Derache, J.R. Mathieu. 1985. Biochem. Biophys. Acta 834, 8-22.

6. Koppenol W.H., F. Levine, T.L. Hatmaker, J. Epp, J.D. Rush. 1986. Arch. Biochem. Biophys. 251, 594-599.

7. Glenn J.K., L. Akileswaran, M.H. Gold. 1986. Arch. Biochem. Biophys. 251, 688-696.

POSSIBILITIES OF DIAGNOSING THE ZINC, MANGANESE, COPPER, IODINE, SELENIUM AND CADMIUM STATUS

M. Anke, B. Groppel, U. Krause, L. Angelow, T. Kośla, A. Regius, T. Masaoka, E. Siegert

Karl-Marx-Universität Leipzig, Wissenschaftsbereich Tierernährungschemie, Dornburger Straße 24, Jena, GDR-6900

Introduction

The diagnosis of the trace element status in humans is extremely difficult since, as a rule, only blood, hair, nails and excretions are available for this purpose. The capacity of indicating the trace element status of several tissues is element-specific. Therefore, not all of them are suited as indicator tissues for the trace element status. Furthermore, it must be taken into consideration that the indicating capacity of organs in the case of deficiency can be different from that in the case of exposure. The investigation of these laws could not be carried out in humans, but in animals. In this way, it is possible to assess the capacity of the tissues available in humans to indicate the trace element status.

1. Zinc

The Zn status of the body and its organs is controlled homeostatically (Kirchgessner et al. 1977). According to the investigations in goats (table 1), the skeleton, i.e. the rib, the testicles, the hair and the cardiac muscle in our particular case, are least subjected to this regulation. On the average, the Zn content in the ribs of goats having died of Zn deficiency was reduced by 28 %. This finding is in good accordance with the results obtained by Berg and Kollmer (1987) who detected that about one third of the Zn was removed from the skeleton of rats in the case of Zn deficiency.
The Zn content in ribs, hair, serum, liver, kidneys and cerebrum was determined in a large number of cattle and sheep used in agriculture and correlations between the concentrations in the individual tissues were looked for (table 2). It was found that, under practical conditions, the Zn content of the ribs of cattle

Trace Element Analytical Chemistry in Medicine and Biology

Table 1: The capacity of 16 parts of the body and of the milk of 31 control and 22 zinc deficient goats to reflect the zinc status (mg/kg dry matter) (Anke and Risch 1979)

organ, tissue	control goats		Zn deficient goats		p	%
	s	$\bar{x}$	$\bar{x}$	s		
rib	16	80	58	14	< 0.001	72
testicles	23	74	55	18	< 0.05	74
hair	22	117	93	31	< 0.01	79
heart	24	75	63	18	< 0.05	84
liver	41	105	89	57	>	85
aorta	24	69	59	19	>	86
colostrum	14	41	36	10	>	88
blood plasma	0.58	1.16	1.06	0.62	>	91
spleen	18	88	80	23	>	91
carpal bone	15	58	54	15	>	93
lungs	19	67	63	15	>	94
cerebrum	15	56	54	9	>	96
ripe milk	7	27	26	10	>	96
bone marrow	8	29	29	12	>	100
kidneys	25	92	97	35	>	105
skeleton muscle	37	172	193	60	>	112
ovary	25	59	66	28	>	112
pancreas	83	86	97	76	>	113

and sheep actually correlated with that of blood serum. Hence it follows that both parts of the body reflect the Zn status, at least to a certain extent. The influence of stress while blood samples are taken must be underlined here (Reinhardt et al. 1986). This factor varies the Zn content of blood serum significantly.

Table 2: Correlations between the zinc content of several organs of cattle and sheep, resp.

organ, species		p	y	r
rib : serum	sheep	< 0.001	-0.14 + 0.014x	0.71
	cattle	< 0.05	0.75 + 0.010x	0.42
rib : cerebrum	sheep	< 0.001	36.2 + 0.20x	0.48
rib : liver	cattle	< 0.05	85.1 + 0.78x	0.35
hair : kidneys	cattle	< 0.01	63.8 + 0.24x	0.47
serum : cerebrum	sheep	< 0.05	43.2 + 9.1x	0.46

It is astonishing that the Zn content of ribs correlates with that of cerebrum and that of ribs with that of kidneys.
It can be summarized that several parts of the skeleton can reflect the Zn status and that there is also a certain correlation between blood serum and Zn status when the samples are not taken under stress conditions. The hair of animals and humans is less suited as indicator of the Zn status than rib and serum.

2. Manganese

The homeostatic regulation of the Mn content of organs is less distinct than that of Zn. The Mn content in the liver of goats having died of Mn deficiency was reduced by more than 50 % (table 3).

Table 3: The capacity of 16 parts of the body of 35 control and 22 manganese deficient goats to reflect the manganese status (mg/kg dry matter) (Anke and Risch 1979)

organ, tissue	control goats s	control goats $\bar{x}$	Mn deficient goats $\bar{x}$	Mn deficient goats s	p	%
liver	3.3	10.1	4.1	2.1	< 0.001	41
hair	1.1	2.1	1.1	0.5	< 0.001	52
kidneys	1.9	5.0	2.6	1.1	< 0.001	52
heart	0.9	2.2	1.2	0.8	< 0.001	54
ovary	2.3	4.6	2.6	1.5	< 0.05	57
bone marrow	1.3	2.5	1.6	0.6	< 0.01	64
aorta	1.5	2.2	1.4	0.8	< 0.05	64
spleen	1.6	3.5	2.4	0.8	< 0.01	69
pancreas	4.3	7.0	5.2	2.4	< 0.05	74
skeleton muscle	1.0	1.6	1.2	0.7	>	75
cerebrum	1.1	3.9	3.1	1.2	< 0.05	79
rib	3.6	9.7	7.9	4.1	>	81
blood plasma mg/l	0.06	0.09	0.08	0.03	>	89
lungs	1.2	2.3	2.1	1.6	>	91
testicles	1.2	3.0	2.8	1.3	>	93
carpal bone	2.5	6.3	6.1	1.3	>	97

The liver was the best indicator of the Mn status. Hair, kidneys, cardiac muscle, ovary, bone marrow, spleen and cerebrum also contained significantly more Mn than the corresponding organs of Mn deficiency goats. No significant differences were found in blood plasma.
The individual correlative calculation of the Mn content of liver, kidneys, blood serum, cerebrum, rib and wool of sheep used in agriculture demonstrated the suitability of the Mn status of liver and kidneys as indicator organs of the Mn status.

Table 4: The correlation between the Mn content of liver and kidneys of sheep

organs	p	y	r
liver : kidneys	< 0.001	$2.54 + 0.20x$	0.54

Liver, kidneys and the hair of animals as well (when the stage of development - anagenous, catagenous, telagenous hair - and the colour and kind of hair are taken into consideration) reflect the Mn status with decreasing reliability. The hair of the head of humans does not react in the same way as the periodically changed hair of animals. The hair of humans is continuously changed, i.e. as a rule, 85 to 90 % of human hair is in the anagenous phase, 1 % in the catagenous phase and 9 % in the telagenous phase. A change in these proportions leads to a great variability of the Mn content.

3. Copper

The Cu status was best reflected by the cerebrum of goats (table 5). As a rule, Cu deficient goats died when they had $<$ 50 % of the normal Cu content in the cerebrum. After the cerebrum, the Cu content of liver, blood plasma and hair reacted to the deficient Cu supply. Just like in cardiac muscle, significant differences between control and Cu deficient animals were detectable in these tissues. All other differences were not significant. The individual correlative calculation of the Cu content of cerebrum, liver, hair, kidneys and ribs of cattle confirmed the suitability of cerebrum, liver and, to a certain degree, of hair as indicator organs of the Cu status (table 6).

Table 5: The capacity of 16 parts of the body of 35 control and 25 copper deficient goats to reflect the copper status (mg/kg dry matter) (Anke and Risch 1979)

organ, tissue	control goats		Cu-deficient goats		p	%
	s	$\bar{x}$	$\bar{x}$	s		
cerebrum	4.0	14	6.4	1.8	< 0.001	46
liver	6.8	10	5.3	2.5	< 0.001	52
blood plasma (mg/l)	0.24	0.67	0.41	0.19	< 0.01	61
hair	1.5	4.7	3.0	1.4	< 0.001	64
testicles	2.9	6.5	4.5	1.5	< 0.05	69
pancreas	2.6	4.6	3.4	2.0	>	74
aorta	2.2	3.7	2.9	1.3	>	78
rib	3.2	7.3	6.1	2.0	>	84
heart	3.4	13	11	2.9	< 0.01	85
spleen	1.6	4.8	4.1	1.4	>	85
lungs	2.4	9.3	8.0	3.1	>	86
kidneys	3.4	11	9.7	3.6	>	88
skeleton muscle	1.3	3.6	3.2	1.3	>	89
carpal bone	2.2	5.4	4.8	2.0	>	89
bone marrow	2.8	7.2	7.9	4.2	>	110
ovary	5.6	10	12	8.7	>	120

When the Cu content of the liver was < 35 mg/kg dry matter the Cu content of liver and cerebrum, of liver and hair and of cerebrum and hair correlated with r 0.68 to 0.63. In animals with a Cu content in the liver > 34 mg/kg dry matter, the dependence of the Cu content became weak or disappeared since the capacity of accumulating Cu is lower in cerebrum and hair than in the liver. The liver of cattle can store up to 1000 mg Cu/kg dry matter. Cerebrum and hair can not at all accumulate such Cu amounts. Therefore, the linear dependence of the Cu content between these 3 indicator organs disappears.
Cerebrum and liver reflect the Cu status under practical conditions. The hair is only suitable as indicator tissue of the Cu status under defined conditions.

Table 6: The correlations between the copper content of several organs of cattle depending on the copper content of the liver (mg/kg dry matter)

organs,	Cu content of the liver (mg/kg dry matter)	p	y	r
liver : cerebrum	< 35 mg	< 0.001	4.33+0.32x	0.68
	> 34 mg	< 0.01	13.0+0.024x	0.30
	all samples	< 0.001	10.4+0.038x	0.49
liver : hair	< 35 mg	< 0.001	3.90+0.14x	0.63
	> 34 mg	> 0.05	-	(0.13)
	all samples	< 0.001	6.22+0.008x	0.32
cerebrum : hair	< 35 mg	< 0.01	3.22+0.27x	0.64
	> 34 mg	> 0.05	-	(0.15)
	all samples	> 0.05	-	(0.10)

4. Iodine

The I content of organs and tissues is not at all regulated homeostatically (table 7). I deficient goats with 40 µg I/kg ration dry matter only excreted via milk 2 % of the I amount secreted by control goats (500 µg I/kg ration dry matter).
All parts of the body reflected the I status significantly. Blood serum and hair are also good indicators of the I status and it is not of importance if TI or PBI is selected as standard.
The I concentration of the blood serum of calves, fattening cattle, lambs and kids used in agriculture correlated with the I content of the organs investigated very closely (0.44 - 0.99). Blood serum, milk and hair indicate the I status reliably.

Table 7: The capacity of 13 parts of the body and the milk of control and iodine deficient goats to reflect the iodine status (µg/kg dry matter) (Groppel and Anke 1983)

organ, tissue	control goats		I deficient goats		p	%
	s	$\bar{x}$	$\bar{x}$	s		
ripe milk	71	247	6	1	< 0.001	2
lungs	248	374	36	26	< 0.001	10
pancreas	272	344	51	22	< 0.05	15
liver	84	180	37	12	< 0.001	21
spleen	279	300	70	45	< 0.001	23
blood serum, TI (µg/l)	28	64	15	5	< 0.001	23
blood serum, PBI (µg/l)	24	52	12	5	< 0.001	23
kidneys	134	261	64	37	< 0.001	25
aorta	299	586	159	65	< 0.001	27
ovary	108	334	101	48	< 0.001	30
heart	66	135	46	17	< 0.001	34
hair	90	192	68	33	< 0.001	35
skeleton muscle	81	202	108	59	< 0.01	53
cerebrum	38	80	50	23	< 0.05	62
rib	24	57	38	18	< 0.05	70

5. Selenium

The Se content of the body and of individual tissues is also less homeostatically regulated than that of Zn, Mn and Cu (table 8). The blood serum only contains about 25 % of the Se amount found in control animals. Cardiac muscle and milk are also good indicators of the Se status. Only the cerebrum did not react to Se deficiency with a significantly decreased Se content.

It was possible to study Se depletion in blood serum and hair over a long period of time (table 9). As early as after 8 weeks of trial, the Se content of hair plateaued and kept this level up to 6 months after the beginning of trials. Just like in hair, the Se content of the blood serum was reduced by two thirds after 2 months and kept on decreasing to about 20 % of the Se amount found in control animals. Hence it follows that the

Table 8: The capacity of 12 parts of the body and the milk of control and selenium deficient goats to reflect the selenium status (µg/kg dry matter) (Angelow et al. 1986)

organ, tissue	control goats		Se deficient goats		p	%
	s	$\bar{x}$	$\bar{x}$	s		
blood plasma	29	130	36	13	< 0.001	28
heart	190	1045	386	106	< 0.001	37
aorta	86	364	135	11	< 0.001	37
ripe milk	92	247	93	32	< 0.001	38
skeleton muscle	85	387	146	28	< 0.001	38
liver	330	1123	464	78	< 0.001	41
rib	40	170	69	18	< 0.001	41
lungs	192	1239	515	72	< 0.001	42
hair	20	350	150	30	< 0.001	43
spleen	275	1648	838	160	< 0.001	51
kidneys	1221	5011	2771	349	< 0.001	55
pancreas	320	1161	690	67	< 0.01	59
cerebrum	112	716	604	111	> 0.05	84

blood serum reacted more sensitively to Se deficiency than hair. Both are suited as indicators of the Se status.

Table 9: The selenium content of the serum and hair of control and selenium deficient goats depending on the length of experiments (µg/l or kg dry matter)

length of experiment,	tissue	control goats		Se deficient goats		p	%
		s	$\bar{x}$	$\bar{x}$	s		
2 months	serum	25	103	38	16	< 0.001	37
	hair	42	353	129	46	< 0.001	39
4 months	serum	17	161	28	2	< 0.001	17
	hair	96	333	129	46	< 0.001	39
6 months	serum	11	126	31	8	< 0.001	25
	hair	114	377	131	17	< 0.001	35

6. Cadmium

The importance of the trace element cadmium is in the toxic range. In this way it is different from the other elements described in this paper. Cd mainly accumulates in kidneys (table 10).

Table 10: The capacity of 15 parts of the body of male and female goats to reflect cadmium exposure (kidneys = 100) (Kronemann and Anke 1983)

organ, tissue	$\bar{x}$	s
kidneys	100	0
liver	30	13
hair	17	11
aorta	8.4	7.1
ovary	5.3	0.0
spleen	5.2	3.6
lungs	5.0	1.9
heart	4.5	4.3
skeleton muscle	3.0	2.1
testicles	1.6	0.0
rib	1.3	0.8
carpal bone	1.1	0.3
blood plasma	1.1	0.0
bone marrow	1.2	1.4
cerebrum	0.6	0.3

The liver accumulates only 30 % of the Cd amount which was found per kg kidney dry matter. The hair incorporated 17 % of the kidney Cd concentration and all other organs stored $<$ 10 % in their tissue. The cerebrum of goats contained least Cd. The individual correlations of the Cd content of kidneys, liver, hair, muscle, cerebrum, metatarsal bone and blood serum was investigated in horses used in agriculture (table 11).
It was found out that the Cd content of kidneys and liver of horses correlated most intensively (r 0.91). Apart from blood serum, it was possible to calculate significant correlations. Hence it follows that the Cd content of kidneys, liver and, to a certain extent, of hair delivers data on a Cd exposure. Hair can also take up exogenous Cd and therefore, it is not as much the indicator of the individual Cd status as of the Cd exposure

Table 11: The correlations between the cadmium content of several organs of horses

organ, tissue	p	y	r
kidneys : liver	< 0.001	-39 + 0.20x	0.91
kidneys : cerebrum	< 0.001	0.281 + 0.0007x	0.78
kidneys : muscle	< 0.001	1.24 + 0.0052x	0.71
kidneys : hair	< 0.001	0.094 + 0.00007x	0.44
kidneys : metatarsus	< 0.05	1.88 + 0.0006 x	0.27
liver : cerebrum	< 0.001	0.391 + 0.003x	0.88
liver : muscle	< 0.001	2.09 + 0.025x	0.79
liver : hair	< 0.001	0.101 + 0.0004x	0.62
liver : metatarsus	< 0.05	2.02 + 0.0032x	0.29
hair : cerebrum	< 0.001	0.211 + 3.27x	0.60
hair : muscle	< 0.001	1.26 + 22x	0.47
hair : metatarsus	< 0.05	1.75 + 3.40x	0.26
muscle : cerebrum	< 0.001	0.32 + 0.083x	0.67
muscle : metatarsus	< 0.01	1.83 + 0.11x	0.36

of the area. This finding is also important for the identification of a Cd exposure at work or in a Cd emission area.

Summary

The Zn status of several tissues is most difficult to register. Various parts of the skeleton (rib) and, to a certain extent, blood serum (and hair) are suitable for that purpose. The Mn status is best reflected by liver, kidneys (and hair). Cerebrum, liver (and serum or hair) deliver data on the Cu status.
The I status is reflected by milk and all parts of the body. The Se status is best indicated by blood serum and the Cd status by kidneys. The blood serum is not suited for the identification of a Cd exposure. A critical analysis of the indicating capacity of individual tissues is necessary.

Literature

1. Angelow, L., M. Anke, U. Krause. 1986. In: M. Anke et al. (eds.) Mengen- und Spurenelemente 6, p. 202, Karl-Marx-Univ. Leipzig, GDR

2. Anke, M., M. Risch. 1979. Haaranalyse und Spurenelementstatus. VEB Gustav Fischer Verlag Jena

3. Berg, D., W. E. Kollmer. 1987. In: L. S. Hurley et al. (eds.) Trace Element Metabolism in Man and Animal 7, p. 87 abstracts

4. Groppel, V., M. Anke. 1983. In: M. Anke et al. (eds.) Spurenelement-Symposium 4, p. 156

5. Kirchgessner, M., W. A. Schwarz, H.-P. Roth. 1977. In: M. Kirchgessner (ed.) Trace Element Metabolism in Man and Animal 3, p. 116

6. Kronemann, H., M. Anke. 1983. In: M. Anke et al. (eds.) Mengen- und Spurenelemente 3, p. 141, Karl-Marx-Univ. Leipzig, GDR

7. Reinhardt, F., H.-Ch. Gabsch, M. Hanefeld. 1986. In: M. Anke et al. (eds.) Mengen- und Spurenelemente 6, p. 161, Karl-Marx-Univ. Leipzig, GDR

ESSENTIALITY OF THE TRACE ELEMENT BROMINE

M. Anke, B. Groppel, W. Arnhold, M. Langer

Karl-Marx-Universität Leipzig, Wissenschaftsbereich Tierernährungschemie, Dornburger Straße 24, Jena, GDR-6900

Introduction

Together with F, Cl and I, Br belongs to the 7th group of the Periodic System of Chemical Elements. It occurs in large amounts in the biosphere. 10 to 515 mg Br/kg were found in the soil of Great Britain; on an average 54 mg/kg. Plants of this area contained between 5 and 157 mg Br/kg dry matter; on an average 45 mg/kg (Wilkins 1978). Similar Br concentrations in soils (81 and 63 mg Br/kg, resp.) and vegetation (30 and 6.4 mg Br/kg, resp.) were found in Japan (Yuita et al. 1982a, b). Compared to I ten times more Br is taken up by the flora from the soil. Br accumulates in plants (Werchowskaja 1958).

Foodstuffs from Central Europe contain Br concentrations which exceed those of Cu and which come closer to those of Zn. Fish can accumulate particularly much Br (Grote 1980, Montag and Grote 1981).

In spite of or due to its abundant occurrence, it has not yet been possible to detect significant essential functions of the trace element Br in microbes, plants, animals and humans. (Nielsen 1986).

Only Oe et al. (1981) found low Br concentrations in serum and cerebrum of patients with haemodialysis. They saw a correlation between the insomnia observed in patients and the low Br values. Br supplementations improved the capacity of sleeping. This finding is not new. As early as 50 years ago, Zondek and Bier (1932) reported the potential existence of a Br-containing sleeping hormone in the pituitary gland of dogs. This hypothesis has not yet been confirmed though Br was used as sleeping draught before barbiturates.

Within the systematic analysis of the essentiality of several trace elements (Ni, As, Li, Cd, F, Al, V), the influence of a Br-poor diet on the prenatal and postnatal growth, the reproduction performance, milk production and mortality of goats has also been investigated now.

Trace Element Analytical Chemistry in Medicine and Biology

Material and methods

The Br deficiency experiments began in July 1986 and were repeated twice with growing, gravid and lactating goats (table 1).

Table 1: Control and experimental goats of the bromine deficiency experiment

year	control goats ash	straw	cellulose	Br deficient goats	remark
1986/87	6	6	6	6	2 Br deficient goats died
1987/88	0	5	5	5	2 Br deficient goats died

Apart from the Br deficiency group, 3 control groups were included in the experiment. The cellulose group was directly compared to the Br deficiency group. Both groups only differed in the Br offer.
The cellulose control goats were given about 20 mg Br/kg dry matter in their ration, the Br deficient goats $<$ 1 mg Br/kg dry matter. Cellulose control animals took in similar Br amounts as herbivores in nature. Straw was used as litter for straw control goats. This straw was eaten by the animals instead of cellulose. Thus, they had an additional supply with all inorganic substances occurring in straw. In addition to the components of the synthetic control ration (table 2), ash control animals were given 1 % lignite ash and furthermore, they had an additional supply with all components of ash. The effect of other inorganic components of the ration whose essentiality has not yet been detected was to be excluded in this way.
Since the digestibility of straw is worse then that of cellulose, straw control goats gained less weight than the cellulose control animals. The ash control goats gained 11 g less weight daily than the cellulose control animals. Thus, we can proceed from the assumption that the cellulose control ration approximately meets the requirements. The goats of all groups took part in the experiment until exitus.
The composition of the semi-synthetic ration is represented in table 2.

Table 2: The composition of the semi-synthetic ration

components	amount in 100 kg	
potato starch	48.30	kg
beet sugar	32.00	"
casein	10.00	"
urea	3.00	"
sunflower oil	3.00	"
KH_2PO_4	1.34	"
$CaCO_3$	1.00	"
NaCl	350.0	g
$Al_2(SO_4)_3$, 60 %	200.0	"
MgO	220.0	"
K_2SO_4	350.0	"
$ZnSO_4 \times 7\ H_2O$	50.0	"
$FeSO_4 \times 7\ H_2O$	45.0	"
$MnSO_4 \times 4\ H_2O$	40.0	"
S	35.0	"
Li_2CO_3	10.6	"
$CuSO_4 \times 5\ H_2O$	4.0	"
KBr	3.0	"
$Ni_2SO_4 \times 7\ H_2O$	3.4	"
$Cr_2(SO_4)_3 \times 18\ H_2O$	600	mg
$(NH_4)\ VO_3$	460	"
NaF	220	"
$NaWO_4 \times H_2O$	180	"
KJ	100	"
$(NH_4)_6\ Mo_7O_{24} \times 4\ H_2O$	92	"
SeO_2	80	"
$CoSO_4 \times 7\ H_2O$	80	"
As_2O_3	40	"
$CdCl_2 \times H_2O$	36	"
vitamin A	2000	"
vitamin D_3	400	"
vitamin E	20000	"

The animals were fed ad libitum every day. Residual feedstuffs were weighed every day. The drinking water of goats was destilled. They were weighed every fortnight, at 7 a. m. Their kids were weighed every 7 days.

Results

1. Growth

The kids of Br deficient goats had a significantly lower birth weight than control goats (table 3). The difference is insignificant compared to the results obtained in other deficiency experiments with Mo, V, As, Ni, F and Cd, they were considerably greater (Anke and Groppel 1988).

Table 3: The influence of bromine deficiency on the pre- and postnatal development of kids

parameter (n)	control kids s	control kids $\bar{x}$	Br deficient kids $\bar{x}$	Br deficient kids s	p	%
first day of life (kg) (30; 9)	0,68	2,7	2,4	0.81	>0.05	89
168 experimental days (g/day) (11;10)	34	109	66	60	>0.05	61

After weaning on the 91st day of life, the kids were given the control and Br deficiency ration, resp. (table 2) and their live weight gain was registered during 168 days (table 3).
Even during their intrauterine development, the kids of Br deficient goats gained 11 % less weight. It was not possible to demonstrate the influence of the Br deficiency on live weight gain during the suckling periode since only 2 kids were still alive on the 91st day of life. The Br deficiency trials were continued with these 2 kids and 9 animals being not depleted intrauterinely. The kids with Br-poor rations gained 39 % less weight during the following 168 days of life. This reduced weight gain corresponds to that of Se deficiency (Anke et al. 1987b).

2. Reproduction

Br-poor rations did not take effect on heat intensity. But it reduced the success of the first insemination and the conception rate of goats significantly (table 4). All control goats became

Table 4: The influence of bromine deficiency on reproduction performance and life expectancy

parameter	control goats	Br deficient goats	p
success of first insemination (%)	87	50	< 0.05
conception rate (%)	100	57	< 0.01
services per gravidity	1.1	1.1	> 0.05
abortions (%)	0	12	> 0.05
kids per goat carrying to terms	1.5	1.3	> 0.05

gravid whereas 43 % of the animals with Br-poor rations were not gravid. The number of services per gravidity was the same in both groups. This finding points to the fact that, on the average of both groups, repeated matings took the same effect on the conception rate.
In the goats which were not gravid after the first service, further matings only led to gravidity in one case. One Br deficient goat aborted. The control goats gave birth to more kids than Br deficient goats. The difference remained insignificant. It can be summarized that Br-poor rations influenced the reproduction performance significantly.

3. Milk and milk fat production

The Br-poor nutrition of goats reduced milk production by 20 % (table 5). The difference between control and Br deficient goats is insignificant. The influence of Br deficiency on the fat content is less remarkable. The difference between the groups only amounts to 8 %. The reduced milk production and the lower milk fat content resulted in a milk fat production of Br deficient goats significantly decreased by 25 %. The influence of Br deficiency on the milk fat production is greater than that found in Se, As and Cd deficient goats (Anke et al. 1987a, 1986a, b). However, the amount of produced milk would have been absolutely sufficient to meet the energy and protein requirements of Br deficient kids.

Table 5: The influence of bromine deficiency on the milk performance of goats during the first 56 days of lactation

data	control goats s	control goats $\bar{x}$	Br deficient goats $\bar{x}$	Br deficient goats s	p	%
milk (g/day)	260	1424	1145	797	> 0.05	80
fat (%)	0.58	3.68	3.20	0.78	> 0.05	87
fat (g/day)	13.1	52.5	39.5	30.3	> 0.05	75

4. Blood picture

It was astonishing that the Br-poor nutrition of goats also influenced the haemoglobin and haematocrit values of the blood (table 6). Compared to the control group and to the experimental

Table 6: The influence of bromine deficiency on the blood picture of female goats

parameter	s	$\bar{x}$	$\bar{x}$	s	p	%
haemoglobin, mmol/l	0.89	6.8	5.7	1.0	< 0.05	84
haematocrit, PCV	0.04	0.32	0.29	0.07	> 0.05	91
MCHC, mmol/l	0.68	21.3	19.9	1.2	< 0.01	93
leucocytes, gp/l	2.7	8.8	9.9	3.2	> 0.05	112

F, Al, Na, Ni, As, Co, Li, Cd and V deficient goats, Br deficient goats had the by far worst haemoglobin and haematocrit level. The blood analysis presented here was carried out in January 1988, in the 3rd or 4th month of gravidity.

Table 7: The influence of age on the haemoglobin level of the blood of control and bromine deficient goats

gravidity (n)	control goats s	control goats $\bar{x}$	Br deficient goats $\bar{x}$	Br deficient goats s	p	%[1]
1.	0.26	7.1	6.2	1.2	> 0.05	87
2.	1.1	6.6	5.3	0.82	> 0.05	80
p		> 0.05	> 0.05			-
%		93	85			

[1] control goats $\hat{=}$ 100 %

As a rule, older goats had a worse haemoglobin status than younger ones (table 7). The reduction of the haemoglobin content of the blood was stronger in 2-year-old Br deficient goats.

5. Mortality

The Br-poor ration of goats did not take a significant effect on the mortality of adult goats (table 8). But it influenced the life expectancy of intrauterinely Br-depleted kids significantly. The majority of kids died during the suckling period without specific deficiency symptoms in a good nutritional state.

Table 8: The influence of bromine deficiency on the mortality of goats and their kids

data	control goats $\bar{x}$	Br deficient goats $\bar{x}$	p
died kids, %	0	67	< 0.001
died adult goats, %	18	27	> 0.05

Discussion

The essentiality of Br has already been investigated repeatedly. Winnek and Smith (1937) could not register an influence of a Br-poor ration on growth, reproduction performance and health in rats over a period of 11 weeks. A small growth response to dietary trace addition of Br has been reported for chicks (Huff et al. 1956) and mice (Bosshardt et al. 1956) fed a semi-synthetic diet containing iodinated casein to produce a hyper-thyroid-induced growth retardation. The Br content of the basal diet was not given, and the indications of growth requirement for Br have apparently not been investigated further.
The experiments with goats fed a Br-poor ration ($<$ 1.0 mg Br/kg ration) which were repeated twice showed an influence on growth, conception rate, milk fat production, the haemoglobin level and the mortality of the offspring. The results obtained up to now do not yet allow to make a statement on the essentiality of Br. They must be confirmed by at least 3 further repetitions and metabolic investigations.
Human dietary intakes of Br are large and variable. Duggan and Lipscomb (1971) obtained an average Br intake of 24 mg/day over

2 years from US diets, while Hamilton and Minski (1972/73) reported a mean intake of 8.4 mg/day from English total diets and Varo and Koivistoinen (1980) calculated an average intake of 4.2 mg/day from Finish diets. Actual intakes will be much higher where organic bromine compounds are used as fumigants for soils and stored grains.

Summary

In experiments with growing, gravid an lactating goats which were repeated twice, a Br-poor nutrition (< 1 mg Br/kg ration dry matter) led to a significantly reduced growth, a worse conception rate, reduced milk fat production, decreased haemoglobin content and a lower life expectancy of the offspring intrauterinely Br-depleted. Further investigations are necessary for the clarification of the essentiality of Br.

Literature

1. Anke, M., L. Angelow, B. Groppel, W. Arnhold. 1987b. In: Anke, M. et al. (eds.) Mengen- und Spurenelemente. 7, p. 440, Karl-Marx-Univ. Leipzig, GDR

2. Anke, M., L. Angelow, B. Groppel, T. Košla, M. Langer. 1987a. In: Anke, M. et al. (eds.) Mengen- und Spurenelemente. 7, p. 431, Karl-Marx-Univ. Leipzig, GDR

3. Anke, M., B. Groppel. 1988. Zbl. Pharmazie, Pharmakotherapie und Laboratoriumsdiagnostik 127, 197

4. Anke, M., B. Groppel, A. Schmidt, H. Kronemann. 1986b. In: Anke, M. et al. (eds.) Spurenelement-Symposium 5, p. 937 Karl-Marx-Univ. Leipzig, Friedrich-Schiller-Univ. Jena, GDR

5. Anke, M., A. Schmidt, U. Krause, B. Groppel, K. Gruhn, G. Hoffmann. 1986a. In: Anke, M. et al. (eds.) Mengen- und Spurenelemente 6, p. 225, Karl-Marx-Univ. Leipzig, GDR

6. Bosshardt, D. K., I. W. Huff, R. H. Barnes. 1956. Proc. Soc. Exp. Biol. Med. 92, 219

7. Duggan, R. E., G. Q. Lipscomb. 1971. J. Dairy Sci. 54, 695

8. Grote, B. 1980. Wellenlängedispersive Röntgenfluoreszenzspektrometrie von Brom und Jod in Lebensmitteln. Diss. Univ. Hamburg, FRG

9. Hamilton, E. I., M. J. Minski. 1972/73. Sci. Total Environ. 1, 375

10. Huff, J. W., D. K. Bosshardt, O. P. Miller, R. H. Barnes. 1956. Proc. Soc. Exp. Biol. Med. 92, 216

11. Montag, A., B. Grote. 1981. Z. Lebensm. Unters. Forsch. 172, 123

12. Nielsen, F. H. 1986. In: W. Mertz (ed.) Trace Elements in Human and Animal Nutrition - Fifth Edition. Academic Press, Inc. Orlando, USA, p. 415

13. Oe, P. L., R. D. Vis, I. H. Meijer, F. van Langevelde, W. Allen, C. v. d. Meer, H. Verkeul. 1981. In: J. McC. Howell (ed.) Trace Element Metabolism in Man and Animals. Australian Academy of Science, Canberra, p. 526

14. Varo, P., P. Koivistoinen. 1980. Acta Agric. Scand., Suppl. 22, 165

15. Werchowskaja, I. N. 1958. In: A. P. Winogradow, M. Trénel (eds.) Spurenelemente in der Landwirtschaft, Akademie-Verlag, Berlin p. 613

16. Wilkins, C. 1978. J. agric. Sci. Camb. 90, 109

17. Winnek, P. S., A. H. Smith. 1937. J. Biol. Chem. 121, 345

18. Yuita, K., Y. Nobusawa, M. Shibuya, S. Aso. 1982a. Soil Sci. Plant Nutr. 28, 315

19. Yuita, K., S. Akabe, M. Shibuya, S. Aso. 1982b. Soil. Sci. Plant Nutr. 28, 499

20. Zondek, H., A. Bier. 1932. Klin. Wochenschr. 11, 759

INTER-INDIVIDUAL VARIABILITY OF THE KINETICS OF LEAD IN MAN

S. Heusler-Bitschy, R. Knutti, Ch. Schlatter
Institute of Toxicology , Federal Institute of Technology & University of Zürich, CH-8603 Schwerzenbach, Switzerland

Abstract

In surveys of blood-lead concentrations in occupationally not lead-exposed adults a high inter-individual variability is generally found, which can not easily be explained just by differences of lead intake with food.

The aim of this investigation was to look for the factors governing the individual steady state concentrations of lead in blood. In a preliminary step, individual blood-lead concentrations measured by graphite furnace atomic absorption spectrometry were shown to be stable both during the course of a single day (CV < 2%) and also throughout the year (CV 6 to 23%).

Eight volunteers with "environmental" blood-lead concentrations ranging from 2.5 µg/dl (volunteer H) to 10 µg/dl (volunteer D) participated in the main part of the study which was carried out with single oral doses of 0.5 and 1.5 mg of isotopically labeled lead (= "experimental" lead, 69.4% ^{204}Pb). Kinetic data were derived from the determination of lead in blood, urine and feces by Isotopic Dilution Mass Spectrometry and can be summarized as follows:

1. The estimated degree of intestinal absorption from the empty stomach varied from 10 to 80%.
2. For seven volunteers the half-life of lead in blood was in the range of 39 to 53 days. Only volunteer H with the lowest environmental blood-lead concentration of 2.5 µg/100ml also had a much lower half life of only 24 days.

Trace Element Analytical Chemistry in Medicine and Biology

3. There was no significant difference of renal clearance with regard to the experimental lead between the two volunteers with the lowest and the highest blood-lead concentration (H: 56±11 µl/min, D: 61±15 µl/min).
4. Fecal clearance was higher in volunteer H (50 to 135 µl/min) than in volunteer D (20 to 40 µl/min).

The fourfold difference in blood-lead concentrations between volunteers H and D can not be accounted for by the differences in kinetic parameters investigated in this study.

Introduction

Whereas quite a lot is known about the average pharmacokinetic characteristics of many xenobiotics - including lead (1) - we know much less about the person to person variability of such parameters. The starting point of this study (2) was a survey on blood-lead concentrations in adult persons without known occupational exposure to lead (Fig. 1). Blood lead levels in members from the author's institute spanned a range from 2.5 to about 10 µg/100 ml. As it seemed unlikely that the full

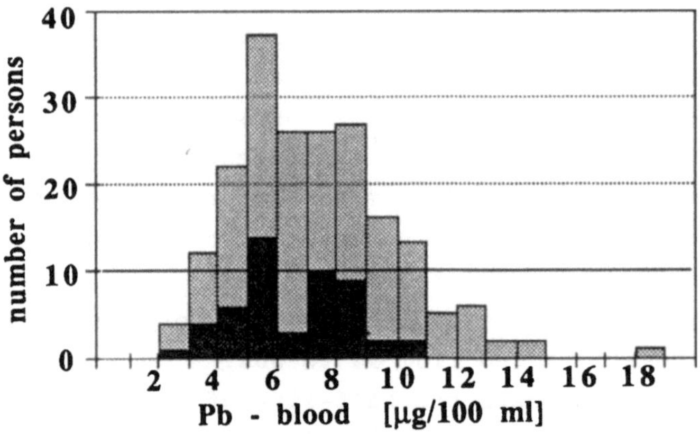

Fig.1 Distribution of blood-lead concentrations in 191 persons without occupational exposure to lead living in the area of Zürich/Switzerland (1983). Dark shaded: members of the authors institute.

range of this variability can be assigned just to differences in lead intake, it was decided to investigate physiological factors playing a role in lead kinetics. Whereas the low blood-lead concentrations found in all our volunteers are considered as safe with respect to health, marked individual differences in susceptibilities might be of importance in persons occupationally exposed to lead and other heavy metals.

In a first step intra-individual variabilities of the "environmental" blood lead concentrations were determined. The main part of the experiments, then, was done with single oral doses of isotopically labeled lead (= "experimental" lead, 69.4% ^{204}Pb). The aspects investigated were (a) half-life of the experimental lead in blood, (b) renal clearance and (c) fecal elimination of absorbed lead (endogenous fecal clearance).

Analytical Methods

Intra-individual diurnal and seasonal variability of the environmental blood-lead concentrations were determined with our standard method (3) using graphite furnace atomic absorption spectrometry (AAS) (Perkin Elmer Model 5000, HGA 500). For the determination of the pharmacokinetik behaviour of single oral doses of experimental lead samples of venous blood (0.5 ml), urine (2 ml) and homogenized and lyophilized feces (0.5 g) were analyzed by Isotopic Dilution Mass Spectrometry (IDMS) (MAT Model 261). For calibration, a known amount of ^{208}Pb was added as an internal standard (2).

Results

Diurnal variability of individual blood-lead concentrations

4 volunteers participated in this part of the study. During a single day venous blood samples were collected at 8:00 am, 2:00 pm, and 5:00 pm. The mean coefficient of variation for the individual was less than 2%, i.e. within the precision of the AAS method and there was no evidence of physiologically caused diurnal variations as found by Lehnert and Szadkowski (4).

Seasonal variability of individual blood-lead concentrations

The determination of blood-lead concentrations in 10 persons throughout a year resulted in individual coefficients of variation in the range of 6 to 23% (Fig. 2). There were no indications of common seasonal trends, confirming data published by Delves et al. (5). The larger variabilities during the 10-month period probably reflect individual variations of lead intake with food.

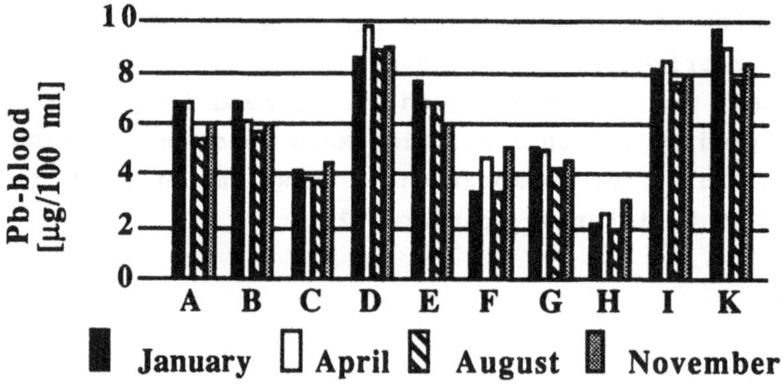

Fig.2 Seasonal variability of individual environmental blood-lead concentrations in ten volunteers from the authors laboratory.

Half-life of the disappearance of experimental lead from blood

Seven volunteers were given 0.5 mg and one 1.5 mg of experimental lead in the form of an aqueous solution of the carbonate salt. This dose was taken in the middle of a 12-hours fasting period, in order to reduce interfering effects of other food constituents with the absorption from the intestinal tract. A large variation was found in the estimated degree of absorption from the intestinal tract ranging from about 10% to about 80%. Interestingly, the highest value is found for volunteer H with the lowest environmental blood lead concentration, whereas the lowest degree of absorption is associated with volunteer D, whose environmental blood-lead concentration is around 10 µg/dl.

For seven of the eight volunteers the half-lives of experimental lead in blood determined from samples taken at days 43 and 87 lie within a narrow range of 39 to 53 days. Only volunteer H, the person with the lowest environmental blood-lead concentration of 2.5 µg/dl exhibits an appreciably faster rate of elimination from blood with a half-life of only 24 days.

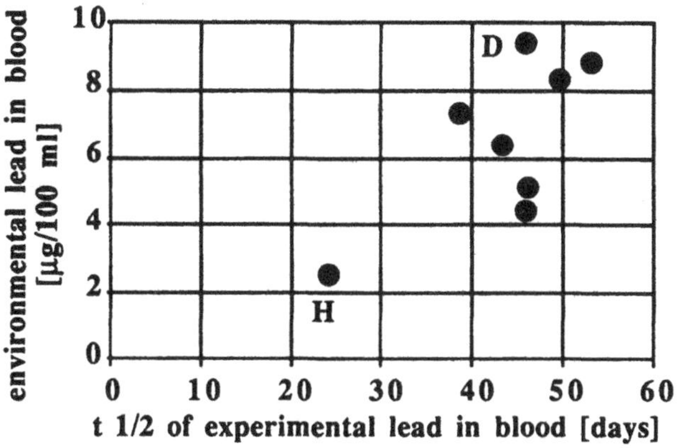

Fig.3 Mean environmental whole-blood concentrations vs. $t_{1/2}$ of experimental lead in blood.

Fig. 3 indicates that the environmental blood-lead concentrations are correlated with the corresponding half-lives of lead in blood.

Renal clearance of experimental and environmental lead

The determination of the renal and the fecal clearance was restricted to the two volunteers with the lowest (H: 2.5 µg/dl) and the highest (D: 10 µg/dl) environmental blood-lead concentration. Twice, with an interval of 8 months, they received a single oral dose of 0.5 mg experimental lead. In both experiments samples of urine and venous blood were taken at three occasions.

Results are shown in Fig. 4. Clearance values with respect to experimental lead are in the range of 40 to 80 µl/min and do not differ for volunteers H and D. Moreover, they are closely reproduced in the two experiments. Our results also agree with data from literature (1). The faster disappearance of lead from blood as found for volunteer H, therefore, can not be explained by differences in renal clearance.

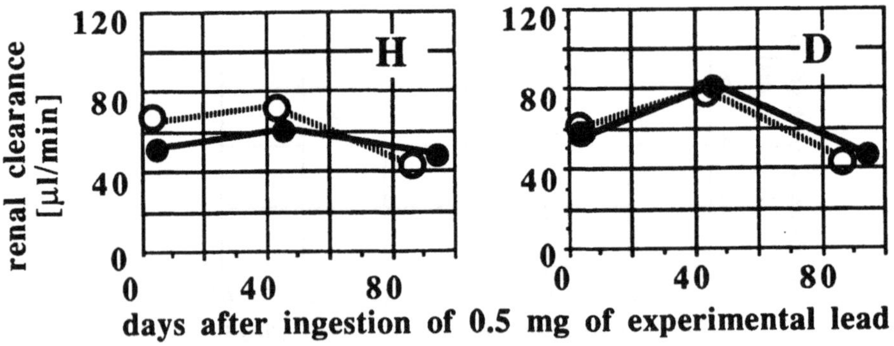

Fig.4 Renal clearance of experimental lead determined at an interval of 8 months with two single oral doses of 0.5 mg experimental lead (volunteers H and D)

Fecal clearance of experimental lead

This kinetic parameter was determined together with the second determination of the renal clearance described above. For volunteer D the fecal clearance is 40 µl/min on day 44 and 20 µl/min on day 87 or about 1/3 to 1/2 of the renal clearance. The values for volunteer H are distintly higher e.i. 135 µl/min on day 44 and 50 µl/min on day 87. However, the high value for volunteer H has to be taken with some precaution as the sampling period was for technical reasons only 15 hours, whereas usually feces was collected over a period of 48 hours.

Conclusions

This study indicates, that besides differences in lead intake with food, the elimination rate from whole blood (but not essentially from the body) is the main factor governing the inter-individual variability of environmental blood-lead concentrations. As the experimentally determined differences of the combined renal and fecal elimination can not explain the factor of two found between the half-lives, additional mechanisms as e.g. differences in net flow of lead from blood to the skeleton or the kidneys have to be considered for further investigations.

Acknowledgement

We thank Prof. V. Koeppel and his team of the Institute of Crystallography and Petrography at the ETH Zürich for the generous help in using their mass spectromectric facility.

References

1. Chamberlain, A.C.. 1985. Proc.R.Soc.Lond.B 224, 149

2. Heusler-Bitschy, S. 1988, Dissertation ETH-Zürich, Nr. 8561.

3. Knutti, R. 1984. In: Fortschritte in der atomspektroskopischen Spurenanalytik, Band 1 (Welz. B., ed.), p. 327

4. G. Lehnert & D. Szadkowski, Die Bleibelastung des Menschen, p.43. Verlag Chemie, Weinheim 1983

5. Delves, H.T., Sherlock, J.C & Quinn, M.J. 1984) Human Toxicol. 3, 279

Subject Index

A

B

C

D

E

F

G

H

I

K

L

M

N

O

P

R

S

T

U

V

W

X

Z

Author Index

E

F

G

H

I

J

K

W

Y

Z

List of Participants

Abou-Shakra, Mr. Fadi R.
Chemistry Department
University of Surrey
GB-GU2 5XH Guildford, Surrey
England

Acartürk, Mr. Esmeray
Medical School, Cardiology Dept.
Cukurova University
Tr-Balcali/Adana
Turkey

Aggett, Mr. P.J.
Dept. of Child Health
University of Aberdeen
GB-Aberdeen AB9 22D
England

Akabane, Mr.Isoko
Tama Chemicals Co. Ltd.
Research Laboratory
5-36-2 Kamata OTA-KU
J-144 Tokyo
Japan

Alt, Mr. Friedrich
Institut für Spektrochemie
Postfach 778
D-4600 Dortmund
FR Germany

Andersson, Mrs. Margareta
Stora Technology
Box 60
S-66100 Saeffle
Sweden

Andrasi, Mr. E.
Institute of Inorganic
and Analytical Chemistry
VIII. Nuzeum Krt.4 B
H-1088 Budapest
Hungary

Anke, Mr. Manfred
Tierernährungschemie
Universität Jena
Domburger Str. 24
DDR-6900 Jena
DDR

Aras, Mr. Namik
Middle East Technical University
Tk-06531 Ankara
Turkey

Asche, Mr. Wolfgang
Chemische Rundschau
Barerstr. 80
D-8000 München 40
FR Germany

Bayer, Mr. Wolfgang
Friedrich-Zandel-Str. 7
D-7000 Stuttgart
FR Germany

Beguin, Mr. Yves
Rue de Selys 2
B-4000 Liège
Belgium

Bellisola, Mr. Guiseppe
Via Valfiorita 4
I- S. Maria die Negrar-Verona
Italy

Benedik, Mr. Kiha
Insitute "Jozef Stefan"
YU-61000 Ljubljana
Yugoslavia

Berthelot, Mr. Alain
Laboratoire de Phys.-Pharmacie
Universitè de Franche-Comtè
Place-Saint-Jacques
F-25030 Besancon-Cedex
France

Bethel, Mrs. Ulrike
Universität Hamburg
Martin-Luther-King-Platz 6
D-2000 Hamburg 13
FR Germany

Bödecker, Mr. Claus-Dieter
E. Merck
Frankfurter Str. 250
D-6100 Darmstadt 1
FR Germany

Boman, Mr. Johan
Physics Dept. 3B
Chalmers University of Technology
S-412 96 Göteborg
Sweden

Bossalles, Mr. Wolfgang
Höchst AG
Klinische Chemie H 811
D-6230 Frankfurt/M. 80
FR Germany

Brätter, Mr. Peter
Hahn-Meitner-Institut
Glienicker Str. 100
D-1000 Berlin 39
FR Germany

Byrne, Mr. Anthony R.
"J. Stefan " Institute
Jamova 39
YU-61000 Ljubljana
Yugoslavia

Caroli, Mr. Sergio
Istituto Superiore di Sanita
Viale Regina Elena 299
I-00161 Roma
Italy

Cavadini, Mr. Claude
Nestec SA Nestle Research Center
P.O.Box 353
Ch-1800 Vevey
Switzerland

Cercasov, Mr. Vadim
Institut für Physik der
Universität Hohenheim
Garbenstr. 30
D-7000 Stuttgart 70
FR Germany

Chaudhri, Mr. M.A.
Austin Hospital
AUS-3084 Melbourne
Australia

Chtchekine, Mr. Konstantin
All-Union Research Institute
of Radiation Engeneering
USSR-Moscow
USSR

Corhay, Mr. Jean Louis
Univ. of Liège, IPNE,BL15
Sart Tilmann
B-4000 Liège
Belgium

Damsgaard, Mrs. Else
Isotope Divison
Risa National Laboratory
DK-4000 Roskilde
Denmark

Delgado, Mrs. Mayorga Veronika
Los Queltehoes
RCH-3555-Santiago
Chile

Dempster, Mr. W.S.
University of Cape Town
Institut of Child Health
Red Cross Childrens Hospital
ZA-7700 Rondebosch, Cape Town
South Africa

Derache, Mr. Philippe
Inserm U-87
2, Rue F. Magendie
F-31400 Toulouse
France

Doerner, Mr. Klaus
Universitäts-Kinderklinik Kiel
Schwanenweg 20
D-2300 Kiel 1
FR Germany

Duflou, Mrs. Hedwig
Institut voor Nucleaire
Wetenschappen
Proeftuinstraat 86
B-9000 Gent
Belgium

Dunemann, Mr. Lothar
Institut für Anorganische
und Analytische Chemie der
TU Clausthal
Paul-Ernst-Str. 4
D-3392 Clausthal-Zellerfeld
FR Germany

Durrant, Mr. Steven
Chemistry Department
University of Surrey
GB-GU2 5XH Guildford, Surrey
England

Ebner, Mr. Heinrich
Piringerhofstr. 13
A-4020 Linz
Austria

Eisenmann, Mr. Thomas
GSF, Inst. f. Ökolog. Chemie
Ingolstädter Landstr. 1
D-8042 Neuherberg
FR Germany

Everett, Mr. George
U.S. Plant. Soil and
Nutr. Laboratory
Tower Rd
Ithaca, NY 14853
USA

Farrar, Mrs. Gillian
Biology Division
Pharmaceutical Sciences
Aston University
GB-B4 7ET Birmingham
England

Frank, Mr. Adrian
National Veterinary Institute
P.O. Box 7o 73
S-750 07 Uppsala
Sweden

Fresenius, Mr. W.
Im Maisel
D-6204 Taunusstein 4
FR Germany

Frosch, Mrs. Doris
Martinstr. 44
D-4000 Düsseldorf 1
FR Germany

Fuchs-Pohl, Mr. Gerald
E. Merck
Frankfurter Str. 250
D-6100 Darmstadt 1
FR Germany

Fuchswans, Mr. Werner
Institut für Med. Physiologie
Universität Wien
Gusenleithnergasse 30/12
A-1140 Wien
Austria

Galgan, Mrs. Vera
National Veterinary Institute
P.O. Box 70 73
S-750 07 Uppsala
Sweden

Gatschke, Mr. Wolfram
Hahn-Meitner-Institut
Glienicker Str. 100
D-1000 Berlin 39
FR Germany

Gawlik, Mr. Dieter
Hahn-Meitner-Institut
Glienicker Str. 100
D-1000 Berlin 39
FR Germany

Gercken, Mr. Bertold
Hahn-Meitner-Institut
Glienicker Str. 100
D-1000 Berlin 39
FR Germany

Greenberg, Mr. Robert
National Bureau of Standards
Gaithersburg, MD 20899
USA

Grill, Mr. Peter
GSF, Institut f. Ökolog. Chemie
Ingolstädter Landstr. 1
D-8042 Neuherberg
FR Germany

Hamm, Mr. Volker
Universität Hamburg
Institut für Anorg. und Angew. Chemie
Martin-Luther-King-Platz 6
D-2000 Hamburg 13
FR Germany

Hansen, Mrs Christine
GSF, Institut für
Biophys. Strahlenforschung
Paul-Ehrlich-Str. 20
D-6000 Frankfurt
FR Germany

Hasse, Mrs. Sylvia
GSF, Institut für Ökolog. Chemie
Ingolstädter Landstr. 1
D-8042 Neuherberg
FR Germany

Havercroft, Mrs. Janet
Chemistry Department
University of Surrey
GB-GU2 5XH Guildford, Surrey
England

Hedrich, Mrs. Martina
Hahn-Meitner-Institut
Glienicker Str. 100
D-1000 Berlin 39
FR Germany

Heil, Mrs. Monika
Department of Pediatrics
University of Vienna
Währingerstr. 38
A-1090 Wien
Austria

Herber, Mr. Robert F.M.
Coronel Laboratory
University of Amsterdam
Meibergdreef 15
NL-1105 AZ Amsterdam
The Netherlands

Hermann, Mrs. Renate
Die Neue Ärztliche
Schleissheimer Str. 295
D-8000 München 40
FR Germany

Himmel, Mrs. Angelika
Klinisch-Chem. Institut
Katharinen Hospital
Kriegsbergstr. 60
D-7000 Stuttgart
FR Germany

Hoffmann, Mr. Jürgen
Eduard Spranger Str. 7
D-8000 München 45
FR Germany

Holler, Mrs. Gabriele
CYANAMID GmbH
Pfaffenriederstr. 7
D-8190 Wolfratshausen
FR Germany

Horvat, Mr. M
Institute "Jozef Stefan"
Jamova 39
YU-61000 Ljubljana
Yugoslavia

Iyengar, Mr. G.V.
National Bureau of Standards
Building 235, B-125
Gaithersburg, MD 20899
USA

Jerons, Mr. Ulrich
Institut für Spektrochemie
D-4600 Dortmund
FR Germany

Johansson, Mr. Erland
Department of Physic. Biol.
P.O. Box 535
S-75121 Uppsala
Sweden

Kaenzig, Mr. Andre
Institut für Toxikologie
der Universität und ETH Zürich
CH-8603 Schwerzenbach
Switzerland

Käkkonen, Mr. Perit
Department of Chemistry
University of Oulo
Linnanmaa
SF-90570 Oulo
Finland

Kalliomäki, Mrs. P.-Liisa
Institute of Occ. Health
Topekuksenleatu 41A
SF-00250 Helsinki
Finland

Kaother, Mr. A.R.S.
Commission, IAEC
P.O.Box 765
IRQ-Baghdad
Iraq

Kirchgessner, Mr. M
Technische Universität München
D-8050 Freising-Weihenstephan
FR Germany

Kitao, Mr. Kensuke
National Institute
of Radiological Sciences
J-260 Chiba-Shi
Japan

Kleff, Mr. Ulrich
Fa. Nattermann
Nattermannallee 1
D-5000 Köln
FR Germany

Klein, Mr. Eckard
Untersuchungsstelle für
Umwelttoxikologie des Landes
Schleswig-Holstein
D-2300 Kiel 1
FR Germany

Knapp, Mr. Günter.
Technische Universität Graz
Techniker Str. 4
A-8010 Graz
Austria

Knutti, Mr. Rudolf
Institut für Toxikologie
der ETH und Univ. Zürich
CH-8603 Schwerzenbach
Switzerland

Kollmer, Mr. Willy
GSF, Institut für Strahlenbiologie
Ingolstädter Landstr. 1
D-8042 Neuherberg
FR Germany

Kreppel, Mr. Helmut
Walter-Straub-Institut
f. Pharmakologie u. Toxikologie
d. Universität München
Nußbaumstr. 26
D-8000 München 2
FR Germany

Kron Tomas
GSF, Institut für Biophysikalische
Strahlenforschung
Paul-Ehrlich-Str. 20
D-6000 Frankfurt
FR Germany

Kruse-Jarres, Mr. J. D.
Klinisch-Chemisches Institut
Katharinenhospital
Kriegsbergstr. 60
D-7000 Stuttgart 1
FR Germany

Kürner, Mr. Hans
Kürner Analysentechnik
Herderstr. 2
D-8200 Rosenheim
FR Germany

La Torre, Mr. Francesco
Istituto Superiore di Sanita
Viale Regina Elena 299
I-00161 Roma
Italy

Lamberty Mr. Andree
JRC-Geel, CBNM
Steenweg op Retie
B-2440 Geel
Begium

Larsen, Mr. Erik
National Food Agency
Moerkhoej Bygade 19
DK-2860 Soeborg
Denmark

Lee Jones, Mr. A.O.
U.S. Food and Drug
Administration
200 C St.S.W.
Washington, D.C. 20204
USA

Lejon, Mr. Mats
Physics Dept. 3A
Calmers Univ. Technology
S-41296 Gothenburg
Sweden

Lieser, Mr. K.H.
TH Darmstadt
Fachbereich Anorganische Chemie
Hochschulstr. 4
D-6100 Darmstadt
FR Germany

Liu Nian, Mr. Qing
INFN Laboratori di Legnaro
I-35020 Legnaro Padova
Italy

Loerinci, Mr. Gabor
GSF, Institut für
Ökologische Chemie
Ingolstädter Landstr. 1
D-8042 Neuherberg
FR Germany

Lombeck, Mrs. Ingrid
Univ. Kinderklinik
Moorenstr. 5
D-4000 Düsseldorf
FR Germany

Lugowski, Mr. Stan
Faculty of Dentistry
University of Toronto
124 Edward St.
Toronto, Ontario M5G 1G6
Canada

Mansour, Mr. M.
GSF, Institut für
Ökologische Chemie
Schulstr. 10
D-8050 Freising-Attaching
FR Germany

Markert, Mr. Bernd
Dept. of System Research
Universität Osnabrück
Artilleriestr. 34
D-4500 Osnabrück
FR Germany

Massey, Mr. Robert
Ministry of Agriculture,
Fisheries and Food
Food Science Laboratory
Queen St.
GB-Norwich NR2 4SX
England

Meier, Mrs. Susanne
GSF, Institut für Strahlenbiologie
Ingolstädter Landstr. 1
D-8042-Neuherberg
FR Germany

Messerschmidt, Mr. Jürgen
Institut für Sprektrochemie
Bunsen-Kirchhoff-Str. 11
D-4600 Dortmund 1
FR Germany

Michel, Mr. Rolf
Zentraleinrichtung für Strahlenschutz
Am Kleinen Felde 30
D-3000 Hannover 1
FR Germany

Monshausen, Mr. Frank
Colora Messtechnik GmbH
Technisches Büro München
Dorfstr. 21
D-8000 München 60
FR Germany

Müller, Robert
GSF, Institut für Ökologische Chemie
Ingolstädter Landstr. 1
D-8042 Neuherberg
FR Germany

Muntau, Mr. H.
IRC-Ispra
Chem. Division
I-21020 Ispra
Italy

Murakami, Mr. Yukio
Tama Chemicals Co. Ltd.
Research Laboratory
5-36-2 Kamata Ota-Ku
J-144 Tokyo
Japan

Nederbragt, Mr. H.
University of Utrecht
Dept. of Pathology
Fac. of Veterinary Science
NL-3584 CL Utrecht
The Netherlands

Negretti de Brätter, Mrs. V.
Hahn-Meitner-Institut
Glienicker Str. 100
D-1000 Berlin 39
FR Germany

Neidhart, Mr. B
Institut für Arbeitsphysiologie
Universität Dortmund
Ardeystr. 67
D-4600 Dortmund
FR Germany

Nepveu, Mr. Francoise
Lab. Chimie Bioinorganique
38, Rue des 36-Ponts
F-31400 Toulouse
France

Müller, Robert
GSF, Institut für Ökologische Chemie
Ingolstädter Landstr. 1
D-8042 Neuherberg
FR Germany

Muntau, Mr. H.
IRC-Ispra
Chem. Division
I-21020 Ispra
Italy

Murakami, Mr. Yukio
Tama Chemicals Co. Ltd.
Research Laboratory
5-36-2 Kamata Ota-Ku
J-144 Tokyo
Japan

Nederbragt, Mr. H.
University of Utrecht
Dept. of Pathology
Fac. of Veterinary Science
NL-3584 CL Utrecht
The Netherlands

Negretti de Brätter, Mrs. V.
Hahn-Meitner-Institut
Glienicker Str. 100
D-1000 Berlin 39
FR Germany

Neidhart, Mr. B.
Institut für Arbeitsphysiologie
Universität Dortmund
Ardeystr. 67
D-4600 Dortmund
FR Germany

Nepveu, Mr. Francoise
Lab. Chimie Bioinorganique
38, Rue des 36-Ponts
F-31400 Toulouse
France

Neve, Mr. Jean
Université Libre de Bruxelles
Institut de Pharmacie
Campus Plaine 205-1
B-1050 Brussels
Belgium

Ochs, Mr. Andreas
Med. Klinik II
Universität Freiburg
D-7800 Freiburg
FR Germany

Pääkö, Mr. Paavo
Dept. of Pathology
University of Oulo
Kajaanintie 52 D
SF-90220 Oulu
Finland

Panday, Mr. Vijay
KFA-Jülich
Abt. Z. Ch.
D-5170 Jülich
FR Germany

Peretz, Mr. A.
University Hospital Sant-Pierre
Dept. of Rheumatology and Physical Medicine
322 Rue Haute
B-1000 Brussels
Belgium

Peronnet, Mr.
Laboratoire Herbaxt
Zi Nord BAT 5
F-Torcy
France

Plantin, Mr. Lars-Olof
Karolinska Institute
Research Center, Huddinge Hospital
S-141 86 Huddinge
Sweden

Quercy, Giquel
Laboratoire Herbaxt
Zi Nord BAT 5
F-77200 Torcy
France

Reichl, Mr. Franz
Walter-Straub-Institut
Pharmakologie u. Toxikologie
LMU München
D-8000 München
FR Germany

Reichlmayr-Lais, Mrs. A.M.
TU München-Weihenstephan
D-8050 Freising
FR Germany

Ribas, Mr.Ozonas Bartolome
Instituto de Bioquimica
CSIC-Complutense Univ.
Facultad de Farmacia
E-280 40 Madrid
Spain

Rösick, Mr. Ullrich
Hahn-Meitner-Institut
Glienicker Str. 100
D-1000 Berlin 39
FR Germany

Ruhstorfer, Mr. Heribert
GSF, Institut für Toxikologie
Ingolstädter Landstr. 1
D-8042 Neuherberg

Runkel, Mr. K.H.
GSF, Institut für Toxikologie
Ingolstädter Landstr. 1
D-8042 Neuherberg
FR Germany

Runow, Mr. Klaus-Dietrich
Institute for Environmental Diseases
D-3501 Emstal
FR Germany

Sabbioni, Mr. E.
IRC, Ispra
I-21020 Ispra
Italy

Salgo, Mr. Laszlo
Dept. of Paediatrics
Medical University
P.O.Box 471
H-6701 Szeged
Hungary

Schelenz, Mr. Rieder
IAEA, Laboratory Seibersdorf
P.O.Box 200
A-1400 Wien
Austria

Schenkel, Mr. H.
Dorotheenweg 1
D-7024 Filderstadt
FR Germany

Schilling, Thomas
GSF, Institut für Ökologische Chemie
Ingolstädter Landstr. 1
D-8042 Neuherberg
FR Germany

Schleich, Mrs. Christine
Universität Trier
Abteilung Chemie
Postfach 38 25
D-5500 Trier
FR Germany

Schmitt, Mr. York
Klin. Chem. Institut
Katharinenhospital
Kriegsbergstr. 60
D-7000 Stuttgart
FR Germany

Schramel, Mr. Peter
GSF, Institut für Ökologische Chemie
Ingolstädter Landstr. 1
D-8042 Neuherberg
FR Germany

Schramel, Mrs. Gisela
GSF, Institut für Ökologische Chemie
Ingolstädter Landstr. 1
D-8042 Neuherberg
FR Germany

Shumpei, Mr. Shimizu
Moses Lakes Industries INC.
780 Randolph Road
Moses Lake, Wash. 98837
USA

Simonoff, Mr. Gabriel
Chimie Nucleaire,
CEN Bordeaux Gradignan
Le Haut Vigneau
F-33170 Gradignan
France

Simonoff, Mrs. Monique
Directeur de Recherche
Chimie Nucleaire
CEN Bordeaux Gradignan
Le Haut Vigneau
F-33170 Gradignan
France

Skreblin, Mrs. Mirjana
Institute "Jozef Stefan"
YU-61000 Ljubljana
Yugoslavia

Steffan, Mrs. Ilse
Institut für Analytische Chemie
Währingerstr. 38
A-1090 Wien
Austria

Steinnes, Mr. Eiliv
Dept. of Chemistry
Univ. of Trondheim
N-7055 Dragvoll
Norway

Stelte, Mr. Wilhelm
Institut für Ernährungswissenschaft
Justus-Liebig-Universität Giessen
Goethestr. 55
D-6300 Giessen
FR Germany

Stoeppler, Mr. M.
KFA, Institut für Chemie (ICH4)
Postfach 19 13
D-5170 Jülich
FR Germany

Stone, Mrs. Susan F.
Center for Analytical Chemistry
National Bureau of Standards
Gaithersburg, MD 20899
USA

Suter, Mr. Paolo
Schweizer Fachpresse Versand
St. Georgenstr. 32
CH-8400 Winterthur
Switzerland

Teuber, Roland
GSF, Institut für Ökologische Chemie
Ingolstädter Landstr. 1
D-8042 Neuherberg
FR Germany

Toelg, Mr. G.
Institut für Sprektrochemie
und Angew. Spektroskopie
Bunsen-Kirchhoff-Str. 11
D-4600 Dortmund
FR Germany

Tomiak, Mr. Andreas
Hahn-Meitner-Institut
Glienicker Str. 100
D-1000 Berlin 39
FR Germany

Tsurahide, Mr. Cho
Tama Chemicals Co. Ltd.
Research Laboratory
5-36-2 Kamata Ota-Ku
J-144 Tokyo
Japan

Van Dael, Mr. Peter
University of Antwerp
Lab. of Bromatology
Universiteitsplein 1
B-2610 Wilryk
Belgium

Van den Hamer, Mr. C.J.A.
Interfaculty Reactor Institute
Mekelweg 15
NL-2629 NB Delft
The Netherlands

Van Renterghem, Mrs. Diana
Faculteit van de Wetenschappen
Proeftuinenstraat 86
B-9000 Gent
Belgium

Van Soest, Mr. P.J.
Dept. of Animal Science
Cornell University
Ithaca, NY. 14853
USA

Varvaritsa, Mr. Vladislav
All-Union Research Institute
of Radiation Engeneering
USSR-Moscow
USSR

Verdingh, Mr. Vital
CEC, JRC-Geel
Steenweg Naar Retie
B-2440 Geel
Belgium

Verheesen, Mr. J.H.H.
Institute of Veterinary Path.
State Univ. of Utrecht
Yalelaan 1
NL-3584 CL Utrecht
The Netherlands

Vermeir, Mrs. Gerda
Lab. of Analytical Chem.
Inst. for Neclear Science
Proeftuinstraat 86
B-9000 Gent
Belgium

Vis, Mr. R.D.
Natuurkundil Lab.
Vrije Universiteit
De Boelelaan 1081
NL-1081 HV Amsterdam
The Netherlands

Voget, Mr. M.
GSF, Institut für Ökolog. Chemie
Schulstr. 10
D-8050 Attaching
FR Germany

Walker, Mrs. Nicola
University of Surrey
Chemistry Department
GB-GU2 5XH Guildford, Surrey
England

Ward, Mr. Neil I.
Department of Chemistry
University of Surrey
GB-GU2 5XH Guildford, Surrey
England

Weber, Mr. Georges Y.J.
I.P.N.E. University of Liège
Sart-Tilman B15
B-4000 Liège
Belgium

Welz, Mr. B.
Bodenseewerk Perkin-Elmer
Postfach 11 20
D-7770 Überlingen
FR Germany

Werk, Mr. Wolfgang
Römerstr. 32 A
D-8023 Pullach
FR Germany

Weser, Mr. U.
Anorganische Biochemie
Physiol.-Chemisches Institut
Hoppe-Seyler-Str. 1
D-7400 Tübingen
FR Germany

Wigfield, Mr. Donald
Department of Chemistry
Carleton University
CDN-KIS SB6 Ottawa, Ontario
Canada

Wittl, Christiane
GSF, Institut für Ökolog. Chemie
Ingolstädter Landstr. 1
D-8042 Neuherberg
FR Germany

Woittiez, Mr. Joost R.W.
Netherlands Energy
Research Foundation (ECN)
P.O.Box 1
NL-1755 ZG Petten
The Netherlands

Yadegarian, Mr. Hadji Abadi
University of Surrey
Chemistry Department
GB-GU2 5XH Guildford, Surrey
England

Yokel, Mr. Robert
College of Pharmacy
University of Kentucky
Rose Street
Lexington, Kentucky 40536-0082
USA

Yoshizako, Mr. Mamoru
Tama Chemicals Co. Ltd.
Research Laboratory
5-36-2 Kamata OTA-KU
J-144 Tokyo
Japan

Yukawa, Mr. Masaen
National Institute of Radiology Sce.
9-1 Anagawa-4Chome
J-260 Chiba-Shi
Japan

Zeisler, Mr. Rolf
National Bureau of Standards
Center for Analytical Chemistry
Gaithersburg, MD 20899
USA

Zettel, Mr. Victor
Chemistry Department
University of Surrey
GB-GU2 5XH Guildford, Surrey
England

Bei Fragen zur Produktsicherheit wenden Sie sich bitte an:
If you have any questions regarding product safety,
please contact:

Walter de Gruyter GmbH
Genthiner Straße 13
10785 Berlin
productsafety@degruyterbrill.com